H.R. Figulla R. Kandolf B. McManus (Eds.)

Idiopathic Dilated Cardiomyopathy

Cellular and Molecular Mechanisms, Clinical Consequences

With 117 Figures

Springer-Verlag
Berlin Heidelberg New York
London Paris Tokyo
Hong Kong Barcelona
Budapest

Prof. Dr. med. HANS-REINER FIGULLA
Medizinische Klinik und Poliklinik
Postfach 3742, W-3400 Göttingen, Fed. Rep. of Germany

Prof. Dr. med. REINHARD KANDOLF
Institut für Pathologie, Universität Tübingen
W-7400 Tübingen and
Max-Planck-Institut für Biochemie
W-8033 Martinsried, Fed. Rep. of Germany

Prof. BRUCE MCMANUS, Ph.D., M.D.
Department of Pathology and Microbiology
University of Nebraska, Medical Center
600 South 42nd Street, Omaha, NE68198-6495, USA

ISBN-13:978-3-642-77893-3 e-ISBN-13:978-3-642-77891-9
DOI: 10.1007/978-3-642-77891-9

Library of Congress Cataloging-in-Publication Data Idiopathic dilated cardiomyopathy : cellular and molecular mechanisms, clinical consequences / H.R. Figulla, R. Kandolf, B. McManus.p.cm. Includes bibliographical references and index.ISBN-13:978-3-642-77893-3 1. Myocardium – Diseases. 2. Heart – Dilatation. I. Figulla, H.R. (Hans-Reiner), 1949– II. Kandolf, R. (Reinhard), 1949– . III. McManus, Bruce M. [DNLM: 1. Cardiomyopathy, Congestive – etiology – congresses. 2. Cardiomyopathy, Congestive – physiopathology – congresses. 3. Cardiomyopathy, Congestive – therapy – congresses. WG 200 I19] RC685.M9I37 1993 616.1'24 – dc20 DNLM/DLC 92-49036

Softcover reprint of the hardcover 1st edition 1993

Typesetting Best-set Typesetter Ltd., Hong Kong
23/3130/SPS-543210 – Printed on acid-free paper

Preface

Idiopathic cardiomyopathy has been categorized for over 10 years by the World Health Organization as dilated, hypertrophic, or restrictive. These categories have somewhat vague boundaries; however, idiopathic dilated cardiomyopathy (IDC) finds its general definition from unexplained four-chamber dilatation, prominent systolic dysfunction with depressed ejection fraction, a marked increase in total myocardial mass without wall thickening, myocyte hypertrophy and nuclear polyploidy, rarity of frank myocardial scars, but increased interstitial connective tissue and myocyte atrophy, all in the setting of anatomically normal heart valves and coronary arteries. Mural thrombi within cardiac chambers and atrioventricular valve regurgitation are common secondary concomitants. Myocardial inflammation is rare, but immunological abnormalities have been suggested. Ultrastructural features include increased mitochondrial numbers, lipid deposits, and myofibrillar loss. Progressive pump failure with or without rhythm disturbances typically leads to ever-intensified medical therapies, and ultimately to transplantation or early death.

The need for better understanding of IDC is represented by incidence rates varying from 5 to 8/100 000 population across several populations, 10 000–20 000 IDC deaths annually in the United States, and the generally increasing prevalence of the disease. More than 50% of heart transplant recipients have IDC as their primary myocardial disease. The half-life of patient survival without transplantation is generally less than 5 years. Treatment with antiarrhythmics, calcium channel blockers, β-adrenergic blockers, and angiotensin-converting enzyme inhibitors is being evaluated in clinical trials, but the role of such treatments in the pathobiology of progressive disease is unclear.

Idiopathic cardiomyopathy – a condition entailing protracted and often fatal heart dysfunction – is on the verge of being found out! Cellular and molecular strategies are being used to dissect the etiogenetic factors in previously indistinguishable diseased hearts. Recent insights into the familial bases of hypertrophic cardiomyopathy and IDC provide hope for the possibility of conquering these diseases. Thus, IDC, as an ill-defined subset of abnormal hearts, is among the groups of diseases being resolved pathobiologically.

Work discussed at an International Symposium on Idiopathic Dilated Cardiomyopathy in Baden-Baden, Federal Republic of Germany, January 29–February 1, 1992, emphasized the relationship among structural features and metabolic, molecular, virological, and immunological processes which may explain the causes and physiological consequences of IDC. The symposium sessions provided many opportunities for discussion and critical assessment of data germaine to IDC. A panel of cell and molecular biologists, physiologists, immunologists, virologists, pathologists, cardiologists, and epidemiologists considered many unanswered questions about IDC:

- What is IDC? How should it be defined?
- Does the definition affect the outcome?
- How might IDC be modified, reversed, or prevented?
- What can be learned about mechanisms of IDC from processes in cardiomyopathy of known causes?
- Are familial cases of IDC determined socially or biologically?
- What role is there for primary abnormalities in genes or gene expression of contractile proteins?
- How are the sarcolemmal calcium-sodium exchanger and calcium pump and sarcoplasmic reticulum involved?
- Which channels are paramount to normal myocyte contractility?
- What is the role of calcium-binding proteins?
- How does one explain the apparent paradox of IDC: markedly increased muscle mass yet progressive failure?
- Thus, how much of the hypertrophy in IDC is primary, and how much is compensatory?
- How do interstitial components contribute to the disease?
- Are microcirculatory abnormalities cause or effect?
- Are virus infections a factor in IDC? If so, how and how often?
- What role does the immune system play in IDC?

In consideration of the many potential pathways to a dilated, poorly contractile heart, the possibility of specific therapy remains a vision. However, interventions which improve cardiac function are targeted to known abnormalities of cardiac myocytes or to injurious agents. Current therapeutic approaches were addressed, as evaluated directly in humans and characterized in animal models. A further topic was the development of better targets for amelioration of the disease.

The goal of the Symposium was to tackle the enigmas of IDC. By mutual efforts, the seemingly insurmountable problems of IDC will eventually succumb to scrutiny and solution in the laboratory and clinic. Progress depends heavily on collegial dialogues such as those which spawned this volume.

We would like to thank everyone who has contributed to this book. It was a pleasure for us to work with Dr. C. Osthoff from the Springer-Verlag, whose efforts were truly professional. We wish also to thank Mrs. R. Soballa, Göttingen, for her tremendous secretarial contribution.

January 1993

HANS R. FIGULLA
REINHARD KANDOLF
BRUCE MCMANUS

Contents

I. Introductory Perspectives

II. Epidemiology and Natural Course of IDC

III. Basic Molecular Concepts Toward Understanding of IDC

IV. Structural Abnormalities in IDC

V. Contraction and Relaxation

VI. Microcirculation and Oxygen Supply in Cardiac Hypertrophy and IDC

VII. Ionic Channels and Signal Gating

VIII. Virus and Cardiomyopathy

IX. Immunological Factors in Cardiomyopathy

X. Therapeutic Options in IDC and Myocarditis

List of Contributors

Addresses are given at the beginning of the respective contribution

I. Introductory Perspectives

Idiopathic Dilated Cardiomyopathy: Current Concepts in Clinical Research

H.R. Figulla

Clinical research aims at answering two "simple" questions:
How can a disease be prevented?
How can a disease be cured?

Prevention of Idiopathic Dilated Cardiomyopathy

Since the etiology of idiopathic dilated cardiomyopathy (IDC) is yet unknown, it is not possible to establish reliable means for its prevention. However, some risk factors for the development of IDC can be elucidated. Firstly, there seems to be a genetic disposition for IDC. During tissue typing of 102 IDC patients, HLA-DR4 was found in 40% compared with only 24% in normal patients; this difference was statistically significant [20]. A familial history was demonstrated in 7% of patients [23]; in a recent study, 20% of IDC patients had first-degree relatives with ventricular dysfunction. Complex segregation analysis of the families reported in these studies suggests a single dominant locus with incomplete penetrance [24].

Secondly, malnutrition and deficiency of selenium or the amino acid taurine as well as alcohol abuse are considered to be risk factors [16, 28, 32]. However, an epidemiological study demonstrating alcohol abuse to be a risk factor is, to my knowledge, still lacking. The so-called alcoholic cardiomyopathy cannot be discriminated morphologically, hemodynamically, or in the clinical course [15]. Abstinence from alcohol and smoking may lower the risk of the disease [14, 25]. Some authors claim that alcohol increases calcium inflow, thus inhibiting glycolysis, as was observed in hamsters [1]. Calcium inflow is inhibited by verapamil [40].

On the whole, the association of these risk factors with IDC is very weak. A primary prevention of the disease is therefore yet impossible.

Department of Cardiology and Pulmonology, Division of Internal Medicine, University of Göttingen, W-3400 Göttingen, Fed. Rep. of Germany

Figulla et al. (Eds.)
Idiopathic Dilated Cardiomyopathy

What Are the Present Concepts for Curing or Improving IDC?

Current Therapy

First of all, we must differentiate between symptomatic and asymptomatic patients. Is there any evidence that an asymptomatic patient with IDC will benefit from any therapy? We recommend alcohol abstinence. However, in ours and in other series of IDC patients the hemodynamic course remained unaffected by the patients' drinking habits' [8, 15]. Cessation of exercise is usually recommended even in the NYHA classes I and II due to the concept of overload in cardiomyopathy [17]. However, there is no evidence that this positively affects outcome. Immunosuppressive drug therapy with prednisone for prevention of an ongoing autoimmune reaction was evaluated in IDC but showed negative results [31].

About 50% of IDC patients die suddenly. Up to now it has been assumed that this is due to ventricular fibrillation. However, in a recent study based on the findings of six stable IDC patients awaiting cardiac transplantation who arrested while on telemetry, four demonstrated sinus bradycardia, one AV block, and one electromechanical dissociation due to pulmonary embolism in combination with a small myocardial infarct at necropsy [21]. In none of these cases was ventricular fibrillation the reason behind the arrest. Ventricular runs can be detected in 31%–60% of patients with IDC. No correlation at all between sudden death and the incidence of spontaneous PVCs was found in six of seven studies [27, 29]. Even with programmed electrical stimulation, no specific risk group was detected, and suppression of ventricular tachycardia with antiarrhymthic drugs was very problematic [5, 33].

A study investigating the effects of implantable debrillators is currently beginning (Defibrilat). There are well-known adverse effects of antiarrhythmic drugs (proarrhythmic effects and negative inotropy); a specific risk group of those patients who die due to ventricular fibrillation cannot be identified. Thus, antiarrhythmic drug therapy in patients without symptomatic arrhythmias cannot be recommended.

In all patients (symptomatic or asymptomatic) with highly reduced ejection fraction (<0.35), antithrombotic therapy is of proven benefit to avoid thromboembolism [12, 18]. Angiotensin-converting enzyme inhibitors (ACEI) such as enalapril reduced mortality in patients with symptoms of congestive heart failure, as shown in the CONSENSUS I and SOLVD I trials [3, 36]. In both trials, 15%–18% of the patients were diagnosed as having dilated cardiomyopathy; ACEI also proved of therapeutic benefit for this subgroup. In patients with a left ventricular ejection fraction less than 0.35 without overt heart failure, ACEI reduced morbidity [41].

Digitalis in pharmacological doses increases cytoplasmatic calcium concentrations by inhibiting the sodium-potassium exchange pump and by stimulating sodium-hydrogen exchange due to a decrease in intracellular pH [35]. This may enhance the lusitropic abnormality (=relaxation abnormality) due to abnormal intracellular calcium handling as seen in patients with IDC

[17, 26]. However, in some studies investigating the effects of digitalis on heart failure patients in sinus rhythm, the drugs showed slight beneficial effects [19] which, aside from its positive inotropic effects, may be predominantly attributable to the sympathoinhibitory responses on the baroreceptors [7]. A beneficial effect of digitalis on survival in IDC has never been demonstrated.

In IDC and other diseases of overloaded hearts with increased wall tension, cyclic AMP production is deficient, probably due to β_1-receptor downregulation and partial uncoupling of the β_1-receptor from adenylate cyclase by means of a reduced ratio of the G protein-stimulating subunit to the G-protein inhibitory subunit [6, 26]. β_1-receptor downregulation seems to be more pronounced in IDC compared with ischemic left ventricular dysfunction, while uncoupling is more pronounced in the latter [2]. Inotropic drugs which elevate cyclic AMP exhibit temporary beneficial hemodynamic effects. In most of the older long-term studies, mortality was elevated, albeit not significantly. The most recent trial (PROMISE trial), a double-blind, placebo-controlled study in which 40 mg milrinone/day was administered to 561 heart failure patients compared with 527 placebo patients (46% of patients had a nonischemic cause of heart failure) showed a 28% increase in mortality [30]. These results support the hypothesis that the decline in production of cyclic AMP reduces energy expenditure in chronic heart failure [4].

Experimental Therapy Under Investigation

Two double-blind, randomized trials are presently underway which address the problem of endogenously high levels of catecholamines and energy expenditure in IDC. The first is the Metropolol in IDC Trial conducted by Waagstein. We are eagerly awaiting the publication of its results this year (for discussion see [34]). The second trial is investigating the effects of the calcium blocker diltiazem in IDC (DiDi Trial) which is being conducted in our clinic. A pilot trial showed beneficial effects of this therapy [10]. The rationale behind the use of calcium blockers in cardiomyopathy is to affect abnormal myocardial calcium handling, alcohol toxicity, inappropiate vasomotility, and to decrease heart rate; the latter reduces energy expenditure, which in turn may entail positive effects on lusitropic abnormalities [11]. Calcium blockers have been shown to prevent microcirculatory disorders in the cardiomyopathic Syrian hamster, thus improving oxygen supply [9]. Microcirculatory abnormalities have also been shown in human IDC [39].

The following hypothesis may be formulated for the pathogenesis of IDC: in an acute or chronic illness, myocytes are diminished due to ongoing necrotization. The ongoing disease may progress or may spontaneously terminate. However, once myocardial destruction has reached a certain degree, heart failure progresses due to the structural and functional abnormalities resulting from wall stress overload [17]. This concept would explain why IDC is morphologically and functionally indistinguishable from other cardiac diseases with myocardial overload.

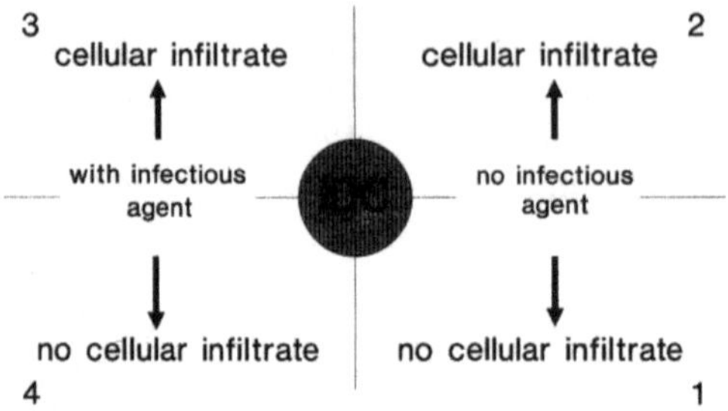

Fig. 1. A suggested new classification of IDC/myocarditis. The term IDC should be used if myocardial dysfunction of unknown cause is detectable by hemodynamic measurements. According to endomyocardial biopsy (EMB) – unless classified cardiomyopathies can be detected – four subsets of IDC can be differentiated: IDC without any infectious agent on EMB – without infiltrate (*1*), with infiltrate (*2*); IDC with an infectious agent – with an infiltrate (*3*), without infiltrate (*4*)

	infectious agent	no infectious agent
cellular infiltrate	3 current therapy + virus-suppressive therapy + cytokine blockers	2 current therapy + cytokine blockers (IL1 ra, TNF blocker)
no cellular infiltrate	4 current therapy + virus-suppressive therapy Ifn a,ß,y	1 - anticoagulants - ACEI - diuretics - (ß-blockers Ca-blockers) } current therapy

Fig. 2. Differential therapy in IDC. Although speculative, the classification suggested in Fig. 1 warrants the option of such a differential therapy. See text for further explanation

In recent years this hypothesis has been supported by the finding of enterovirus genomes in about 25% of endomyocardial biopsies in IDC patients. Additionally, the spirochete *Borrelia burgdorferi* has been found in the myocardial biopsies of some IDC patients [37]. A myocyte-necrotizing agent was therefore characterized (see [22] for review). However, the mechanisms behind the enterovirus and *Borrelia* persistence are unclear; a model of recurrent viral infection is being discussed [22]. It would be appropriate, of course, to eliminate the virus within the myocardium. First results of interferon-alpha treatment in patients with a positive enterovirus endomyocardial biopsy are encouraging [38].

Future Therapy

Profound suppression of cardiac contractile function can accompany inflammatory myocardium and cardiac allograft rejection despite minimal necrosis of the myocardium. Cytokines such as interleukin-1, tumor necrosis factor and others have been found to suppress left ventricular function due to partial uncoupling of the β-receptor from the adenylcyclase [13]. The availability of interleukin-1 receptor antagonists (IL1 ra) as well as tumor necrosis factor binding proteins will extend the possibilities for clinical research in the coming years [42].

Summary: Current and Future Concepts in Therapy

To summarize current and future concepts of clinical research, I would suggest giving up the terms "myocarditis" and "idiopathic dilated cardiomyopathy;" these terms may only describe both sides of the same coin. Instead, I would suggest the incorporation of modern diagnostic tools for differentiation of IDC according to the detection of infectious agents in myocardial tissue. Secondly, patients with cellular infiltrates should be distinguished from those without them (Fig. 1). This classification would allow for a more differential therapy. Those patients without any infectious agents and without noncellular infiltrates should receive unspecific heart failure therapy such as diuretics, ACEI, anticoagulants, possibly β-blockers or calcium blockers, pacemakers, and heart transplantation. If a virus is detectable, heart failure therapy in combination with a virus-suppressive agent may be more appropriate. If a cellular infiltrate in combination with an infectious agent and severe left ventricular dysfunction is found, unspecific therapy in combination with a virus-suppressive agent and cytokine blockers may be effective. In the case of an infectious agent without cellular infiltrates, unspecific therapy and virus-suppressive therapy should be used (Fig. 2).

References

1. Auffermann W, Wu ST, Parmley WW, Wikman-Coffelt J (1990) Glycolysis in heart failure: a ^{31}P-NMR and surface fluorometry study. Basic Res Cardiol 85:342–357
2. Bristow MR, Anderson FL, Port JD, Skerl L, Hershberger RE, Larrabee P, O'Connell JB, Renlund DG, Volkman K, Murray J, Feldman AM (1991) Differences in β-adrenergic neuroeffector mechanisms in ischemic versus idiopathic dilated cardiomyopathy. Circulation 84:1024–1039
3. CONSENSUS (Cooperative North Scandinavian Enalapril Survival Study) Trial Study Group (1987) Effects of enalapril on mortality in severe congestive heart failure. N Engl J Med 316:1429–1435
4. Curfman GD (1991) Inotropic therapy for heart failure – an unfulfilled promise. N Engl J Med 325:1509–1510
5. Das SK, Morady F, DiCarlo L Jr, Baerman J, Krol R, De Buittleir M, Crevey B (1986) Prognostic usefulness of programmed ventricular stimulation in idiopathic dilated cardiomyopathy without symptomatic ventricular arrhythmias. Am J Cardiol 58:998–1000

6. Feldman MD, Copelas L, Gwathmey JK, Phillips P, Warren SE, Shoen FJ, Grossman W, Morgan JP (1987) Deficient production of cyclic AMP. Pharmacologic evidence of an important cause of contractile dysfunction in patients with end-stage heart failure. Circulation 75:331–339
7. Ferguson DW, Berg WJ, Sanders JS, Roach PJ, Kempf JS, Kienzle MG (1989) Sympathoinhibitory responses to digitalis glycosides in heart failure patients. Circulation 80:65–77
8. Figulla HR, Rahlf G, Nieger M, Luig H, Kreuzer H (1985) Spontaneous hemodynamic improvement or stabilization and associated biopsy findings in patients with congestive cardiomyopathy. Circulation 71:1095–1104
9. Figulla HR, Vetterlein F, Glaubitz M, Kreuzer H (1987) Inhomogenous capillary flow and its prevention by verapamil and hydralazine in the cardiomyopathic Syrian hamster. Circulation 76:208–216
10. Figulla HR, Rechenberg J, Wiegand V, Soballa R, Kreuzer H (1989) Beneficial effects of long-term diltiazem treatment in dilated cardiomyopathy. J Am Coll Cardiol 13:653–658
11. Figulla HR, Kreuzer H (1993) The role of calcium channel blockers in idiopathic dilated cardiomyopathy. In: Figulla HR, Kandolf R, McManus B (eds) Idiopathic dilated cardiomyopathy, molecular and structural mechanisms, clinical consequences. Springer, Berlin Heidelberg New York, pp 382–389
12. Fuster V, Gersh BJ, Giuliani ER, Tajik AJ, Brandenburg RO, Frye LR (1981) The natural history of idiopathic dilated cardiomyopathy. Am J Cardiol 47:525–531
13. Gulick T, Chung MK, Pieper IY, Lang LG, Schreiner GF (1989) Interleukin-1 and tumor necrosis factor inhibit cardiac myocyte β-adrenergic responsiveness. Proc Natl Acad Sci USA 86:6753–6757
14. Hartz AJ, Anderson AJ, Brooks HL, Manley JC, Parent GT, Barboriak JJ (1984) The association of smoking with cardiomyopathy. N Engl J Med 311:1201–1206
15. Ikram M, Williamson HG, Won M, Crozier IG, Wells EJ (1987) The course of idiopathic dilated cardiomyopathy in New Zealand. Br Heart J 57:521–527
16. Johnson RA, Palacios I (1982) Dilated cardiomyopathies of the adult. N Engl J Med 307:1051–1058
17. Katz AM (1990) Cardiomyopathy of overload. A major determinant of prognosis in congestive heart failure. N Engl J Med 322:100–110
18. Kyrle PA, Korninger C, Gössinger H, Glogar D, Lechner K, Niessner H, Pabinger I (1985) Prevention of arterial and pulmonary embolism by oral anticoagulants in patients with dilated cardiomyopathy. Thromb Haemost 54:521–523
19. Lee DC-S, Johnson RA, Bingham JB, Leahy M, Dinsmore RE, Goroll AH, Newell JB, Strauss W, Haber E (1982) Heart failure in outpatients. N Engl J Med 306:699–705
20. Lima CJ, Lima SC (1989) HLA antigens in idiopathic dilated cardiomyopathy. Br Heart J 62:379–383
21. Luu M, Stevenson WG, Stevenson LW, Baron K, Walden J (1989) Diverse mechanisms of unexpected cardiac arrest in advanced heart failure. Circulation 80:1675–1680
22. McManus BM, Kandolf R (1991) Evolving concepts of cause, consequence and control in myocarditis. Curr Opin Cardiol 6:418–427
23. Mestroni L, Miani D, Lenarda AD, Silvestri F, Bussani R, Filippi G, Camerini F (1990) Clinical and pathologic study of familial dilated cardiomyopathy. Am J Cardiol 65:1449–1453
24. Mechels VV, Moll PP, Miller FA, Tajik J, Chu JS, Driscoll DJ, Burnett JC, Rodeheffer RJ, Chesebro JH, Tazelaar HD (1992) The frequency of familial dilated cardiomyopathy in a series of patients with idiopathic dilated cardiomyopathy. N Engl J Med 326:77–82
25. Molgaard H, Kristenson BO, Baandrup U (1990) Importance of abstention from alcohol in alcoholic heart disease. Int J Cardiol 26:373–375
26. Morgan JP, Earny RE, Allen PD, Grossman W, Gwathmey JK (1990) Abnormal intracellular calcium handling, a major cause of systolic and diastolic dysfunction in ventricular myocardium from patients with heart failure. Circulation 81 [Suppl III]:21–32
27. Ometto R, de Maria R, Gavazzi A, Biagini A (1991) Prevalence and prognostic significance of complex ventricular arrhythmias in dilated cardiomyopathy. Eur Heart J 12 [Suppl]:77 (abstr)

28. Oster O, Prellwitz W, Kasper W, Meinertz T (1983) Congestive cardiomyopathy and the selenium content of serum. Clin Chim Acta 128:125–132
29. Packer M (1985) Sudden unexpected death in patients with congestive heart failure. A second frontier. Circulation 72:681
30. Packer M, Carver JR, Rodeheffer RJ, Ivanhoe RJ, DiBianco R, Zeldis SM, Hendrix GH, Bommer WJ, Elkayam U, Kukin ML, Mallis GI, Sollano JA, Shannon J, Tandon PK, DeMets DL (1991) Effect of oral milrinone on mortality in severe chronic heart failure. N Engl J Med 325:1468–1475
31. Parrillo JE, Cunnion RE, Epstein S, Parker MM, Suffredini AF, Brenner M, Schaer GL, Palmeri ST, Cannon RO III, Alling D, Wittes JT, Ferrans VJ, Rodriguez ER, Fauci AS (1989) A prospective, randomized, controlled trial of prednisone for dilated cardiomyopathy. N Engl J Med 321:1061–1068
32. Pion PD, Kittleson MD, Rogers QR, Morris JG (1987) Myocardial failure in cats associated with low plasma taurine. A reversible cardiomyopathy. Science 237:764–768
33. Poll DS, Marchlinski EF, Buxton AE, Doherty JU, Waxman HL, Josephson ME (1984) Sustained ventricular tachycardia in patients with idiopathic dilated cardiomyopathy. Electrophysiologic testing and lack of response to antiarrhythmic drug therapy. Circulation 70:451–456
34. Shanes JG (1987) β-Blockade-rational or irrational therapy for congestive heart failure. Circulation 76:971–973
35. Smith TW (1988) Digitalis. Mechanisms of action and clinical use. N Engl J Med 318: 358–365
36. SOLVD Investigators (1991) Effect of enalapril on survival in patients with reduced left ventricular fractions and congestive heart failure. N Engl J Med 325:293–302
37. Stanek G, Klein J, Bittner R, Glogar D (1990) Isolation of *Borrelia burgdorferi* from the myocardium of a patient with longstanding cardiomyopathy. N Engl J Med 322:249–252
38. Stille-Siegener M, Heim A, Klingel K, Kandolf R, Mall G, Kreuzer H, Figulla HR (1993) Interferon therapy in enterovirus-associated idiopathic dilated cardiomyopathy. In: Figulla HR, Kandolf R, McManus B (eds) Idiopathic dilated cardiomyopathy, molecular and structural mechanisms, clinical consequences. Springer, Berlin Heidelberg New York, pp 369–73
39. Treasure CB, Vita JA, Cox DA, Fish D, Gordon JB, Mudge GH, Colucci WS, Sutton MGSJ, Selwyn AP, Alexander RW, Ganz P (1990) Endothelium-dependent dilatation of the coronary microvasculature is impaired in dilated cardiomyopathy. Circulation 81:772–779
40. Wu S, White R, Wikman-Coffelt J, Sievers R, Wendland M, Garrett J, Higgins CB, James T, Parmley WW (1987) The preventive effect of verapamil on ethanol-induced cardiac depression: phosphorus-31 nuclear magnetic resonance and high-pressure liquid chromatic studies of hamsters. Circulation 75:1058–1064
41. Yusuf S (1991) SOLVD II. Lecture at the American Heart Association Meeting, 11 November 1991, Anaheim, California
42. Hannum CH, Wilcox CJ, Arend WP, Joslin FG, Dripps DI, Heimdal PL, Armes LG, Sommer A, Eisenberg SP, Thompson RC (1990) Interleukin-1 receptor antagonist activity of a human interleukin-1 inhibitor. Nature 343:336–340

II. Epidemiology and Natural Course of IDC

Epidemiology of Idiopathic Dilated Cardiomyopathy

H. Kuhn, F. Gietzen, G. Beer, and K. Pethig

The present paper provides an overview of the epidemiology of idiopathic dilated cardiomyopathy (IDC) with special reference to clinical aspects. It considers general problems in defining epidemiologic features of IDC, special clinical problems in the epidemiology of IDC, prognostic features as manifested by annual mortality rate, the frequency of familial IDC, the incidence and prevalence of IDC, and special costs entailed by the therapy of IDC.

General Problems

General problems in defining epidemiologic features of IDC include the fact that incidence and prevalence estimates are commonly based on data from selected population subgroups such as patients referred to cardiac units, general hospitals, or major medical institutions, cases identified in autopsy subsets or hospital discharge, mortality statistics, cross-sectional survey data, or industrial health screening programs [1, 14, 27]. Attempts to describe the epidemiologic characteristics of IDC are thus limited by a number of factors. There is only one study in the United States which is based on a review of the complete medical records of all patients diagnosed in a delineated community [14]. No such studies are available in Europe or other countries.

Special Clinical Problems

The main special clinical problem is the fact that all epidemiologic studies available (see "Incidence and Prevalence") are based on patients selected in a very advanced stage of the disease, i.e., all patients with mild to moderate, well-compensated heart failure are ruled out by the selection criteria. This of course leads to a pronounced underestimation of the incidence and prevalence of IDC, because the population of patients with a less advanced stage of the disease is rather high [22, 45, 51, 53]. Patients presenting with a very early stage of disease account for at least 20% in series of patients with

Department of Internal Medicine and Cardiology, Bielefeld Hospital, Teutoburgerstraße 50, W-4800 Bielefeld 1, Fed. Rep. of Germany

Figulla et al. (Eds.)
Idiopathic Dilated Cardiomyopathy

IDC (in this study, patients were identified by a normal cardiothoracic ratio). However, there is no doubt that the frequency of patients with an early clinical stage of IDC also includes patients with an already enlarged heart on X-ray. This is indicated by analysis of heart failure of any origin using a scoring system for the detection of early stage heart failure [22, 23]. In this study, the subset of patients with an early stage (independent of heart size on X-ray) accounted for nearly 50% of all patients with heart failure. The Framingham study differentiates between patients with major, (i.e. neck vein distension and acute pulmonary edema), and minor (i.e., hepatomegaly and angle edema) signs of congestive heart failure. However, from a clinical point of view, there is no doubt that both subsets include an already advanced stage of IDC [40].

Regardless of these considerations, the epidemiologic evaluation is complicated by the fact that there is no universally accepted definition of heart failure, particularly as far as early stages (latent heart failure) are concerned [21, 22]. The same holds true for other definitions. Mainly because of a lack of universally accepted definitions of alcoholic heart disease [5, 20, 46, 55, 57, 90], myocarditis [4, 34, 50, 53, 54, 60, 74], hypertensive heart disease, and left ventricular hypertrophy [52], different diagnostic criteria of IDC are used. This gives rise to different epidimeologic features, i.e., patients are predominantly ruled out or are included with regard to alcoholic heart disease, myocarditis, or moderate arterial hypertension (the common link being the dilated poorly contracting left ventricle; see "Incidence and Prevalence").

The percentage of patients with a history of alcohol abuse (at least 100 g/day) accounts for about 20% in series of consecutive patients [51]. In our opinion, such patients should be excluded because the habit of alcohol consumption may profoundly influence the clinical course of IDC. The intake of alcohol may deteriorate the functional class of patients with IDC, and the withdrawal of alcohol may lead to clinical and hemodynamic improvement or even complete regression of IDC [5, 46, 73, 90]. Consequently the results of epidemiologic trials may be markedly influenced by different proportions of patients with IDC, in whom alcohol abuse has ceased or persists during follow-up (alcohol may trigger the clinical course of genetically transmitted IDC; see "Familial IDC").

The same considerations do not hold true for myocarditis because, at least as far as histologic criteria of myocarditis are concerned, there is no proof for the prediction of the clinical course of IDC. In addition, there is much debate on the definition and on the clinical and the scientific criteria of myocarditis [4, 8, 10, 12, 31, 34, 50, 53, 59, 60, 74]. The titles of recently published papers show this; for example, Peters and Poole-Wilson use the title "Myocarditis: Continuing Clinical and Pathologic confusion" [74]. Even very critical and emotional wording has been applied for the subject; for instance, Lie, the cardiac pathologist of the Mayo clinic, wrote in 1988 of the "pernicious nature of the highly subjective and irreproducibly quantitative criteria for biopsy diagnosis of myocarditis" and "follies in biopsy diagnosis" [59, 60]. He was referring to the so-called Dallas criteria of myocarditis and to the extreme interobserver variability in the judgement of single criteria of myocarditis [60, 83] used by the panel which published the Dallas criteria of myocarditis [4].

Recently, a blinded study (i.e., no information for the cardiac pathologist) was finished showing the same proportion of histologic changes compatible with the diagnosis of borderline myocarditis (according to the Dallas criteria) in subsets of patients with completely different myocardial disorders [7]. Therefore, epidemiologic considerations to date should not consider the possibility of myocarditis being present among patients with IDC by using histologic differentiations. Similar to latent heart failure, which is a major concealed epidemiologic source of IDC among patients with heart failure, unrecognized early-stage IDC may be also present in epidemiologic studies of complete common left bundle branch block (LBBB) [35, 79]. In these studies no heart catheterization has in general been performed to diagnose IDC unequivocally, and patients were classified into those suffering from heart failure of unknown cause, angina pectoris, arterial hypertension, or primary conduction defects. In 20%–40% of patients with IDC, the ECG shows LBBB [45, 51] (Table 1). LBBB may preexist for many years before the clinical manifestation of congestive heart failure of IDC, i.e., up to 22 years [44, 45].

To characterize the epidemiology of IDC, it must be established whether patients with preexisting LBBB combined with normal cardiothoracic ratio on X-ray show an already dilated, poorly contracting left ventricle. It can be speculated that patients with overt IDC are born with a clinically still inapparent, slightly dilated left ventricle combined with a reduced ejection fraction. On the other hand, cases are reported in whom primarily completely normally contracting left ventricle together with LBBB was present before IDC was diagnosed [18, 70].

Recently, we have seen three such patients studied by ECG and coronary angiography; one of them was most interesting (Fig. 1; unpublished data). In this patient an ECG was performed because of uncharacteristic pains of the chest. LBBB was detected in combination with a completely normal one- and two-dimension echocardiogram. Two years later, ejection fraction of the left ventricle was slightly reduced, as demonstrated by ECG and left heart catheterization (angiogram). An IDC which persisted was now diagnosed. Acute myocarditis demonstrated by flulike fever of 39.6°C combined with transient total AV block and transient sinus arrest together with slightly increased creatine phosphokinase was diagnosed 9 months later. Now the patient developed severe cardiac decompensation with pulmonary edema. The severe reduction in ejection fraction and dilatation of the left ventricle persisted. Heart transplantation was needed 3 years later. The patient is now in a very good condition.

This clinical course demonstrates that viral infection may trigger preexisting IDC causing severe deterioration (see "Familial IDC"). Furthermore, this clinical course demonstrates that early stage IDC may develop from completely normal heart function in patients with LBBB. However, it should be noted that irrespective of normal function at rest pronounced

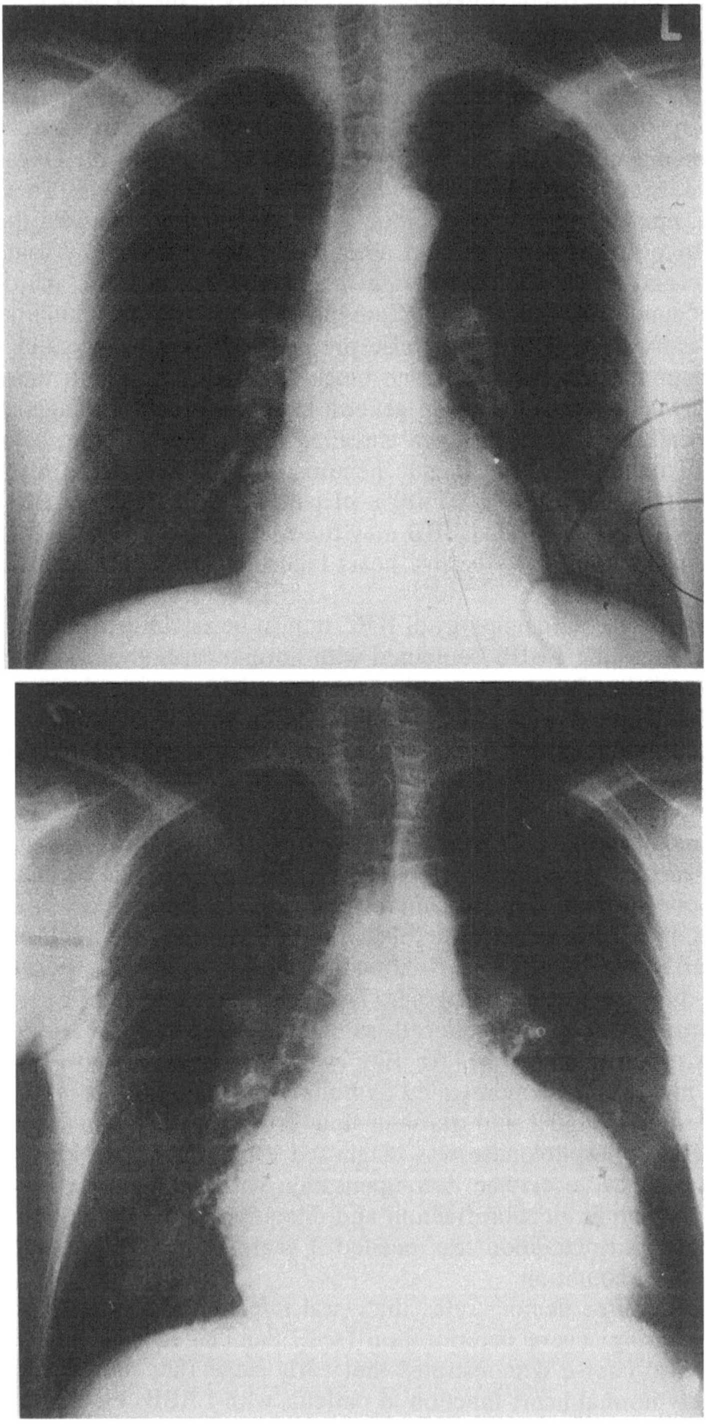

a

b

alterations may be present such as focal light and electron microscopic changes, metabolic alterations (negative arterial coronary venous lactate extraction in atrial pacing), myocardial perfusion defects demonstrated by thallium-201 scintigraphy, or disturbed left ventricular function at exercise in patients with LBBB and no disturbed contraction of the left ventricle at rest [44, 45, 48, 49, 50, 61, 70]. The term latent cardiomyopathy has therefore been suggested for this myocardial disorder [45, 49, 50].

According to the studies of Curtius and coworkers, the functional deterioration rate of patients with early-stage IDC and concomitant LBBB is about 20% per 4 years [19]. However, no reliable epidemiologic data are available that enable the frequency of IDC developing from normal hearts with LBBB to be evaluated, although new LBBB and subsequent congestive heart failure has been described [18, 44, 45, 70, 79, 80].

In summary, there is strong evidence for profound epidemiologic underestimation of IDC, as has been also stressed by other authors [1, 16, 27, 55, 86, 93].

Mortality of IDC

Annual mortality rates of different studies as expressed by cumulative survival rates of patients with IDC are listed in Table 1. As can be seen, there has been no tendency to a better prognosis since 1973. However, recent analysis of the mortality by Camarini and coworkers (this volume) indicates that the prognosis has possibly improved somewhat in recent years. Different selection criteria used (predominantly different stages of IDC) may, again, explain the pronounced prognostic difference shown.

Familial IDC

The frequency of familial IDC in series of patients with IDC accounts for 2%–63% of cases (Table 2). In 489 consecutive patients diagnosed in our institution (Düsseldorf and Bielefeld) the rate of familial IDC was 6% [54]. The low rates published in the literature are based on retrospective studies and not on systematic echocardiographic studies in first- and second-degree relatives of index patients (who are often completely asymptomatic although IDC is demonstrated by ECG) [65, 66]. High rates include putative cases. Analysis of pedigree reveals different modes of inheritance (autosomal

←

Fig. 1a,b. Development of IDC in a 38-year-old patient with complete common LBBB. **a** Heart size and heart function was normal; however, LBBB was present. **b** Six years later IDC had developed "triggered by myocarditis." Cardiac transplantation was needed. For further explanation see text

Table 1. Annual mortality in patients with IDC (cumulative survival rates)

Reference	Year	*n*	Annual mortality
Kuhn [43]	1973	21	20%
McClellan [63]	1978	123	15%
Bory [11]	1978	30	11%
Koide [39]	1980	36	7%
Shirey [84]	1980	113	8%
Hess [37]	1980	70	16%
Kunkel [56]	1980	68	15%
McArthur [2]	1981	146	10%
Fuster [25]	1981	104	17%
Kuhn [51]	1982	221	10%
Unverferth [87]	1984	69	35%
Meinertz [64]	1984	74	26%
Figulla [24]	1985	56	15%
Schwarz [81]	1987	68	15%
Hammond [34]	1987	79	13%
Regitz [77]	1988	102	15%
Hagege [32]	1988	116	8%
Komajda [42]	1990	201	15%
		1697	7%–35% (range)

Table 2. Frequency of familial IDC in series of patients with IDC

Reference	Year	Frequency
Sekiguchi [82]	1978	31%
Ando [3]	1978	52%
Yamaguchi [94]	1978	63%
Fuster [25]	1981	2%
Pongpanich [75]	1986	10%
Griffin [30]	1988	25%
Kuhn [54]	1988	6%
Valantine [92]	1989	9%
Mestroni [65]	1990	7%
Csanády [17]	1991	23%
Michels [66]	1992	20%

dominant [17, 29, 72], recessive [3, 92, 94], X-linked [9, 92], or single dominant locus with incomplete penetrance [66]. A genetic defect has also been indicated by other studies [41, 68, 69, 88, 89].

Irrespective of different selection criteria and methods applied, these studies strongly indicate that, similar to hypertrophic cardiomyopathies, IDC is a genetically transmitted, often asymptomatic myocardial disorder. This may support the theory that on the basis of a primarily myocardial defect the clinical course of IDC may be accelerated or the clinical manifesta-

Table 3. Annual incidence of IDC (patients per 100000 inhabitants)

Reference	Publication year	Annual incidence	Year(s) covered
Torp [86]	1981	5.0	1970–1977
Torp [86]	1981	10.0	1980
Kawai [38]	1980	2.6[a]	1974
Baggar [6]	1984	0.73	1980–1981
Codd [14]	1989	6.0	1975–1984
Codd [14]	1989	9.9[b]	1975–1984

[a] Predominantly IDC.
[b] Adjusted for United States white (male) population

tion or preexisting IDC caused by several secondary factors, such as alcohol abuse, endocrine peripartum influences, viral infections, or merely by normal or elevated arterial blood pressure [45, 47, 54].

Incidence and Prevalence of IDC

The mean age of patients with IDC is about 45 years. In 489 consecutive patients of our institution (Düsseldorf and Bielefeld) it was 46.6 years. Our oldest patient was 86 years old and our youngest under 1 year of age [36, 58, 75, 78]. Confirmed postnatally, even in utero IDC has been detected by fetal echocardiogram [91]. There is a pronounced preponderance of men, who account for 74% of cases ([51, 54] and literature in Table 1). The differences between the incidence rates shown in Table 3 can be explained by different definition and selection criteria (see "Special Clinical Problems").

In a carefully analyzed study published recently by Codd and coworkers [14], the adjusted annual incidence rate for the United States was 9.9 patients/100000 for men and 2.9 for women, and the corresponding prevalence rate was 58.0 for men and 19.4 for women (adjusted rates, white population in 1980). Much higher incidence rates are observed in subsets (e.g., the incidence rate among men 45–54 years old is 26.5 patients/100000 per year, and the corresponsing prevalence rate is 198.3 patients/100000 inhabitants). This study and other trials, however, illustrate the problem of adjusted rates, because the data calculated are based on the observation of only 45 patients with IDC observed between 1975 and 1984 in Olmstedt County, Massachusetts [14] (population-based study; see "Special Clinical Problems"). Based on questionnaires sent to general practitioners the prevalence in two regions of England was 8.3/100000 [93].

Thus the true incidence and prevalence rates are unknown. All authors agree that the published data underestimate the occurrence of the disease. In addition to arguments mentioned above, higher rates of prevalence and incidence of IDC can be also assumed by considering patients with IDC who are commonly excluded from epidemiologic investigations, for example,

Table 4. Annual incidence of congestive heart failure (patients per 100000 inhabitants)

Reference	Year	Population	Annual incidence
McKee [40]	1971	Men	370
McKee [40]	1971	Women	250
Chesebro [13]	1987	-	175
Eriksson [24]	1987	Men, 50–67 years	590
Eriksson [24]	1987	Women 50–60 years	290

Table 5. Prevalence of congestive heart failure

Reference	Year	Age (years)	Prevalence
Logan [62]	1958	45–64	200
Gibson [26]	1966	45–64	1000
McKee [40]	1971	-62	300
Packer [71]	1987	—	1300[a]
Eriksson [23]	1987	50–60	2900

[a] Adjusted.

patients with mild or moderate arterial hypertension combined with severe reduced ejection fraction, patients with other forms of heart disease which does not explain the severely depressed contraction of left ventricle (valvular heart disease or coronary heart disease). These patients account for about 35% in series of patients with unequivocally confirmed or suspected IDC [51].

To check the data of direct epidemiologic IDC studies (Table 3) they can be compared with results obtained from epidemiologic studies of patients with LBBB or congestive heart failure. Using epidemiologic data of LBBB published by Hardarson and coworkers [35] (prevalence of LBBB for men 430 and for women 280 per 100000; corresponding annual incidence rates of 32 and 37 per 100000, age 33–71 years; frequency of IDC 14%) a prevalence rate of IDC of 200 patients (men) and 130 patients (women) per 100000 inhabitants and an annual incidence rate of men of 15/100000 inhabitants (men) and 17/100000 inhabitants (women) can be calculated (assumptions: LBBB patients account for 30% of all IDC patients [45, 51]).

Assuming a frequency of advanced IDC of 35% among patients with congestive heart failure (Tables 4, 5) [13, 28, 31] a corresponding prevalence rate of IDC (derived by calculation from epidemiologic data of congestive heart failure) of about 70–455/100000 and an annual incidence rate of about 60–130/100000 can be estimated. In addition, there is strong evidence for an increasing frequency of congestive heart failure [22, 71, 95]. These data derived from subsets of epidemiologic studies of LBBB or congestive heart failure again indicate a profound underestimation of the incidence of IDC.

Table 6. Epidemiology of IDC

	Deaths/year	New patients/year	Occurrence
USA	12000	20000	115000
Europe	35000	59000	347000
FRG	4000	6000	39000

In approximative terms, they may reflect an up to tenfold higher incidence of IDC compared with values obtained from direct epidemiologic IDC studies. Using the minimum rates of IDC as described above, the deaths/year, new patients/year, and occurrence of IDC for the United States, Europe, and Federal Republic of Germany can be calculated (Table 6; assumptions: 80% men, 20% women with IDC, annual mortality 10%, prevalence 50.3, incidence 8.5; population of United States 230 million, Europe 691 million, Federal Republic of Germany 78 million).

Special Costs

Special costs engendered by the therapy of IDC may be of major importance for health care systems in some countries. These costs are caused first by the tremendous increase in the implantation of automatic internal cardioverter-defibrillator (AICD). The reason is the generally accepted fact that sudden cardiac death in IDC (30%–50%) [15, 33, 42, 45, 51, 64, 67], and, as a consequence of the alarming results of the CAST study [76], that the medical possibilities for preventing sudden cardiac death are very limited, and that the implantation of AICD has become relatively easy (transvenous approach).

The second reason is the increase in heart transplantation and the fact that IDC accounts for more than 50% of patients who need heart transplantation [85].

References

1. Abelmann WH (1985) Incidence of dilated cardiomyopathy. Postgrad Med J 61:1123–1124
2. McArthur CGC, Binbrek AS, Chetty S, Sanderson JE, Fitchet DH, Goodwin JF, Oakley C (1981) The effect of left ventricular function, arrhythmias and evidence of previous viral infection on congestive cardiomyopathy. In: Goodwin JF, Hjalmarson A, Olsen EGJ (eds) Congestive cardiomyopathy. Hässle, Mölndal, p 236
3. Ando M, Takao A, Mori K, Sekiguchi M (1978) The mode of inheritance in idiopathic dilated cardiomyopathy. In: Sekiguchi M, Olsen EGJ (eds) Cardiomyopathy. University of Tokyo Press, Tokyo, p 445
4. Aretz HT, Billingham ME, Edwards WD, Factor S, Fallon JT, Fenoglio JJ, Olsen EGJ (1986) Myocarditis: a histopathologic definition and classification. J Cardiovasc Pathol 1:3
5. Askanas A, Udoshi M, Sadjadi SA (1980) The heart in chronic alcoholism: a noninvasive study. Am Heart J 99:9

6. Bagger JP, Baandrup U, Rasmussen K, Moller M, Versterlund T (1984) Cardiomyopathy in western Denmark. Br Heart J 52:327
7. Beer G, Frenzel H, Hort W, Kuhn H (1990) Clinical value of endomyocardial biopsy: a prospective blinded study in patients with various myocardial disorders. Eur Heart J 11 [Suppl]:410 (abstr)
8. Bender JR (1991) Idiopathic dilated cardiomyopathy, an immunologic, genetic or infectious disease, or all of the above? Circulation 83:704–706
9. Berko BA, Swift M (1987) X-linked dilated cardiomyopathy. N Engl J Med 316:1186–1191
10. Bouhour JB, Petitier H, De Lajartre AY, Almazor M, Nicolas G, Horeau J (1976) La biopsie myocardique dans les myocardiopathies congestives en apparence primitives. Arch Mal Coeur 69:485
11. Bory M, Manchini JC, Dijane P, Sacerdote P, Mostefa-Cara S, Egre A, Serradimigne A (1978) Critèries de prognostic des myocardiopathies apparement primitives hypocinétiques (étude prospective de 60 cas). Arch Mal Coeur 71:870
12. Calquist JF, Menlove RL, Murray MB, O'Connell JB, Anderson JL (1991) HLA class II (DR and DQ) antigen associations in idiopathic dilated cardiomyopathy – validation study and meta-analysis of published HLA association studies. Circulation 83:515–522
13. Chesebro JH, Burnett JC (1987) Cardiac failure: characteristics and clinical manifestations. In: Brandenburg RO, Fuster V, Giuliano ER, McGoon DC (eds) Congestive heart failure. Year Book Medical Publishers, Chicago, pp 645–665
14. Codd MB, Sugrue DD, Gersh BJ, Melton LJ (1989) Epidemiology of idiopathic dilated and hypertrophic cardiomyopathy – a population-based study in Olmstedt County, Minnesota, 1975–1984. Circulation 80:564–572
15. Constanzo-Nordin MR, O'Connel JB, Engelmeier RS, Moran JF, Scanlon PJ (1986) Dilated cardiomyopathy: functional status, hemodynamics, arrhythmias and prognosis. Cathet Cardiovasc Diagn 11:445
16. Coughlin SS, Szklo M, Baughman K, Pearson TA (1990) The epidemiology of idiopathic dilated cardiomyopathy in a biracial community. Am J Epidemiol 131:48–56
17. Csanády M, Faragó M, Forster T, Högye M, Piros G (1991) Study of the course of inheritance of dilated familial cardiomyopathy. Eur Heart J 12 [Suppl]:191 (abstr)
18. Curtius JM, Stechern V, Kuhn H, Loogen F (1984) Follow up of patients with latent cardiomyopathy. Z Kardiol 73:695–700
19. Curtius JM, Welslan K, Breuer HW, Loogen F (1985) Follow up of patients with mild dilated cardiomyopathy. Z Kardiol 74:369–373
20. Dancy M, Bland JM, Leech G, Gaitonde MK, Maxwell JD (1985) Preclinical left ventricular abnormalities in alcoholics are independent of nutritional status, cirrhosis and cigarette smoking. Lancet 1:1122–1125
21. Denolin H, Kuhn H, Krayenbühl AP, Loogen F, Reale A (1983) The definition of heart failure. Eur Heart J 4:445–448
22. Eriksson H, Wilhelmsen L, Caidahl K, Svärdsudd K (1991) Epidemiology and prognosis of heart failure. Z Kardiol 80 [Suppl 8]:1–6
23. Eriksson H, Svärdsudd K, Larsson B, Ohlson LO, Welin L, Tibblin G, Wilhelmsen L (1987) Dyspnoea in a cross-sectional and longitudinal study of middle-aged men: a study of men born in 1913 and 1923. Eur Heart J 8:1015–1023
24. Figulla HR, Rahlf G, Nieger M, Luig H, Kreuzer H (1985) Spontaneous hemodynamic improvement or stabilisation and associated biopsy findings in patients with congestive cardiomyopathy. Circulation 71:1095–1104
25. Fuster V, Gersh BJ, Giuliani ER, Tajik AJ, Brandenburg RO, Freye RL (1981) The natural history of dilated cardiomyopathy. Am J Cardiol 47:525
26. Gibson TC, White KL, Klainer LM (1966) The prevalence of congestive heart failure in two rural communities. J Chronic Dis 19:141–152
27. Gillum RF (1986) Idiopathic cardiomyopathy in the United States, 1970–1982. Am Heart J 111:752–755
28. Gorlin R (1983) Incidence, etiology and prognosis of heart failure. Cardiovasc Rev Rep 4:765–770

29. Graber HL, Unverferth DV, Baker PB, Ryan JM, Baba N, Wooley CF (1986) Evolution of a hereditary cardiac conduction and muscle disorder: a study involving a family with six generations affected. Circulation 74:21–35
30. Griffin ML, Hernandez A, Martin TC, Goldring D, Bolman RM, Spray TL, Strauss AW (1988) Dilated cardiomyopathy in infancy and children. J Am Coll Cardiol 11:139–144
31. Haber RH, Le Jemtel T, Sonnenblick EH (1988) The pathophysiologic profile of congestive heart failure. Cardiovasc Drugs Ther 2 [Suppl 1]:397–400
32. Hagege A, Desnos M, Fernandez F, Cristofini P, Guenot O, Bourmayan C, Marcantoni JP, Guerot C (1988) Prognostic factors in dilated cardiomyopathy. Arch Mal Coeur 81: 1473–1479
33. Haissaguerre M, Bonnet J, Le-Goff G (1986) Prevalence, significance and prognosis of ventricular arrhythmias in 236 cases of dilated cardiomyopathy. Arch Mal Coeur 79:32–38
34. Hammond EH, Menlove RL, Anderson JL (1987) Predictive value of immunofluorescence and electron microscopic evaluation of endomyocardial biopsies in the diagnosis and prognosis of myocarditis and idiopathic dilated cardiomyopathy. Am Heart J 114:1055
35. Hardarson T, Arnason A, Eliasson GJ, Palsson K, Eyjolfsson K, Sigfusson N (1987) Left bundle branch block: prevalence, incidence, follow up and outcome. Eur Heart J 8:1075–1079
36. van der Hauwaert LG, Boudewin D, Dumoulin M (1983) Long term echocardiographic assessment of dilated cardiomyopathy in children. Am J Cardiol 52:1066
37. Hess OM, Turina J, Goebel NH, Grob P, Krayenbühl HP (1977) Prognostic evaluation of congestive cardiomyopathy. Z Kardiol 66:351–360
38. Kawai C (1980) Studies on cardiomyopathy in Japan. In: Sekiguchi M, Olsen EGJ (eds) Cardiomyopathy. University of Tokyo Press, Tokyo, p 3
39. Koide T, Kato A, Takabatake Y, Iizuka M, Ushida Y, Ozeki K, Morooka S, Kakihana M, Serizawa T, Tanaka S, Ohja T, Momomura S, Murao S (1980) Variable prognosis in congestive cardiomyopathy. Jpn Heart J 21:451
40. McKee PA, Castelli WP, McNamara PM (1971) The natural history of congestive heart failure: the Framingham study. N Engl J Med 285:1441–1446
41. Koike S, Xawa S, Yabu K (1987) Familial dilated cardiomyopathy and human leucocyte antigen. A report of two family cases. Jpn Heart J 28:941–945
42. Komajda M, Jais JP, Reeves F, Goldfarb B, Bouhour JB, Juillieres Y (1990) Factors predicting mortality in idiopathic dilated cardiomyopathy. Eur Heart J 11:824–831
43. Kuhn H, Kübler W, Loogen F, Gleichmann U (1973) Die Kongestive Kardiomyopathie. Med Welt 24:996
44. Kuhn H, Breithardt G, Knieriem HJ, Loogen F (1978) Endomyocardial catheter biopsy in heart disease of unknown etiology. In: Kaltenbach M, Loogen F, Olsen EGJ (eds) Cardiomyopathy and myocardial biopsy. Springer, Berlin Heidelberg New York, p 121
45. Kuhn H, Breithardt G, Knieriem HJ, Köhler E, Lösse B, Seipel L, Loogen F (1978) Prognosis and possible presymptomatic manifestations of congestive cardiomyopathy. Postgrad Med J 54:451
46. Kuhn H, Loogen F (1978) The effect of alcohol on the heart, including alcoholic cardiomyopathy. STIAR 32:2–12
47. Kuhn H, Krelhaus W, Breithardt G, Knieriem HJ, Loogen F (1978) Behandlungsmöglichkeiten und Prognose der idiopathischen Kardiomyopathien. Med Welt 29:621–627
48. Kuhn H, Lösse B, Hort W, Loogen F (1979) Studies for the early detection of congestive cardiomyopathy in patients with left bundle branch block. Circulation 59/60 [Suppl II]:37 (abstr)
49. Kuhn H, Lösse B, Hort W (1980) Studies in patients with abnormal electrocardiogram of unknown etiology. In: Kübler W, Tillmanns A, Cebe H (eds) Microcirculation of the heart, theoretical and clinical problems. Springer, Berlin Heidelberg New York, p 288
50. Kuhn H, Lösse B, Boch H, Becker R, Hort W (1981) Prognosis of patients with congestive cardiomyopathy (CCM) – therapeutic, hemodynamic, morphologic and metabolic aspects.

In: Goodwin JF, Hjalmarson A, Olsen EGJ (eds) Congestive cardiomyopathy. Hässle, Mölndal, p 213
51. Kuhn H, Becker R, Fischer J, Curtius JM, Lösse B, Hort W, Loogen F (1982) Studies on the etiology, the clinical course and the prognosis of patients with dilated cardiomyopathy (DCM). Z Kardiol 71:497–508
52. Kuhn H (1982) The definition of ventricular hypertrophy. Eur Heart J 3 [Suppl A]:5–8
53. Kuhn H (1983) The natural history of dilated cardiomyopathy and the problem of myocarditis. In: Just H, Schuster HP (eds) Myocarditis – cardiomyopathy. Springer, Berlin Heidelberg New York, p 13
54. Kuhn H (1988) Critical evaluation of clinical trials. Symposium: dilated cardiomyopathy, update 1988. Xth congress of the European Society of Cardiology, Vienna
55. Kuhn H (1991) Epidemiology, clinical course and prognosis of dilated cardiomyopathy. Z Herz Thorax Gefäßchir 5:1–11
56. Kunkel B, Kober G, Lapp H, Kaltenbach M (1978) Correlations between clinical and morphologic findings and natural history in congestive cardiomyopathy. In: Kaltenbach M, Loogen F, Olsen EGJ (eds) Cardiomyopathy and myocardial biopsy. Springer, Heidelberg Berlin New York, p 271
57. Levi GF, Quadri A, Ratti S, Basagni M (1977) Preclinical abnormality of left ventricular function in chronic alcoholics. Br Heart J 39:35–77
58. Lewis AB, Neustein HB, Takahashi M, Lurie PR (1985) Findings on endomyocardial biopsy in infants and children with dilated cardiomyopathy. Am J Cardiol 55:143–145
59. Lie JT (1988) Myocarditis and endomyocardial biopsy in unexplained heart failure: a diagnosis in search of a disease. Ann Intern Med 109:525–528
60. Lie JT (1989) Myocarditis and endomyocardial biopsy. Ann Intern Med 110:165–166
61. Lösse B, Kuhn H, Krönert H, Rafflenbeul E, Köhler E, Schicha H, Feinendegen LE (1979) Hemodynamic and radiocardiographic exercise studies in patients with complete left bundle branch block and normal-sized heart. Z Kardiol 68:304–312
62. Logan WPD, Cushion AA (1958) Morbidity statistics from general practice. 1. Studies on medical and population subjects (no 14). HM Stationery Office, London
63. McClellan J, Waller B, Segal J, Stapleton J, Harvey P (1978) Prognostic features of idiopathic congestive cardiomyopathy. Am J Cardiol 41
64. Meinertz T, Hofmann T, Kasper W, Just H (1987) Prognostische Bedeutung ventrikulärer Arrhythmien bei dilativer Kardiomyopathie. Internist 28:164
65. Mestroni L, Miani D, Di Lenarda A, Silvestri F, Bussani R, Filippi G, Camerini F (1990) Clinical and pathologic study of familial dilated cardiomyopathy. Am J Cardiol 65: 1449–1453
66. Michels VV, Moll PP, Miller FA, Tajik AJ, Chu JS, Driscoll DJ, Burnett JC, Rodeheffer RJ, Chesebro JH, Tazelaar HD (1992) The frequency of familial dilated cardiomyopathy in a series of patients with idiopathic dilated cardiomyopathy. N Engl J Med 326:77–82
67. von Olshausen K, Schäfer A, Mehmel HC, Schwarz F, Senges J, Kübler W (1984) Ventricular arrhythmias in idiopathic dilated cardiomyopathy. Br Heart J 51:195
68. Onishi S, Wada A (1985) Heretofamilial factors as a cause of myocardial dammage. In: Sekiguchi M, Olsen EGJ, Goodwin JF (eds) Myocarditis and related disorders. Springer, Berlin Heidelberg New York, p 34
69. O'Connel JB, Fowles RE, Robinson JA (1983) Clinical and pathologic findings of myocarditis in two families with dilated cardiomyopathy. Am Heart J 107:127–135
70. Opherk D, Schuler G, Wetterauer K, Manthey J, Schwarz F, Kübler W (1989) Four-year follow-up study in patients with angina pectoris and normal coronary arteriograms ("syndrome X"). Circulation 80:1610–1616
71. Packer M (1987) Prolonging life in patients with congestive heart failure: the next frontier. Circulation 75 [Suppl IV]:1–3
72. de Paepa A, Kluyskens Y, van Durme JP, Naudts K, Claeys R, de Wagter X (1991) Familial idiopathic dilated cardiomyopathy. Acta Cardiol 46:577–582
73. Pavan D, Nicolosi GL, Lestuzzi C, Burelli C, Zardo F, Zanuttini D (1987) Normalization of variables of left ventricular function in patients with alcoholic cardiomyopathy after cessation of excessive alcohol intake: an echocardiographic study. Eur Heart J 8:535–540

74. Peters NS, Poole-Wilson PA (1991) Myocarditis-continuing clinical and pathologic confusion. Am Heart J 121:942–947
75. Pongpanich B, Isaraprasart S (1986) Congestive cardiomyopathy in infants and children. Clinical features and natural history. Jpn Heart J 27:11–15
76. Preliminary Report (1989) Effect of ecainide and flecainide on mortality in a randomized trial of arrhythmia suppression after myocardial infraction. N Engl J Med 321:406
77. Regitz V, Schug AL, Schüler S, Yanka C, Hetzer R, Fleck E (1988) Herzinsuffizienz bei dilatativer Kardiomyopathie und koronarer Herzerkrankung. Dtsch Med Wochenschr 113:781
78. Schmaltz AA, Apitz J, Hort W (1987) Dilated cardiomyopathy in childhood. Problems of diagnosis and long-term follow up. Eur Heart J 8:100
79. Schneider JF, Emerson TH, Kreger BE, McNamara PM, Kannel WB (1978) Newly aquired left bundle branch block: the Framingham Study. Ann Intern Med 90:303
80. Schuler G, Opherk D, Schwarz F, Görlich J, Mathey J (1986) Prognostic significance of concomitant left bundle branch block in patients with reduced coronary dilatory capacity with syndrome X. Circulation [Suppl] II:449
81. Schwarz F, Mall G, Cebe H, Schmitzer E, Manthey J, Scheurlen H, Kübler W (1984) Determinants of survival in patients with congestive cardiomyopathy: quantitative morphologic findings and left ventricular hemodynamics. Circulation 70:923–928
82. Sekiguchi M, Hasegawa A, Ando M (1978) A proposal for analyzing the familial occurrence of cardiomyopathy in the clinical situation. In: Sekiguchi M, Olsen EGJ (eds) Cardiomyopathy. University of Tokyo Press, Tokyo, p 445
83. Shanes JG, Ghali J, Billingham ME, Ferrans VJ, Fenoglio JJ, Edwards WD, Tsai CC, Saffits JE, Isner J, Furner S, Subramanian R (1987) Interobserver variability in the pathologic interpretation of endomyocardial biopsy result. Circulation 75:401
84. Shirey EK, Proudfit WL, Hawk WA (1980) Primary myocardial disease. Correlation with clinical findings, angiographic and biopsy diagnosis. Am Heart J 99:198
85. Solis E, Kaye MP (1986) The registry of the international society for heart transplantation: third official report – June 1986. J Heart Transplant 5:2
86. Torp A (1981) Incidence of congestive cardioymopathy. In: Goodwin JF, Hjalmarson A, Olsen EGJ (eds) Congestive cardiomyopathy. Hössle, Mölndal, p 18
87. Unverferth DV, Magorien RD, Moeschberger ML, Baker BT, Fetters JK, Leier CV (1984) Factors influencing the one year mortality of dilated cardiomyopathy. Am J Cardiol 54:147
88. Urie PM, Billingham ME (1988) Ultrastructural features of familial cardiomyopathy. Am J Cardiol 62:325–327
89. Voss EG, Reddy CVR, Detrano R, Virmani R, Zabrinskie JB, Fotino M (1984) Familial dilated cardiomyopathy. Am J Cardiol 54:456
90. Wang R, Mallon J, Alterman AI, McLellan AT (1987) Alcohol and dilated cardiomyopathy: incidence and correlation with clinical outcome. J Subst Abuse Treat 4:209–213
91. Webber S, Sandor G, Farquharson D, Talor G, Jamieson S (1991) Management and prognosis of dilated cardiomyopathy detected by foetal echocardiography. J Am Coll Card 17:257a (abstr)
92. Valantine HA, Hunt SH, Fowler MB, Billingham ME, Schroeder JS (1989) Frequency of familial nature of dilated cardiomyopathy and usefulness of cardiac transplantation in this subset. Am J Cardiol 63:959–963
93. Williams DG, Olsen EGJ (1985) Prevalence of overt dilated cardiomyopathy in two regions of England. Br Heart J 54:153–155
94. Yamaguchi M, Toshima H (1978) A genetic analysis of idiopathic cardiomyopathies in Japan. In: Sekiguchi M, Olsen EGJ (eds) Cardiomyopathy. University of Tokyo Press, Tokyo, pp 421–428
95. Yusuf S, Thom T, Abbot RD (1989) Changes in hypertension treatment and in congestive heart failure mortality in the United States. Hypertension 13 [Suppl 1]:74–79

Natural History of Idiopathic Dilated Cardiomyopathy

F. Camerini, A. Di Lenarda, G. Lardieri, L. Mestroni, A. Perkan, B. Pinamonti, A. Salvi, and G. Sinagra

Introduction

Idiopathic dilated cardiomyopathy (IDC) continues to be an etiological and pathogenetic enigma. Many possible causes have been considered and are being studied at present. Genetic factors, virus infections, autoimmunity alcohol, pregnancy, and slight hypertension may play a role. However the etiology continues to be unknown in a large percentage of patients. The possibility of different etiological factors and of various pathogenetic mechanisms may explain the heterogeneity of clinical presentation, functional status, complications, and characteristics of evolution and response to therapy of the disease [22].

In the past the prognosis of IDC was usually ominous, with a high mortality (nearly 50%) within the first 2 years after diagnosis [24]. However, more recent studies have indicated that in addition to the patients who present heart failure with a progressive downhill course, or who die suddenly and unexpectedly, there are others whose course remains stable for years, and who may also show improvement, sometimes surprising (Table 1).

A better understanding of the natural history of the disease and the need to identify prospectively patients at higher risk for early mortality have become increasingly important. There is the need to inform patients correctly about their future, to initiate programs of active treatment of heart failure and sometimes prevention of sudden death, and to schedule cardiac transplantation in time. Moreover, a better knowledge of the prognosis and course of the disease is necessary whenever pharmacological trials are planned and carried out in these patients.

Limitations of Studies

Many authors have drawn attention to the evolution of the disease and the factors which may influence it. However, these studies are impaired by many important limitations. First, the concept of "natural history" itself cannot be fully accepted because patients who have been studied in more recent years are frequently subjected to complex pharmacological interven-

Department of Cardiology, Ospedale Maggiore and University, Trieste, Italy

Figulla et al. (Eds.)
Idiopathic Dilated Cardiomyopathy

Table 1. Studies of mortality in IDC

Reference	Year	*n*	Mortality			Sudden death (%)
			1 year (%)	2 years (%)	5 years (%)	
Cuzzato [14]	1978	271	0	3	22	–
Kuhn [45]	1978	58	8	10	35	–
Convert [12]	1980	132	22	31	44	–
Koide [43]	1980	36	8	10	12	–
Lengyel [50]	1981	98	24	41	66	4
Fuster [24]	1981	104	32	46	62	–
Kuhn [46][a]	1982	258	8	14	29	–
Franciosa [23][a]	1983	87	23	48	–	50
Gavazzi [25]	1984	137	14	28	55	22.7
Unverferth [67]	1984	69	35	–	–	–
Figulla [21]	1985	56	17	26	39[c]	–
Diaz [15]	1987	169	28	41	57	8.6
Ikram [38]	1987	72	9	15	40	50
Ogasawara [55][a]	1987	111	3	14	47	–
Likoff [51]	1987	74	43[b]	–	–	–
Stevenson [62]	1987	28	45	75	–	–
Juilliere [39]	1988	111	10	26	50	19
Hofmann [34][a]	1988	110	–	34	–	64
Keogh [41]	1988	79	28	40	–	–
Razzolini [59]	1989	77	22	32	51	–
Varani [68]	1989	79	16	31	47	24
Ciaccheri [7][a]	1990	138	6	12	58	15
Gavazzi [26]	1990	178	5	9.5	–	39.2
Kelly [40][a]	1990	53	10	28	74	30
Komajda [44]	1990	201	0	4	16	28.5

[a] Prospective or partially prospective studies.
[b] 6 months.
[c] 4 years.

tions which may modify survival and evolution of the disease. Moreover, in some cardiological departments patients all stages of IDC are studied [44] while in tertiary referral centers patients are frequently admitted in a more advanced phase of the disease, frequently with potential indications for heart transplantation. Secondly, the great majority of studies are retrospective, with only few exceptions [7, 23, 34, 40, 46]; in Ogasawara et al.'s paper [55] part of the patients were analyzed retrospectively and part prospectively. The limitations of retrospective studies are well known: criteria of inclusion and exclusion may change with time, follow-up may not be rigorous, and loss of information may be high. For these reasons in some studies nearly 10% of patients had to be excluded because data were incomplete [15] while in others the cause of death was known in only few cases [59]. The problem is particularly relevant when aspects such as sudden death are considered.

Another problem which must be considered is that of methodology. The importance of a coronary angiography in these patients was stressed by Likoff et al. [51] and Franciosa et al. [23], who observed that in a population with congestive heart failure the subgroup affected by ischemic cardiomyopathy had a higher mortality than patients affected by IDC. In many published papers coronary angiography was not carried out systematically or is not even mentioned [12, 14, 15, 24, 25, 40, 41, 43, 50, 68]. It is therefore possible that some series may not be homogeneous for studies on mortality. A similar problem concerns endomyocardial biopsy. Although the role of this invasive procedure in diagnosis and prognosis is still debated [4], the diagnosis of active myocarditis (and sometimes of specific heart muscle diseases) can only be made using this technique. A review of the literature regarding this aspect shows that in the majority of studies endomyocardial biopsy was not mentioned or not carried out [7, 12, 14, 23–25, 34, 40, 43, 44, 50, 59, 68]. In some of them, only a subgroup was histologically studied [15, 21, 39, 41, 51] while in a minority of studies only [38, 45, 55, 62, 67] was endomyocardial biopsy systematically performed.

The study of factors which may identify patients at risk of sudden death shows that frequent complex ventricular arrhythmias (associated with depressed left ventricular function) [13, 34, 53, 60] may identify patients at risk of sudden death. For these studies a 24- or 48-h Holter monitoring is necessary. Again, Holter monitoring was not performed (or was not mentioned) in many series, mainly those dating back to 20 years ago [12, 14, 24, 25, 39, 43, 44, 59, 68]. In others only subgroups of patients were studied [7, 15, 55] while in a minority a systematic study was carried out in all cases [34, 51]. Only exceptionally was continuous monitoring performed in the coronary care unit [67].

Goodwin [29], discussing the possible etiological factors of IDC, notes a number of conditions related to the syndrome of IDC, such as alcohol, systemic hypertension, and pregnancy. Alcohol abuse was considered in many studies as a "risk factor" or a condition which may be related to or increase the severity of the disease [12, 15, 24, 25, 43]; however, in other studies alcoholic patients were considered separately [46], or ongoing alcohol abuse was a cause of exclusion [34, 40, 41]. Also different were the levels above which alcohol intake was considered dangerous: at least "2 quarts of beer" or "8 ounces of hard liquor" for the preceding 10 years [24], or "40 g daily" [41], "100 g/die" [34], ">1 g/kg/die" [7] etc. Other authors were more "optimistic," such as Alexander [1], whose definition of excess alcohol consumption was at least 4 pints of beer or two measures of whisky a day, and Brigden and Robinson [3], who defined alcoholism as a daily intake of 15 pints of beer or a 750-ml bottle of spirits. These data stress the difficulty in defining alcoholic heart disease as a specific entity and reflect our uncertainties about the etiological, potentiating, or conditioning role of alcohol in IDC and the quantity required to damage the myocardium. Similar considerations can be made regarding systemic hypertension. Oakley [54] mentioned the possible relation between high blood pressure and the natural history of the disorder, considering high blood pressure as a cofactor of congestive failure in a heart which has been damaged by other agents. While "severe hypertension" is considered a cause of exclusion, and in this case a dilated dysfunctioning heart is usually classified as hypertensive heart disease, moderate hypertension was considered by some authors as a cofactor in the etiology of the disease. For example, only blood pressure of at least 170/100 mmHg or above 180/100 mmHg or "treatment with at least two drugs" were causes of exclusion in Ciaccheri et al.'s [7] and Juillière et al.'s papers [39]. Of the 69 patients in the series of Unverferth et al. 23 [67] had a history of blood pressure over 140/90 mmHg, while Ikram et al. [38] considered hypertension as a cause of exclusion. Knowledge of the values of blood pressure in patients with IDC is important because the presence of a history of slight hypertension or the detection of values of blood pressure higher than normal (between 140/90 and 170/100 mmHg) was found to be one of the factors predictive of improvement [17]. Similarly, Hatle et al. [33] reported a favorable course in patients who developed systemic hypertension after successful treatment for heart failure.

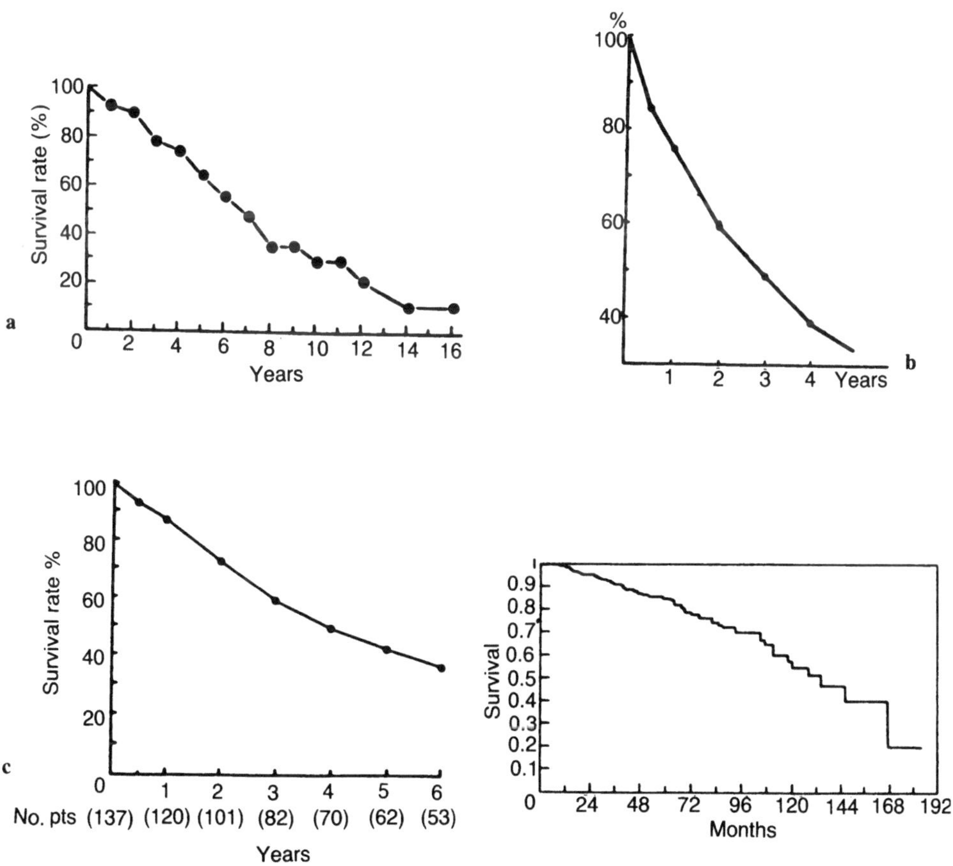

Fig. 1a–d. Examples of survival curves with constant and progressive slopes in the course of follow-up. **a** Kuhn et al. [45]; **b** Lengyel and Kökeny [50]; **c** Gavazzi et al. [25]; **d** Komajda et al. [44]

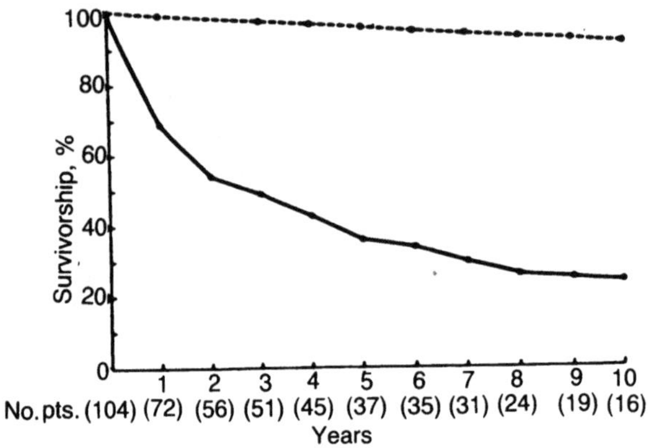

a

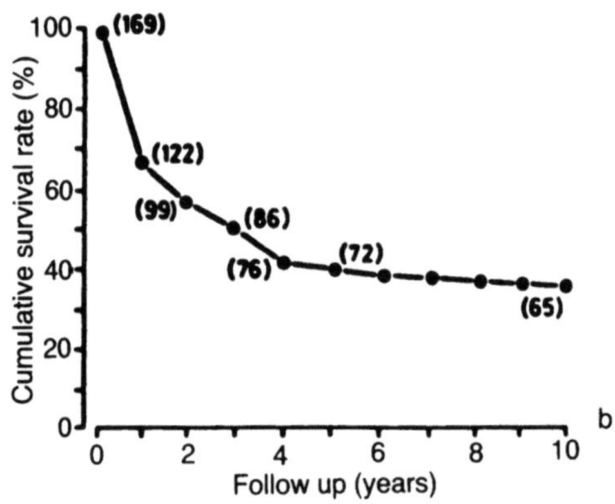

b

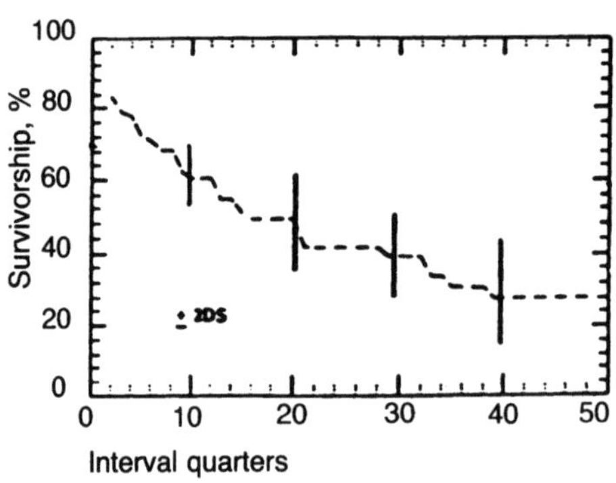

c

Fig. 2a–c. Examples of survival curves characterized by high mortality in the first years. Later mortality was reduced and the remaining showed stabilization or improvement. **a** Fuster et al. [24]; **b** Diaz et al. [15]; **c** Razzolini et al. [59]

Survival and Mortality

The analysis of the survival curves of the series published in the past 14 years indicates the presence of two phenomena: the different shapes and slopes of the survival curves and the highly variable incidence of mortality after 1, 2, and 5 years. In some studies [25, 38, 44, 45, 50] mortality was homogeneously progressive with constant slopes of the survival curves (Fig. 1) while in others [15, 24, 59] (Fig. 2) mortality was very high in the first years, with a steep descent of the curve. Later, after 4–5 years, mortality was reduced, and the remaining patients showed a stabilization or even a striking improvement. A similar evolution of the disease was observed by Diaz et al. [15], who noticed that patients who survived 5 years "were unlikely to die thereafter," while Figulla et al. [21] suggested that the pathogenesis of the disease may not be unique and that IDC can be divided into progressive and nonprogressive forms. All these data suggest not only a clinical and hemodynamic involvement of variable degree, but also the possibility that different etiological factors or cofactors may be operating in different patients.

The highly variable rates of mortality among different studies can be explained by various factors such as the changing degree of severity of the disease and the possibility of an earlier diagnosis. Moreover, the survival rates may be clearly different if the earliest manifestations of the disease, the first appearance of subjective symptoms, or the development of cardiac failure are considered [14, 43]. Data available 15 years ago suggested that the course of the disease is usually steadily downhill, and that death commonly ensues within 6 months–5 years after the onset of symptoms [29]. However, more recent observations, with earlier detection of the disease and the use of new treatments [angiotensin-converting enzyme (ACE) inhibitors, beta-blockers] suggest the possibility of changes in mortality and evolution of IDC [16].

In a further extension of some preliminary observations we studied the course of the disease in 131 patients affected by IDC. The patient population was divided into three groups: group 1 (July 1 1971–June 30 1981; $n = 23$), group 2 (July 1 1981–June 30 1986; $n = 48$), and group 3 (July 1 1986–June 30 1989; $n = 60$). The end of follow-up was July 1990. Twenty-nine clinical, noninvasive and invasive parameters obtained at first examination were analyzed. A survival study of the three groups showed a significant improvement over time (1 year). Indeed, while survival of patients of group 1 was similar to that observed by Fuster et al. [24], group 2 and especially group 3 patients showed a significant reduction in mortality. Moreover, cardiac transplantation further increased survival in patients with end-stage IDC. Patients of group 3 showed a significantly lower mean age, NYHA functional class, cardiothoracic ratio, and a shorter duration of symptoms lower incidence of heart failure, mitral regurgitation, first-degree AV block, and left bundle branch block. In addition, they were more frequently treated with ACE inhibitors and beta-blockers. These data thus suggest that a different selection of patients and new treatments may have modified the

Table 2. Studies of prognostic factors in DCM: predictors of poor outcome

	Predictive value present	Predictive value absent
Age	[24, 38, 43]	[8, 13, 18, 19, 21, 23, 34, 39, 44, 45, 50, 52, 59, 64, 67]
Sex	[43, 68]	[18, 21, 38, 39, 44, 52, 67]
Symptom duration	[43, 44[a]]	[13, 21, 23, 45, 50, 62, 64, 67]
Onset dis (flu-like)	–	[15, 50, 67]
NYHA class	[12, 13[a], 39, 42[a], 44, 50, 59, 68[a]]	[18, 21, 34, 52, 60, 62]
Alcohol	–	[15, 21, 39, 44, 50, 67]
S3 gallop	–	[15, 67]
Mitral rigurg.	[43]	[15, 21, 67]
LV hypetr (ECG)	–	[12, 13]
Heart rate ↑	[68[a]]	[21, 23]
AF	[34[a], 38, 60, 67]	[12, 13, 15, 18, 42, 43, 44, 51]
AV-IV cond. delay	[12, 16, 57, 67[a], 68]	[13, 21, 34, 39, 44, 50, 51, 60]
VE arrhythmias	[13[a], 18, 34[a], 37[a], 53, 60[a], 62[a], 67[a]]	[15, 39, 42, 51, 56, 65]
CTR	[18[a], 24, 38, 39, 43, 44, 45]	[21]
LA	[50]	–
E point	[50]	–
LV wall thickness	[50]	[19, 67]
LVEDD	[19[a], 34, 44[a]]	[13, 18, 21, 41, 50, 64, 67]
LVESD	[19, 44]	[18]
Wall motion abnormalities	[12]	–
Δ%	[18, 67]	[19, 42, 50]
LV filling pattern	[8[a], 58]	–
LV geometry	[19]	–
LVEDV	[15, 34, 44, 45, 59]	[8, 12, 21, 60]
LVESV	[15, 44[a], 59]	[8, 21, 60]
LVEF	[8[a], 12, 13, 15, 34[a], 39[a], 44, 45, 51[a], 59[a], 61[a], 67]	[18, 21, 38, 42, 52, 60, 64]
RAP	[44, 59, 60, 64, 67[a]]	[62]
LA-PCWP	[12, 13, 18, 23, 31, 42, 43, 52, 59[a], 64, 67[a]]	[62]
LVEDP	[12, 15, 18, 24, 34, 44, 59, 61]	[13, 60]
RVEDP	[60]	–
SVI	[12, 62[a]]	[15, 59, 67]
LVSWI	[12, 23, 30]	–
PAP	[12, 44· 60]	[21, 39, 42]
PVR	–	[41, 58]
SVR	[23]	–
SAP	[40]	–
CI	[13[a], 23, 24, 34[a], 43, 44, 59, 62, 64]	[12, 18, 42, 52, 60, 67]
Histological parameters	[21, 32, 45, 61]	[61[a], 67]
Exercise hemod	[28, 31]	–
VO_2 max	[47, 51[a], 52[a], 66]	–
Perfusion defects	[18]	–
Neurohorm. fact.	[42[a], 64[a]]	–

AF, Atrial fibrillation; CTR, cardiothoracic ratio; LA, left atrium; LVEDD, left ventricular end-diastolic diameter; LVESD, left ventricular end-systolic diameter; Δ%, shortening fraction; LVEDV, left ventricular end-diastolic volume; LVESV, left ventricular end-systolic volume; LVEF, left ventricular ejection fraction; RAP, right atrial pressure; LA-PCWP, left atrium-pulmonary capillary wedge pressure; LVEDP, left ventricular end-diastolic pressure; RVEDP, right ventricular end-diastolic pressure; SVI, stroke volume index; LVSWI, left ventricular stroke work index; PAP, pulmonary artery pressure; PVR, pulmonary vascular resistance; SVR, systemic vascular resistance; SAP, systemic arterial pressure; CI, cardiac index; VO_2 max, oxygen consumption.

[a] Multivariate analysis.

baseline characteristics and survival curve of patients affected by IDC studied in recent years.

Predictors of Poor Outcome

Several authors have tried to identify prognostic factors that predict a poor outcome in IDC in an attempt to select patients for cardiac transplantation or for more aggressive forms of treatment. There are no univocal data regarding the prognostic value of the different parameters considered (Table 2), but some factors emerge as major and useful prognostic indicators.

Functional Data

Functional status evaluated by the NYHA class [12, 13, 39, 42, 44, 50, 59, 68] and particularly by peak O_2 consumption during exercise is an important prognostic indicator [51, 52, 66]. Mancini et al. [52] recently identified patients with the worst prognosis or those eligible for heart transplantation if peak VO_2 during exercise was under 14 ml/kg per minute, or even more so (the outcome was even worse) if this value was under 10 ml/kg. Other papers [28, 31] suggest that a more complex assessment may be based on hemodynamic evaluation during exercise.

Left Ventricular Dimensions, Geometry, and Function

In some studies a simple parameter such as cardiothoracic ratio showed prognostic value [18, 24, 38, 39, 43–45]. These data were confirmed by studies which evaluated left ventricle dimensions or volumes [15, 19, 34, 44, 45, 59]. With few discrepancies there is agreement on the poor outcome of patients with an enlarged left ventricle (end diastolic diameter over 7.6 cm) [19] or with severe impairment of left ventricular function (left ventricular ejection function below 35%) [12, 39, 41]. In this context left ventricular ejection fraction emerges not only as an independent prognostic indicator [8, 12, 13, 15, 34, 39, 44, 45, 51, 59, 61, 67] but also as an important decision-making factor for the timing of heart transplantation [41].

In Stevenson et al. study's [62], which included only patients with left ventricular ejection fraction below 25%, 1-year mortality was 64%. Keogh et al. [41] demonstrated that patients referred for heart transplantation and with left ventricular ejection fraction greater than 20% had a relatively low priority for transplant (no patient died within 6 months) while the mortality after 6 months was very high in patients with 19% or less ejection fraction. Recently, two interesting prognostic factors were studied: the first regarding the prognostic value of left ventricular architecture [19] and the second the role of right ventricular function [60]. Douglas et al. [19] found that alterations of the ellipsoid left ventricular cavity and an abnormal distribution of afterload were associated with a higher mortality rate. Keogh et al. [42] and Romeo et al. [60], on the other hand, showed a shortened survival and higher incidence of sudden death in patients with impaired right hemodynamics. This late aspect suggests that right ventricular dysfunction might play also a role on the genesis of dysrhythmias [60].

Conduction Delay and Arrhythmias

Left ventricular conduction delay has been reported to be a major determinant of prognosis [12, 17, 57, 67, 68], but most of the studies failed to confirm these data [13, 15, 21, 34, 39, 44, 50, 51, 60]. The role of atrial fibrillation is controversial, and with the exception of the study of Convert et al. [12], in which atrial fibrillation was an indicator of better prognosis, in most studies this factor was prognostically indifferent [12, 13, 15, 18, 42–44] or a significant predictor of poor prognosis in others [34, 38, 60, 67].

Assessment of the prognostic role of complex ventricular arrhythmias is definetely more problematic. In this context methodological hetereogeneity of the studies and different definitions adopted are important factors that may explain the wide range of results. Whereas some authors reported a significant correlation between frequency and degree of complexity of ventricular arrhythmias and incidence of sudden cardiac death [34, 35, 53], others failed to demonstrate such correlation [15, 36, 39, 42, 51, 56, 65]. Hofmann et al. [34] and Meinertz et al. [53] showed that complex ventricular arrhythmias (more than 20 couplets/24 h and/or more than 20 episodes/24 h of nonsustained ventricular tachycardia in Meinertz et al.'s study [53]) predict a high probability of sudden death in presence of impaired left ventricular function (LVEF less than 40%). Similar results (importance of multiform ectopic ventricular beats) in patients with reduced cardiac index were obtained by Costanzo Nordin et al. [13]. In this study a correlation between complex ventricular arrhythmias and sudden death was observed, but unlike with Meinertz et al. [53], ventricular tachycardia did not show a prognostic significance based on univariate and multivariate analysis. Finally, Romeo et al. [60] using multivariate analysis identified the association of complex ventricular arrhythmias (Lown 4), right ventricular hemodynamic impairment, and atrial fibrillation as independent prognostic indicator of sudden death, but they did not confirm the concomitant prognostic role of left ventricular dysfunction [53].

Histological Parameters

Recently some studies have suggested that a quantitative approach based on morphometric analysis obtained by electron or light microscopy could help in prognostic stratification [21, 32, 45, 61]. Moreover Figulla et al. [21] and Hammond et al. [32] showed that myofibrillar volume loss represents an appropriate prognostic factor predictive of hemodynamic deterioration and death.

Hemodynamic Parameters

Many studies have shown different hemodynamic parameters to have prognostic value: cardiac index [13, 23, 24, 34, 43, 44, 59, 62, 64], pulmonary capillary wedge pressure [12, 13, 18, 23, 31, 42, 43, 52, 59, 64, 67], end-diastolic left ventricular pressure [12, 13, 15, 18, 24, 34, 44, 59, 61], right atrial pressure [44, 59, 60, 64, 67], and pulmonary artery pressure [12, 44, 59, 60]. Stevenson et al. [63] recently emphasized the potential prognostic role of medical treatment and clearly demonstrated that survival in a group of patients referred for heart transplantation was not predicted by the severity of hemodynamic compromise at the moment of the referral but by hemodynamic pattern after individually tailored medical treatment.

Recent Advances

Left Ventricular Filling Pattern

Different left ventricular filling patterns may be observed with echo Doppler and radionuclide ventriculography [2, 8, 48]. Some patients show a predominant filling during atrial contraction while others show a "restrictive" filling pattern, characterized by rapid early diastolic filling, rapid deceleration of flow (short E deceleration time with Doppler), and reduced filling during atrial contraction (high E/A ratio wtih Doppler) [2] (Fig. 3). We were able to demonstrate [58] that a "restrictive" filling pattern is frequent in dilated cardiomyopathy (38/79 patients, 48%) or is associated with a more

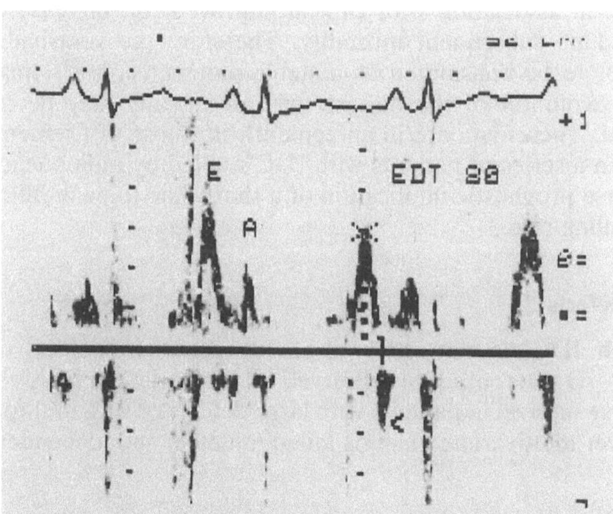

Fig. 3. An example of transmitral Doppler curve with restrictive filling pattern in a patient with dilated cardiomyopathy. The curve is characterized by predominant E wave and short E deceleration time (*EDT*)

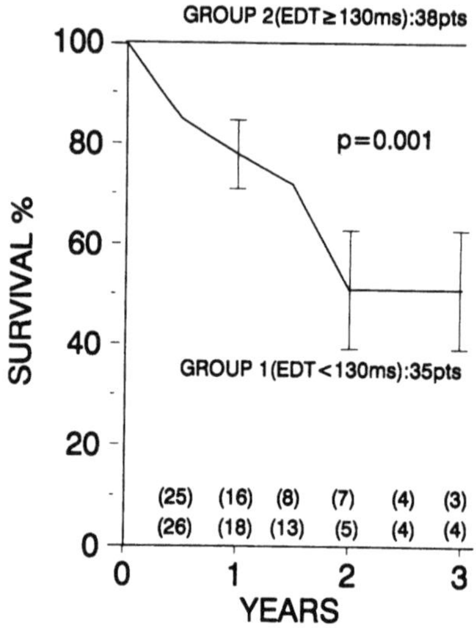

Fig. 4. Actuarial survival curves of two groups of patients with dilated cardiomyopathy. Survival is significantly lower in patients of group 1. with restrictive left ventricular filling pattern at Doppler (E deceleration time, *EDT*, <130 ms)

severe disease (Fig. 4). Moreover, the persistence during follow-up of a restrictive filling pattern was a prognostic marker of high incidence of events such as cardiac transplantation or cardiac death. Considering the prognostic value, it must be noted that during a mean follow-up of 15 months, all 11 patients of this series who died or required cardiac transplantation had shown a persistent "restrictive" filling pattern. However, a subgroup of 15 patients with restrictive filling pattern at diagnosis, showed a normalization during the follow-up, in association with clinical improvement, increase of ejection fraction, and no subsequent mortality. Therefore the ventricular filling pattern has not to be considered as a stable parameter, but it may vary in response to evolution of the disease and presumably may be influenced by treatment. These data are in agreement with those of Clements et al. [8] who found in a series of patients with IDC studied by radionuclide angiography an adverse prognostic implication of a short time to peak filling rate and high early filling rate.

Thallium Perfusion Defects

Usually patients with IDC do not show reversible perfusion defects on Tallium scintigraphy. Doi et al. [18] observed at univariate analysis a significant lower 3-year survival in patients with large defects or with multiple small defects. However multivariate analysis failed to detect an independent prognostic value.

Neurohormonal Assessment

The activation of neurohormonal systems in IDC could contribute to accelerate the deterioration of left ventricular function. As demonstrated in several studies, the most marked activation of the sympathetic nervous system (elevated plasma norepinephrine levels) and of the renin-angiotensin system (hyponatriemia) significantly contribute to higher mortality [9, 49, 64]. Moreover, attention was recently devoted to vasodilator hormones (prostaglandine, atrial natriuretic peptide), and Gottlieb et al. [30] and Keogh et al. [42] demonstrated the independent prognostic value of a high atrial natriuretic peptide concentration.

Comments

The following conclusions can be drawn from the existing data.

- It is unlikely that a single parameter can represent a major or unique determinant of prognosis.
- Different methodological approaches, selection criteria, statistical methods adopted and duration of follow-up may affect results.
- pooling all the literature data shows that a comprehensive approach (combining history, physical examination, invasive and noninvasive parameters) allows a reliable prognostic stratification.
- Data on age, sex, onset modality, and symptom duration are of limited value.
- The prognostic role of atrial fibrillation and atrioventricular and intraventricular conduction delay remain controversial.
- Ventricular dimensions and function, hemodynamic compromise, particularly if persistent after individually tailored medical treatment, represent important prognostic factors.
- Complex ventricular arrhythmias if associated with severely impaired left and right ventricular function have a predictive value, particularly regarding sudden death.
- Morphometric demonstration at endomyocardial biopsy of a reduced myofibrillar volume fraction may have prognostic relevance.
- Research in the field of neurohormonal assessment, evaluation of right ventricular function, left ventricular architecture and afterload, functional capacity, metabolic and perfusion patterns, left ventricular filling patterns, and late potentials can contribute to a better prognostic stratification and timing for heart transplantation.

Improvement in IDC

Some studies [15, 24, 25, 59] have shown that the majority of deaths occur within 2–3 years after the beginning of symptoms, and that survival curves after a rapid slope during the first years tend to become flat and sometimes almost parallel to those of general population [15, 24]. This aspect may suggest that patients who survive 3–4 years may have a stable course or even a sustained improvement [15]. Some 20%–40% of patients with DCM may do so [15, 21, 24, 25, 41, 62]. The probability of clinical improvement may be reduced in the very advanced stage of the disease. Nevertheless about 20% of patients referred for cardiac transplantation [41, 62] or with NYHA functional class 3 or 4 [25, 41] or very depressed left ventricular ejection fraction [21, 41] may benefit from a stabilization or improvement of symptoms. However, only in two of these studies was the classification of severity of patients based on ejection fraction [21, 41].

There are few studies analyzing the factors predictive of improvement. A short disease history (<7 months [62]; <12 months [41]) and a NYHA functional class 1 or 2 are more frequently associated with a good prognosis. According to Figulla et al. [21] improvement in dilated cardiomyopathy is associated to a myofibril volume fraction on endomyocardial samples higher than 60%. Therapeutic agents such as ACE inhibitors may also improve survival, ejection fraction, and quality of life [5, 6, 10, 11]. Moreover, beta-blockers seem to influence improvement in about 60% of patients with dilated cardiomyopathy [20, 27, 69]. These preliminary data must be confirmed by the ongoing multicenter trial on metoprolol in IDC.

In a recent analysis [17] of our study population we tried to identify the patients who improve and which factors are associated with improvement in IDC. A total of 94 patients, followed for 24 ± 6 months, were classified according to a clinical and instrumental score, including factors related to changes in left ventricular function and dimension, NYHA functional class, physical examination, cardiothoracic ratio and exercise time. Two groups of patients emerged: 36 patients (38.3%) improved; 27 patients (28.7%), stable or worsened, were considered together with the 31 patients (33%) who underwent cardiac transplantation (8 cases, 8.5%) or died (23 cases, 24.5%) within the same period. At the last follow-up visit the symptomatic state according to NYHA classification of the 36 improved patients was significantly better, and no patient showed clinical signs of heart failure. Mean left ventricular end-diastolic diameter decreased from 70 to 62 mmHg, and mean ejection fraction improved from 25% to 43% (a highly significant increase of 70%). More than one third of these patients (13) showed a normalization of ejection fraction, and 5 of them (13.9%) had an apparent complete recovery. Ejection fraction did not play a major role in improvement. In fact, in some cases a significant clinical improvement was also seen in patients with ejection fraction less than 15%. In a discriminant analysis indicators of improvement (in order of significance) were: beta-blocker therapy, previous history or blood pressure between 140/90 and 170/100 mmHg, ACE inhibitors therapy, left ventricular end-systolic volume below <105 ml/m^2, NYHA class 1 or 2, and history of disease of 12 months or less before diagnosis.

On the basis of these data we defined three groups: (a) high probability of improvement: 27 patients with the presence of at least four of the six screened parametes; (b) moderate probability of improvement: 49 patients with the presence of two or three independent predictors; and (c) low probability of improvement: 18 patients with only one or more of these parameters. Patients of the first group did very well, with about 90% having estimated probability of sustained improvement. On the other hand, no patient of the third group improved during follow-up. In conclusion it was possible to confirm that a subgroup of patients with dilated cardiomyopathy show sustained improvement during follow-up, and that therapeutic agents such as beta-blockers and ACE inhibitors may significantly influence the course of the disease.

Conclusions

Studies of the natural history of IDC are characterized by many methodological limitations; the majority are retrospective, and differing etiological and diagnostic criteria have been considered. Mortality, very high in the oldest studies, is at present significantly reduced, and in more than one third of patients improvement is possible. In a small number an apparent complete recovery may be observed. Predictors of poor outcome are at present well known, although the role of some factors, such as complex ventricular arrhythmias, is not completely elucidated. A better knowledge of natural history of IDC is important for rational planning of treatment and for correct information to patients.

References

1. Alexander CS (1966) Idiopathic heart disease. I. Analysis of 100 cases with special reference to chronic alcoholism. Am J Med 41:213–228
2. Appleton CP, Hatle LK, Popp RL (1988) Demonstration of restrictive ventricular physiology by Doppler echocardiography. J Am Coll Cardiol 11:757–768
3. Bridgen WW, Robinson JF (1964) Alcoholic heart disease. BMJ ii:1283–1289
4. Camerini F, Salvi A, Sinagra G (1991) Endomyocardial biopsy in dilated cardiomyopathy and myocarditis: which role? Int J Cardiol 31.1–8
5. Captopril Multicenter Research Group (1983) A placebo-controlled trial of captopril in refractory chronic congestive heart failure. J Am Coll Cardiol 24:755–763
6. Captopril-Digoxin Multicenter Research Group (1988) Comparative effects of therapy with captopril and digoxin in patients with mild to moderate heart failure. JAMA 259:539–544.
7. Ciaccheri M, Castelli G, Nannini M, Santoro G, Troiani V, Di Lenarda A, Miani D, Sinagra G, Mestroni L, Risoli A, Palli D, Dolara A, Camerini F (1990) Valutazione prognostica della cardiomiopatia dilatativa: follow up di 138 pazienti. G Ital Cardiol 20: 645–650
8. Clements IP, Brown ML, Zinsmeister AR, Gibbons RJ (1991) Influence of left ventricular diastolic filling on symptoms and survival in patients with decreased left ventricular systolic function. Am J Cardiol 67:1245–1250
9. Cohn JN, Levine TB, Olivari MT, Garberg V, Lura D, Francis GS, Simon AB, Rector T (1984) Plasma norepinephrine as a guide to prognosis in patients with chronic congestive heart failure. N Engl J Med 311:819–823
10. Cohn JN, Johnson G, Ziesche S, Cobb F, Francis G, Tristani F, Smith R, Dunkman WB, Loeb H, Wong M, Bhat G, Goldman S, Fletcher RD, Doherty J, Hughes CV, Carson P, Cintron G, Shabetai R, Haakenson C (1991) A comparison of enalapril with hydralazine-isosorbide dinitrate in the treatment of chronic congestive heart failure. N Engl J Med 325:303–310
11. CONSENSUS Trial Study Group (1987) Effects of enalapril on mortality in severe congestive heart failure. Results of the Cooperative North Scandinavian Enalapril Survival Study (CONSENSUS). N Engl J Med 316:1429–1435

12. Convert G, Delaye J, Beaune J, Biron A, Gonin A (1980) Etude pronostique des myocardiopathies primitives non obstructives. Arch Mal Coeur 3:227–237
13. Costanzo-Nordin MR, O'Connel JB, Engelmeier RS, Moran JF, Scanlon PJ (1985) Dilated cardiomyopathy: functional status, hemodynamics, arrhythmias, and prognosis. Cathet Cardiovasc Diagn 11:445–453
14. Cuzzato V, Giommi L, Frigo G, Candiani L, Travagli E, Rispoli T, Feder G (1978) La storia naturale delle cardiomiopatie primitive. Proceedings of the IX congress of the ANMCO, pp 123–130
15. Diaz RA, Obasohan A, Oakley CM (1987) Prediction of outcome in dilated cardiomyopathy. Br Heart J 58:393–399
16. Di Lenarda A, Lardieri G, Mestroni L, Miani D, Pinamonti B, Salvi A, Sinagra G, Camerini F (1990) Dilated cardiomyopathy: changing survival in the last 20 years. Circulation 82 [Suppl II]:387
17. Di Lenarda A, Lardieri G, Mestroni L, Miani D, Pinamonti B, Salvi A, Sinagra G, Camerini F (1991) Natural history in dilated cardiomyopathy: is improvement possible? Eur Heart J 12 [Suppl]:77 (abstr)
18. Doi YL, Chikamori T, Takata S, Yonezawa Y, Poloniecki JD, Ozawa T, McKenna WJ (1991) Prognostic value of thallium-201 perfusion defects in idiopathic dilated cardiomyopathy. Am J Cardiol 67:188–193
19. Douglas PS, Morrow R, Ioli A, Reichek N (1989) Left ventricular shape, afterload and survival in idiopathic dilated cardiomyopathy. J Am Coll Cardiol 13:311–315
20. Engelmeier RS, O'Connell JB, Walsh R, Rad N, Scanlon PJ, Gunnar RM (1985) Improvement in symptoms and exercise tolerance by metoprolol in patients with dilated cardiomyopathy: a double-blind, randomized, placebo-controlled trial. Circulation 72:536–546
21. Figulla HR, Rahlf G, Nieger M, Luig H, Kreuzer H (1985) Spontaneous hemodynamic improvement or stabilization and associated biopsy findings in patients with congestive cardiomyopathy. Circulation 71:1095–1104
22. Fowles RE (1990) Natural history and prognosis of overt dilated cardiomyopathy. In: Baroldi G, Camerini F, Goodwin JF (eds) Advances in cardiomyopathies. Springer, Berlin Heidelberg New York, pp 337–346
23. Franciosa JA, Wilen M, Ziesche S, Cohn JN (1983) Survival in men with severe chronic left ventricular failure due to either coronary heart disease or idiopathic dilated cardiomyopathy. Am J Cardiol 51:831–836
24. Fuster V, Gersh BJ, Giuliani ER, Tajik AJ, Brandenburg RO, Frye RL (1981) The natural history of idiopathic dilated cardiomyopathy. Am J Cardiol 47:525–531
25. Gavazzi A, Lanzarini L, Cornalba C, Desperati M, Raisaro A, Angoli L, De Servi S, Specchia G (1984) Dilated (congestive) cardiomyopathy. G Ital Cardiol 14:492–498
26. Gavazzi A, De Maria R, Caroli A, Di Lenarda A, Veritti U, Miani D, Sinagra G, Luvini M, Ometto R, Borgia M, Ciaccheri W, Campana C, Graziano G, Lanzarini L, Gronda E, Camerini F (1990) Complex ventricular arrhythmias in dilated cardiomyopathy: a multicenter italian experience. In: Baroldi G, Camerini F, Goodwin JF (eds) Advances in cardiomyopathies. Springer, Berlin Heidelberg New York, pp 354–367
27. Gilbert EM, Mestroni L, Anderson JL, Woadley SL, Bristow MR (1989) Can response to beta-blocker therapy in idiopathic dilated cardiomyopathy be predicted by baseline parameters? Circulation 80:II–428
28. Goldrach S, Haissaguerre M, Broustet JP (1987) Valeur pronostique de l'exploration hemodynamique de repos et de l'épreuve d'effort dans 103 cas de myocardiopathies primitives dilatées. Arch Mal Coeur 9:1333–1340
29. Goodwin JF (1978) Congestive cardiomyopathy. In: Hurst JW, Logue RB, Schlant RC, Wenger NK (eds) The heart. McGraw-Hill, New York, pp 1567–1580
30. Gottlieb SS, Kukin ML, Ahern D, Packer M (1989) Prognostic importance of atrial natriuretic peptide in patients with chronic heart failure. J Am Coll Cardiol 13:1534–1539
31. Griffin BP, Shah PK, Ferguson J, Rubin SA (1991) Incremental prognostic value of exercise hemodynamic variables in chronic congestive heart failure secondary to coronary artery disease or to dilated cardiomyopathy. Am J Cardiol 67:848–853

32. Hammond EH, Menlove RL, Anderson JL (1987) Predictive value of immuno-fluorescence and electron microscopic evaluation of endomyocardial biopsies in the diagnosis and prognosis of myocarditis and idiopathic dilated cardiomyopathy. Am Heart J 114:1055–1065
33. Hatle L, Orjaviko O, Storstein O (1976) Chronic myocardial disease. Acta Med Scand 199:399–405
34. Hofmann T, Meinertz T, Kasper W, Geibel A, Zehender M, Hohnloser S, Stienen U, Treese N, Just H (1988) Mode of death in idiopathic dilated cardiomyopathy: multivariate analysis of prognostic determinants. Am Heart J 116:1455–1463
35. Holmes J, Kubo SH, Cody RJ, Kligfield P (1985) Arrhythmias in ischemic and nonischemic dilated cardiomyopathy: prediction of mortality by ambulatory electrocardiography. Am J Cardiol 55:146–151
36. Huang SK, Messer JV, Denes P (1983) Significance of ventricular tachycardia in idiopathic dilated cardiomyopathy: observations in 35 patients. Am J Cardiol 51:507–512
37. Ikegawa T, Chino M, Hasegawa H, Usuba F, Suzuki S, Ookura M, Nishikawa K (1987) Prognostic significance of 24-hour ambulatory electrocardiographic monitoring in patients with dilative cardiomyopathy: a prospective study. Clin Cardiol 10:78–82
38. Ikram H, Williamson HG, Won M, Crozier IG, Wells EJ (1987) The course of idiopathic dilated cardiomyopathy in New Zeland. Br Heart J 57:521–527
39. Juilliere Y, Danchin N, Brianson S, Khalife K, Ethevenot G, Balaud A, Gilgenkrantz JM, Pernot C, Cherrier F (1988). Dilated cardiomyopathy: long-term follow-up and predictors of survival. Int J Cardiol 21:269–277
40. Kelly TL, Cremo R, Nielsen C, Shabetai R (1990) Prediction of outcome in late-stage cardiomyopathy. Am Heart J 119:1111–1121
41. Keogh AM, Freund J, Baron DW, Hickie JB (1988) Timing of cardiac transplantation in idiopathic dilated cardiomyopathy. Am J Cardiol 61:418–422
42. Keogh AM, Baron DW, Hickie JB (1990) Prognostic guides in patients with idiopathic or ischemic dilated cardiomyopathy assessed for cardiac transplantation. Am J Cardiol 65: 903–908
43. Koide T, Kato A, Takabatake Y, Iizuka M, Uchida Y, Ozeki K, Morooka S, Kakihana M, Serizawa T, Tanaka S, Ohya T, Momomura S, Murao S (1980) Variable prognosis in congestive cardiomyopathy. Role of left ventricular function, alcoholism and pulmonary thrombosis. Jpn Heart J 21:451–463
44. Komajda M, Jais JP, Reeves F, Goldfarb B, Bouhour JB, Juilliere Y, Lanfranchi J, Peycelon P, Geslin Ph, Carrie D, Grosgogeat Y (1990) Factors predicting mortality in idiopathic dilated cardiomyopathy. Eur Heart J 11:824–831
45. Kuhn H, Breithardt G, Knieriem HJ, Kohler E, Losse B, Seipel L (1978) Prognosis and possible presymptomatic manifestations of congestive cardiomyopathy (COCM). Postgrad Med J 54:451–459
46. Kuhn H, Becker R, Fisher J, Curtius JM, Losse B, Hort W, Loogen F (1982) Untersuchungen zur Atiololgie, zum Verlauf und zur Prognose der dilatativen Kardiomyopathie (DCM). Z Kardiol 71:497–508
47. Latham RD, Thornton JW, Mulrow JP (1987) Cardiovascular reserve in idiopathic dilated cardiomyopathy as determined by exercise response during cardiac catheterization. Am J Cardiol 59:1375–1379
48. Lavine SJ (1991) Left ventricular diastolic function in idiopathic cardiomyopathy: Doppler hemodynamic correlations. Echocardiography 8:151–161
49. Lee WH, Packer M (1986) Prognostic importance of serum sodium concentration and its modification by converting enzyme inhibition in patients with severe heart failure. Circulation 73:257–267
50. Lengyel M, Kokeny M (1981) Follow-up study in congestive (dilated) cardiomyopathy. Acta Cardiol 36:35–48
51. Likoff MJ, Chandler SL, Kay HR (1987) Clinical determinants of mortality in chronic congestive heart failure secondary to idiopathic dilated or to ischemic cardiomyopathy. Am J Cardiol 59:634–638

52. Mancini DM, Eisen H, Kussmaul W, Mull R, Edmunds LH, Wilson JR (1991) Value of peak exercise oxygen consumption for optimal timing of cardiac transplantation in ambulatory patients with heart failure. Circulation 83:778–786
53. Meinertz T, Hofmann T, Kasper W, Treese N, Bechtold H, Stienen U, Pop T, Leither EV, Andresen D, Meyer J (1984) Significance of ventricular arrhythmias in idiopathic dilated cardiomyopathy. Am J Cardiol 53:902–907
54. Oakley C (1978) Diagnosis and natural history of congested (dilated) cardiomyopathies. Postgrad Med J 54:440–447
55. Ogasawara S, Sekiguchi M, Hiroe M, Komazawa T, Kawai Y, Fujita N, Nakamura K, Kasanuki H, Tanaka E, Kobayashi K, Ohta M, Hirosawa K (1987) Prognosis of dilated cardiomyopathy: for a retrospective to a prospective study employing multivariate analysis. Jpn Circul J 51:699–706
56. Olshausen K, Schafer A, Mehmel HC, Schwarz F, Senges J, Kubler W (1984) Ventricular arrhythmias in idiopathic dilated cardiomyopathy. Br Heart J 51:195–201
57. Olshausen KV, Stienen U, Schwarz F, Kubler W, Meyer J (1988) Long-term prognostic significance of ventricular arrhythmias in idiopathic dilated cardiomyopathy. Am J Cardiol 61:146–151
58. Pinamonti B, Di Lenarda A, Sinagra G, Lardieri G, Rustja D, Camerini F (1991) Restrictive left ventricular filling pattern in dilated cardiomyopathy: clinical, hemodynamic and prognostic implications. Circulation 84 [Suppl II]:327
59. Razzolini R, Boffa GM, Stritoni P, Budano S, Ramondo A, Rampazzo C, Chioin R, Dalla Volta S (1989) Storia naturale delle miocardiopatie dilatative. G Ital Cardiol 19:114–120
60. Romeo F, Pelliccia F, Cianfrocca C, Cristofani R, Reale A (1989) Predictors of sudden death in idiopathic dilated cardiomyopathy. Am J Cardiol 63:138–140
61. Schwarz F, Mall G, Zebe H, Schmitzer E, Manthey J, Scheurlen H, Kubler W (1984) Determinants of survival in patients with congestive cardiomyopathy: quantitative morphologic findings and left ventricular hemodynamics. Circulation 70:923–928
62. Stevenson LW, Fowler MB, Schroeder JS, Stevenson WG, Dracup KA, Fond V (1987) Poor survival of patients with idiopathic cardiomyopathy considered too well for transplantation. Am J Med 83:871–876
63. Stevenson LW, Dracup KA, Tillisch JH (1989) Efficacy of medical therapy tailored for severe congestive heart failure in patients transferred for urgent cardiac transplantation. Am J Cardiol 63:461–464
64. Stevenson LW, Tillisch JH, Hamilton M, Luu M, Chelimsky-Fallick C, Moriguchi J, Kobashigawa J, Walden J (1990) Importance of hemodynamic response to therapy in predicting survival with ejection fraction ≤20% secondary to ischemic or nonischemic dilated cardiomyopathy. Am J Cardiol 66:1348–1354
65. Stewart RAH, Adams K, McKenna WJ, Oakley CM (1988) Arrhythmia and prognosis in dilated cardiomyopathy. Br Heart J 59:113
66. Szlachcic J, Massie BM, Kramer BL, Topic N, Tubau J (1985) Correlates and prognostic implication of exercise capacity in chronic congestive heart failure. Am J Cardiol 55:1037–1042
67. Unverferth DV, Magorien RD, Moeschberger ML, Baker PB, Fetters JK, Leier CV (1984) Factors influencing the one-year mortality of dilated cardiomyopathy. Am J Cardiol 54:147–152
68. Varani E, Magelli C, Bassein L, Truncellito L, Ribani MA, Rubino I, Fini M, Ferretti RM, Magnani B (1989) Caratteristiche cliniche e fattori prognostici dello scompenso cardiaco cronico secondario a miocardiopatia dilatativa primitiva ed a cardiopatia ischemica. G Ital Cardiol 19:1–6
69. Waagstein F, Hjalmarson A, Swedberg K, Wallentin I (1983) Beta-blockers in dilated cardiomyopathies: they work. Eur Heart J 4 [Suppl A]:173

III. Basic Molecular Concepts Toward Understanding of IDC

Molecular Analysis of Cardiac Growth and Development in Transgenic Mouse Model Systems

K.R. Chien, K.U. Knowlton, K. Lee, R.S. Ross, H.A. Rockman, and T.X. O'Brien

Introduction

In the past several years a great deal of work has focused on the potential biochemical and cellular mechanisms which regulate cardiac growth, hypertrophy, and failure (for review see [1]). Although previous studies in a variety of model systems has provided a valuable catalogue of the alterations in cardiac structure and function during the onset of these distinct physiological states, relatively little is known concerning the molecular switches which mediate compensatory cardiac hypertrophy and the transition to heart failure. The current challenge is to identify the precise pathways which mediate these responses through a critical examination of the cause-effect relationships between potential signaling molecules and the acquisition of well-defined phenotypic features of the hypertrophied and/or failing heart.

Studies of the growth and development of neuronal and skeletal muscle cells have documented the utility of genetic approaches to identify the pathways which control the growth and development of terminally differentiated cell types. Until recently, these studies were primarily confined to relatively simple organisms, in which genetic manipulation is feasible and the propagation times are relatively short, such as fruit flies (*Drosophila*), and worms (*C. elegans*). However, the recent development of transgenic technology in mice and other mammalian species has provided the opportunity to utilize genetic approaches to study growth and development in models which closely resemble in vivo physiology in man (for review see [2, 3]). Several of these recent studies have relied on the ability to genetically manipulate the mouse embryo, resulting in the development of lines of mice which harbor transgenes whose expression is targeted to specific tissues [2, 3]. Accordingly, the influence of the overexpression of normal or mutant genes in specific tissue types can now be studied in the in vivo setting, allowing an assessment of the role of candidate genes in the regulation of the growth and development of differentiated cell types. In addition, genetic recombination can be utilized to generate lines of mice in which specific

Department of Medicine, Center for Molecular Genetics, and the American Heart Association-Bugher Foundation Center for Molecular Biology, UCSD School of Medicine, La Jolla, CA 92093

Figulla et al. (Eds.)
Idiopathic Dilated Cardiomyopathy

genes have been disrupted, resulting in the targeted ablation of genes which may be required to activate growth or development of terminally differentiated cells [4].

Recent studies from a variety of laboratories have begun to harvest the power of transgenic technology to study the role of individual candidate genes in the regulation of cardiac structure and function [5–7]. For the purpose of the present brief review, our discussion focuses on the utility of transgenic approaches and the rat cardiac myosin light chain (MLC-2) gene as a model system to identify the molecular switches which regulate ventricular muscle growth and development. Transgenic studies hold great promise for furthering our understanding of the genetic mechanisms which activate compensatory hypertrophy of the myocardium and the subsequent transition to overt cardiac failure.

The Rat Cardiac MLC-2 Gene as a Model System to Study Cardiac Growth and Development

Recent studies from our laboratory have utilized the rat cardiac MLC-2 as a genetic marker for the activation of cardiac growth and development [8–15]. The utility of the MLC-2 gene as a model system is based upon its muscle-specific pattern of expression and new advances in uncovering the molecular regulation of this cardiac muscle gene. The rat cardiac MLC-2 gene encodes the regulatory, phosphorylatable myosin light chain that is expressed exclusively in cardiac and slow-twitch skeletal muscle, (soleus) [7]. In addition to serving as a model for the study of mechanisms which regulate cardiac muscle specificity, the rat cardiac MLC-2 gene has also been valuable in identifying the signals which augment the expression of constitutively expressed contractile protein genes during the course of cardiac muscle cell hypertrophy (Table 1). During in vivo pressure overload hypertrophy, there is an up-regulation of a panel of myofibrillar protein genes, leading to a quantitative increase in the contractile protein content of individual cardiac muscle cells. α-Adrenergic receptor-mediated hypertrophy of cultured ventricular muscle cells is accompanied by the transcriptional activation of the MLC-2 gene [8], while in vivo pressure overload results in a several

Table 1. A classification of genetic programs during myocardial cell hypertrophy

Induction of immediate early genes	Induction of embryonic genes	Induction of constitutive contractile protein genes	Lack of induction of cardiac-specific genes
c-*fos*	ANF	MLC-2	Cardiac Na^+ Channel
c-*jun*	Skeletal α-actin	Cardiac α-actin	SR Ca^{2+} ATPase
Egr-1	β-MHC		
jun-B			
Nur-77			

fold increase in MLC-2 gene expression (K. Lee, KR Chien, unpublished observations). Taken together, these results suggest that the up-regulation of the MLC-2 gene can serve as a genetic marker for the quantitative increase in constitutively expressed contractile protein genes, which is a central feature of cardiac hypertrophy [1]. Our underlying premise is that the cellular signals which activate MLC-2 gene expression are also likely to play a larger role in the activation of other features of the hyper-

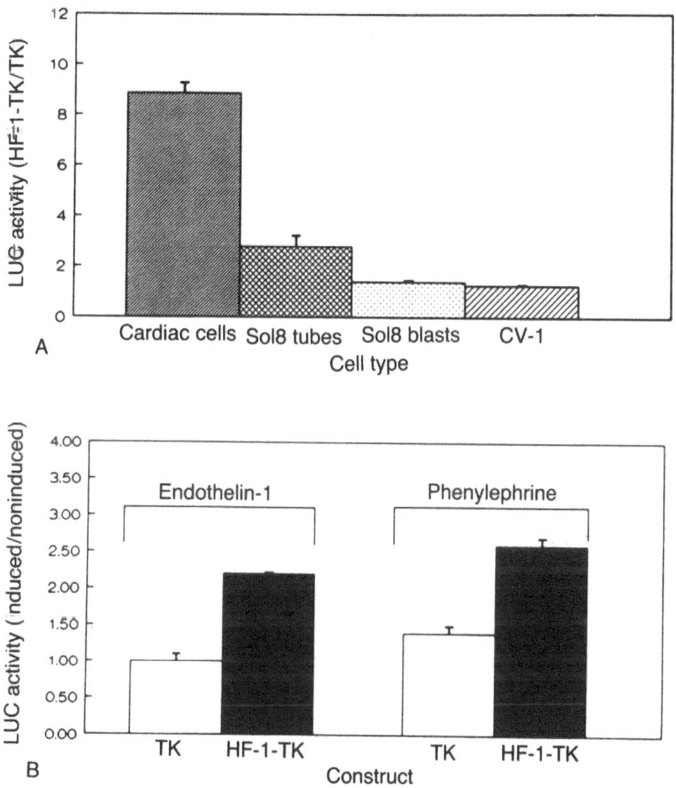

Fig. 1A,B. A conserved 28-bp element, HF-1, confers cardiac specific and inducible expression to a neutral promoter construct. A duplex oligonucleotide containing HF-1 sequences (5′-GCCAAAAGTGGTCATGGGGTTATTTTTA-3′) was inserted into the thymidine kinase luciferase vector PT109 and subsequently analyzed for its ability to confer cardiac specific expression in transient expression assays of neonatal rat myocardial cells, *So*18 myoblasts, *So*18 myotubes, and CV-1 cells (**A**) or on inducible expression during stimulation with either phenylephrine or endothelin (**B**) In each case, a CMV β-galactosidase vector was cotransfected to control for the efficiency of transfection. Results represent the mean ± standard error of the mean of four experiments, with four separate plasmid preparations. The relative activity of the above constructs was normalized to RSV/LUC in various cell types. As a reference, the relative activity of PT109 was 1.8% ± 0.1% of RSV/LUC in cardiac cells; 8.9% ± 0.5% in the *So*18 tubes; 5.8% ± 0.1% in *So*18 blasts; and 8.7% ± 0.5% in CV-1 cells. Inducibility assays were performed as previously described [11]. Results represent three separate experiments with three separate plasmid preparations and are expressed as the mean ± standard error of the mean (from [11] with permission)

trophic response in cardiac muscle cells. Of particular interest would be the identification of factors which bind to the critical regulatory sequences within the MLC-2 gene and thereby ultimately control MLC-2 expression during cardiac growth, hypertrophy, and development.

Toward this goal, recent studies have employed transient assays in cultured myocardial cells to identify a 250-bp fragment of the MLC-2 promoter which can confer cardiac muscle specific expression, indicating that the sequences within the MLC-2 gene which are required to maintain cardiac muscle specificity can be localized to this small gene fragment [11]. This fragment contains two conserved regulatory sequences (HF-1 and HF-2; Fig. 1) which not only are required to maintain cardiac muscle specificity of the MLC-2 promoter fragment but also play an important role in inducible expression during cardiac muscle cell hypertrophy [13]. The conserved 28-bp HF-1 element appears to be of particular importance since a single copy

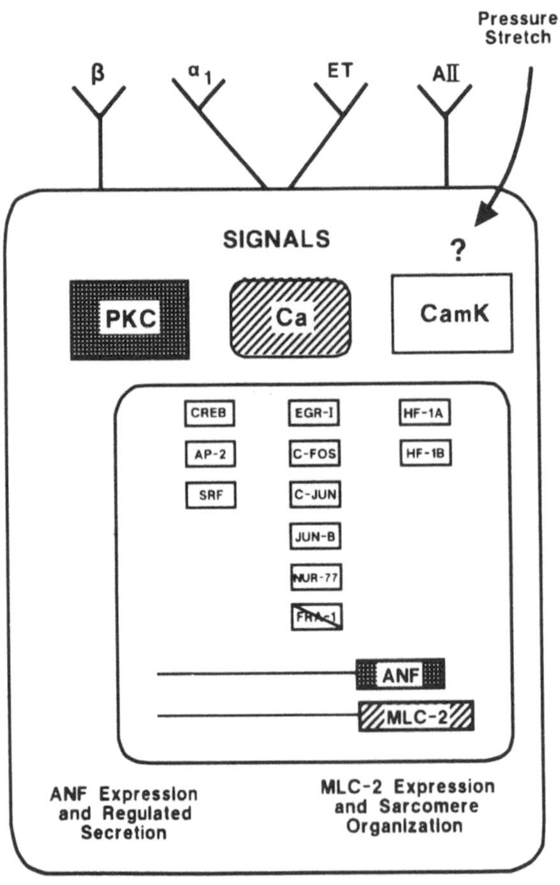

Fig. 2. Candidate signaling mechanisms which may activate growth and hypertrophy of myocardial cells (from [1] with permission)

of the HF-1 element can confer cardiac muscle specificity to an otherwise neutral promoter [13]. The conserved 28-bp HF-1 element binds to a ubiquitous factor (HF-1a) and a muscle factor (HF-1b/MEF2) which act to confer cardiac muscle specificity [13]. Recent studies have documented that the HF-1a and HF-1b factors, previously observed to be important for cardiac muscle specific expression, also appear to have a dual role in the activation of MLC-2 gene expression in the setting of cardiac muscle cell hypertrophy, suggesting that shared mechanisms may regulate the MLC-2 gene during cardiac growth and development. Utilizing the MLC-2 gene as a genetic marker for the activation of the hypertrophic response in cardiac muscle cells, a number of candidate regulatory signaling molecules have been identified in the signaling pathway (Fig. 2), including extracellular stimuli (α-adrenergic agonists, angiotensin II, endothelin-1) [1, 14]; G proteins (H-RAS) [15]; cytosolic kinases (protein kinase C) [16], and transcriptional factors (HF-1a, HF-1b/MEF2, etc.) [1]. With the rapidly growing number of potential signaling molecules it has become essential to critically assess the individual role of these genes in the activation of the hypertrophic program in vivo.

Limitations of In Vitro Models

While in vitro model systems have played an important role in the identification of candidate signaling pathways which might regulate the MLC-2 gene, there are clearly inherent limitations to these studies in cultured myocardial cells. In cultured myocyte models it is difficult to examine the mechanisms which restrict expression to individual cardiac compartments (atrial/ventricular/conduction system), and by necessity the experiments are limited to a defined point in cardiac development, i.e., neonatal. It would be extremely valuable to be able to monitor the effects of specific mutations

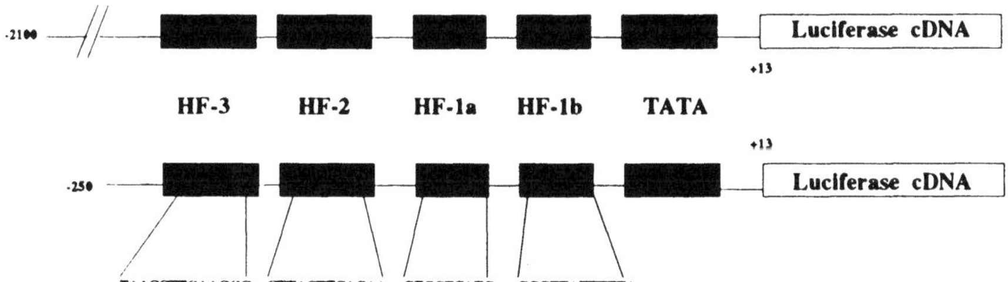

Fig. 3. Organization of the MLC-2 luciferase fusion gene utilized for generation of transgenic mice. A 250-bp rat MLC-2 promoter fragment was ligated into the promoterless luciferase expression vector pSVOALΔ5′. The HF-1a, HF-1b, HF-2, and HF-3 conserved sequences are depicted within the 250-bp rat cardiac MLC-2 promoter (11). (from [7] with permission)

within the 250-bp MLC-2 promoter fragment on chamber-specific expression, particularly as a function of normal embryonic development. Since the rat cardiac MLC-2 gene is also expressed at high levels in slow-twitch skeletal muscle (soleus) [7], it would also be of interest to determine whether divergent mechanisms might dictate muscle specificity in these two distinct but closely related striated muscle subtypes. Toward this goal, recent studies in our laboratory have harnessed transgenic approaches to identify potential signalling pathways which control MLC-2 gene expression during cardiac growth and development in vivo.

Generation and Breeding of MLC-2 Luciferase Transgenic Mice

Transgenic mice harboring a rat cardiac MLC-2-luciferase fusion gene have been obtained by standard approaches [17]. Since previous studies have documented that the 250-bp MLC-2 luciferase fusion gene (Fig. 3) is sufficient for cardiac specific and inducible expression in cultured cardiac muscle cells during α-adrenergic receptor mediated hypertrophy, independent mouse lines were generated which harbor this transgene, thereby allowing a direct analysis as to the fidelity of the studies in cultured cells to the in vivo physiologic context. MLC-2 luciferase transgenic mice were identified by Southern or PCR analysis of tail genomic DNA. To ensure that the pattern of expression of the transgene was not due simply to the influence of copy number and/or the site of transgene integration, at least two independent lines were examined in these studies, which were subjected to further analysis. (For a more complete description of the generation of the transgenic mice see [7]).

Ventricular-Specific Expression of the MLC-2 Luciferase Gene in Transgenic Mice

In two independent lines, a 250-bp MLC-2 promoter fragment can confer cardiac muscle-specific expression of the luciferase reporter gene in transgenic mice [7]. Although the endogenous mouse cardiac MLC-2 gene is expressed as an abundant transcript in both cardiac and slow-twitch skeletal muscle, luciferase activity was detected exclusively in cardiac tissue. Similar results were obtained with mice harboring a 2100-bp MLC-2 luciferase fusion gene. These results indicate that the 250-bp MLC-2 promoter fragment, while sufficient to direct cardiac muscle-specific expression, lacked the necessary sequences to allow expression of the gene in the slow skeletal muscle context, thereby providing direct evidence that distinct molecular mechanisms may have evolved to control muscle specific expression of this gene in these two closely related but distinct striated muscle subtypes. Interestingly, expression of the endogenous MLC-2 gene, as well as the MLC-2 luciferase fusion gene, was restricted to the ventricular compartment of the heart, with background levels in the atrium (Table 2). Thus, the 250-

Table 2. Regional specificity of luciferase activity (mean light units/μg protein ± SE) in various cardiac compartments of adult MLC250LUC transgenic mice

Tissue	MLC250L-3 (n = 4)	MLC250-31 (n = 5)
Left venticle	370.0 ± 73.2	11.5 ± 3.0
Right ventricle	39.4 ± 5.0[a]	9.9 ± 9.7
Right atrium	0.4 ± 0.1[a]	0.6 ± 0.5[a]
Left atium	0.1 ± 0.1[a]	0.2 ± 0.1[a]

[a] $p < 0.001$ versus LV.

bp MLC-2 promoter fragment can not only confer cardiac muscle-specific expression but can target expression exclusively to the ventricular compartment. The ventricular-specific expression of the MLC-2 gene is quite intriguing since it suggests that the MLC-2 gene may be an excellent model system to study the growth and development of ventricular muscle cells in transgenic model systems. In addition, it may serve as a valuable tool to target the expression of any candidate gene into the ventricular chamber, which should allow a direct examination of its consequent effects on normal ventricular growth and development. Studies are currently in progress to identify the cardiac muscle factors which are responsible for ventricular specification during the earliest stages of embryonic cardiac development in these transgenic mice, through the generation of lines which harbor point mutations in critical regulatory elements (HF-1a, HF-1b, HF-2, etc.) within the 250-bp MLC-2 promoter fragment.

Regulated Expression of the MLC-2 Luciferase Fusion Gene During In Vivo Murine Hypertrophy in Normal and Transgenic Mice

The previously mentioned studies in transgenic mice have examined the regulation of the MLC-2 gene strictly as a function of normal development. Since HF-1a and HF-1b play an important role in both cardiac specific and inducible expression in cultured myocardial cells, it has become apparent that the pathways which regulate MLC-2 expression during cardiac development and hypertrophy may be closely related. To examine this question, it became imperative to develop a reproducible model of in vivo pressure overload hypertrophy in normal mice, with eventual application to the MLC-2 luciferase transgenic mouse lines [18]. Given the 1-mm diameter of the mouse aorta (Fig. 4), the difficulty of establishing this murine model cannot be overstated. To circumvent the problems related to the minimal size of the mouse vessels, microsurgical approaches were adopted to produce a constriction between the right innominate and left carotid arteries,

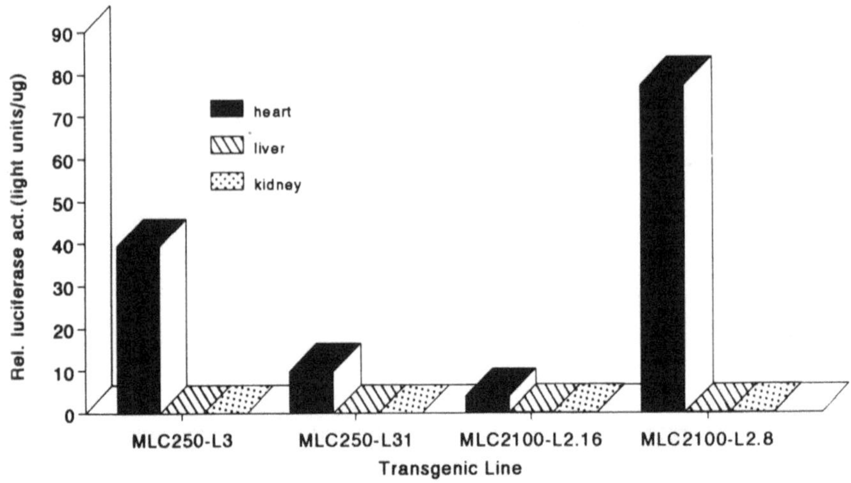

Fig. 4. Distribution of luciferase activity in various muscle and nonmuscle tissues of founder mice containing the MLC250-L fusion gene or progeny of the MLC 2100-bp luciferase lines. The heart, liver, and kidney were harvested from the transgenic mice of three MLC250L lines. Activity in soleus, liver, and kidney were at background levels. Luciferase activity is expressed in light units per microgram protein. (from [7] with permission)

resulting in a stable 35-mmHg gradient across the transverse aorta of the mouse, as assessed by the direct monitoring of the gradient via carotid cannulation (0.6-mm diameter) in this mouse preparation (Fig. 5). The degree of hemodynamic overload was stable after 6 h, and the mortality (less than 10%) and success rate (greater than 80%) compared favorably with larger animal models. Following 7–10 days of banding, all of the phenotypic markers for hypertrophy are present in the ventricular myocardium, including an increase in heart weight and heart weight/body weight ratio, myocyte area/volume, induction of a conserved program of immediate early gene expression, and the activation of embryonic markers (ANF) and constitutively expressed contractile protein genes (MLC-2). Thus, a reproducible in vivo murine model of pressure overload hypertrophy has been developed. This model system should now allow the direct analysis of the role of the precise elements within the MLC-2 promoter fragment which are required for the induction of this gene, as well as an embryonic marker gene (ANF) during the delivery of a bona fide in vivo mechanical stimulus for hypertrophy. In addition, it may become feasible to directly assess the role of individual candidate genes in the transduction of a mechanical stimulus, in this case pressure overload, into biochemical signals which activate transcription of the MLC-2 and ANF genes in the in vivo setting. These studies are currently in progress.

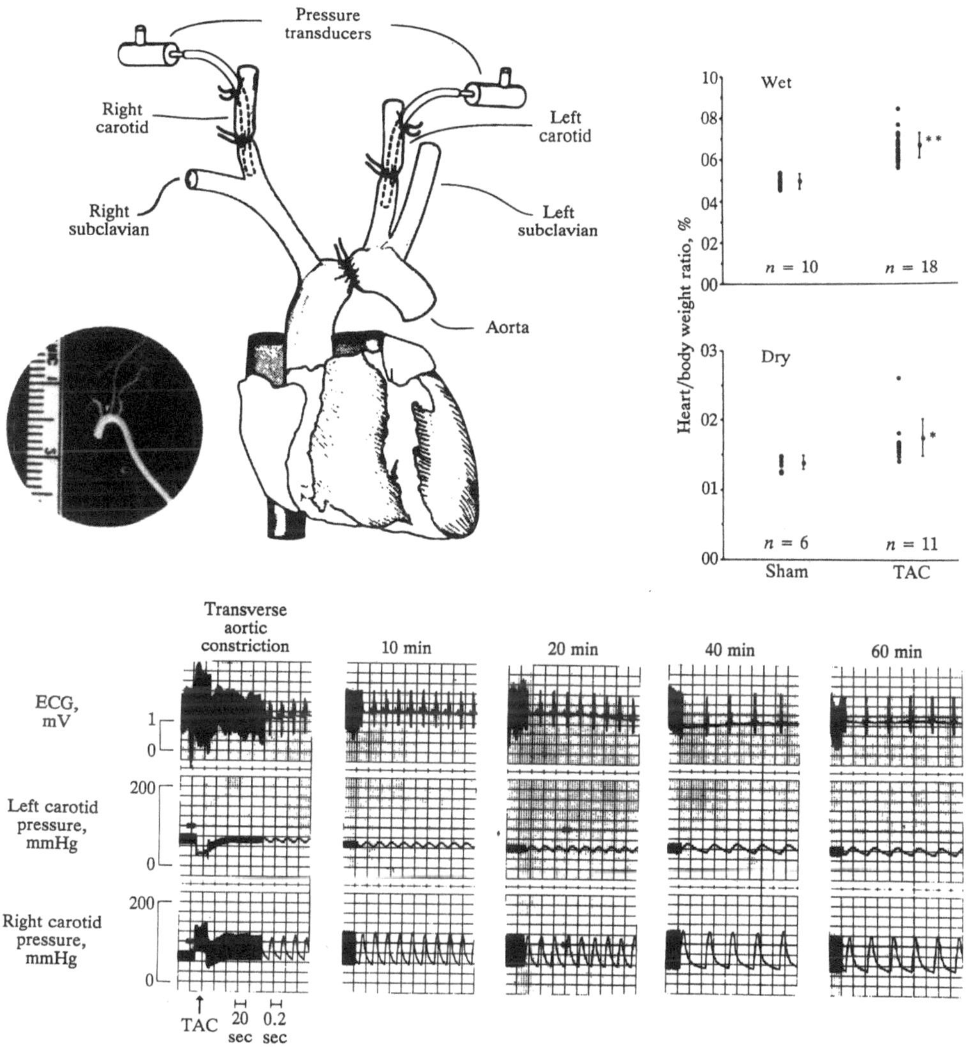

Fig. 5. Induction of in vivo pressure overload in murine myocardium following transverse aortic constriction. *Upper left panel*, Schematic presentation of the surgical preparation. *Inset*, Silastomer cast of the mouse aortic tree. *Lower panel*, Electrophysiologic and hemodynamic effects of thoracic aortic banding. *Upper right panel*, Effects of chronic transverse aortic constriction on heart weight/body weight ratio. For the assessment of heart weight, the mice were divided into two groups, sham operated and transverse aortic constricted (*TAC*). Following 7 days of TAC, a 33% increase in wet weight and a 24% increase in dry heart weight/body weight ratio was observed. Data are expressed as individual experiments and mean ± standard deviation. $^{*}p < 0.05$; $^{**}p < 0.001$ (from [18] with permission)

Summary/Future Directions

In summary, transgenic approaches hold great promise for the study of the molecular pathways which activate cardiac growth and development. The ability of the MLC-2 promoter fragment to target expression exclusively into ventricular muscle cells at the earliest stage of cardiac development should be of value in targeting the expression of candidate genes which influence ventricular growth and development. In addition, since the MLC-2 luciferase transgenic mice which harbor normal and mutant MLC-2 promoter constructs have integrated the transgene into their germ line cells, it should now be possible to mate these animals with separate lines of mice that harbor transgenes that influence MLC-2 expression, providing a means to examine transcriptional regulation in the in vivo context. In an analogous fashion, it is anticipated that genetically based transgenic models which develop cardiac hypertrophy and failure may also be potentially feasible. The ability to ablate specific genes through homologous recombination and embryonic stem cell technology offers the possibility of generating mice which lack the critical molecular switches which regulate normal cardiac growth and development, which may lead to the development of novel model systems of human cardiac disorders. As can be seen, the rat cardiac MLC-2 gene and studies of its regulated expression in transgenic mice represents a valuable experimental approach for increasing our understanding of the mechanisms which regulate the expression of contractile protein genes during cardiac growth and development.

References

1. Chien KR, Knowlton KU, Chien, S (1991) Regulation of cardiac gene expression during myocardial growth and hypertrophy: molecular studies of an adaptive physiologic response. FASEB J 5:3037–3046
2. Hanahan D (1989) Transgenic mice as probes into complex systems. Science 246:1265–1275
3. Field LJ (1991) Cardiovascular research in transgenic animals. Trends Cardiovasc Med 1:141–146
4. Thomas KR, Capecchi MR (1987) Site-directed mutagenesis by gene targeting in mouse embryo-derived stem cells. Cell 51:503–512
5. Field LJ (1988) Atrial natriuretic factor-SV 40 T antigen transgenes produce tumors and cardiac arrythmias in mice. Science 239:1029–1033
6. Jackson T, Allard MF, Sreenan CM, Doss LK, Bishop SP, Swain JL (1990) The cmyc-proto-oncogene regulates cardiac development in transgenic mice. Mol Cell Biol 10:3709–3716
7. Lee KJ, Ross RS, Rockman HA, Harris A, O'Brien TX, Van-Bilsen M, Evans SM, Zhu H, Chien KR (1992) Myosin light chain 2-luciferase transgenic mice reveal distinct regulatory programs for cardiac and skeletal muscle specific expression of a single contractile protein gene. J Biol Chem (in press)
8. Lee H, Henderson S, Reynolds R, Dunnmon P, Yuan D, Chien KR (1988): α-1 adrenergic stimulation of cardiac gene transcription in neonatal rat myocardial cells: effects on myosin light chain-2 gene expression. J Biol Chem 263:7352–7358
9. Henderson SA, Xu Y, Chien KR (1988) Nucleotide sequence of full length cDNAs encoding rat cardiac myosin light chain-2. Nucleic Acid Res 16:4722
10. Iwaki K, Sukhatme VP, Shubeita HE, Chien KR (1990) α and β adrenergic stimulation induce distinct patterns of immediate early gene expression in neonatal rat myocardial cells:

fos/jun expression is associated with sarcomere assembly; Egr-1 induction is primarily an α_1 mediated response. J Biol Chem 265:13809–13817
11. Zhu H, Garcia A, Ross RS, Evans SM, Chien KR (1991) A conserved 28bp element (HF-1) within the rat cardiac myosin light chain-2 gene confers cardiac specific and α adrenergic inducible expression in neonatal rat myocardial cells. Mol Cell Biol 11:2273–2281
12. Knowlton KU, Baracchini E, Ross RS, Harris AN, Henderson SA, Evans SM, Glembotski CC, Chien KR (1991) Co-regulation of the atrial natriuretic factor and cardiac myosin light chain-2 genes during α-adrenergic stimulation of neonatal rat ventricular cells. J Biol Chem 266:7759–7768
13. Navankasattusas S, Zhu H, Garcia A, Evans SM, Chien KR (1992) A ubiquitous factor (HF-1a) and a muscle factor (HF-1b/MEF-2) forms an E-box independent pathway for cardiac muscle specific expression. Mol Cell Biol (in press)
14. Shubeita HE, McDonough PM, Harris A, Knowlton KU, Glembotski C, Brown JH, Chien KR (1990) Endothelin induction of sarcomere assembly and cardiac gene expression in ventricular myocytes: a paracrine mechanism for myocardial cell hypertrophy. J Biol Chem 265:20555–20562
15. Thorburn A, Thorburn J, Powers S, Shubeita HE, Feramisco JR, Chien KR (1992) HRAS dependent pathway for cardiac muscle hypertrophy (submitted)
16. Shubeita HE, Martinson E, Van-Bilsen M, Chien KR, Brown JH (1992) Expression of a constitutively active protein kinase C gene activates the transcription of the ANF and MLC-2 genes in cultured myocardial cells. Proc Natl Acad Sci USA 89:1305–1309
17. Hogan B, Costantini F, Lacy L (1986) Manipulating the mouse embryo: a laboratory manual. Cold Spring Harbor Laboratory, New York
18. Rockman HA, Ross RS, Harris AN, Knowlton KU, Steinhelper ME, Field L, Ross J, Chien KR (1991) Segregation of atrial specific and inducible expression of an ANF transgene in an in vivo murine model of cardiac hypertrophy. Proc Natl Acad Sci USA 88:8277–8281

The Promoter of the Human Cardiac β-Myosin Heavy-Chain Gene Is Controlled by an Upstream Enhancer

A. Wettstein and H.-P. Vosberg

Introduction

The two human cardiac myosin heavy chains (MHC), designated α and β, are members of a multigene family comprising ten or more sarcomeric myosin genes. The two cardiac genes are located on chromosome 14; the skeletal muscle genes are on chromosome 17 [6, 14, 15]. The cardiac α-MHC is fast with respect to ATPase activity and speed of contraction; in contrast, β-MHC is slow. The two isoforms are distinctively distributed within the heart: α is found primarily in the atria and β in the ventricles. Both proteins are also expressed in certain skeletal muscles, β in slow twitch fibers and α in the masseter muscle [1, 16]. Due to the differences in their contractile properties, these two myosins have profound effects on the function of cardiocytes in which they are present. Their discrete distribution in different tissues implies that separate control mechanisms exist for the transcription of these genes. Some of the components involved in regulation have been identified. However, the picture is far from complete. Particularly well established is the role of the thyroid hormone (T_3) and that of retinoic acid [7, 13]. Both compounds stimulate the α and repress the β gene. Investigations of cardiac myosin HC genes are of growing interest since it is known that certain forms of familial hypertrophic cardiomyopathy, a dominantly inherited disorder of the heart accompanied by primary hypertrophy and an elevated risk of sudden cardiac death [9, 17].

Most functional studies of muscle genes, even of those which are either transiently or predominantly cardiac genes, have been carried out in primary or established skeletal muscle cells. Thus little is known about the modes by which the cardiac phenotype is maintained. The regulation of skeletal muscle genes is to a large extent controlled by myogenic lineage determining factors such as myoD and related proteins [18, 21]. None of these proteins seems to play a role in cardiocytes. Hence, a major question about the control of cardiac genes is related to the unknown factors required for their expression.

Max-Planck-Institut für physiologische und klinische Forschung, Abteilung für Experimentelle Kardiologie, Benekestr. 2, W-6350 Bad Nauheim, Fed. Rep. of Germany

Figulla et al. (Eds.)
Idiopathic Dilated Cardiomyopathy

To approach this question we have studied the regulation of the β-gene by analyzing the promoter and promoter-related *cis*-acting DNA sequences. Methods applied were based on DNA-mediated gene transfer involving a reporter gene in cultured myoblasts and in nonmuscle cells. The goal was to delineate the promoter of the β-gene as well as additional regulatory sequences in the region preceding the promoter. Functional studies were carried out with myoblasts. However, the experiments were extended to protein binding studies with nuclear extracts of cardiac tissue from rabbit to identify and ultimately to characterize a factor or factors which bind to β-specific regulatory sites.

Materials and Methods

For mapping of the β-myosin HC promoter and other regulatory elements within the sequenced 5′ flanking region of the β-gene either random partial digestion with exonuclease III or targeted PCR amplification of selected subregions was applied. DNA fragments were cloned in an expression vector comprising the gene for chloramphenicol acetyltransferase [10] (CAT; vector designation pCAT-O) and transfected into myoblasts derived from embryonic chicken muscle (musculus pectoralis major). Growth and fusion protocols were as reported earlier [2]. Nuclear proteins were extracted from rabbit heart tissue and liver and from human fibroblasts. The isolation was as reported in the literature [5]. Target sequences for electrophoretic mobility shift assays were either amplified by PCR or chemically synthesized. Double-stranded wild-type and mutated oligonucleotides were obtained by renaturation of single-stranded DNA. The 5′ end labeling was achieved by T4 polynucleotide kinase and gamma-[^{32}P]ATP. Analytical binding reactions and DNase I protcction assays were carried out according to standard protocols.

Results and Discussion

The human β-MHC gene has been cloned and sequenced in its entire length [4, 11, 12]. The gene consists of 40 exons including two untranslated exons at the 5′ end. The mRNA has a length of 6008 bases and codes for 1945 amino acids. The β-protein sequence has a high degree of homology with other human and nonhuman vertebrate MHCs of cardiac and skeletal muscle. Differences between myosins appear to be clustered in certain subregions of the protein which are presumably important for the isotype specific properties of the β-chain [11].

The transcription start point which is identical with the 5′ end of the nontranslated exon 1 has been assigned with a remaining ambiguity of a few basepairs. All nucleotide distances given in the text below refer to this point. Expression control sequences are located in the region immediately preceding the gene.

```
-389
GGCCAGAAGTGCCCTGCGGGAGAGCCAGGACATTGGCTGCCTGTGGTCTT

                   -316             A
GGTGGTCGTGGTCAGTTCCCTCTCCTGCCAGCT[GTGGAATG]TGAGGCCTG

                 -268                           -240
GCCTGGGAGATATTTTTGCTGCACTTTGAGCACCCCGCCCCCTGGAACT

                  -217                -202
CAGACCCTGCACAGTCCATGCCATAACAATGACGACCACTTCCAATTGTT

TCCTAGCTGGAGAGGCGGGGAGGGGAGCACTGTTTGGGAAGGGGGGGAGC

                                        B
CTGGGGGAAATGCTTCTAGTGACAACAGCCCTTT[CTAAAT]CCGGCTAGGG

                                            C
ACTGGGTGCCGTTGGGGGTGGGGGTGCCCTGCTGCCCCA[TATATA]CAGCC

                                       +1
CCTGAGACCAGGTCTGGCTCCACAGCTCTGTCCTGCTCTG
                        ▲
```

Fig. 1. The upstream sequence of the *β*-gene is shown between position −389 and +1. The boxed sequence motifs are the *β*-enhancer (*A*), the CTAAAT site (*B*), and the TATATA site of the basal promoter (*C*). The *arrowhead* in position −15 points to an alternative determination of the transcription start point [19]. The positions −316, etc. designate cutoff points of deletion constructs used in the identification of the enhancer

Two basal promoter signals, TATATA and CTAAAT, have been identified in position −50 and −105 within the flanking region, respectively. Figure 1 depicts this region from position −389 to the presumed position +1.

Transcription Control Experiments

The identification of a DNA signal stimulating the *β*-promoter was achieved by gene transfer experiments with muscle and nonmuscle cells. For this purpose upstream sequences were systematically deleted from 5′ to 3′ within the 5′ flanking region of the *β*-gene and ligated to a CAT expression vector without promoter. In these experiments a 316-bp-long segment within the upstream region was determined harboring in addition to the promoter a *cis*-acting element (an enhancer or an enhancerlike element) required for activation of the promoter. Induction of CAT reporter gene activity was measured in embryonic chicken myoblasts and, as a control, in HeLa cells. The control of the *β*-gene is muscle cell specific and depends on the differentiation state. An inhibitor of muscle genes (bromodeoxyuridine) suppresses enhancement of the *β*-promoter. The basal promoter alone is extremely weak even in muscle cells.

Removal of DNA 3′ to −316 resulted in a gradual decrease in promoter activation. Truncation to position −268 was still compatible with CAT activity; however, induction was markedly reduced. With a segment beginning at −202 CAT activity was almost not detectable. These results suggested that an essential part of the regulatory control system was located between −316 and −268.

Protein Binding Studies

The regulation of a gene by *cis*-acting DNA signals requires the interaction of *trans*-acting binding factors. In the search for proteins recognizing a sequence (or sequences) between −316 and −268 nuclear extracts were isolated from cardiac muscle. Since heart muscle of human origin was not available, rabbit heart tissue was used instead. This decision seemed justified since extensive homology exists between the human and rabbit β-MHC upstream sequences (about 80%).

Protein binding was monitored by electrophoretic mobility shift assays with radioactively labeled DNA fragments as probes. DNA-protein complexes were formed with a DNA fragment extending from −362 to −249. If the same fragment was applied in an unlabeled form as competitor DNA in a 100-fold molar excess radioactive DNA-protein complexes were undetectable. If, however, competing DNA was selected from an overlapping

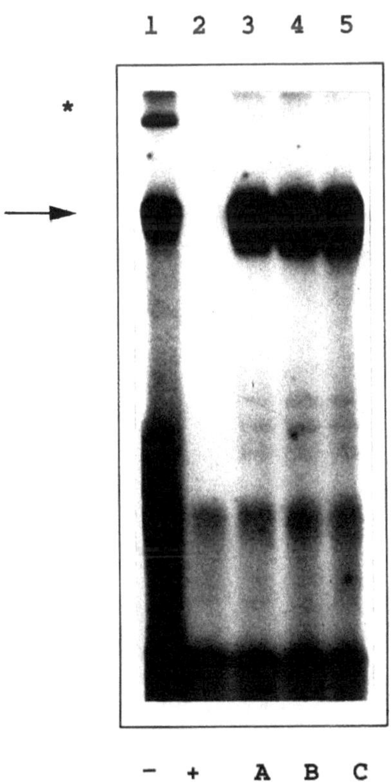

Fig. 2. Mobility shift experiments with the enhancer region as DNA test sequence and protein extracts of rabbit heart nuclei. The DNA target was in all assays a double-stranded 26mer radioactively labeled at the 5′ end of the coding strand. The lanes show: *1*, complex formation with radioactive DNA substrate; *2*, the same in the presence of a 100-fold molar excess of unlabeled 26mer; *3–5*, with 100-fold molar excess of unrelated DNA sequences: *A*, a 131-bp fragment of intron 24 of the β-gene; *B*, a double-stranded 32mer of intron 24; *C*, a 115-bp cDNA fragment representing exon 16–18 sequences. The *arrow* on the left points to the complex; *asterisk*, an unspecific DNA-protein complex

sequence (−298 to −193) very little competition was seen (not shown). Hence, it was concluded that the relevant sequence involved in promoter activation might be located between −298 and −316. We have previously designated this activating sequence "upstream regulatory element", implying that it functions as a β-specific enhancer [20].

To characterize the target region involved in DNA-protein complex formation in more detail we have synthesized a double-stranded oligonucleotide (a 26mer) extending from position −316 to −291. This sequence exhibited complex formation with cardiac nuclear extracts (Fig. 2). No competition was seen if DNA fragments containing β-gene sequences unrelated to the promoter region (a sequence from intron 24) were applied

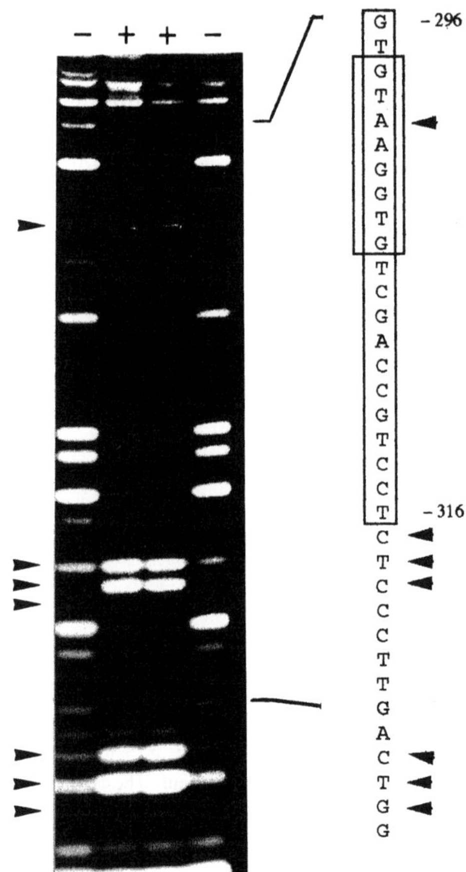

Fig. 3. DNase I protection assay with a 5′ flanking DNA segment including the enhancer in position −305 to −298. Protein extracts of rabbit heart nuclei were incubated with amplified 124-bp DNA fragments labeled at the 5′ ends of the coding strands and treated with DNaseI. *Minus, plus lanes* show digestion products in the absence or in the presence of protecting nuclear protein, respectively. The protected sequence is written on the right margin. *Box*, the predominantly protected segment; *double-boxed* the enhancer within this segment; *arrowheads*, sites of preferred cleavage in the presence of nuclear protein

in excess over the target 26mer, suggesting specificity of the binding. This conclusion was corroborated by a DNase I protection test using the same extracts of heart nuclei and a DNA fragment extending from position −362 to −249. The protected area included the region between −316 and −296 (Fig. 3).

Delineation of the upstream enhancer was refined further by mobility shift assays using either normal (wild-type) or altered sequences between −316 and −291 as binding targets. These experiments were originally designed to investigate the possibility that a hexameric sequence motif CAGCTG within this region was responsible for activation of the β-promoter. This motif seemed appealing since it matched the consensus sequence CANNTG, also designated as E-box. Sequences of this type are regulatory target elements for the action of myogenic differentiation factors such as myoD and related proteins (for review see [21]). They have been identified as control elements preceding a number of muscle genes. None of the myogenic factors has ever been detected in cardiocytes. We speculated, however, that an E-box like sequence in a cardiac gene might be recognized by a heart specific regulatory protein possibly related to the known myogenic factors, but nevertheless of distinct character.

A sequence specific mutation was introduced into the 26mer by changing a triplet preceding the hexamer and the first triplet within this motif. Thus TGCCAG was converted to GTAACT. The presumed E box read in the altered state ACTCTG. This mutation did not affect the binding of nuclear protein isolated from rabbit heart to the target sequence. However, changed patterns of protein binding were seen if sequence mutations were introduced at the 3′ end and adjacent to the 3′ end of the presumed E box (Table 1).

β-MHC gene specific regulatory upstream sequences are known from three different mammalian species: man, rabbit, and rat [3, 16, 19, 20]. In all three species a control element has been identified in roughly homologous regions relative to the transcription start points of the respective genes. A sequence comparison shows identity over a length of eight nucleotides, reading GTGGAATG (in the human gene from position −305 to −298; Fig. 4). Site-directed mutations leading to loss of protein binding fall within the frame of this octamer. We believe therefore that the principal regulatory site responsible for the β-gene is this octamer and not the CANNTG-related motif discussed previously.

The successful demonstration of a protein binding to a DNA region involved in the control of the β-MHC promoter was taken as an opportunity to initiate the isolation of this protein from rabbit cardiac tissue in quantities suitable for biochemical analysis. The strategy was based on DNA affinity

Table 1. Wild-type and mutated enhancer sequences of the β-MHC gene

WT	T C C T G C C A G C T G T G G A A T G T G A G G C C
Mut. 1	T C C T G C C A G C *g t c t* G A A T G T G A G G C C
Mut. 2	T C C T G C C A G C T G T G G A A *c a c* G A G G C C
Mut. 3	T C C *g t a a c t* C T G T G G A A T G T G A G G C C

```
                     -305               -298
                    |                      |
Man       C A - G C T | G T G G A A T G | T G A G G
            *   * * * | * * * * * * * * | *   * * *
Rat       G A - G C T | G T G G A A T G | T A A G G
            *   * *   | * * * * * * * * |     * * *
Rabbit    C A G G C A | G T G G A A T G | C G A G G
```

Fig. 4. Comparison of enhancer sequences of β-MHC genes of man, rat, and rabbit

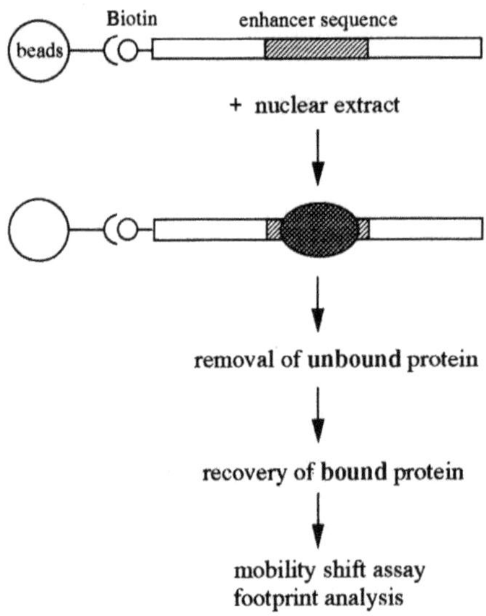

Fig. 5. Purification scheme for enhancer binding protein by DNA affinity beads. See text for the components of the system

binding. The DNA ligand was an amplified 124-bp fragment harboring the enhancing octamer. For PCR a coding primer was used which carried a biotin residue at its 5′ end. This residue mediated immobilization of the DNA fragment on the surface of magnetic beads coated with streptavidin [8]. After mixing of total nuclear protein with a suspension of DNA bound to the beads, proteins attached to the DNA were separated magnetically together with the beads in the tip of the vial. Supernatant was subsequently removed, beads were washed, and bound protein was detached in a buffer containing high salt (800 m*M* KCl). The procedure is outlined in Fig. 5. Proteins enriched by this protocol were able to induce a mobility shift and a DNase I protection pattern identical with that obtained with total extract (details to be reported elsewhere).

This protein binds specifically to the enhancer, but alone it is probably not sufficient to activate the promoter. The reason to believe that additional factors are required is based on the observation that the enhancer binding protein is not tissue specific. We have found a similar or identical protein in nuclei of cultured human fibroblasts and also in rabbit liver nuclei.

In summary, by investigating the transcriptional regulation of the human β-MHC gene we have determined an octameric sequence in position −305 to −298 which is required for activating the β-promoter in a muscle-specific manner. This sequence motif is involved in binding of a protein which was isolated from the nuclei of rabbit heart.

The functional tests in this study were done with myoblasts of chicken. Strictly speaking we have identified a *cis*-acting element involved in promoter stimulation in skeletal muscle but not in cardiac muscle. However, in an independent report it was shown recently that this sequence is necessary for expression in cardiac cells as well [19]. Hence, we consider this octamer as an essential component in the activation of the β-MHC gene.

The role of the nuclear binding protein in the control of expression requires further analysis. Since this protein is found in muscle cells and in a related or identical form in nonmuscle cells, it is possible that it is necessary but probably not sufficient to induce β-specific transcription. Other, presumably cardiac specific factors might attach to this protein, or alternatively might be directed to adjacent DNA binding sequences. We anticipate that a characterization of the octamer binding protein will significantly improve our understanding of the mechanism by which an essential cardiac gene is controlled.

References

1. d'Albis A, Janmot C, Mira JC, Couteaux R (1991) Characterization of a ventricular V1 myosin isoform in rabbit masticatory muscles. Developmental and neural regulation. Basic Appl Myol 1:23–34
2. Blau HM, Webster C (1981) Isolation and characterization of human muscle cells. Proc Natl Acad Sci USA 78:5623–5627
3. Cribbs LL, Shimizu N, Yockey CE, Levin JE, Jakovcic S, Zak R, Umeda PK (1989) Muscle-specific regulation of a transfected rabbit myosin heavy chain β gene promoter. J Biol Chem 264:10672–10678
4. Diederich KW, Eisele I, Ried T, Jaenicke T, Lichter P, Vosberg HP (1989) Isolation and characterization of the complete human β-myosin heavy chain gene. Hum Genet 81: 214–220
5. Dignam JD, Lebovitz RM, Roeder RG (1983) Accurate transcription initiation by RNA polymerase II in a soluble extract from isolated mammalian nuclei. Nucleic Acids Res 11:1475–1489
6. Edwards YH, Parker M, Povey S, Wets LF, Parrington MJ, Solomon E (1985) Human myosin heavy chain genes assigned to chromosome 17 using a human cDNA clone as a probe. Ann Hum Genet 49:101–109
7. Everett AW, Sinha AM, Umeda PK, Jakovcic S, Rabinowitz M, Zak R (1984) Regulation of myosin synthesis by thyroid hormone: relative change in the α- and β-myosin heavy chain mRNA levels in rabbit heart. Biochemistry 23:1596–1599
8. Gabrielsen OS, Hornes E, Korsnes L, Ruet A, Oeyen TB (1989) Magnetic DNA affinity purification of yeast transcription factor tau – a new purification principle for the ultrarapid isolation of near homogeneous factor. Nucleic Acids Res 17:6253–6267
9. Geisterfer-Lowrance AAT, Kass S, Tanigawa G, Vosberg HP, McKenna W, Seidman CE, Seidman JG (1990) A molecular basis for familial hypertrophic cardiomyopathy: a beta cardiac myosin heavy chain gene missense mutation. Cell 62:991–998

10. Gorman CM, Moffat GT, Howard BH (1982) Recombinant genomes which express chloramphenicol acetyltransferase in mammalian cells. Mol Cell Biol 2:1044–1051
11. Jaenicke T, Diederich K, Haas W, Schleich J, Lichter P, Pfordt M, Bach A, Vosberg HP (1990) The complete sequence of the human β-myosin heavy-chain gene and a comparative analysis of its product. Genomics 8:194–206
12. Liew CC, Sole MJ, Yamauchi-Takihara K, Kellam B, Anderson DH, Lin L, Liew JC (1990) Complete sequence and organization of the human cardiac β-myosin heavy-chain gene. Nucleic Acids Res 18:3647–3651
13. Rottmann JN, Thompson WR, Nadal-Ginard B, Mahdavi V (1990) Myosin heavy chain gene expression: interplay of *cis* and *trans* factors determines hormonal and tissue specificity. In: Pette D (ed) The dynamic state of muscle fibers. de Gruyter, Berlin, pp 3–16
14. Saez LJ, Gianola KM, McNally EM, Feghali R, Eddy R, Shows TB, Leinwand LA (1987) Human cardiac myosin heavy chain genes and their linkage in the genome. Nucleic Acids Res 13:5443–5459
15. Schwartz K, de la Bastie D, Mercadier JJ, Swynghedauw B, Lompré AM (1990) In: Swynghedauw B (ed) Cardiac hypertrophy and failure. INSERM/Libbey, London, pp 105–135
16. Shimizu N, Dizon E, Zak R (1992) Both muscle-specific and ubiquitous nuclear factors are required for muscle-specific expression of the myosin heavy-chain β gene in cultured cells. Mol Cell Biol 12:619–630
17. Tanigawa G, Jarcho J, Kass S, Solomon SD, Vosberg HP, Seidman JG, Seidman CE (1990) A molecular basis for familial hypertrophic cardiomyopathy: an alpha/beta cardiac myosin heavy chain hybrid gene. Cell 62:991–1006
18. Tapscott SJ, Weintraub H (1991) MyoD and the regulation of myogenesis by helix-loop-helix proteins. J Clin Invest 87:1133–1138
19. Thompson WR, Nadal-Ginard B, Mahdavi V (1991) MyoD1-independent muscle-specific enhancer controls the expression of the β-myosin heavy-chain gene in skeletal and cardiac muscle cells. J Biol Chem 266:22678–22688
20. Vosberg HP, Horstmann-Herold U, Wettstein A (1992) The regulation of the human β myosin heavy-chain gene. In: Hasenfuss G, Holubarsch C, Just H, Alpert N (eds) Cellular and molecular alterations in the failing human heart. Steinkopff, Darmstadt, pp 161–173
21. Weintraub H, Davis R, Tapscott S, Thayer M, Krause M, Benezra R, Blackwell TK, Turner D, Rupp R, Hollenberg S, Zhuang Y, Lassar A (1991) The *myoD* gene family: nodal point during specification of the muscle cell lineage. Science 251:761–766

Growth Factor Expression in the Cardiac Myocyte

P. Cummins

Introduction

The processes involved in both normal and abnormal growth of the heart are poorly understood. Enlargement of the heart following haemodynamic overload or as a result of congenital heart disease in both the young and the adult mammal may result from cellular and molecular mechanisms which have common features. Protein growth factors are known to play a critical role in the cycle of normal cell division, differentiation and growth of cells from a variety of organs and tissues. In the heart, their effects on endothelial cells and role in the processes of angiogenesis are the subject of intense interest. However, relatively less is known about the possible role of growth factors in the proliferation and enlargement of the cardiac myocyte. Recent studies, both in vitro and in vivo, suggest that myocytes are the target of selected growth factors, and that they could be implicated in both normal and abnormal hypertrophic growth.

Proliferative Capacity of Cardiac Myocytes

Our currently poor understanding of the processes involved in cardiac enlargement during neonatal development, congenital hypertrophy, adaptive hypertrophy subsequent to haemodynamic overload and deficient growth following myocardial infarction imposes severe limits on the development of new approaches to the treatment of heart disease. Fundamental to this problem is the apparent inability of cardiac myocytes to proliferate. Ventricular myocytes are considered to be terminally differentiated cells, incapable of mitotic division and cytokinesis [33]. Soon after birth, myocytes cease hyperplasia and further muscle growth is by cellular hypertrophy. Injury to cardiac muscle results in hypertrophy or scar formation by non-muscle cells which can further compromise cardiac performance.

Despite this major limitation, which has become a central dogma of cardiovascular research, there are a number of indications which suggest

Director, British Heart Foundation Molecular Cardiology Unit and Senior Lecturer in Physiology, Department of Physiology, The Medical School, University of Birmingham, Birmingham, B15 2TJ, UK

Figulla et al. (Eds.)
Idiopathic Dilated Cardiomyopathy

that the inability of cardiac myocytes to undergo mitotic division is a property that may be repressed and not irreversibly lost [2]. First and foremost is the fact that myocytes proliferate in the developing heart although the rate of division declines rapidly during the late fetal and early postnatal stages [33]. In the early 1960s Rumyantsev demonstrated that adult atrial myocytes might be triggered into DNA synthesis, nuclear division and limited cytokinesis by ventricular overload subsequent to experimental myocardial infarction [24]. In the amphibian, both ventricular and atrial myocytes undergo mitosis 15–20 days following wounding of the myocardium, even to the extent of regeneration of the tip of the excised ventricle in the newt [19]. Isolated adult mammalian myocytes in culture have also been stimulated to divide by inserting the gene for the large tumour antigen from the SV40 virus into the myocyte genome [2]. In vivo the same antigen gene has been coupled to the promoter for the gene for atrial natriuretic factor and injected into fertilised mouse eggs. Some of the transgenic mice developed tumours of the atrium due to uncontrolled division of myocytes which in most other respects appeared normal [10]. While these are isolated findings, they suggest that it is necessary to look beyond the limitations posed by past thinking. Indeeed they all point to a real need to understand the molecular processes involved in the proliferation and growth of the cardiac myocyte.

Role of Peptide Growth Factors

Relatively little is known of the signalling molecules in the heart that might coordinate growth. Such molecules would need to communicate growth signals both within and between cells, from cell surface sarcolemmal receptors to the nucleus, interact with genomic DNA to initiate cellular growth, regulate the synthesis of muscle proteins, and modify the ability of cardiac myocytes to proliferate and enlarge. Potential candidates as regulators of myocardial growth and differentiation which could fulfil the above roles are the multifunctional polypeptide growth factors. Two classes of growth factors, namely the transforming and fibroblast growth factors, have been implicated in signalling processes involving the cardiac myocyte [18, 26]. Attention has focussed in particular on transforming growth factor-β (TGF-β) and acidic (aFGF) and basic (bFGF) fibroblast growth factors.

Transforming and Fibroblast Growth Factors

TGF-β is the prototype of a family of proteins that regulate cell growth and division [23]. Other members of the family include the activins, inhibins, müllerian inhibiting substance in mammals and the decapentaplegic gene complex transcript which controls morphogenesis in *Drosophila*. The range of cell types that respond to TGF-β and the variety of cellular responses elicited are wide. It can either stimulate or inhibit cell proliferation and growth in addition to other seemingly unrelated actions [28]. TGF-β is a potent regulator of myogenesis in in vitro models using cell lines exhibiting properties of both skeletal and smooth muscle, reversibly inhibiting fusion and expression of muscle-specific genes such as creatine kinase, acetylcholine receptors and transition from β and γ to α-actin [11]. These effects are also seen in embryonic and skeletal muscle myoblasts. It appears to act at an early stage of commitment to differentiation in mouse and rat myoblasts rather than inhibiting expression of differentiated functions per se. [11, 12]. It has either no effect or possibly suppresses cell proliferation of myoblasts [12].

aFGF and bFGF are heparin-binding growth factors that are part of a larger family of structurally related trophic factors including the *hst*/K-*fgf* gene product found in Kaposi's sarcoma, the product of the *int*2 gene in embyonic carcinoma cell lines and FGF-5 which is expressed in transformed and normal tissues [31]. aFGF and bFGF were originally shown to influence the proliferation of cells of mesodermal and neuroectodermal origin [31]. It is now clear that they are present in a wide range of organs, solid tissues, tumours and cultured cells and are now considered as potent mitogens for diverse cell types. In this respect aFGF and bFGF seem to differ only in their potency. Both aFGF and bFGF elicit proliferation of isolated skeletal muscle myoblasts and block myogenic differentiation, probably between the G_0 and G_1 transition [26]. Interestingly, the action of FGFs on skeletal muscle can be inhibited by TGF-β, suggesting a plausible model for regulation of myoblast proliferation in vivo [11].

Transforming and Fibroblast Growth Factors in the Heart

A number of growth factors are present in non-myocyte cells in the heart. In particular, the fibroblast growth factors are known to be mitogenic for capillary endothelial cells in vitro and to induce angiogenesis in picomolar quantities in vivo [13]. As peptide growth factors can act in both a paracrine and an autocrine fashion, synthesis or storage in non-myocyte cells does not exclude actions on the cardiac myocyte. Indeed this may well be one mode of controlling their effects on such cells. Nonetheless, the evidence for the presence of transforming and fibroblast growth factors in the myocyte has been accumulating in the last 3 years and is pointing to a possible role in the regulation of cell proliferation, differentiation and growth.

TGF-β protein and mRNA expression has been investigated in the mouse heart [30]. Using different polyclonal antibodies raised to the amino-terminal 30 residues of TGF-β, strong intracellular immunohistochemical staining of cardiac myocytes or myocyte extracellular matrix (ECM) has been observed. The staining, which was present in both adult and neonatal tissues and was somewhat stronger in atria than ventricles, depended on the particular antisera employed. Mutually exclusive staining of either ECM or myocytes with different antisera was obtained, suggesting that differential distribution of TGF-β isoforms might be involved. Some support for this arises from the finding that a 2.4 kb transcript of TGF-β mRNA was detected in both adult and neonatal whole mouse heart extracts but additional 0.8 and 1.1 kb transcripts were present in the adult [30].

In contrast in the rat, immunohistochemical staining for TGF-β has been shown to be intense around intramyocardial blood vessels, with only weak radially diffuse staining in the surrounding myocardium [9]. mRNA for TGF-β was detected only in the non-myocyte fraction of mixed isolated cardiac myocyte and non-myocyte cells [9]. These studies have therefore left unclear the precise cellular location of TGF-β. Interestingly, it has been demonstrated that TGF-β mediates the transformation of epithelial to mesenchymal cells during cardiac embryogenesis in the developing endocardial cushion and valves of embryonic mice and chicks [21, 30].

Studies of the effects of TGF-β on isolated cardiac myocytes have also been reported. TGF-β inhibits proliferation of neonatal rat cardiac myocytes [14] and provokes the re-expression of "fetal" isoforms of myosin heavy chain and actin contractile proteins at a time when these cells are normally expressing the adult isoforms [20]. These transitions are similar to those following haemodynamic overload. This finding suggests that cardiac protein isoforms which we have shown to undergo well-defined transitions during both development and haemodynamic overload [5, 7] may be linked in some way to growth factor expression, perhaps by sharing regulatory gene elements involved in cell proliferation and growth [26].

aFGF and bFGF expression has also been reported in the heart and isolated myocyte. bFGF has been localised by immunofluorescence to nuclei and intercalated discs in adult bovine hearts, with more accumulation around atrial than ventricular fibres [15], but to endothelial cells and cardiac myocytes in the human ventricle [4]. Few studies on aFGF localisation in relation to the cardiac myocyte have been reported. In isolated adult rat myocytes, both aFGF and bFGF have been detected by radioimmunoassay and/or immunoblotting [27]. In isolated neonatal rat myocytes there is controversy as to whether aFGF is present in the myocyte cytoplasm [25] or is localised to the ECM [32]. This raises an important point regarding the FGFs. Their strong heparin-binding properties in vitro almost certainly reflect their in vivo affinities for ECM heparan sulphate proteoglycans and glycosoaminoglycans. If FGFs are concentrated and stored in the ECM, they could be released for action on adjoining and adjacent cells. However the FGFs do not have classical amino-terminal leader sequences typical of secreted proteins that are extracellularly active [32]. It is thus not clear at present which cells they are synthesised in, how they exit from cells, and which, if any, cell ECM is involved.

It is nonetheless clear that high levels of both aFGF and bFGF protein are present in the heart. Between 100 and 500 μg purified aFGF/kg has been isolated from adult porcine, canine or bovine hearts [22, 25] and 6–13 μg purified bFGF/kg from bovine heart [22]. These may not, however, reflect the absolute tissue levels. A recent study from our own laboratory has indicated that bFGF content of the adult bovine myocardium is nearer 300 μg/kg, and that in turn even this may be an underestimate [8].

In contrast to the high protein levels of both aFGF and bFGF, it has proved difficult to detect mRNA for either growth factor by conventional techniques because of the relatively low copy numbers that are available in any tissue in vivo [22]. Recently, mRNA transcripts of 3.8 kb and 7.1 kb for aFGF and bFGF, respectively, which are comparable to those present in other tissues, have been identified in the bovine heart [6].

The effects of FGFs on isolated cardiac myocytes in vitro also point to a role in myocyte proliferation and cardiac specific protein gene expression. bFGF stimulates proliferation and DNA synthesis in neonatal rat myocytes which can be inhibited by TGF-β, once again suggesting a joint role for these growth factors [14]. Like TGF-β, both aFGF and bFGF also induce expression of "fetal" contractile protein genes in neonatal myocytes [20]. However the two FGFs differ in their effects on striated alpha-actin genes with bFGF alone being identical to TGF-β.

Overall therefore, while there is a body of evidence that links the expression of TGF-β, aFGF and bFGF to the cardiac myocyte, there are many inconsistencies in the findings to date. One important factor that may partly explain the contradictions in some of the above studies is that the precise form of each of the three growth factors expressed in the heart has yet to be established. TGF-β exists as at least three characterised forms (TGF-β_1, -β_2, -β_3) [28], with studies in the heart focussed mainly on TGF-β_1 and to a much lesser extent TGF-β_2. Given that the three isoforms are each located on a different chromosome, that different isoforms are expressed in various cells, and that physiological mechanisms of activation of their respective precursor latent forms may be specific [1, 27], this could have important implications for studying specific isoforms in the heart.

Likewise, both aFGF and bFGF have been characterised in several different isoforms, many of them as potential post-translational cleavages of various lengths from the amino-terminal domains [3]. The 16- to 18-kDa forms of both FGFs are known to exist as higher molecular weight (20–26 kDa) versions [3]. It is not clear yet whether these are partly the result of random proteolysis, or whether tissue or developmentally regulated specific post-translational modifications are involved. Support for the latter comes from recent preliminary evidence on bFGF expression in the rat which suggests that different isoforms of bFGF may be developmentally and hormonally regulated [17]. Clearly further studies on TGF-β or the FGFs in the heart should use transcript or isoform specific DNA probes or antibodies respectively to investigate mRNA and protein expression.

Finally, there is evidence that the above growth factors may play an important role in wound healing and cardioprotection in the heart. The role of the FGFs may be intricately linked to their potent angiogenic influences on endothelial cells. However TGF β has also been shown to be up-regulated at both the protein and mRNA level in myocytes at the margin of experimentally infarcted areas between 24 and 48 h after coronary artery ligation [29] and to prevent severe cardiac injury when injected intravenously immediately before or after experimental myocardial ischaemia [16].

Taken overall, the above studies strongly suggest that cardiac myocytes are targets for the action of TGF-β, aFGF and bFGF in vivo and may play an important role in myocyte hyperplasia, differentiation and growth. While the limited available evidence suggests that these growth factors are present during cardiac development, it also suggests that they are present in the adult myocardium. If their role during cardiac development is linked in some way to myocyte proliferation, it leaves open the question as to what further role they may play in the adult. In particular, are they involved in growth processes in general, irrespective of myocyte proliferation? Moreover, the wider role of these growth factors in the myocardium, in particular in the normal, haemodynamically overloaded and damaged heart is unknown.

Future Directions

A number of important questions require investigation and should point the way for future research in this area:

1. What is the precise cellular or extra-cellular location of TGF-β, aFGF, and bFGF in the developing and adult human heart with respect to the myocyte in vivo?
2. Are the expressions of TGF-β, aFGF and bFGF modified in the haemodynamically overloaded and hypertrophied adult heart?
3. Are the expressions of TGF-β, aFGF and bFGF modified in cases of congenital cardiac hypertrophy and is this dependent on a primary or secondary stimulus to growth?
4. How is the expression of these growth factors, whether at the protein or mRNA level, both during development and in the adult, linked to the expression of their respective receptors? What is the cellular localisation of these receptors?
5. Are different isoforms of these growth factors, in particular bFGF, localised in different subcellular compartments, and is this a means of regulating their expression?
6. Are different isoforms of TGF-β, aFGF or bFGF expressed in the cardiac myocyte, and are these developmentally regulated?
7. Are mRNA and protein expression of TGF-β, aFGF and bFGF closely linked temporally and spatially in the normal developing and overloaded heart.
8. Does the pattern of expression of TGF-β, aFGF and bFGF differ in the atria and ventricles during development to reflect the different potentials for DNA synthesis and perhaps mitotic division of their respective myocytes?
9. Can the different isoforms of these growth factors be isolated and tested in vitro for their effects on isolated neonatal and adult cardiac myocytes?
10. Is the expression of these growth factors modified in the adult heart subsequent to experimental myocardial infarction? If so, is the expression confined to the immediate post-infarction phase and/or the latter compensated growth phase?
11. Are there links in vivo, as demonstrated during development in vitro, between the expression of TGF-β and the FGFs and the transitions in cardiac specific contractile protein isoforms in the adult hypertrophied heart?

References

1. Abraham JA, Whang JL, Tumolo A et al. (1986) Human basic fibroblast growth factor nucleotide sequence and genomic organisation. EMBO J 5:2523–2528
2. Barnes DM (1988) Joint Soviet–US attack on heart muscle dogma. Science 242:193–195
3. Burgess WH, Maciag T (1989) The heparin-binding fibroblast growth factor family of proteins. Ann Rev Biochem 58:575–606
4. Casscels W, Speir E, Sasse J et al. (1990) Isolation, characterisation and localization of heparin-binding growth factors in the heart. J Clin Invest 85:433–441
5. Cummins P (1982) Transitions in human atrial and ventricular myosin light-chain isoenzymes in response to cardiac pressure overload-induced hypertrophy. Biochem J 205:195–204
6. Cummins P, Beestone S, Chilton D (1991) Fibroblast growth factor mRNA and protein expression in the developing heart. Eur Heart J 12 [Suppl]:210
7. Cummins P, Lambert J (1986) Myosin transitions in the human and bovine heart: a developmental and anatomical study of heavy and light chain subunits in the atrium and ventricle. Circ Res 58:846–858
8. Cummins P, Logan A, Cummins B (1991) Basic fibroblast growth factor in the developing bovine heart. Biochem Soc Trans 19:79S.
9. Eghbali M (1989) Cellular origin and diistribution of transforming growth factor-β1 in the normal rat myocardium. Cell Tissue Res 256:553–558
10. Field LJ (1988) Atrial natriuretic factor-SV 40 T antigen transgenes produce tumors and cardiac arrhythmias in mice. Science 239:1029–1033
11. Florini JR (1987) Hormonal control of muscle growth. Muscle Nerve 10:577–598
12. Florini JR, Magri KA (1989) Effects of growth factors on myogenic differentiation. AM J Physiol 25:C701–C711
13. Folkman J, Klagsbrun M (1987) Angiogenic factors. Science 235:442–447
14. Kardami E (1990) Stimulation and inhibition of cardiac myocyte proliferation in vitro. Mol Cell Biochem 92:129–135
15. Kardami E, Frandrich RR (1989) Basic fibroblast growth factor in atria and ventricles of the vertebrate heart. J Cell Biol 109:1865–1875
16. Lefer AM, Tsao P, Aoki N et al. (1990) Mediation of cardioprotection by transforming growth factor beta. Science 249:61–64
17. Liu L, Kardami E (1991) Hypothyroidism favors expression of high molecular weight basic FGF in the heart. J Cell Biochem [Suppl] 15C:171
18. Long CS, Kariya K, Karns L, Simpson PC (1990) Trophic factors for cardiac myocytes. J Hypertension 8 [Suppl 7]:S219–S224
19. Oberpriller JO (1988) Nuclear characteristics of cardiac myocytes following the proliferative response to the mincing of the myocardium in the adult newt, *Notophthalmus viridescens*. Cell Tissue Res 253:619
20. Parker TG, Schneider MD (1990) Peptide growth factors can provoke "fetal" contractile protein gene expression in rat cardiac myocytes. J Clin Invest 85:507–514
21. Potts JD, Runyan RB (1989) Epithelial-mesenchymal cell transformation in the heart can be mediated, in part, by transforming growth factor β. Dev Biol 134:392–401
22. Quinkler W, Maasberg M, Bernotat-Danielowski S (1989) Isolation of heparin-binding growth factors from bovine, porcine and canine hearts. Eur J Biochem 181:67–73
23. Roberts A, Sporn MB (1990) The transforming growth factor-betas. In: Sporn MB, Roberts AB (eds) Peptide growth factors and their receptors. Springer, Berlin Heidelberg New york, pp 419–472 (Handbook of experimental pharmacology, vol 95, part 1)
24. Rumyantsev PP (1977) Interrelations of the proliferation and differentiation processes during cardiac myogenesis and regeneration. Int Rev Cytol 51:187–273
25. Sasaki H, Hoshi H, Hong Y-M et al. (1989) Purification of acidic fibroblast growth factor from bovine heart and its localization in the cardiac myocytes. J Biol Chem 264:17606–17612
26. Schneider MD, Parker TG (1990) Cardiac myocytes as targets for the action of peptide growth factors. Circulation 81:1443–1456

27. Speir E, Yi-Fu Z, Lee M et al. (1988) Fibroblast growth factors are present in adult cardiac myocytes, in vivo. Biochem Biophys Res Commun 157:1336–1340
28. Sporn MB, Roberts AB, Wakefield LM et at. (1987) Some recent advances in the chemistry and biology of transforming growth factor beta. J Cell Biol 105:1039–1045
29. Thompson NL, Bazoberry F, Speir EH et al. (1988) Transforming growth factor beta-1 in acute myocardial infarction in rats. Growth Factors 1:91–99
30. Thompson NL, Flanders KC, Smith JM et al. (1989) Expression of transforming growth factor-beta 1 in specific cells and tissues of adult and neonatal mice. J Cell Biol 108:661–669
31. Waterfield MD (ed) (1989) Growth factors. Br Med Bull 45:317–604
32. Weiner HL, Swain JL (1989) Acidic fibroblast growth factor mRNA is expressed by cardiac myocytes in culture and the protein is localized to the extracellular matrix. Proc Natl Acad Sci USA 86:2683–2687
33. Zak R (1974) Development and proliferative capacity of cardiac muscle cells. Circ Res 34–35 [Supp II]:17–25

IV. Structural Abnormalities in IDC

The Cytoskeleton and Extracellular Matrix in Human Hearts with Dilated Cardiomyopathy

J. Schaper, B. Speiser, and T. Brand

Introduction

Dilated cardiomyopathy is a disease that in its final stage can be treated only by transplantation. The explanted diseased hearts enable the study of the structural composition of viable tissue that can be optimally fixed and prepared for different morphological methods, as well as the investigation of large tissue samples as opposed to the small samples provided by biopsy removal during catheterization of the heart.

In ultrastructural and immunocytochemical studies of the intracellular changes in dilated cardiomyopathy [8], an enlargement of the extracellular space, and the presence of an increased amount of fibrotic material have been noted. The extracellular matrix of normal myocardium contains numerous proteins including fibronectin, laminin, various collagens, and different types of proteoglycans. In a recent study we investigated the composition of the extracellular matrix in normal human myocardium using monoclonal antibodies and immunocytochemical techniques [10], and the present work is a continuation of this study in diseased human myocardium. The cytoskeleton is closely connected with the extracellular matrix via talin–α-actinin–vinculin [3] and the integrins [1], and therefore a systematic study was undertaken to determine the structure and composition of both the matrix and cytoskeleton of diseased human myocardium.

Material and Methods

Explanted hearts from patients with end-stage dilated cardiomyopathy (ejection fraction < 20%) were obtained at the time of transplantation, and tissue samples were immediately fixed in either liquid nitrogen for immunocytochemistry or in 3% glutaraldehyde for electron microscopy. The frozen tissue was stored at $-70°$C, and the samples for electron microscopy were embedded in Epon following routine procedures.

Department of Experimental Cardiology, Max-Planck-Institute, W-6350 Bad Nauheim, Fed. Rep. of Germany

Figulla et al. (Eds.)
Idiopathic Dilated Cardiomyopathy

Immunofluorescent Staining. Commercially available antibodies against the following matrix proteins were used in this study: fibronectin (ICN Immuno Biologicals, Lisle, UK), laminin (Dianova, Hamburg, FRG), vimentin (Dianova, Hamburg, FRG), collagen I and III (Bioscience Products, Switzerland), collagen VI (Telios, San Diego, USA).

Antibodies that were against cytoskeletal proteins were vinculin (Biomakor, Rehovot, Israel) and desmin (Boehringer, Mannheim, FRG). Frozen tissue sections were air-dried and fixed for 10 min in acetone at −20° C. After rinsing in phosphate-buffered saline the tissue was incubated with the first antibody. The sections were washed 3 × 2 min with phosphate-buffered saline, incubated with biotinylated (anti-mouse, anti-rat, or anti-rabbit IgG; Amersham, UK) as the second antibody for 1 h, then rinsed and treated with fluorescein isothiocyanate labeled streptavidin. Nuclei were counterstained with 0.002% propidium iodide (modified from [4]). The sections were examined in an Olympus Vanox T fluorescence microscope, and the results were recorded by photography on Kodak 200 ASA color film.

Electron Microscopy. Sections of 50-nm thickness were prepared, stained with uranylacetate and lead citrate, and viewed and photographed in a Philips CM 10 electron miscroscope.

In Situ Hybridization. After appropriate pretreatment, frozen sections were incubated overnight with radioactively labeled cRNA probes sense, and antisense that was prepared from a polymerase chain reaction amplification product of the coding region from the human desmin sequence (we are grateful to Dr. H. Sharma from our laboratory for preparing the probe). The sections were covered with nuclear track photographic emulsion, exposed for 10 days, and then viewed in the light microscope and photographed.

Results and Discussion

Electron microscopy was used as a basic technique to determine the degree of hypertrophy and degeneration in the tissue samples. In contrast to normal myocardium, which shows regular arrangement of myofilaments and mitochondria, the hypertrophied myocytes from diseased hearts exhibited numerous alterations, including enlargement of nuclei, lack of contractile material, accumulation of mitochondria of different sizes and bizarre shapes and the occurrence of lipid droplets and myelin figures. Many cells showed sequestration of particles in the extracellular space, which was widened and filled with cellular debris, active fibroblasts and macrophages, and single collagen fibers or bundles (Fig. 1).

In normal myocardium, fibronectin, laminin, and different collagens are present in the space between myocytes and surrounding blood vessels [10]. These elements comprise the normal scaffolding of the cardiac tissue and ensure both the stability of the cellular arrangement and the resistance to mechanical forces due to contractile activity of the muscle. Under normal

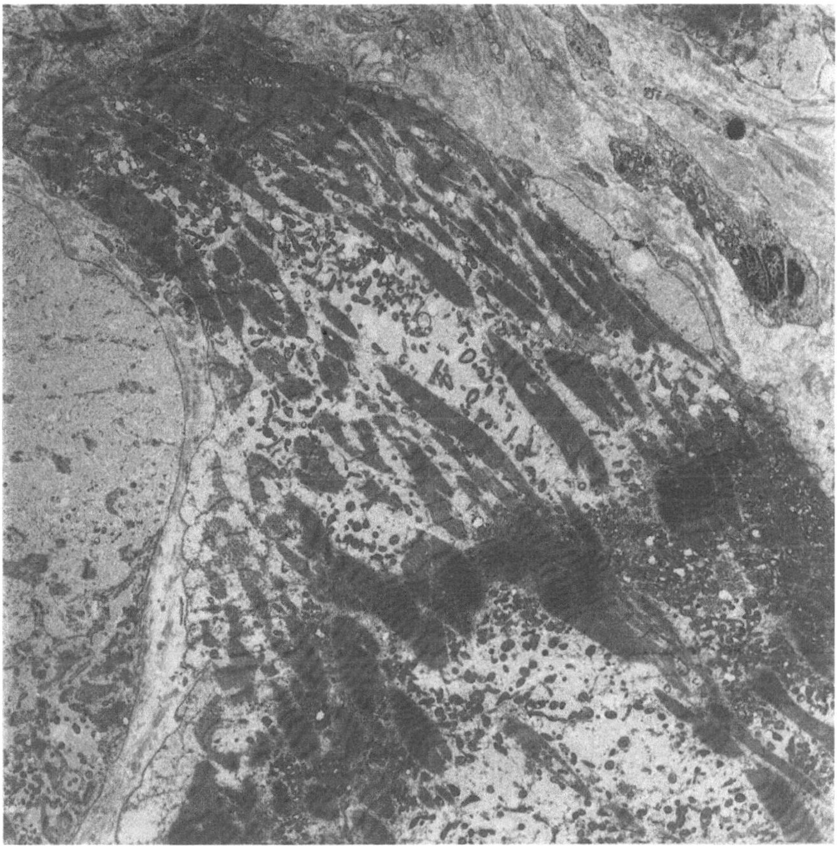

Fig. 1. Electron micrograph of diseased human myocardium. The cell is hypertrophied, the contractile material is reduced in the myocytes, and the extracellular space contains cellular debris and collagen bundles. ×4500

conditions, immunofluorescence of these proteins shows fine septa separating the cells. However, in diseased human myocardium the extracellular proteins were significantly increased in the widened extracellular space and exhibited a fluorescence more intense than in normal tissue. The increase in fibronectin was especially prominent, as it was observed in increased amounts that sometimes formed thick sheaths which encapsulated and isolated the myocardial cells (Fig. 2). Intracellular labeling for fibronectin was evident in myocardial cells, indicating the existence of a dilated and prominent T tubular system of the myocytes as observed in the electron microscope.

Laminin in normal myocardium is localized as a fine line representing the basal laminae surrounding myocytes, endothelial cells, and smooth muscle cells. It was greatly increased in amount and not only occurred in the

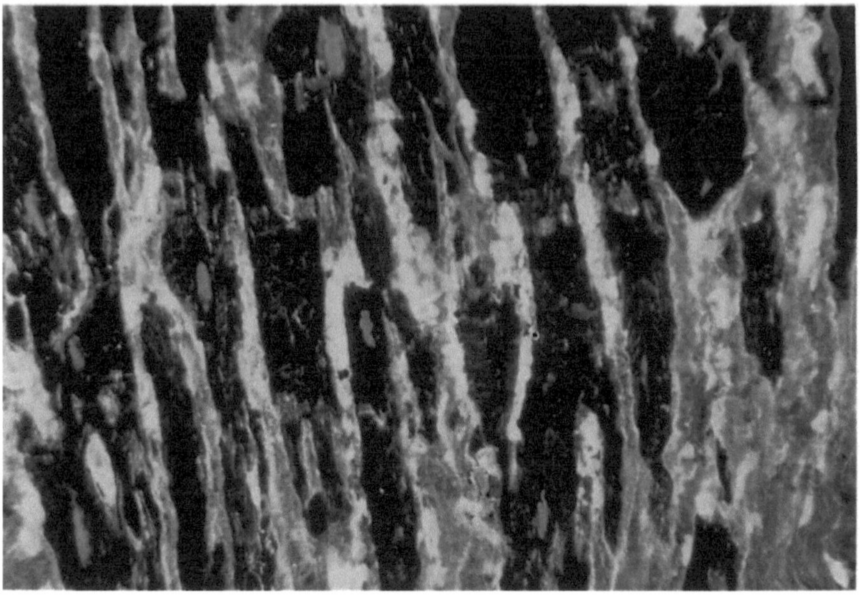

Fig. 2. Immunocytochemical localization of fibronectin. *Green fluorescence*, fibronectin abundantly distributed in the widened extracellular space. There is a fine labeling of the T tubular system in the otherwise dark, unlabeled myocytes. Nuclei are stained *red* in myocytes and *yellow* in the interstitium

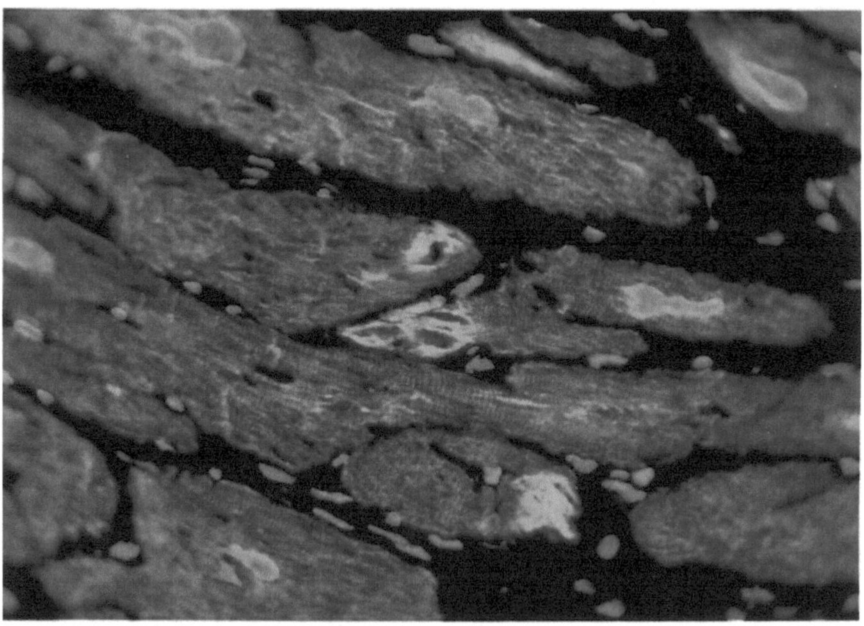

Fig. 3. Immunocytochemical localization of desmin. The labeling pattern is normal and/or moderately disturbed. The yellow fluorescence is typical of a completely disarranged desmin localization

basement membranes but was also widely distributed in the extracellular space. Likewise, all collagens were increased, and vimentin was present in numerous fibroblasts. The findings in the extracellular matrix have been previously described in greater detail [9].

As mentioned above, the extracellular matrix proteins, especially fibronectin and laminin, are connected to the cytoskeleton of the cells via talin–α-actinin–vinculin, and this connection may be important for the function of the cell. In normal myocardium vinculin labeling was observed in the intercalated discs and sarcolemma, which is in accordance with earlier reports [7]. In cardiac tissue from diseased human hearts, vinculin was abundantly present not only at all myocyte membranes but also within the myocytes [8]. It is therefore hypothesized that in dilated cardiomyopathy the loss of contractile proteins is compensated by an increase in the different elements of the cytoskeleton, thereby maintaining cellular stability and perhaps producing an increased stiffness of the cardiac cells.

The immunocytochemical localization of desmin further strengthens this hypothesis. Desmin, specific for muscle tissue, belongs to the family of intermediate filaments that are typically 10 nm thick [5, 6]. Desmin is localized in normal cardiac cells at the level of the Z lines where it con-

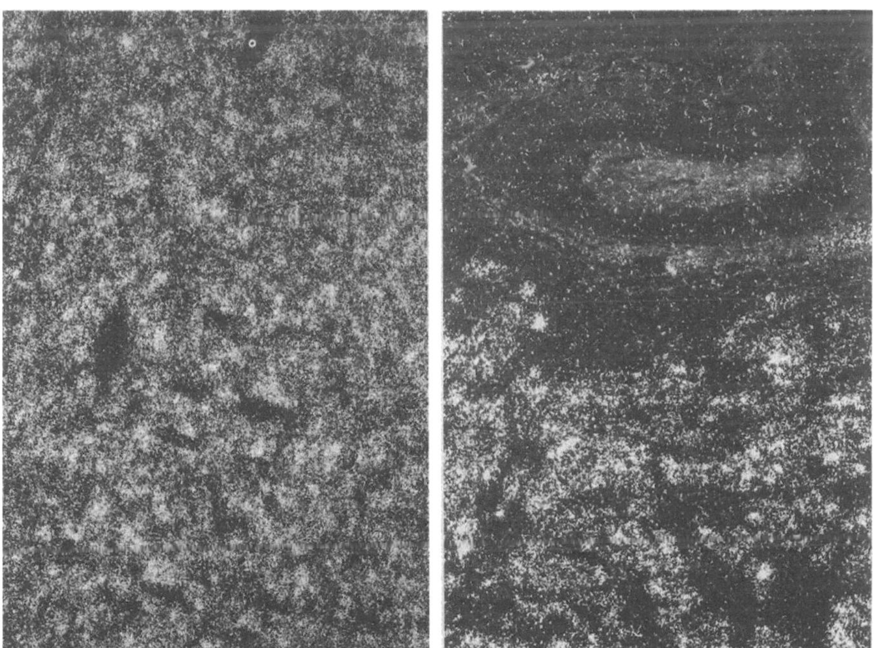

a b

Fig. 4a,b. In situ hybridization for mRNA of desmin. **a** In normal myocardium the labeling pattern is regularly distributed, indicating the presence of mRNA in all myocardial cells. **b** In diseased myocardium the distribution of the mRNA is inhomogeneous over myocytes and shows clustering of the label, and it is absent over fibrotic tissue (*bottom*) nects the sarcomeres with each other and to the lateral sarcolemma [2]. In diseased cardiac myocytes the localization of desmin is often disturbed. Some myocytes show a normal localization of desmin with distinct cross striations, other cells show a slight disarrangement, but in many cells a complete disorganization combined with an increase in the amount of desmin can be observed (Fig. 3). These findings on the protein level were confirmed by the localization of the mRNA for desmin using a specific probe (Fig. 4). It is evident from Fig. 4 that in normal tissue the mRNA label is homogeneously distributed, but that there are great differences in the quality of labeling in diseased myocardium. The absence of desmin mRNA in some areas is due to the presence of fibrotic tissue; however, this does not explain the altered labeling pattern showing an irregular distribution and clustering. The latter most probably is the equivalent of the accumulation of the protein desmin in individual cells.

The morphological findings described here may be summarized as follows: the contractile machinery of the myocardial cells as well as other subcellular organelles such as mitochondria and the T tubular system are degenerated in dilated cardiomyopathy, whereas the myocytes are hypertrophied. The cytoskeleton, on the other hand, shows proliferation of the components desmin and vinculin, and the extracellular matrix proteins are more abundantly expressed compared to normal myocardium. All these alterations may contribute to the reduced functional properties of the myocardium that result finally in heart failure.

References

1. Albelda S, Buck A (1990) Integrins and other cell adhesion molecules. FASEB J 4: 2868–2880
2. Altmannsberger M, Weber K, Droste R, Osborn M (1985) Desmin is a specific marker for rhabdomyosarcomas of human and rat origin. Am J Pathol 118:85–95
3. Hynes R, Destree A, Wagner D (1982) Relationships between fibronectin, actin, and cell substratum adhesion. Cold Spring Harb Symp Quant Biol 46:659–670
4. Jones K, Kniss D (1987) Propidium iodide as a nuclear counterstain for immunofluoresence studies of cells in culture. J Histochem Cytochem 35:123–125
5. Lazarides E (1982) Biochemical and immunological characterization of intermediate filaments in muscle cells. Methods Cell Biol 25:333–357
6. Osborn M, Geisler N, Shaw G, Sharp G, Weber K (1982) Intermediate filaments. Cold Spring Harb Symp Quant Biol 46:413–429
7. Pardo J, Siliciano J, Craig S (1983) Vinculin is a component of an extensive network of myofibril-sarcolemma attachment regions in cardiac muscle fibers. J Cell Biol 97:1081–1088
8. Schaper J, Froede R, Buck A, Bleese N (1988) Impaired myocardial ultrastructure and cytoskeleton in cardiomyopathic human myocardium. In: Schultheiß H (ed) New concepts in viral heart disease. Springer, Berlin Heidelberg New York, pp 295–302
9. Schaper J, Speiser B (1992) The extracellular matrix in the failing human heart. Basic Res Cardiol 87:303–309
10. Speiser B, Rieß C, Schaper J (1991) The extracellular matrix in human myocardium. I. Collagen I, III, VI, and IV. Cardioscience 2:225–234

Macro- and Micropathology of Dilated Cardiomyopathy

E.G.J. Olsen

The WHO/ISFC recommendation for definition and classification of cardiomyopathy is followed here [1].

Macroscopy

When a patient dies in the late phases of the disease the heart is grossly enlarged, with all cardiac chambers severely dilated. Although measurement of the myocardial wall may be normal (or less than normal), hypertrophy is invariably present, and heart weights of double the normal values are not unusual [2]. The heart muscle is flabby and may be pale [3]. The enlarged cardiac chambers usually show endocardial thickening haphazardly distributed (Fig. 1). Thrombus is present at the apex or at other sites in over 50% of hearts [4, 5]. On cutting the myocardium, areas of fine fibrosis are not infrequent, and areas of fibrous replacement may be evident. The coronary arteries, with rare exceptions, are normal [6]. Although the natural history of dilated cardiomyopathy is not yet fully understood, patients may die at any stage during the disease process. This implies that the above features may not be evident, and only some hypertrophy with or without dilatation may be found.

Microscopy

Upon histological examination, the myocardial fibres are in normal regular arrangement. If dilatation has been pronounced, as in long-standing cases, attenuation can be striking, resulting in normal or less than normal myocardial fibre diameters. Nuclear changes of hypertrophy in the form of polar blunting, hyperchromasia or vesicular changes, however, are evident. The interstitium may show varying degrees of an increase in collagen tissue, occasionally to a severe degree. In addition, areas of fibrous replacement of the myocytes can be found. The intramyocardial arterioles are usually

Department of Cardiovascular Pathology, Royal Brompton National Heart and Lung Hospital, Sydney Street, London SW3 6NP, UK

Figulla et al. (Eds.)
Idiopathic Dilated Cardiomyopathy

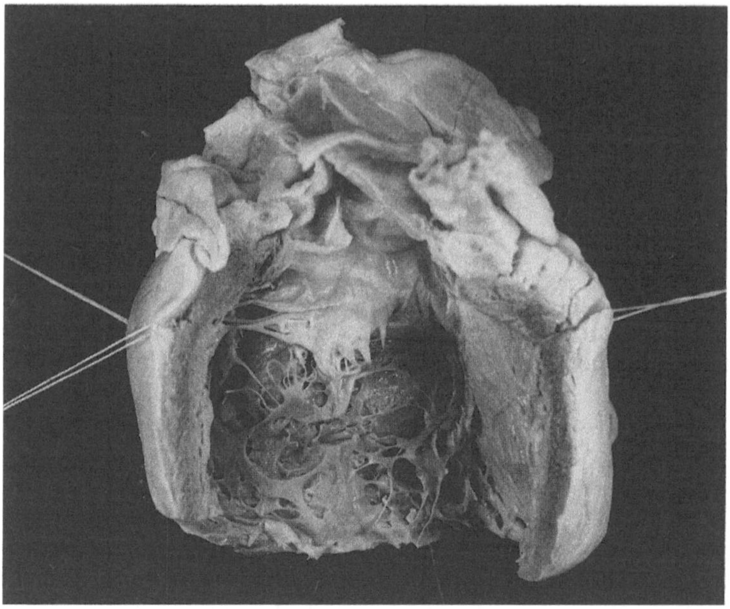

Fig. 1. The left ventricle has been opened to show the dilated cavity. Note the thickened endocardium haphazardly distributed. The wall of the left ventricular wall is within normal limits despite the hypertrophy; this is due to the dilatation

normal, but some vessels may show fibrous intimal changes narrowing the lumina by about 50%. These vessels are usually located in areas of fibrosis [7, 8]. The endocardium shows varying degrees of thickening, and if dilatation has been present for at least 6/52 foci of smooth muscle become pronounced (Fig. 2), which in extreme cases may be confluent.

Morphometric Studies

Hypertrophy has been assessed by a variety of techniques including experimental studies [9]. In man, correct interpretation is fraught with difficulty because of dilatation. Recording of total heart weight is also unreliable, but a technique for removing epicardial fat, valves, and the great vessels has been suggested [10], closely aproximating correct ventricular weights. Attempts to correlate morphological features to clinical parameters, including length of history, LEVDP and ejection fraction have yielded contradictory results. Some have shown correlation; others have not [11–13]. Partial correlation has been reported by Figulla and coworkers [14]. Prognosis has also been assessed [12, 15, 16] at light and electron microscopic levels of investigation.

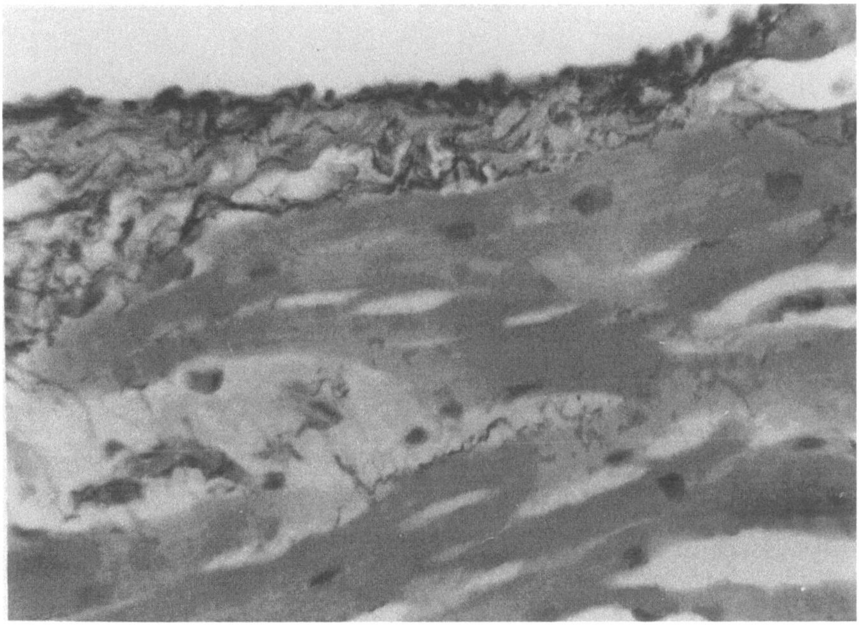

Fig. 2. High-power view of the slightly thickened endocardium clearly shows smooth muscle hyperplasia (between the two elastic laminae). The myocardial fibres are hypertrophied and show evidence of attenuation. Miller's elastic van Gieson ×400

Neuronal Counts

These have been assessed by selecting right atrial strips of tissue between the venae cavae, cutting sections at 50-μm intervals (to avoid counting neurons twice). A significant reduction was found in cases of dilated cardiomyopathy [17].

Endomyocardial Biopsy and Myocarditis

With the aid of the bioptome, patients have been biopsied in earlier phases of the disease process, and myocarditis has been found in a significant number of patients, varying from 0% to 65% [18]. In my own series of 2000 patients thus examined, a result of 23% has been obtained. To achieve uniformity of approach, a group of pathologists met in Dallas in 1984 and defined the condition and established also diagnostic recommendations which were based on the authors' original concepts [19].

Active or acute myocarditis (Fig. 3) usually consists of a chronic inflammatory cell infiltrate flooding the interstitium and resulting in necrosis of

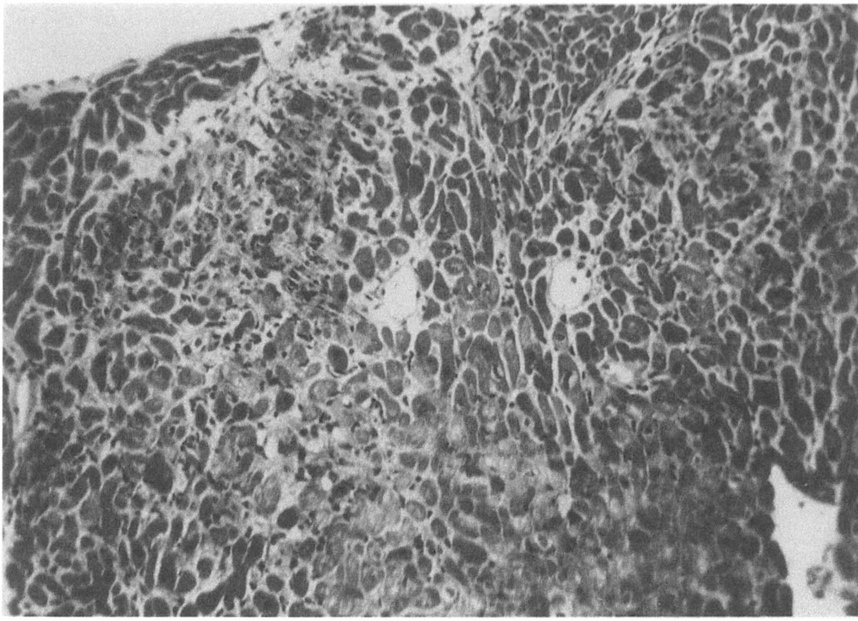

Fig. 3. Low-power view of active myocarditis (focally distributed). Note the chronic inflammatory infiltrate and the necrosis changes of adjacent myocardial fibres. Haematoxylin and eosin ×250

adjacent myocardial fibres. Interstitial fibrous tissue may be increased, but in our experience this is usually not pronounced.

On-going myocarditis is a term used when the next sequential biopsy is at least as severe as the most recent biopsy.

Resolving or healing myocarditis is recognisable when the severity of the inflammatory cell infiltrate has decreased, and although some inflammatory cells may still be in intimate contact with adjacent myocytes, necrosis is usually not widespread and may be totally absent. Interstitial fibrous tissue may be increased.

Resolved or healed myocarditis is the final stage. The inflammatory cell infiltrate has abated, and replacement of myocytes by fibrous tissue may be striking. If this is the patient's first biopsy, no previous myocarditic process can be inferred.

Although chronic inflammatory cells have been used in the description, other inflammatory cells such as eosinophils or neutrophils may dominate the cellular infiltrate.

Staging of the myocarditic process is important, as it may have therapeutic implications. Molecular biology has contributed greatly to understanding the causation of myocarditis, and particularly in implication of viruses. The interpretation of myocarditis is very difficult, both from a clinical point of view and pathologically. Correct interpretation of biopsy samples is important but is full of pitfalls [20].

Histochemical parameters such as glycogen and succinic dehydrogenase may be evaluated in endomycardial biopsy specimens. These substances may be normal, increased or decreased depending on the degree of dilatation and whether or not heart failure has supervened.

It has also been possible to use electron microscopy on these tissue samples. The myocardial fibrils are in regular, parallel alignment with an increase in mitochondria to more than 1 per 2 sarcomeres. Infolding of nuclear membranes consistent with hypertrophy is also found. Varying degrees of degeneration and cellular infiltrate can also be defined [21–23].

Differential Diagnosis

Alcoholic Heart Muscle Disease. Identical changes, including myocarditis, have been documented in dilatation of the heart due to alcohol abuse. It has been reported that if 7 g alcohol is imbibed per day, these changes become evident [24]. Biochemical analysis of endomyocardial biopsy does, however, permit distinction.

Peripartal Heart Disease. According to an expert committee of the World Health Organisation, patients suffering from dilatation and/or heart failure in the third trimester and at any stage following delivery for 3 months are included [25]. Morphologically identical changes to those described under dilated cardiomyopathy are encountered, including myocarditis, described in a joint study from East Africa [26].

Keshan's Disease. Selenium deficiency resulting in changes similar to those in dilated cardiomyopathy has been reported from China [27]. Recognition and treatment of the problem has led to reversal of the condition.

Conclusion

The macro- and micropathology of dilated cardiomyopathy is non-specific, and diagnosis can be achieved only by excluding those conditions that may give rise to a hypertrophied, dilated heart. The recognition of myocarditis is very important, and endomyocardial biopsies at the histological, histochemical and electron microscopic levels serve as a basis upon which other investigations, such as molecular studies, biochemistry and virological analysis, can proceed.

References

1. WHO/ISFC (World Health Organisation/International Society and Federation of Cardiology) (1980) Report of the Task Force on Definition and Classification of Cardiomyopathies. Br Heart J 44:672–673

2. Zeek PM (1942) Heart weight. I. The weight of the normal human heart. Arch Pathol 34:820–832
3. Roberts WC, Siegel RJ, MacManus BM (1987) Idiopathic dilated cardiomyopathy: analysis of 152 necropsy patients. Am J Cardiol 60:1340–1355
4. Olsen EGJ (1972) Cardiomyopathies. Cardiovasc Clin 4:240–261
5. Olsen EGJ (1981) Pathology of congestive cardiomyopathy. In: Goodwin JF, Hjalmarson A, Olsen EGJ (eds) Congestive cardiomyopathy. Hassel, Molndal, pp 66–74
6. Gau GT, Goodwin JF, Oakley CM, Olsen EGJ, Rahimtoola SH, Raphael MJ, Steiner RE (1972) Q waves and coronary arteriography in cardiomyopathy. Br Heart J 34:1034–1041
7. James TN (1964) An etiologic concept concerning the obscure cardiomyopathies. Prog Cardiovasc Dis 7:43–64
8. Factor SM, Cho S, Sonnenblick EH (1981) Verapamil treatment of cardiomyopathy Syrian hamsters: effects on the micro circulation and the extent of myocardial necrosis. Fed Proc 40:758
9. Olsen EGJ (1980) The pathology of the heart, 2nd edn. Macmillan, London, pp 41–55
10. Fulton RM, Hutchinson ECE, Morgan Jones A (1952) Ventricular weight in cardiac hypertrophy. Br Heart J 14:413–420
11. Baandrup U, Florio RA, Rehahn M, Richardson PJ, Olsen EGJ (1981) Critical analysis of endomyocardial biopsies from patients suspected of having cardiomyopathy. II. Comparison of histological clinical haemodynamic information. Br Heart J 45:487–493
12. Baandrup U, Florio RA, Roter F, Olsen EGJ (1981) Electron microscopic investigation of endomyocardial biopsy samples in hypertrophy and cardiomyopathy. A semiquantitative study in 48 patients. Circulation 63:1289–1298
13. Kunkel B, Schneider M, Kober G, Bussmann WD, Hopf R, Kaltenback M (1982) Myocardial biopsy: morphological analysis and clinical implications. Z Cardiol 71:787–794
14. Figulla HR, Rahlf G, Nieger M, Luig H, Kreuzer H (1985) Spontaneous haemodynamic improvement of stabilization and associated biopsy findings in patients with congestive cardiomyopathy. Circulation 71:1095–1104
15. Tangenelli P, Di Lenarda A, Bianciardi G, Salvi A, Silvestri F, Mestroni L, Camerini F (1989) Correlation between histomorphometric findings on endomyocardial biopsy and clinical findings in idiopathic dilated cardiomyopathy. Am J Cardiol 64:504–506
16. Hammond EH, Menlove RL, Anderson JL (1987) Predictive value of immunofluorescence and electron microscopy in the diagnosis and prognosis of myocarditis and idiopathic dilated cardiomyopathy. Am Heart J 114:105–106
17. Amorim DS, Olsen EGJ (1982) Assessment of heart neurons in dilated (congestive) cardiomyopathy. Br Heart J 47:11–18
18. Olsen EGJ, Meany BT, Richardson PJ (1988) The role of biopsy in the diagnosis and follow-up of myocarditis: a critical review. In: Schultheiss H-P (ed) New concept in viral disease. Springer, Berlin Heidelberg New York, pp 285–294
19. Aretz HT, Billingham ME, Edwards WD, Factor SM, Fallon JT, Fenoglio JJ, Olsen EGJ, Schoen FJ (1987) Myocarditis: a histopathologic definition and classification. Am J Cardiovasc Pathol 1:3–13
20. Atkinson JB, Virmani R (1991) The endomyocardial biopsy: techniques, indications, and limitations. In: Virmani R, Atkinson JB, Fenoglio JJ (eds) Cardiovascular pathology. Saunders, London, pp 203–219 (Major problems in pathology, vol 23)
21. Olsen EGJ (1978) Special investigations of COCM: endomyocardial biopsies (morphological analysis). Postgrad Med J 54:486–490
22. Kunkel B, Lapp H, Kober G, Kaltenback M (1978) Ultrastructural evaluations in early and advanced congestive cardiomyopathies. In: Kaltenback M, Loogen F, Olsen EGJ, Bussmann W-D (eds) Cardiomyopathy and myocardial biopsy. Springer, Berlin Heidelberg New York, pp 87–99
23. Ferrans VJ (1989) Pathological anatomy of dilated cardiomyopathies. Am J Cardiol 64:9C–11C
24. Richardson PJ, Atkinson L (1981) The aetiological separation of the congestive cardiomyopathies using the measurement of myocardial enzyme activities in biopsy samples. In: Goodwin JF, Hjalmarson A, Olsen EGJ (eds) Congestive cardiomyopathy. Hassel, Molndal, pp 103–110

25. Report of WHO Expert Committee (1984) Peripartal heart disease. In: Cardiomyopathies technical report series 697. World Health Organisation, Geneva, pp 47–50
26. Sanderson JE, Olsen EGJ, Gatei D (1986) Peripartum heart disease. An endomyocardial biopsy study. Br Heart J 56:285–291
27. Report of a WHO Expert Committee (1984) Keshan disease (in China). In: Cardiomyopathies technical report series 697. World Health Organisation, Geneva, pp 55–57

Detection of Disruption of Sarcolemmal Membrane by Indium-111 Labeled Antimyosin Antibody Scan

T. Yasuda

Introduction

We evaluated patients with a relatively recent onset of symptoms (under 1 year) of congestive heart failure who were suspected to have myocarditis for presence or absence of myocardial cell necrosis with antimyosin scan. To our astonishment, we discovered that 50%–60% of patients with idiopathic dilated cardiomyopathy showed acute myocardial cell necrosis based on antimyosin scan results [1, 2]. Patients with myocarditis may present with congestive heart failure, chest pain, fatigue, dizzy spells, and even syncope or cardiac arrest. In our daily practice these symptoms are not specific, and we may misdiagnose them as coronary disease, idiopathic dilated cardiomyopathy, or other noncardiac disorders. This can be exemplified by the following case.

Case History. A 36-year-old, previously healthy man was admitted because of sudden, unexpected cardiac arrest. He had suffered from a bad cold 1 month prior to this episode. After recovery from the cold, he continued to work every day. On the day of admission, he was playing softball with friends. He hit a single, ran to first base, and then collapsed. The paramedics arrived. He was found to be in ventricular fibrillation. He received DC counter-shock treatment and was brought to Massachusetts General Hospital. On admission his heart rate was 84 and blood pressure 120/80. Physical examination showed him unconscious, with no obvious focal neurological signs. Cardiac examination showed prominent apical impulse, but there was no thrill. There was a loud ventricular third sound. There was no heart murmur, no rub, and the abdomen was flat. The spleen and liver were not palpable. His legs were well perfused without varicose vein or edema. There were no pathological reflexes. Chest X-ray showed left ventricular failure with slight dilatation of the left ventricle. ECG at the time of admission showed a regular sinus rhythm rate of 80/m. There was no evidence of acute myocardial infarction, and acute myocardial infarction was ruled out by cardiac enzymes. Because of his young age, no previous medical problems, and the unresolved cause of sudden cardiac arrest, the

Cardiac Unit, Department of Medical Services, and Nuclear Medicine, Department of Radiology, Massachusetts General Hospital, Harvard Medical School, Boston, MA, USA

Figulla et al. (Eds.)
Idiopathic Dilated Cardiomyopathy

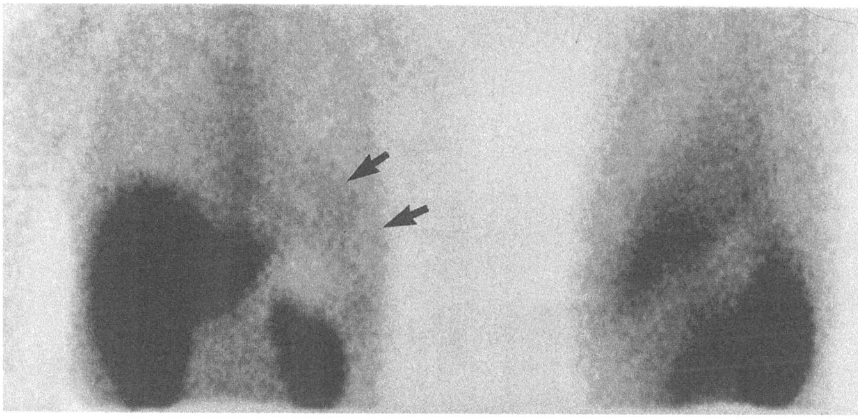

Fig. 1. There is an area of increased uptake in a region of the heart. Single-photon emission computed tomography confirms myocardial uptake. *Arrows*, regions of uptake

patient underwent cardiac catheterization and coronary angiogram. The results were all normal; however, the left ventricle was slightly dilated and moderately hypokinetic. There was no regional wall motion abnormality. Global diffuse hypokinesis was noted on left ventriculogram, and ejection fraction measured by gated scan was 40%. At this point, right ventricular biopsy and antimyosin scan were performed for a diagnosis of myocarditis. The biopsy showed no evidence of myocarditis based on the Dallas criteria. Antimyosin scan demonstrated clear uptake of antibody in the region of the heart (Fig. 1). The patient was treated with digoxin 0.25 mg a day, captopril 25 mg three times a day, propranold 30 mg three times a day, and an automatic implantable cardiac defibrillator.

This history suggests a possible diagnosis of myocarditis. The antimyosin scan showed definite evidence of recent myocardial necrosis. This raises the question of what really had happened to this patient. The patient was not treated with steroids or immunosuppressants because of it "not being myocarditis." He gradually improved, and after 6 months, the left ventricular ejection fraction had returned to a normal of 52%. The global hypokinesis disappeared completely, and the patient returned to work. One year later, LVEF dipped to 40%, and the left ventricle was moderately dilated. He was able to work.

Why did it happen? What does this antimyosin scan mean? What did he have, myocarditis? Surely he had reversible left ventricular dysfunction and recent myocardial cell necrosis.

Antimyosin Scintigraphy

Antimyosin scan is a cardiac imaging procedure using a compound of indium-111 labeled monoclonal antimyosin antibody. It is injected intra-

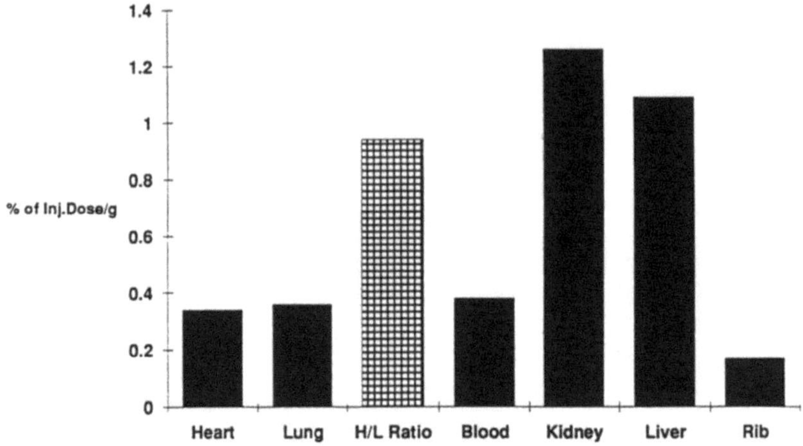

Fig. 2. Biodistribution of indium-111 labeled antimyosin in mice

venously and imaged 48 h after injection. The antibody is Fab fragment, and it was made against the heavy chain of myosin protein [3]. The injection dose of indium-111 is 2 mCi for a patient. Our labeling efficiency, measured by instant thin layer chromatography (Na citrate solvent), showed on average 93%–95%. Antimyosin antibody R11D10 was supplied by Centocor (Malvern, PA, USA).

At 48 h the highest uptake is seen in the kidneys and then the hepatic regions. Some activities are present in the spleen and bone marrow. However, at 24 h the blood pool activities are too high for imaging for myocarditis; it competes with the myocardial uptake, and contrast would not be optimum. The plasma half-life of indium-111 labeled antimyosin antibody Fab is 5 h, and the plasma concentration of antimyosin antibody activities at 24 h drops to 20% and at 48 h to below 10% of initial postinjection blood activities [4, 5].

Biodistribution and Cardiac Uptake

The general biodistribution of antimyosin antibody labeled with indium-111 was published previously by Kishimoto et al. [6] (Fig. 2). This was done in C3H/He mice in an acute stage of an encephalomyocarditis virus infection. The mice were then followed into the chronic stage. The highest uptake was seen in the kidneys, then the liver. After infection with encephalomyocarditis virus, the mice showed significant myocardial uptake on day 10. However, it quickly subsided in 1 month. Of interest, a few of them maintained active uptake, and some appeared to remain so for months. This suggests that myocardial cell necrosis may continue in some for a long time and in others for only a short time. The heterogeneity of the disease is well demonstrated in this homogeneous animal model.

Antimyosin antibodies attach to a myosin heavy-chain component through the broken cell membrane [5, 7] and emit gamma rays, which we image. The binding of antigen and antibody must still have a well-preserved myosin molecular structure. Intensity in uptake of antimyosin antibody correlates with the number of dead cells. The greater the number of dead cells, the clearer the images are. This is best demonstrated in patients with acute myocardial infarction [1–5] who represent a sudden massive myocyte necrosis, versus those with myocarditis in whom the degree of myocyte necrosis is much less.

Most patients with dilated cardiomyopathy do not show intense uptake but rather a faint and vague uptake, as seen in Fig. 1. The distribution of uptake is also heterogeneous and usually involves the entire heart. The sensitivity, specificity, and predictive value of antimyosin scan for a diagnosis of myocarditis in patients with idiopathic detated cardiomyopathy were studied and published previously [1, 2]. The sensitivity is approximately 90%–100%, specificity 50%, and negative predictive value 90%. This means that antimyosin scan can detect myocarditis and also can exclude myocarditis when compared to the Dallas criteria [8]. The large percentage of scan-positive but biopsy-negative patients have been under investigation.

Quantitative Analysis of Antimyosin Scan

Antimyosin uptake can be analyzed quantitatively. The uptake ratio in the cardiac region and right lung region can objectively demonstrate the intensity of antimyosin uptake. When results of the scan and the biopsy were negative, the uptake ratio was approximately 1.3. Positive results on antimyosin scan showed the much higher uptake of 1.69 ($p < 0.001$). When the patients with positive antimyosin scan results were further divided into right ventricular biopsy positive and negative, the difference in uptake ratio was not significant (1.69 versus 1.69). This finding may indicate that the biopsy does not discriminate myocarditis well, or that the antimyosin scan is not specific. However, patients with positive results on antimyosin scans are likely to show improved ejection fractions regardless of the right ventricular biopsy results as compared to patients with negative results on antimyosin scan [9]. In other words, scan(+)/biopsy(+) and scan(+)/biopsy(−) groups are likely to have improved ejection fraction as compared to scan(−)/biopsy(−). The two scan(+) groups would be very likely to present a similar pathological background.

Diagnostic Accuracy of Planar Images and Planar Combined with Single-Photon Emission Computed Tomography

Planar images are simple and diagnostic. They demonstrate 100% sensitivity as compared to 90%–100% sensitivity with a combination of planar and single-photon emission computed tomography (SPECT). The specificity is

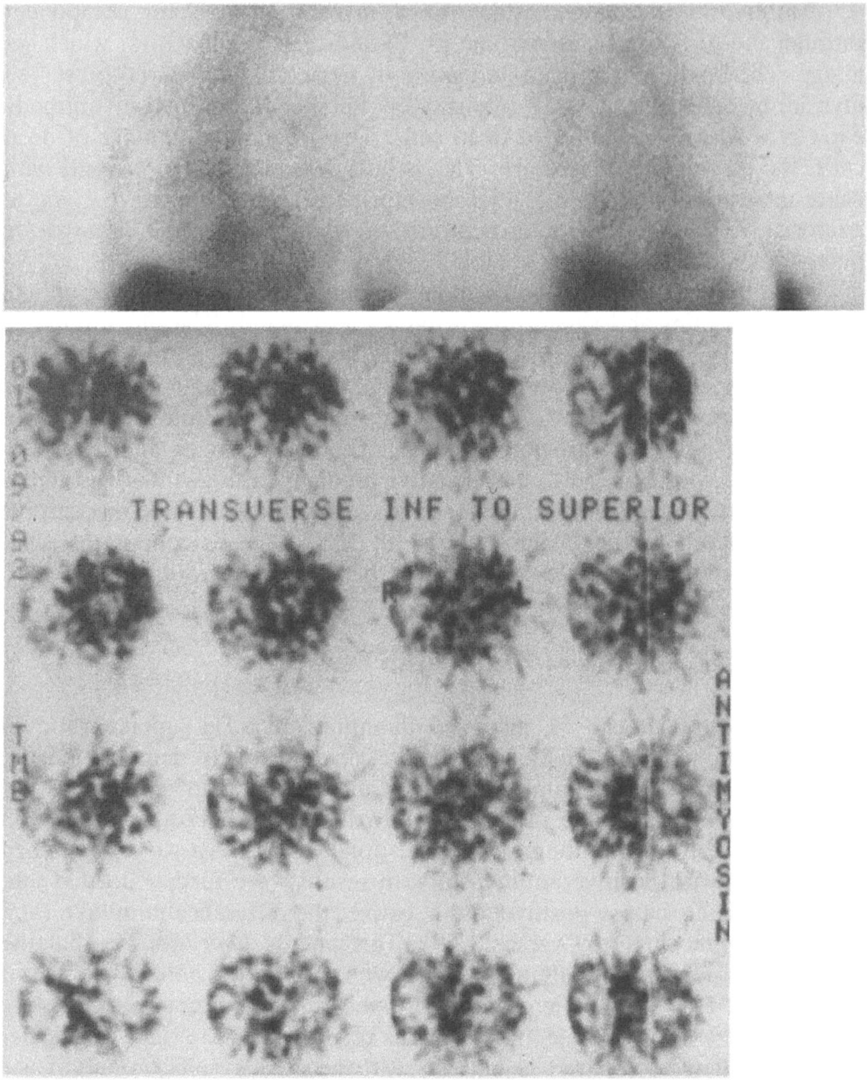

Fig. 3. Intense uptake in the cardiac region. Heart/lung ratio was 1.8. Single-photon emission tomography shows clear ring-like uptake in multiple slices. This was imaged 48 h after injection of antibody

31% as compared to 50% for planar plus SPECT. The planar images are faster and easier to obtain and give less discomfort to the patients. In addition, planar imaging can be done in an intensive care unit by a mobile camera. On the other hand, SPECT is wonderful for clarity in localization of a vague activity and when the activities are not in the myocardium. Using planar imaging it is often difficult to separate the vague and diffuse activities of the blood pool from the myocardial activities. If SPECT is performed

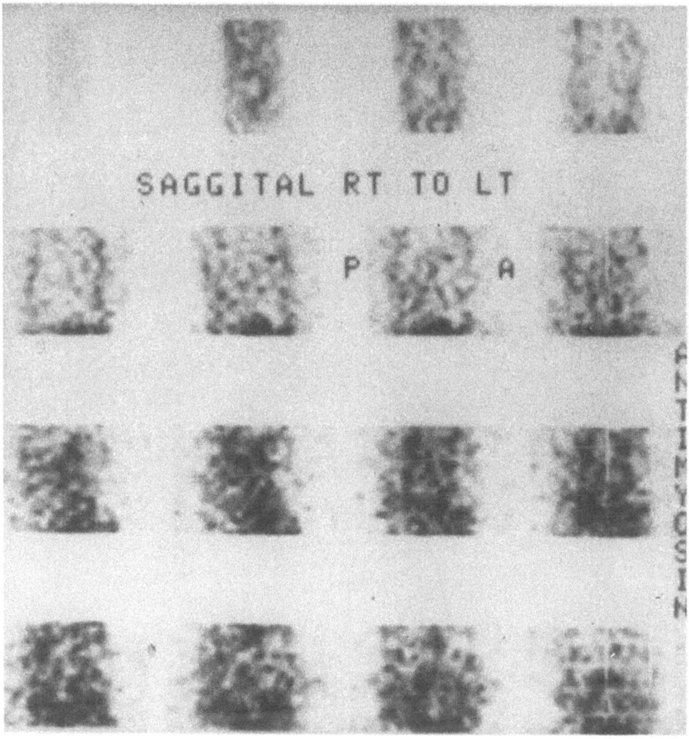

Fig. 3. *Continued*

properly, confidence in reading will increase significantly. If the quality of SPECT is poor, the images are confusing and may not be helpful at all. SPECT should be performed 30–48h after antibody injection. Imaging at 72h is not optimal since we loose activities of uptake from the heart and are not able to collect enough information to process SPECT. Planar images at 72h are clean and easier to read. However, if it is not clean, and/or any question arises from the picture, it is too late to perform SPECT, and the opportunity is missed for SPECT. I personally prefer imaging at 48h; planar performed at 10min in anterior and left anterior oblique 60° then SPECT is performed. The SPECT images are taken in 64 × 64 matrix frame every 6° for 360°, 25s per projection. Since antimyosin images are faint and heterogeneous in uptake, it would be easier to orient oneself in a larger field of SPECT. It should include half of the thorax and a superior part of the liver, which enables localization of the activities.

The three cases presented below demonstrate the clinician's use of this imaging procedure. Results in the first two patients were positive and those in the third negative.

Case 1. A 72-year-old white woman was known to have myocarditis in 1983. She was treated medically for congestive heart failure and maintained her

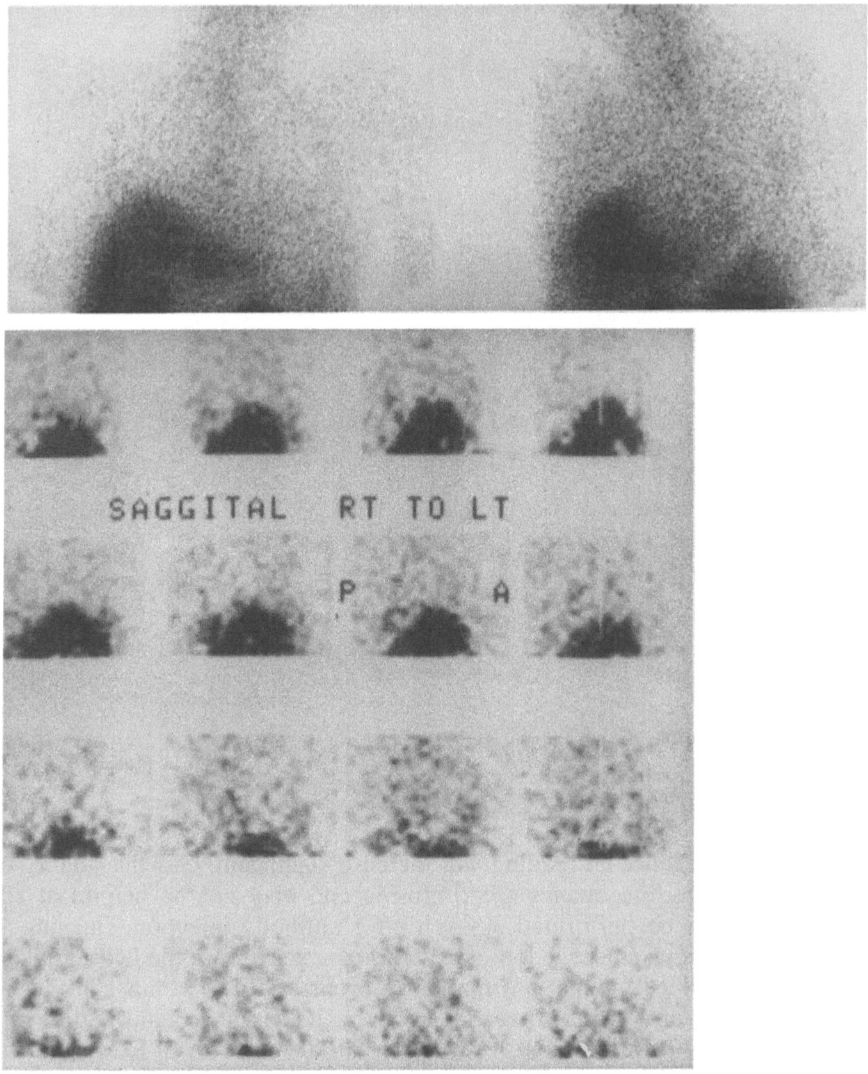

Fig. 4. "Spotty" uptake in the region of left lower chest region. This can be extracardiac. However, single-photon emission tomography confirms ring-like myocardial uptake. The uptake involves the whole left ventricle. This was imaged at 72 h

life-style without difficulty. Her condition was considered to be stable without evidence of myocarditis for years. She played six to nine holes of golf a day with her husband and was able to socialize with her friends. However, for the last month she had felt easily fatigued, and it was found that her LVEF had dropped from 35% to 10%. Also, multiple ventricular ectopic beats were found. A right ventricular biopsy was performed which did not reveal myocarditis. There was myocyte hypertrophy, and a moderate degree of fibriosis was observed. After negative results on biopsy, antimyosin scan was performed to evaluate possibility of myocarditis. Clinically she was suspected to have myocarditis. The scan showed intense uptake in the region of the heart in the anterior and left anterior oblique views. SPECT showed involvement of the entire left ventricle and some of the right ventricle as well (Fig. 3). One week after the scan, the patient had syncope and was admitted for management. This case illustrates the importance of SPECT. SPECT clearty showed the circular uptake in a series of sections. The planar images were convincing, but I could not imagine the extent of involvement as shown by SPECT. SPECT, clearly and with certainty, indicated the location of antimyosin antibody uptake. Another point of this case is that antimyosin scans positivity. It showed intense uptake, and this suggested that rapid cell necrosis has taken place in this patient. Subsequently, she had an episode of syncope and was empirically placed on prednisone.

Case 2. A 43-year-old man had severe upper respiratory infection in October 1991. This was associated with excessive nasal drainage and severe sinus congestion. In December 1991 he was found to have cardiomegaly with moderate shortness of breath. Physical examination showed left ventricular failure. He had no history of serious medical problems in the past. He underwent coronary angiogram, which was normal; however, it confirmed the dilated hypokinetic left ventricle. Results of subsequent right ventricular biopsy were also nondiagnostic for myocarditis. Clinically myocarditis was suspected, and antimyosin scan was performed to detect myocyte necrosis. This was clearly positive (Fig. 4), although it was not as intense as in case 1. This was taken 72 h after injection, and image quality was somewhat reduced, particularly in SPECT. This illustrated an importance of counts for image quality. The count density was so low that planar images were not clear. When SPECT was performed, vague circular activities are demonstrated. This would have been of much better quality if it had been imaged at 48 h. This case also illustrates the clinician's frustration. There was sudden deterioration of left ventricular function without clear cause.

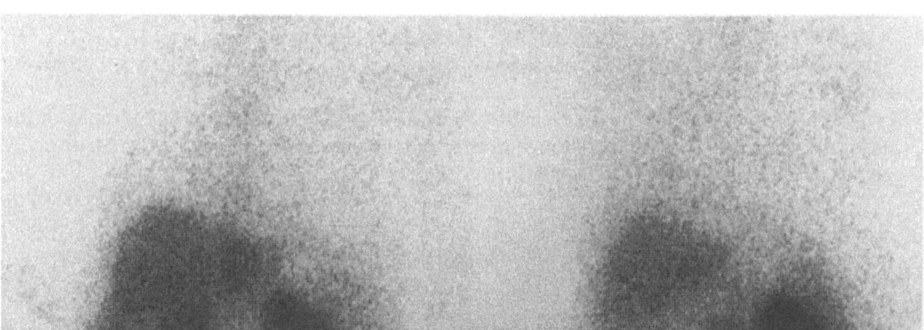

Fig. 5. A negative scan. There is no evidence of increased uptake in the cardiac region. Myocyte necrosis was excluded

Antimyosin scan provided information of active myocyte necrosis whether patient was suffering from myocarditis or not.

Case 3. A 21-year-old man was referred for evaluation of new-onset cardiomegaly with the posibility of myocarditis or myopathy. This study was performed to determine whether he needed right ventricular biopsy (Fig. 5). He was diagnosed as not having acute myocyte cell necrosis by antimyosin scan, and right ventricular biopsy was not performed.

In summary, antimyosin scan is able to detect underlying myocyte necrosis even if the amount of necrosis is small and difficult to confirm by biopsy. If the patients are clinically suspected to have myocarditis, antimyosin scan can be useful to classify the patients into "likely" and "less likely" groups. Accordingly, right ventricular biopsy can be planned.

References

1. Yasuda T, Palacios IF, Dec GW, Fallon JT, Gold HK, Leinbach RC, Strauss HW, Khaw BA, Haber E (1987) Indium 111-monoclonal antimyosin antibody imaging for the diagnosis of acute myocarditis. Circulation 76:306–311
2. Dec GW, Palacios IF, Yasuda T, Fallon JT, Khaw BA, Strauss HW, Haber E (1990) Antimyosin antibody cardiac imaging: its role in the diagnosis of myocarditis. Circulation 16:98–104
3. Khaw BA, Mattis JA, Mekincoff G, Strauss HW, Gold HK, Haber E (1984) Monoclonal activity to cardiac myosin: scintigraphic imaging of experimental myocardial infarction. Hybridoma 3:11–23
4. Khaw BA, Yasuda T, Gold HK, Leinbach RC, Johns J, Kanke M, Barlai-Kovach M, Strauss HW, Haber E (1987) Acute myocardial infarction imaging with indium labeled monoclonal antimyosin Fab. J Nucl Med 28:1671–1678
5. Khaw BA, Fallon J, Beller G, Haber E (1979) Specificity of localization of myosin specific antibody fragments in experimental myocardial infarction: histological, histochemical, autoradiographic and scintigraphic studies. Circulation 6:1527–1531
6. Kishimoto C, Hung G-L, Ishibashi M, Khaw BA, Kolodny GM, Abelman WH, Yasuda T (1991) Natural evolution of cardiac function, cardiac pathology and antimyosin scan in a murine myocarditis model. J Am Coll Cardiol 17:821–827
7. Khaw BA, Scott J, Fallon JR, Cahill SL, Haber E, Homcy (1982) Myocardial injury: quantitation by cell sorting initiated with antimyosin fluorescent spheres. Science 217:1050–1053
8. Aretz HT, Billingham ME, Edward WD, Factor SM, Fallon JT, Fenoglio JJ, Osen EGJ, Schoen FJ (1986) Myocarditis: a histologic definition and classification. Am J Cardiovasc Pathol 1:3–14
9. Haber E, Yasuda T, Palacios IF, Dec GW, Khaw BA (1987) Antimyosin scintigraphy in the diagnosis of acute myocarditis pathogenesis of myocarditis and cardiomyopathy. University of Tokyo Press, Tokyo, pp 277–280

Extracellular Matrix Expression, Organization, and Interaction with Heart Myocytes During Development and Disease

W. Carver, M.L. Burgess, R. Jyring, L. Terracio, and T.K. Borg

Introduction

Background and Perspectives

The extracellular matrix (ECM) of the heart is intimately associated with cardiac function in development and disease. The heart ECM is composed of the interstitial collagens, including predominantly type I and type III collagen, various proteoglycans and noncollagenous glycoproteins such as laminin and fibronectin [2, 7, 22]. These ECM components are organized into an elaborate three-dimensional network which is associated with the structure and physiology of the heart.

A number of studies have described the dynamics of ECM components, including their expression, accumulation, and organization during developmental and disease processes in the heart [7, 22, 49, 50]. Through a combination of structural, biochemical, and molecular techniques the underlying mechanisms by which the ECM influences the organization and physiology of the heart are beginning to be elucidated. The following discussion describes several aspects of ECM components in the heart and the role that these may play in development and disease.

Myocardial ECM Organization

Early reports described the connective tissue of the heart as being composed of collagenous fibers that form an elastic, stress-tolerant network [6, 30, 36, 37]. The arrangement of the collagen fibers was shown to be altered with movement of the muscle, and this led to the hypothesis that collagen is arranged in a network or weave attached to the surface of myocytes and is important in muscle movement [12]. Studies using high-resolution scanning electron microscopy confirmed this arrangement and illustrated that at least three levels of organization exist in the collagen network [6, 36]. A dense region of connective tissue in the subendocardium connects to thick bundles of collagen making up the perimysium. The perimysial collagen bundles connect the different layers of cardiac muscle, thus allowing movement as

Department of Developmental Biology and Anatomy, University of South Carolina School of Medicine, Columbia, SC 29208, USA

Figulla et al. (Eds.)
Idiopathic Dilated Cardiomyopathy

the heart expands to accommodate increased demands. The perimysium terminates at the endomysium, which consists of a weave network surrounding individual myocytes. The endomysium ensheathes the individual myocytes and derives its elasticity from the organization of the ECM components into a weave rather than the resilience of the individual fibers. Also apparent at the level of the individual myocytes are bundles of collagen up to 120 nm in diameter that attach adjacent myocytes and myocytes to capillaries. This arrangement of collagenous connective tissue forms a three-dimensional network that can transmit mechanical tension to different regions of the heart.

ECM in Development

The precisely regulated spatial and temporal distributions of ECM components have only recently begun to be elucidated during development of the heart [7, 22, 24, 39]. The components of the ECM have long been thought to play a passive or structural role in cardiac development and function. However, recent data have shown important roles for the ECM in cell proliferation, differentiation, and migration during embryogenesis [8, 16, 38]. The specific distribution of ECM components and the interaction of cardiac cells with these components may be essential to a variety of cellular processes during development of the heart as well as during adaptation of the heart to physiological or pathological stress.

ECM components have been shown to be critical for early morphogenesis of the heart as well as the formation and organization of the myocardium. For instance, the cardiac jelly, an immense ECM in the embryonic heart, is essential for the initial formation of mesenchymal cells which later populate the heart valves [25, 26, 34]. Also, studies have illustrated that inhibition of collagen cross-linking during developmental or hypertrophic growth results in the formation and rupture of aneurysms in the ventricular wall [4].

ECM in Neonatal Development

Structural and, more recently, molecular analyses have illustrated that collagen mRNAs and proteins accumulate rapidly during neonatal development [6, 11]. This corresponds to a period of development when heart rate, pressure, and passive tension increase [54]. The up-regulation of collagen in the myocardium is potentially an adaptive response to these altered physiological conditions.

Interactions between heart cells and interstitial collagen are mediated, at least in part, by integrins of the β-1 subfamily [41, 43]. The integrins are composed of an α and a β chain to form a functional, transmembrane receptor for specific ECM components [1, 9, 20]. Several members of the β-1 integrin subfamily are thought to be capable of binding interstitial collagen. Recent data have illustrated that β-1 integrins are, similar to collagen, found abundantly in the valves of the neonatal heart [5]. These data suggest that ECM components and corresponding receptors are expressed in similar spatial patterns.

The role of cell-matrix interactions and, more specifically, of particular cell surface components is currently an area of intense investigation in the heart. It is currently not clear why multiple cell surface components are capable of binding individual matrix molecules. For instance, the α-1 and α-2 integrins of the β-1 subfamily are capable of binding collagen as well as other ECM components [1, 9, 20]. It is possible that this multiplicity in integrin-matrix interactions could provide for a wider diversity of signals transduced to the cell from the ECM. However, this has not been conclusively demonstrated in vitro or in vivo.

Basement Membrane

The basement membrane is a specialized region of the extracellular matrix that appears in conjunction with cytodifferentiation of particular cell types, including epithelial and endothelial cells as well as myocytes. This specialized ECM is composed largely of laminin, collagen type IV, heparan sulfate proteoglycan, and entactin/nidogen, and other matrix components [25, 27, 44–46, 52]. Constituent molecules of the basement membrane play a role in various cellular processes including cell migration and differentiation [44, 45]. Immunohistochemical and in situ hybridization studies have illustrated that laminin and collagen type IV are expressed in similar temporal and spatial patterns in the heart [15, 22, 35]. These components are found abundantly in endothelial cells and myocytes of the neonatal heart and are largely absent from connective tissue cells at this developmental stage.

Laminin initially appears during embryonic development as discrete focal patches which later coalesce into the homogeneous basement membrane of the myocytes. It is hypothesized that these early focal sites of laminin function in providing positional information among the developing cells. Immunoelectron-microscopic analysis of cardiac myocytes has illustrated that the distribution of laminin along the myocyte surface is altered during development [33]. That is, in the adult heart, laminin is relatively homogeneous along the surface of heart myocytes. However, in the developing heart, laminin staining is more abundant at the level of the Z disk of the myocyte surface. As discussed below, collagen fibers attach to the myocyte surface at the level of the Z disk. Preliminary evidence also shows that β-1 integrins are abundant at this region of the myocyte surface (unpublished observations), providing further evidence of the coordinated expression and accumulation of matrix and cell surface components.

Northern analysis has illustrated that laminin and collagen type IV mRNA levels are increased in the neonatal heart relative to the levels seen in the late fetal heart [35]. These changes in the organization and synthesis of basement membrane components in the neonatal heart are concurrent with increased pressure and tension during this period. Increased expression of basement membrane components also accompanies increased interstitial collagen in the neonatal heart. These observations are in agreement with those of Chapman et al. [13], who demonstrated increased collagen type IV and interstitial collagen expression in the pressure-overloaded heart. These data provide evidence for the role of mechanical stimulation on expression of ECM components.

Myocyte-ECM Interactions at Specialized Sites

Structural studies showed that collagen fibers attach at specific sites along the surface of cardiac myocytes [3, 7]. The localization of interestitial collagen, other ECM components, including laminin and collagen type IV, and integrins has led to the hypothesis that a specialized point of attachment forms early in the development of the heart and persists throughout the life of the myocyte. This specialized site is located just lateral to the Z disk. It is characterized by several features including: (a) increased staining for basement membrane components laminin and type IV collagen, (b) localization of specific receptors (integrins) for interstitial collagen and laminin, and (c) attachment sites in the myocyte for cytoskeletal components including vinculin and α-actinin.

The functions of this specialized site are: (a) to serve as a location for the attachment of ECM components to the myocyte; (b) to be the site of integration of mechanical stimulation from the ECM to the internal milieu of the cell; and (c) to provide lateral positional information with adjacent myocytes. In order for the ECM to act as a stress-tolerant network, it must be attached to specific sites to allow the expansion and contraction of the network (Fig. 1). With this arrangement, the network can move in synchrony with the contraction cycle of the heart. In order to transmit mechanical stimulation such as stretch resultant from increases in volume and/or pressure, these attachment sites must be firmly anchored to both the

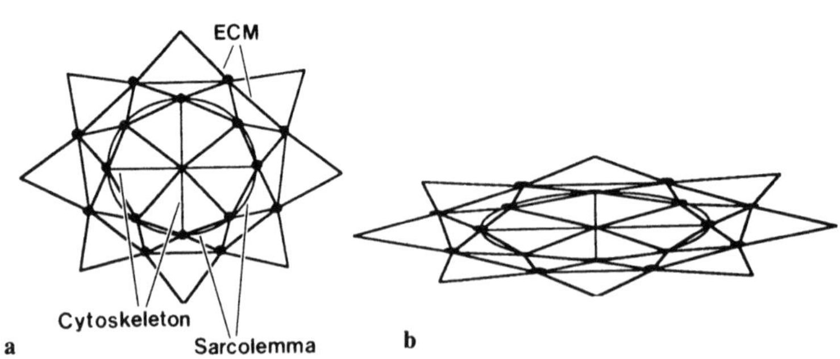

Fig. 1. Schematic illustration of a myocyte with the cytoskeleton and ECM under normal tension in the heart (**a**) and under increased lateral tension (**b**). The repeating series of subunits is affected by extracellular tension exerted on the myocyte. This tension is transmitted through a series of connections to the internal cytoskeletal network

ECM and cytoskeleton. This interaction allows for the force associated with movement of the collagen network to be transferred via the integrin-ECM receptor to the cytoskeletal components. Conversely, during the contraction cycle, where the myocyte shortens and thickens during systole, the force of contraction can be transmitted laterally via the cytoskeleton at the Z disks to the integrins to the ECM.

ECM and Myofibrillogenesis

Myofibrillogenesis is the orderly process by which the contractile fibers are generated. During development, myofibrils are formed in association with elements of the cytoskeleton and with the sarcolemma. Immunofluorescent staining for cytoskeletal components such as vinculin and α-actinin show a clear relationship to the forming myofibrils in heart myocytes [14, 17, 19, 47]. While there is obviously an ordered process of myofibril formation, the role of individual cytoskeletal components has not been determined.

Integrins in the sarcolemma can also influence the pattern of assembly of myofibrils [19]. These receptors, which communicate with the components of the ECM outside of the cell and the cytoskeleton inside of the cell, are critical in the formation of costameres of vinculin [41, 42]. Since the costameres are integral to the formation and alignment of myofibrils, it would appear that integrins may also function in spatial organization of myofibrils. The data from in vitro and limited in vivo studies suggest that myofibrillogenesis is an assembly process which occurs through the co-operativity of the cytoskeleton and integral proteins of the sarcolemma, some of which may be integrins. It is possible that the myocyte cytoskeleton forms an internal stress-tolerant network that is analogous to the ECM

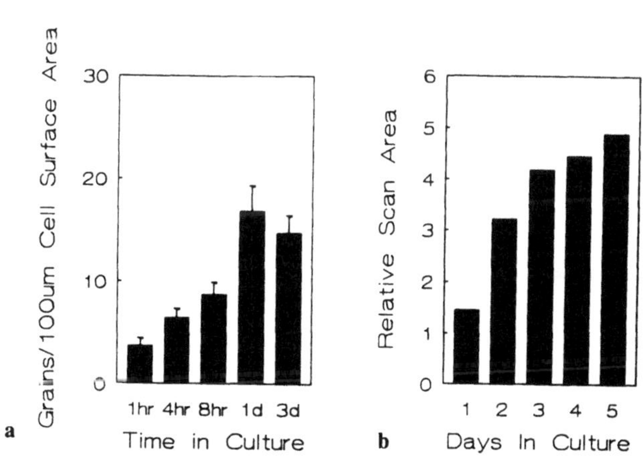

Fig. 2. a Relative mRNA levels of laminin B1 as neonatal myocytes progress into culture. mRNA levels are represented as silver grains/100 μm^2 cell surface area as determined by image analysis of in situ hybridization with a cRNA probe. **b** Relative amounts of immunoprecipitated, metabolically labeled laminin from neonatal heart myocytes at various time points after isolation

network surrounding cells. The information from each of these specialized networks would be transmitted via the transmembrane receptors such as integrins connecting the ECM to the cytoskeleton (see Fig. 1).

While the mechanisms are not well understood, it is clear that cell-matrix interactions play a significant role in myofibril formation. To further understand this process, recent experiments from this group have focused on the expression and distribution of basement membrane components laminin and type IV collagen during the period of myofibrillogenesis in vitro. These studies illustrated that basement membrane mRNA (Fig. 2a) and protein synthesis (Fig. 2b) are increased in isolated cells as they adapt to culture. Laminin (Fig. 3a) and type IV collagen are initially randomly distributed in isolated cells. As cells adapt to culture and myofibrillogenesis proceeds, localization of laminin and collagen type IV are found in a banding pattern reminiscent of Z disk staining (Fig. 3b). These data further demonstrate a specialized region on the myocyte surface that contains cytoskeletal, cell surface, and ECM components and a concerted organization of these components during myofibril formation.

ECM in Disease

Cardiac hypertrophy usually follows an increase in workload that is imposed on the heart. A number of changes have been noted in the heart in response to increased pressure including alterations in myocyte structure and function as well as significant changes in the ventricular ECM [23, 28, 29, 40, 49–51]. Transduction of the force generated by the cardiomyocytes into efficient pumping of the heart is dependent on the relation of the myocyte and the surrounding ECM. In cardiac disease processes, inefficient pumping could arise as a result of alterations in interactions of the myocyte with the ECM producing myocyte slippage and misalignment of myocytes within the ven-

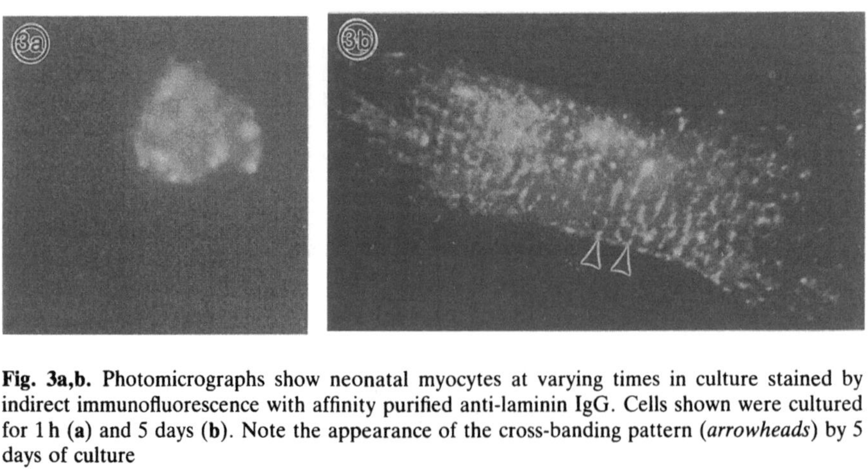

a b

Fig. 3a,b. Photomicrographs show neonatal myocytes at varying times in culture stained by indirect immunofluorescence with affinity purified anti-laminin IgG. Cells shown were cultured for 1 h (**a**) and 5 days (**b**). Note the appearance of the cross-banding pattern (*arrowheads*) by 5 days of culture tricular walls. Thus, it is critical to determine how the ECM is altered in response to abnormal stress in the heart, and how these adaptations influence heart function.

The Collagenous Network in Disease

As described previously, the collagenous network potentially plays several roles in the heart, including the maintenance of capillary patency, transmission of force between myocytes, and coordination of myocytes to produce a forceful contraction of the myocardium. It has been noted that collagen concentration is increased [28], and that there is extensive remodeling of the collagenous fibers during left ventricular hypertrophy [48–50]. Molecular studies have demonstrated a rapid increase in type I and type III collagen mRNAs in response to increased pressure in hypertrophy [13]. Furthermore, recent studies have illustrated that accumulation of collagen type III increases relative to that of collagen type I in the early phases of hypertrophy [7]. As hypertrophy continues, the relative levels of type III to type I collagen decrease. Together these data illustrate that precise changes occur in collagen expression and organization in response to increased pressure in the myocardium.

It is critical to undertake a multifaceted approach as described above in the examination of ECM components, particularly interstitial collagen. No single analytical technique has sufficient sensitivity to describe the ECM components accurately. In addition, single-technique approaches are often misinterpreted since collagen accumulation is regulated at several levels, including synthesis, deposition, and degradation (Table 1). In order to clearly define alterations in the heart ECM in response to chemical or physical factors, it is important to define changing patterns of expression, accumulation, degradation, and organization of the matrix. In this way, it is possible to understand more clearly how the ECM is altered in response to these factors, and what significance this may have in heart function.

It is currently not clear whether the changes seen in ECM expression and accumulation in development or hypertrophy are a direct or indirect result of mechanical forces [10]. A number of chemical and physical factors

Table 1. Current methods for analyzing interstitial collagen

Method	Advantages	Disadvantages
Hydroxyproline	Total collagens	Not specific to types
SDS-PAGE	Specific types	Extraction and analysis
Antibodies	Western blots, localization, semiquantitation	Specificity of antibodies
Histochemistry	Temporal/spacial data	Masking; not specific to types; reproducibility
Molecular (northern)	Expression of specific types	May lack correlation to protein
Molecular (in situ)	Temporal/spacial localization; analysis of small samples	May lack correlation to protein

may influence collagen synthesis, including transforming growth factor-β, glucocorticoids, and cell density. In addition, recent results suggest that the exposure of the heart to elevated levels of vasoactive peptides may have a direct effect on collagen synthesis during renovascular hypertension. That vascular smooth muscle cells treated with angiotensin II increase collagen synthesis has also been shown in vitro [21]. Cardiac fibroblasts also show an enhanced ability to contract collagen gels when treated with angiotensin II, which may be related to the synthetic phenotype of the cells (unpublished observations). How these results relate to the regulation of collagen synthesis in the heart, where a multitude of chemical and physical factors may be present simultaneously, remains to be determined.

Myocyte-Matrix Interactions

Several alterations in gene expression have been described in cardiac myocytes during hypertrophy. Included in these are increased expression of particular oncogenes and cytoskeletal components [32]. As described previously, the cytoskeleton is in communication with the surrounding matrix components through the sarcolemma via receptors such as the integrins [1, 9, 20]. Recent studies from this group illustrated altered expression of specific integrins in cardiac myocytes during hypertrophy [43]. The functional significance of altered integrin expression in heart disease remains to be elucidated. Obviously, disruption of the connections between ECM, integrin, and cytoskeleton could lead to abnormal contraction and force generation in the heart. Increased integrin accumulation on the cardiac myocyte may be a significant adaptive response to the early phases of pressure overload. If too few matrix receptors are present, or if the existing receptors are saturated by the increases in interstitial collagen, further collagen which is deposited would not attach to the myocyte surface. Collagen fibers which normally attach to the myocyte would potentially form complex weave networks around the myocytes that could alter compliance. On the other hand, altered attachment of the cytoskeleton to the cytoplasmic portion of the integrin complex could result in misaligned contractile fibers similar to those seen by ultrastructural examination of severe hypertrophy and other cardiomyopathies. The sum of these structural alterations is a less efficient contracting myocyte.

Metalloproteinases and ECM Remodeling

The presence of extracellular proteases associated with development and disease of the heart have only recently been established ([31] and unpublished observations). There are four classes of extracellular proteases: (a) cysteine, (b) aspartic, (c) serine, and (d) metalloproteinases. In development and disease, the role of these proteases has been proposed to be modification and/or remodeling of the ECM. While the substrates for these enzymes are known to be ECM components, the characterization of the specificity and expression of these is not well understood, particularly during heart disease.

The action of the metalloproteinases on ECM components during disease could result in the appearance of biologically active fragments of ECM molecules or the liberation of growth factors that may influence cell behavior. It is well documented that the ECM can bind growth factors that remain inactive until needed; however, it is not clear how these growth factors are activated or released from the ECM. As ECM accumulation and organization have a profound influence on heart function, clearly the role of molecules which modify the ECM in response to physiological or pathological stress need to be extensively investigated.

Conclusions

The authors have presented this chapter not as an extensive review of all aspects and potential roles of the heart ECM, but to provide several characteristics of the myocardial ECM which appear to be important during development and disease. Therefore the discussion has focused primarily on the areas of expertise of this research group, including ECM structure and interaction with heart cells, primarily myocytes, during specific periods of development and disease. This is not to diminish the importance of the matrix interactions with other cell types in the heart or the role of cell-cell interactions in developmental or disease processes. Several conclusions can be drawn regarding the heart ECM and its interaction with cells during development:

1. ECM components are attached to specialized regions of the myocyte sarcolemma, just lateral to the Z disk. These specialized sites are proposed to play a critical role in the efficient contraction of the myocyte.
2. The integrin receptors for ECM components provide a structural linkage between the ECM and cytoskeleton and may thus be mediators of information from the ECM to the internal milieu of the cell. This information may lead to a variety of cellular responses, including myocyte hypertrophy and altered gene expression or positioning of the contractile elements within the cell.
3. The ECM components and integrins are precisely regulated during development and disease of the heart. Particularly significant is the regulation of collagen synthesis by the cardiac fibroblasts since this is the major component of the heart interstitium and is produced largely by this cell type. A critical balance may be necessary between the accumulation of collagen and collagen-binding integrins on the myocyte surface to provide for efficient myocyte contraction.
4. Muscle pathology may be a result of abnormal interactions at the specialized sites of the sarcolemma with either the ECM surrounding the cell or the cytoskeleton in the cell.

References

1. Albelda SM, Buck CA (1990) Integrins and other cell adhesion molecules. FASEB J 4:2868–2880
2. Borg TK, Gay R, Johnson LD (1982) Changes in the distribution of fibronectin and collagen during development of the neonatal heart. Coll Relat Res 2:2111–2118
3. Borg TK, Johnson LJ, Lill PH (1983) Specific attachment of collagen to cardiac myocytes: in vivo and in vitro. Dev Biol 97:417–423
4. Borg TK, Klevay RE, Gay RE, Siegel R, Bergin ME (1985) Alteration of the connective tissue network of striated muscle in copper deficient rats. J Mol Cell Cardiol 17:1173–1183
5. Borg TK, Rasso DS, Terracio L (1990) Potential role of the extracellular matrix in postseptation development of the heart. Ann NY Acad Sci 58:87–95
6. Borg TK, Caulfield JB (1979) Collagen in the heart. Tex Rep Biol Med 39:321–333
7. Borg TK, Terracio L (1990) Interaction of the extracellular matrix with cardiac myocytes during development and disease. In: Robinson T (ed) Issues in biomedicine. Karger, Basel, pp 113–129
8. Bornstein P, Sage H (1980) Structurally distinct collagen types. Annu Rev Biochem 49:957–1003
9. Buck CA, Horwitz AF (1987) Cell surface receptors for extracellular matrix molecules. Annu Rev Cell Biol 3:179–205
10. Carver W, Nagpal M, Nachtigal M, Borg TK, Terracio L (1991) Collagen expression in mechanically stimulated cardiac fibroblasts. Circ Res 69:116–122
11. Carver W, Borg TK, Terracio L (1992) Collagen expression and accumulation in the neonatal rat heart: quantitative analysis of mRNAs and proteins (submitted)
12. Caulfield JB, Borg TK (1979) The collagen network of the heart. Lab Invest 40:354–371
13. Chapman D, Weber KT, Egbahli M (1990) Regulation of fibrillar collagen types I and III and basement membrane type IV collagen gene expression in pressure overloaded rat myocardium. Circ Res 67:787–794
14. Dlugosz AA, Antin PB, Nachmias VT, Holtzer H (1984) The relationship between stress fiber-like structures and nascent myofibrils in cultured cardiac myocytes. J Cell Biol 99:2268–2278
15. Drake CJ, Davis LA, Walters L, Little CD (1990) Avian vasculogenesis and the distribution of collagens I, IV, laminin, and fibronectin in the heart primordia. J Exp Zool 255:309–322
16. Ekblom P, Vestweber D, Kemler R (1986) Cell-matrix interactions and cell adhesion during development. Annu Rev Cell Biol 2:27–47
17. Hilenski LL, Terracio L, Sawyer R, Borg TK (1989) Effects of extracellular matrix on cytoskeletal and myofibrillar organization in vitro. Scanning Microsc 3:535–548
18. Hilenski LL, Terracio L, Borg TK (1991) Myofibrillar and cytoskeletal assembly in neonatal rat cardiac myocytes cultured on laminin and collagen. Cell Tissue Res 264:577–587
19. Hilenski LL, Xuehui M, Vinson N, Terracio L, Borg TK (1992) The role of $\beta 1$ integrin in speading and myofibrillogenesis in neonatal rat cardiomyiocytes in vitro. Cell Motil Cytoskeleton (in press)
20. Humphries MJ (1990) The molecular basis and specificity of integrin-ligand interactions. J Cell Sci 97:585–592
21. Kato H, Suzuki S, Tajima S, Ogata Y, Tominaga T, Sato A, Saruta T (1991) Angiotensin II stimulates collagen synthesis in cultured vascular smooth muscle cells. J Hypertension 9:17–22
22. Kitten GT, Markwald RR, Bolender DL (1987) Distribution of basement membrane antigens in cryopreserved early embryonic hearts. Anat Rec 217:379–390
23. Kozlovskis PL, Fieber LA, Pruitt DK, Bailey RB, Smets MJD, Bassett AL, Kimura S, Myerburg RJ (1987) Myocardial changes during the progression of left ventricular pressure-overload by renal hypertension or aortic constriction: myosin, myosin ATPase and collagen. J Mol Cardiol 19:105–114

24. Krug EL, Mjaatvedt CH, Markwald RR (1987) Extracellular matrix from embryonic myocardium elicits an early morphogenetic event in cardiac endothelial differentiation. Dev Biol 120:348–355
25. Laurie GW, Bing JT, Klumman HK, Hassell JR, Aumailly M, Martin GR, Feldman RJ (1986) Localization for binding sites for laminin, heparan sulfate proteoglycan and fibronectin on basement membrane (type IV) collagen. J Mol Biol 189:205–216
26. Markwald RR, Krug EL, Runyan RB (1985) Proteins in cardiac jelly which induce mesenchyme formation. In: Ferrans V, Rosenquist G, Weinstein C (eds) Cardiac morphogenesis. Liss, New York, pp 60–69
27. Martin GR, Timpl R (1987) Laminin and other basement membrane components. Annu Rev Cell Biol 3:57–85
28. Medugorac I, Jacob R (1983) Characterisation of left ventricular collagen in the rat. Cardiovasc Res 17:15–21
29. Morgan HE, Baker KM (1991) Cardiac hypertrophy: mechanical, neural and endocrine dependence. Circulation 83:13–25
30. Nagel A (1935) Die mechanischen Eigenschaften von Perimysium internum and Sarkolemm bei der quergestreiften Muskelfaser. Z Zellforsch Mikrosk Anat 22:694–706
31. Nakagawa M, Terracio L, Carver W, Bikedal-Hansen H, Borg TK (1992) Expression of metalloproteinase and IL-1α in developing rat hearts (submitted)
32. Parker TG, Schneider MD (1991) Growth factors, proto-oncogenes, and plasticity of the cardiac phenotype. Annu Rev Physiol 53:179–200
33. Price RL, Nakagawa M, Terracio L, Borg TK (1992) Ultrastructural localization of laminin in in vivo embryonic, neonatal, and adult rat cardiac myocytes and in early rat embryos raised in whole embryo culture. J Histochem Cytochem (in press)
34. Potts JD, Dagle JM, Walder JA, Weeks DL, Runyan RB (1991) Epithelial-mesenchymal transformation of embryonic cardiac cells is inhibited by a modified antisense oligodeoxynucleotide to transforming growth factor β3. Proc Natl Acad Sci USA 88:1516–1520
35. Rasso D, Carver W, Borg TK, Terracio L (1992) Localization of basement membrane components, laminin and collagen type IV, in the developing heart (submitted)
36. Robinson TF, Chohen-Gold L, Factor SM (1983) Skeletal framework of mammalian heart muscle: arrangement of inter- and pericellular connective tissue structures. Lab Invest 49:482–498
37. Robinson TF, Factor SM, Capasso JM, Wittenberg BA, Blumenfeld OO, Seifter S (1987) Morphology, composition, and function of struts between cardiac myocytes of rat and hamster. Cell Tissue Res 249:247–255
38. Sanes JR (1989) Extracellular matrix molecules that influence neural development. Annu Rev Neurosci 12:491–516
39. Sinning AR, Lepera RC, Markwald RR (1988) Initial expression of type-I procollagen in chick cardiac mesenchyme is dependent upon myocardial stimulation. Dev Biol 130:167–174
40. Spinale FG, Zellner JL, Tomita M, Crawford FA, Zile MR (1991) Relation between ventricular and myocyte remodeling with the development and regression of supraventricular tachycardia-induced cardiomyopathy. Circ Res 69:1058–1067
41. Terracio L, Gullberg D, Rubin K, Craig S, Borg TK (1989) Expression of collagen adhesion proteins and their association with the cytoskeleton in cardiac myocytes. Anat Rec 223:62–71
42. Terracio L, Simpson DG, Hilenski L, Carver W, Decker RS, Vinson N, Borg TK (1990) Distribution of vinculin in the Z-disk of striated muscle: analysis by laser scanning confocal microscopy. J Cell Physiol 145:78–87
43. Terracio L, Rubin K, Balog E, Jyring R, Gullberg D, Carver W, Borg TK (1991) Expression of collagen binding integrins during cardiac development and hypertrophy. Circ Res 68:734–744
44. Timpl R (1985) Molecular aspects of basement membrane structure. Prog Clin Biol Res 171:63–74

45. Timpl R (1989) Structure and biological activity of basement membrane proteins. Eur J Biochem 180:487–502
46. Timpl R, Dziadek M (1986) Structure, development and molecular pathology of basement membranes. Int Rev Exp Pathol 29:1–112
47. Toyuyasi KT, Maher PA (1987) Immunocytochemical studies of cardiac myofibrillogenesis in early chick embryos. II. Generation of α-actinin dots within titin spots at the time of first myofibril formation. J Cell Biol 105:2795–2801
48. Weber KT, Janicki JS, Pick R, Abrahams C, Shroff SG, Bashey RI, Chen RM (1987) Collagen in the hypertrophied, pressure-overloaded myocardium. Circulation 75I:40–47
49. Weber KT, Janicki JS, Shroff SG, Pick R, Chen RM, Bashey RI (1988) Collagen remodeling of the pressure-overloaded, hypertrophied nonhuman primate myocardium. Circ Res 62:757–765
50. Weber KT, Pick R, Janicki JS, Gadodia G, Lakier JB (1988) Inadequate collagen tethers in dilated cardiopathy. Am Heart J 116:1641–1646
51. Wilkman-Coffelt J, Parmley WW, Mason DT (1979) The cardiac hypertrophy process. Analysis of factors determining pathological vs. physiological development. Circ Res 45:697–708
52. Yurchenco PD, Schittny JG (1990) Molecular architecture of basement membranes. FASEB J 4:1577–1590
53. Zak R (1974) Development and proliferative capacity of cardiac muscle cells. Circ Res [Suppl] II:17–26

V. Contraction and Relaxation

Ventricular Remodeling in Idiopathic Dilated Cardiomyopathies

E.H. Sonnenblick[1], J. Capasso[2], and P. Anversa[2]

Introduction

Idiopathic dilated cardiomyopathies form one of the most complex and least understood problems leading to heart failure and death. Although accounting for 40% of all cases of severe heart failure, their primary etiology is largely unknown. Most studies of the problem have done little to explain the primary events which produce myocardial dysfunction but rather are descriptive of epiphenomena and later blood flow limitations of organ function. These systemic changes largely reflect alterations in neurohumoral factors and the peripheral circulation that ultimately result in the syndrome of congestive heart failure [42, 43]. Moreover, it is uncertain whether the end-stage pathology of idiopathic dilated cardiomyopathies evolves from a single or multiple initial processes.

Myocyte Loss and Ventricular Remodeling

In this discussion two considerations are addressed which are central to the evolution of cardiomyopathies. First, independent of a known cause of the phenomenon, myocyte loss is a primary event in the initial causation of many types of myocardial dysfunction that ultimately evolve into heart failure. This myocyte loss may be segmental as in acute myocardial infarction, or it may be diffuse and focal, as associated with viral disease, toxins, or perhaps with microvascular spasm leading to focal ischemia and necrosis. This latter phenomenon has been described in the Syrian hamster with an inherited cardiomyopathy, alcoholic cardiomyopathy [13], or hypertension associated with diabetes [18]. In addition, normal aging is associated with diffuse myocyte loss [1, 3, 35] upon which any other pathology is superimposed. Second, once primary myocyte loss has occurred, ventricular remodeling ensues with myocyte elongation and thickening, myocyte "slippage," and connective tissue re-organization. Initially, this occurs as an adaptation. Contractile behavior of hypertrophied myocardium is altered,

[1] Division of Cardiology, Department of Medicine, The Albert Einstein College of Medicine, Bronx, NY, USA
[2] Division of Cardiology, New York College of Medicine, Valhalla, NY, USA

Figulla et al. (Eds.)
Idiopathic Dilated Cardiomyopathy

but force production is preserved. With the continued abnormal stresses of ventricular dilatation and further cell loss the process becomes a maladaptation, with terminal events of decreased myocyte contractility, mitral insufficiency, and decreased pump function. Moreover, progression of heart failure is a time-dependent-event and depends on continued ventricular remodeling with increased end-diastolic ventricular volume, augmented ventricular wall stress, and reduced ejection fraction. Such ventricular changes may produce few symptoms. Rather, with neurohumoral factors and restrictions to cardiac output, symptoms of congestive heart failure occur in relation to an altered peripheral circulation and sodium accumulation [44].

In general, ventricular failure ensues from an initial work overload. This overload may result from increased systolic pressure or from a sustained volume load (abnormal valves). Alternatively, the work overload can result from loss of myocytes, whether diffuse (cardiomyopathies or aging) or segmental (myocardial infarction) [42]. This work overload results in myocyte hypertrophy [2] producing contractile alterations characterized by decreased speed of contraction, prolonged duration of contraction, maintained peak force development and extent of shortening, and delayed relaxation [10]. Such mechanical alterations are paralleled by prolongation of the action potential [11], decreased rates of Ca^{2+} accumulation by isolated sarcoplasmic reticulum, and a decrease in actomyosin ATPase in rodents on the basis of a V_1-V_3 isoenzyme shift [45]. These mechanical and biochemical alterations which are characteristic of laterally expanding sarcomeres but not sarcomeres added in series may be viewed as adaptive to the load since they are reversible within certain limits [10, 11]. Further, the same alterations are seen with experimental alloxan diabetes in rats and rabbits in the absence of a work overload [21, 23]. Here again, these mechanical and biochemical changes are reversible with insulin [24]. Although both the biochemical and mechanical changes induced by renal hypertension and diabetes are reversible when the inciting factors are removed, when renal hypertension is superimposed on alloxan-induced diabetes, massive myocyte necrosis occurs, with the development of a dilated cardiomyopathy that results in death of the animal [22]. Microvascular spasm has been observed in such animals and an ischemic component to this myocyte loss suggested [18, 20]. From these studies and a review of the literature in man, the concept of a diabetic-hypertensive dilated cardiomyopathy as a real entity has been strongly supported [23].

Aging leads to the same mechanical and biochemical alterations as a pressure overload hypertrophy [12, 15] but here the load is created by a time-dependent loss of myocytes resulting in reactive myocyte hypertrophy [1]. Further, the effects of aging and hypertension on the mechanics of contraction and myocyte biochemistry are additive [12]. Indeed these findings in animals help explain the diastolic dysfunction that may be seen in the elderly, which is characterized by delayed ventricular relaxation with preserved systolic shortening capacity and ventricular emptying (i.e., normal ejection fraction). While the systolic overload of hypertension is reversible, the effects of aging with myocyte loss are not. This may also help to explain the irreversible end stage of sustained renal hypertension [17].

A relatively compensated state of the ventricle is maintained with these work overloads as long as ventricular mass/volume ratios are maintained [34, 41]. However, if myocyte loss is too great and/or continues for secondary reasons such as ischemia and/or aging, progression of ventricular remodeling occurs. These secondary factors may thus combine to produce a dilated cardiomyopathy in which a single initial cause cannot be defined. Rather, summated causes of myocyte loss combine to perpetuate the process.

Mechanisms of Ventricular Remodeling

Myocardial infarction provides a controlled and quantitative model of the process of ventricular remodeling which we believe generates a dilated cardiomyopathy. With occlusion of the left anterior descending coronary artery in the rat, a variably sized infarction of the left ventricular wall ensues [34–36]. Very rapidly after the infarction there are major changes that occur in the remaining, normally perfused myocardium. Within 2 days, in the presence of a large infarction, ventricular wall remodeling begins, with myocyte elongation and thickening, accompanied by "slippage" of myocytes in the wall leading to a reduction in the number of myocytes across the wall [32]. Some elongation of sarcomeres occurs physiologically, i.e., about 10%, as the end-diastolic filling pressure rises. Further elongation of the myocyte necessary to accommodate in increased volume requires addition of sarcomeres in series. These changes relate to infarct size [34, 38]. In small infarcts, myocyte mass is restored with little elongation of myocytes or change in ventricular volume [35]. However, in large infarcts, diastolic filling pressure rises substantially, augmenting diastolic wall stress several fold due to the LaPlace equation [34].

This results in myocyte elongation, both immediately by physiological sarcomere lengthening and structurally over extended time by the synthesis of new sarcomeres in series. The latter may be accompanied by myocyte "slippage" to accommodate a rapidly induced increase in volume since there are limits to the rate and extent of myocyte elongation. The result of myocyte slippage is a reduction in myocytes across the wall of the ventricle that adds to the load per cell. The contractile behavior of the myocardium undergoing reactive hypertrophy following infarction resembles that of tissue responding to a systolic overload. Contraction rates are slowed as is relaxation with prolongation of contraction [9]. Similarly, myofibrillar ATPase is reduced [26]. Despite the biochemical and mechanical alterations that resemble a pressure overload hypertrophic response, the structural changes are consistent with concentric hypertrophy [8]. It is possible that such increases in diastolic stress may in and of itself also lead to further myocyte damage and cellular death. Taken as a whole, noncompensated eccentric hypertrophy results from this process, as characterized by a reduced mass/volume ratio and a LaPlace load that cannot be normalized [34]. In this case, a vicious cycle is generated in which increased volume generates augmented stress in both systole and diastole, which cannot be balanced by proportionate growth of myocytes [7]. Pump function is further compromised by volume-dependent functional mitral insufficiency. Further pump dysfunction may result from the loss of elastic recoil in the dilated ventricle so that early ventricular filling is impaired. Moreover, such abnormalities in the left ventricle directly affect the right ventricle via a common septum.

Other ischemic events can also mediate remodeling of the left ventricle without major myocyte necrosis. With repetitive periods of 5 min of ischemia followed by 10 min of reflow repeated 12 times in the left anterior coronary artery of the dog, contractile activity of the anterior ventricular wall is rapidly lost and remains so despite return of normal coronary blood flow, a process termed "stunning" [47]. Segmental ventricular bulging occurs in the ischemic/reflow region appropriate to the systolic stress. Structurally, extensive damage occurs to the connective tissue interconnecting myocytes although myocyte necrosis is absent. Such changes in the wall could in and of themselves produce ventricular dilatation and contribute to the course of ventricular remodeling.

Ischemic Stunning

These alterations in connective tissue may also be pertinent to studies of the dilated cardiomyopathy seen in the Syrian hamster [18]. In this inherited cardiomyopathy, focal myocytolitic lesions are seen early after birth with subsequent healing by fibrosis. Later reactive myocyte hypertrophy occurs, with eventual ventricular dilatation and death. Microfil studies have suggested microvascular spasm in the hearts of these animals [19]. Further, it is intriguing that the focal necrotic lesions are surrounded by much larger areas of damage to connective tissue that could be due to ischemia produced by the microvascular spasm. That this microvascular spasm is integral to the disease process is further supported by the finding that verapamil prevents the lesion and the development of the disease process [16]. Mechanical studies of isolated papillary muscles from these hearts have shown initial loss of force development with only late depression of speed of contraction [25]. These initial changes are consistent with initial myocyte loss while the later alterations in mechanics are consistent with reactive hypertrophy. A major continuing concern has been on how a relatively small amount of myocyte loss (5%) can result in major ventricular remodeling. We would speculate that the focal ischemia resulting from microvascular spasm creates a central area of necrosis surrounded by a larger area of ischemic stunning, which as in the dog results in destruction of intermyocyte connective tissue. Such destruction of interconnections between cells might help impair contractile force [14] and also contribute to progressive ventricular dilatation. In such a manner the effect of focal lesions could be magnified. Studies of this

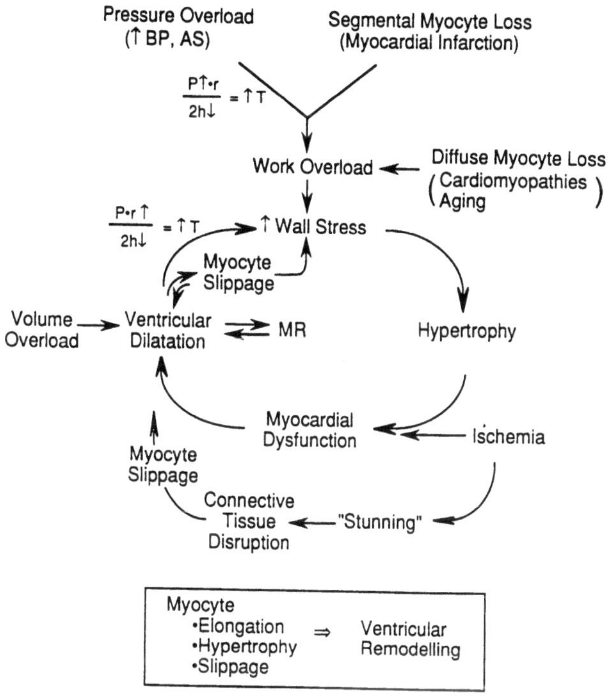

Fig. 1. Ventricular remodeling. Multiple primary factors producing a work overload for the ventricle. This results in an augmental wall stress that results in myocyte hypertrophy. If this cannot compensate for the load, ventricular dilatation ensues which augments wall tension via the LaPlace relation. Moreover, acute elevation of diastolic tension results in myocyte slippage which decreases the number of myocytes across the wall and ruther augments the tension per myocyte. "Stunning" may produce dilatation by damaging intermyocyte connective tissue. *AS*, Aortic stenosis; *BP*, blood pressure;

$$\Delta T = \frac{\mathrm{p \cdot r}}{\mathrm{2h}},$$

with T = wall tension, p = pressure in the venticle, r = radius of the ventricle, h = wall thickness. (From [43])

component of the ventricular wall in dilated cardiomyopathies [39] warrants further study.

Remodeling in Idiopathic Dilated Cardiomyopathies

These multiple experimental models of cardiac dysfunction have provided an overall framework for understanding the evolution of dilated cardiomyopathies in general (Fig. 1). Except for acute infarction and/or aging, the cause of initial loss of myocytes is largely undefined. Depression of contractility in the myocyte itself may or may not occur early. Clearly, following myocardial infarction the remaining myocytes are normal initially. With time their mechanical behavior is altered appropriate to the reactive hypertrophy, but this involves a slowing of contraction and relaxation and not a reduction in force development of shortening capacity. Whether or when a true deficit in force development of the myocyte develops is not at all clear. In some situations Ca^{2+} delivery to the intracellular myoplasm is reduced in late hypertrophy [40] and the troponin-tropomyosin system may be altered [27, 28] so that the translation of Ca^{2+} activation into force may be limited. How or whether this is the case in idiopathic cardiomyopathies in man awaits definition.

Microvascular spasm with focal myocyte necrosis is difficult to delineate while the patient is still alive. Focal stunning with loss of contractile activity and loss of connective tissue could amplify the phenomenon. This may be central to the dilated cardiomyopathy seen in the diabetic hypertensive.

Once initial myocyte loss has occurred or systolic overload persists, the problem becomes ventricular wall remodeling. Increased systolic load results in lateral hypertrophy of myocytes (concentric hypertrophy) while increased volume leads to myocyte elongation by series addition of sarcomeres (eccentric hypertrophy). If the mass can increase to balance the load, mass/volume ratio is preserved with a steady-state maintained [41]. However, with massive cell loss, normal mass/volume relations cannot be preserved. Elevated diastolic pressures and hence tensions lead to further ventricular wall deformation characterized by myocyte slippage, leading to a reduction of the number of myocytes across the wall and further augmenting load per myocyte leading to further myocyte hypertrophy. If hypertrophy with preserved mass/volume is present, the ejection fraction is normal, but abnormal function may reflect delayed relaxation (diastolic dysfunction).

Questions

Described above are initial events in cardiomyopathies that may go unrecognized. Major questions remain relative to these models. As noted above, when does force development per myocyte occur and become a limiting factor? If and when it occurs, is it alterable without affecting further myocyte damage, i.e., energy limitations or Ca^{2+}-induced damage? Myocyte hyperplasia occurs in the myocardium especially late in the process of myocyte loss and reactive hypertrophy [6, 8]; how does this affect the process? Ventricular remodeling occurs in response to initial events, but is it reversible, and if so to what extent? Does ventricular dilatation itself lead to further myocyte loss?

While idiopathic and ischemic cardiomyopathies have a common end-pathway, are they fundamentally different in terms of myocyte loss? Differences in response to beta-adrenergic blockade therapy suggest this to be the case. The mechanism or mechanisms that mediate myocyte loss in both normal (aging) and abnormal conditions have yet to be defined. More knowledge is needed about what controls myocyte viability, and can it be modified to the benefit of the heart.

Acknowledgements. This research was supported by grants PP HL37412-04 and HL # 1572.

References

1. Anversa P, Hiler B, Ricci R, Guideri G (1986) Myocyte cell loss and myocyte hypertrophy in the aging rat heart. J Am Coll Cardiol 8:1441–1448
2. Anversa P, Ricci R, Olivetti G (1986) Quantitative structural analysis of the myocardium during physiologic growth and induced cardiac hypertrophy: a review. J Am Coll Cardiol 7:1140–1149
3. Anversa P, Puntillo E, Nikitin P, Olivetti G, Capasso JM, Sonnenblick EH (1989) Effects of age on mechanical and structural properties of myocardium of Fischer 344 rats. Am J Physiol 256:H1440–1449
4. Anversa P, Capasso JM, Puntillo E, Sonnenblick EH, Olivetti G (1989) Morphometric analysis of the infarcted heart. Pathol Res Pract 185:544–550
5. Anversa P, Capasso JM, Sonnenblick EH, Olivetti G (1990) Mechanisms of myocyte and capillary growth in the infarcted heart. Eur Heart J II [Suppl B]:123–132
6. Anversa P, Palackal T, Sonnenblick EH, Olivetti G, Meggs LG, Capasso JM (1990) Myocyte cell loss and myocyte cellular hyperplasia in the hypertrophied aging rat heart. Circ Res 67:871–885
7. Anversa P, Sonnenblick EHS (1990) Ischemic cardiomyopathy: pathophysiologic mechanisms. Prog Cardiovasc Dis 33:49–70
8. Anversa P, Palackal T, Sonnenblick EH, Olivetti G, Capasso JM (1990) Hypertensive cardiomyopathy: myocyte nuclei hyperplasia in the mammalian rat heart. J Clin Invest 85:994–997
9. Bing OHL, Brooks WW, Conrad CH, Weinstein KB, Spadaro J, Radvany P (1983) Myocardial mechanics of infarcted and hypertrophied non-infarcted myocardium following experimental coronary artery occlusion. Cardiac adaptation to hemodynamic overload, training and stress. Steinkopff, Darmstadt, pp 235–244
10. Capasso JM, Strobeck JE, Malhotra A, Scheuer J, Sonnenblick EH (1982) Contractile behavior of rat myocardium after reversal of hypertensive hypertrophy. Am J Physiol 242 [Heart Circ Physiol 11]:H882–H889
11. Capasso JM, Aronson RS, Sonnenblick EHS (1982) Reversible alterations in excitation-contraction coupling during myocardial hypertrophy in rat papillary muscle. Circ Res 51:189–195
12. Capasso JM, Malhotra A, Scheuer J, Sonnenblick EH (1986) Myocardial biochemical, contractile, and electrical performance after imposition of hypertension in young and old rats. Circ Res 58:445–460
13. Capasso JM, Li P, Guideri G, Anversa P (1991) Left ventricular dysfunction induced by chronic alcohol ingestion in rats. Am J Physiol 261 [Heart Circ Physiol 30]:H212–219
14. Capasso JM, Robinson TF, Anversa P (1989) Alterations in collagen cross-linking impair myocardial contractility in the mouse heart. Circ Res 65:1657–1664
15. Capasso JM, Palackal T, Olivetti G, Anversa P (1990) Severe myocardial dysfunction induced by ventricular remodeling in aging rat hearts. Am J Physiol 259 [Heart Circ Physiol 28]:H1086–H1096
16. Capasso JM, Sonnenblick EH, Anversa P (1990) Chronic calcium channel blockade prevents the progression of myocardial contractile and electrical dysfunction in the cardiomyopathic Syrian hamster. Circ Res 67:1381–1393
17. Capasso JM, Palackal T, Olivetti G, Anversa P (1990) Left ventricular failure induced by long-term hypertension in rats. Circ Res 66:1400–1412
18. Factor SM, Sonnenblick EH (1985) The pathogenesis of clinical and experimental congestive cardiomyopathies. Recent concepts. Prog Cardiovasc Dis 27:395–420
19. Factor SM, Minase T, Cho S, Dominitz R, Sonnenblick EH (1982) Microvascular spasm in the cardiomyopathetic Syrian hamster: a preventable cause of focal myocardial necrosis. Circulation 66: 342–354

20. Factor SM, Minase T, Cho S, Fein F, Capasso JM, Sonnenblick EH (1984) Coronary microvascular abnormalities in the hypertensive-diabetic rat: a primary cause of cardiomyopathy? Am J Pathol 116:9–20
21. Fein FS, Miller-Green B, Sonnenblick EH (1985) Altered myocardial mechanics in diabetic rabbits. Am J Physiol 248:H729–H736
22. Fein FS, Capasso MH, Aronson RS, Cho S, Nordin C, Miller-Green B, Sonnenblick EH, Factor SM (1984) Combined renovascular hypertension and diabetes in rats: a new preparation of congestive cardiomyopathy. Circulation 70:318–330
23. Fein FS, Sonnenblick EH (1985) Diabetic cardiomyopathy. Prog Cardiovasc Dis 4:255-270
24. Fein FS, Miller-Green B, Zola B, Sonnenblick EH (1986) Reversibility of diabetic cardiomyopathy with insulin in rabbits. Am J Physiol 250 [Heart Circ Physiol 19]:H108–H113
25. Forman R, Parmley WW, Sonnenblick EH (1972) Myocardial contractility in relation to hypertrophy and failure in myopathic Syrian hamster. J Mol Cell Cardiol 4:203–211
26. Geenen DL, Malhotra A, Liang D, Scheuer J (1991) Ventricular function and contractile proteins in the infarcted overloaded rat heart. Cardiovasc Res 25:330–336
27. Malhotra A (1990) Regulatory proteins in hamster cardiomyopathy. Circ Res 66:1302–1309
28. Malhotra A, Scheuer J (1990) Troponin-tropomyosin abnormalities in hamster cardiomyopathy. J Clin Invest 86:286–292
29. Michel JB, Lattion AL, Salzmann JL, de Lourdes Cerol M, Philippe M, Camilleri JP, Corvol P (1988) Hormonal and cardiac effects of converting enzyme inhibition in rat myocardial infarction. Circ Res 62:641–650
30. Olivetti G, Lagrasta C, Ricci R, Sonnenblick EH, Capasso JM, Anversa P (1989) Long-term pressure-induced cardiac hypertrophy: capillary and mast cell proliferation. Am J Physiol 257 [Heart Circ Physiol 26]:H1766–H1772
31. Olivetti G, Ricci R, Lagrasta C, Maniga E, Sonnenblick EH, Anversa P (1988) Cellular basis of wall remodeling in long-term pressure overload-induced right ventricular hypertrophy in rats. Circ Res 63:648–657
32. Olivetti G, Capasso JM, Sonnenblick EH, Anversa P (1990) Side-to-side slippage of myocytes participates in ventricular wall remodeling acutely after myocardial infarction in rats. Circ Res 67:23-34
33. Olivetti G, Capasso JM, Sonnenblick EH, Ricci R, Puntillo E, Anversa P (1990) Differences in the temporal effects of aging on the structure and function of rat myocardium. Cor Art Dis 1:240–250
34. Olivetti G, Capasso JM, Meggs LG, Sonnenblick EH, Anversa P (1991) Cellular basis of ventricular remodeling after myocardial infarction in rats. Circ Res 68:856–869
35. Olivetti G, Melissari M, Capasso JM, Anversa P (1991) Cardiomyopathy of the aging human heart, myocyte loss and reactive cellular hypertrophy. Circ Res 68
36. Pfeffer MA, Pfeffer JM, Fishbein MC, Fletcher PJ, Spadaro J, Kloner RA, Braunwald E (1979) Myocardial infarct size and ventricular function in rats. Circ Res 44:503–512
37. Pfeffer MA, Braunwald E (1990) Ventricular remodeling after myocardial infarction, experimental observations and clinical implications. Circulation 81:1161–1172
38. Pfeffer JM, Pfeffer MA, Fletcher PJ, Braunwald E (1991) Progressive ventricular remodeling in rat with myocardial infarction. Am J Physiol 260 {Heart Circ Physiol 29]:H1406–H1414
39. Schaper J, Froede R, Hein S, Buck A, Hashizume H, Speiser B, Fried A, Bleese N (1991) Impairment of the myocardial ultrastructure and changes of the cytoskeleton in dilated cardiomyopathy. Circulation 83:504–514
40. Siri FM, Krueger J, Nordin C, Ming Z, Aronson RS (1991) Depressed intracellular calcium transients and contraction in myocytes from hypertrophied and failing guinea pig hearts. Am J Physiol 261 [Heart Circ Physiol 30]:H514–H530
41. Sonnenblick EH, Strobeck JE, Capasso JM, Factor SM (1983) Ventricular hypertrophy: models and methods. NIH workshop on hypertrophy, 21–22 Sept 1981, Bethesda, Maryland. In: Tarazi RC, Dunbar JB (eds) Perspectives in cardiovascular research, vol 8. Raven, New York, pp 13–20
42. Sonnenblick EH, Factor SM, Eng C, LeJemtel TH, Anversa P (1988) The primary etiology of heart failure: myocyte loss, reactive hypertrophy, dynamic ischemia, and ventricular wall

remodeling. In: Dollery CT, Sherwood L (eds) Cardiac and renal failure: an expanding role for the ACE inhibitors. Hanley and Belfers, Philadelphia, pp 89–99

43. Sonnenblick EH, LeJemtel T, Anversa P (1991) Heart failure: etiological models and therapeutic challenges. In: Lewis BS, Kimchi A (eds) Heart failure – mechanisms and management. Springer, Berlin Heidelberg New York, pp 34–41
44. Sonnenblick EH, LeJemtel TH, Eng C, Factor SM, Anversa P (1989) The relationship of dynamic ischemia, ventricular wall remodeling and reactive hypertrophy to the development of cardiomyopathies: pathophysiological and therapeutic considerations. In: Sonnenblick EH, Laragh JH, Lesch M (eds) New frontiers in cardiovascular therapy: focus on angiotensin converting enzyme inhibition. Excerpta Medica, Amsterdam
45. Swynghedauw B (1989) Remodeling of the heart in response to chronic mechanical overload. Eur Heart J 10:935–943
46. Weisman HF, Bush DE, Mannisi JA, Weisfeldt ML, Healy B (1988) Cellular mechanisms of myocardial infarct expansion. Circulation 78:186–201
47. Zhao M, Zhang H, Robinson TF, Factor SM, Sonnenblick EH, Eng C (1987) Profound structural alterations of the extracellular collagen matrix in postischemic dysfunctional but viable myocardium. J Am Coll Cardiol 10:1322–1334

Mechanisms of Deficient cAMP Generation in Heart Failure

M. Böhm, H. Lensche, and E. Erdmann

Introduction

Heart failure is characterized by the inability of the heart to supply blood to peripheral organs in amounts sufficient to meet the demands of the organism for oxygen and substrates. In order to maintain the perfusion pressure to vital organs, the sympathetic nervous system [26, 46] and renin-angiotensin-aldosterone system [33] are activated, leading to an increase in the circulatory blood volume and to an increase in peripheral resistance. As a consequence of sympathoadrenal activation, an increase in circulating norepinephrine levels has been observed [23, 25, 47], which are closely related to the prognosis of the syndrome of heart failure [25]. The increased circulating and local levels of norepinephrine produce an enhanced stimulation of cardiac β-adrenoceptors. This pathophysiological mechanism has important consequences for the cAMP system of the myocardial cell and, thus, for cardiac contractility.

Under physiological conditions, myocardial contractility is regulated by the activity of the sympathetic nervous system [46]. On the cellular level, the myocardial force of contraction is dependent on the activity of the adenylate cyclase system. A schematic model of the cardiac adenylate cyclase is given in Fig. 1. The adenylate cyclase system is regulated dually by stimulatory β_1-and β_2-adrenoceptors, which couple through a heterotrimeric stimulatory guanine nucleotide binding protein ($\alpha_s\beta y$) to the catalytic subunit of the adenylate cyclase, which catalyzes the formation of cAMP from ATP [32]. Inhibitory m-cholinoceptors and A_1-adenosine receptors [2, 37] couple through inhibitory guanine nucleotide binding proteins ($\alpha_i\beta y$) and mediate a decrease in adenylate cyclase activity [35, 37] and antiadrenergic effects on the myocardial force of contraction [1, 7]. Upon stimulation with agonist-occupied receptors, the α subunits binds GTP and dissociates from the βy subunits to activate their respective effector (e.g., the catalyst of the adenylate cyclase) [32, 41]. Following prolonged stimulation of the adenylate cyclase with agonists, the enzyme becomes desensitized [34, 40]. From the complex regulation it is evident that desensitization to agonist stimulation could involve numerous components of this enzyme complex. This report

From the Medizinische Klinik I der Universität München, Klinikum Großhadern, Marchioninistr. 15, W-8000 München 70, Fed. Rep. of Germany

Figulla et al. (Eds.)
Idiopathic Dilated Cardiomyopathy

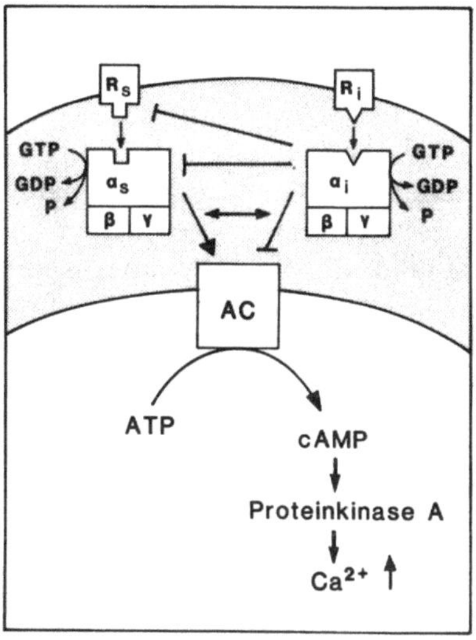

Fig. 1. Scheme on the regulation of the adenylate cyclase activity by stimulatory (R_s) and inhibitory (R_i) receptors and guanine nucleotide binding proteins ($\alpha_s\beta y$, $\alpha_i\beta y$). Following agonist stimulation of inhibitory (A_1-adenosine receptors, m-cholinoceptors, α_2-adrenoceptors) or stimulatory (β_1- and β_2-adrenoceptors, H_2-receptors, glucagon receptors) receptors, the G protein α subunits binds GTP, which leads to the dissociation of the $\alpha\beta y$ complex and the activation of the catalyst of the adenylate cyclase (*AC*). The AC produces cAMP from ATP and leads via activation of the cAMP-dependent protein kinase to phosphorylation of functional membrane proteins, thereby increasing the slow Ca^{2+} inward current. The mechanism of inhibition of AC by α_i is not entirely elucidated. Experimental evidence exists that α_i influences coupling of R_s, reducing the activity of α_s by producing a sink for α_s by releasing βy subunits, or that it directly inhibits AC

reviews the experimental observations on human failing heart muscle which contribute to desensitization of cardiac adenylate cyclase presumably initiated by persistent sympathetic stimulation in heart failure.

Myocardial β-Adrenoceptors

It is well known that prolonged agonist stimulation of β-adrenoceptors leads to densensitization of the adenylate cyclase system [34, 40]. Since stimulation with elevated levels of circulating catecholamines occurs in heart failure [46], it is likely that this could produce a desensitization of cardiac adenylate cyclase in this condition. The first experimental evidence in human failing myocardium was provided by Bristow et al. [12]. These investigators observed a reduction of the number of β-adrenoceptors by 60%. Consistently, a decline in the positive inotropic response of the β-adrenoceptor agonist isoproterenol was observed, while the increase in force of contraction by elevation of the extracellular Ca^{2+} concentration was unchanged. In the following years these findings were confirmed by a large number of investigators (for review see [3–6, 14, 16, 18, 19, 21, 51]).

There is experimental evidence that in the human myocardium β_1- and β_2-adrenoceptor subtypes coexist. This has been shown in right atria [17, 18, 20] and left ventricular myocardium [18, 39, 53]. Both receptor subtypes mediate a stimulation of adenylate cyclase and force of contraction [39]. Bristow et al. [14] reported a β_1-selective downregulation of receptor number and agonist response in failing hearts with dilated cardiomyopathy. This subtype-selective β_1-adrenoceptor downregulation was also observed by other investigators in dilated cardiomyopathy [6, 21, 52]. The latter observation can be readily explained by the 10–30 times higher selectivity of the elevated norepinephrine to stimulate the β_1-adrenoceptor subtype than β_2-adrenoceptors [31]. However, a selective β_1-adrenoceptor downregulation appears not to be a general finding in heart failure. Brodde et al. [21] and Steinfath et al. [52] recently observed a reduction also in the β_2-adrenoceptor number in myocardium from patients with mitral valve disease and ischemic cardiomyopathy, respectively. In addition, Bristow et al. [15] reported an uncoupling of β_2-adrenoceptors from the adenylate cyclase, which could contribute to the reduced effects of β_2-adrenoceptor agonists on force of contraction [6]. Taken together, β_1- and β_2-adrenoceptor downregulation as well as β-adrenoceptor uncoupling is one mechanism which contributes to the reduced cAMP formation in heart failure.

Adenylate Cyclase System

If the β-adrenoceptor downregulation were the only mechanism to reduce cAMP formation, one would expect that agents bypassing β-adrenoceptors

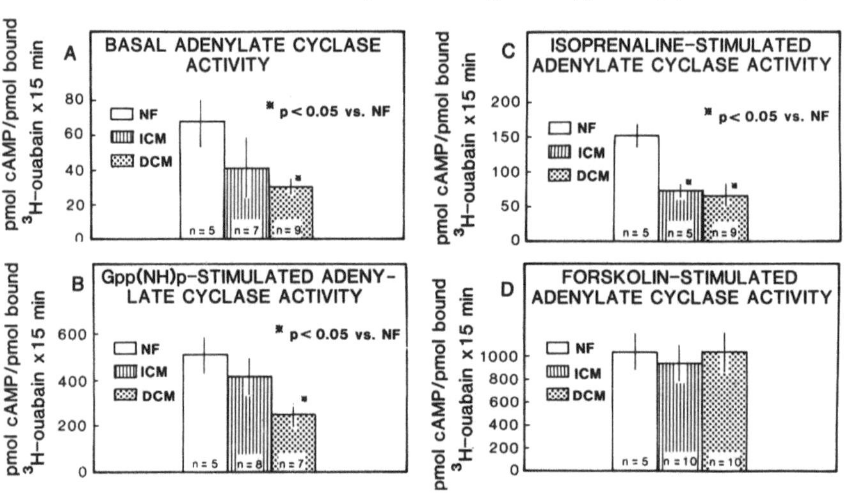

Fig. 2. Activity of basal (**A**), and of Gpp(NH)p (**B**; 100 μmol/l), isoprenaline (**C**; 10 μmol/l), and forskolin (**D**; 10 μmol/l) stimulated adenylate cyclase activity in membranes from the left ventricles of nonfailing hearts (*NF*) and from hearts of patients with ischemic (*ICM*) or dilated (*DCM*) cardiomyopathy. The effects of forskolin and investigated without guanine nucleotides. (From [7])

would elicit sustained inotropic responses. Interestingly, the effect of phosphodiesterase inhibitors such as milrinone [4, 5, 28], isobutylmethylxanthine [3], and pimobendan [9, 49] was also observed to produce diminished inotropic response in failing myocardium. The activity of the soluble crude cAMP phosphodiesterase [4], soluble cAMP phosphodiesterase isoenzymes [49, 55], and sarcoplasmic reticulum-associated cAMP phosphodiesterase [44] have been reported to be unchanged. Thus, one explanation for the functional observations could be a diminished basal cAMP formation. Figure 2 shows the activity of the adenylate cyclase activity in failing and nonfailing myocardium. There was not only a decline in isoprenaline-stimulated enzyme activity but also a reduction in the basal and Gpp(NH)p-stimulated adenylate cyclase. Stimulation of cAMP formation in intact preparations with forskolin [28] or isoprenaline [4] restored the inotropic response to milrinone in failing myocardium. Thus, the reduced basal adenylate cyclase activity is likely to possess importance for the mechanical function of the myocardium. The effect of forskolin on adenylate cyclase [7, 13, 29] (also shown in Fig. 2) and force of contraction [3, 13, 28] is unchanged, indicating that the catalyst is not altered in failing myocardium. However, the reduced basal and Gpp(NH)p-stimulated adenylate cyclase provided evidence for alterations on the level of the signal transducing G proteins.

Guanine Nucleotide Binding Proteins

Stimulatory Guanine Nucleotide Binding Protein

Stimulatory guanine nucleotide binding protein (Gsα) mediates the effects of β-adrenoceptors to the catalyst of the adenylate cyclase [22]. Gsα is a single-gene product with four in RNA isoforms due to alternative splicing [11]. Two protein isoforms have been characterized with the electrophoretic mobility of approximately 45-kDa and 52-kDa proteins [32]. Cholera toxin is able to ADP-ribosylate an arginine residue located in the GTPase site of Gsα [43]. By using [^{32}P]NAD as substrate, one can identify Gsα by autoradiography following electrophoretic separation of membrane proteins. Figure 3 shows [^{32}P]ADP ribosylation in membranes from failing and nonfailing myocardium. HL60 cells and S49 mouse lymphoma cells are also shown. The 45-kDa substrate is lacking in the S49 cyc-variant, which is deficient in Gsα. It can be seen that no difference occurs in the amount of cholera toxin substrates in failing myocardium [29, 50]. Insel and Ransnäs [38, 48] reported that only 5% of immunodetectable Gsα is substrate for cholera toxin labeling. Thus, immunochemical detection is necessary to quantify the amount of Gsα. Figure 4 shows that the synthetic C terminus of Gsα RMHLRQYELL is able to inhibit immunostaining of Gsα by an antiserum raised in rabbits against the same peptide. Figure 5 shows that immunochemical quantification of Gsα is not different in the amount of Gsα. Finally, Feldman et al. [29] reconstituted Gsα from nonfailing and

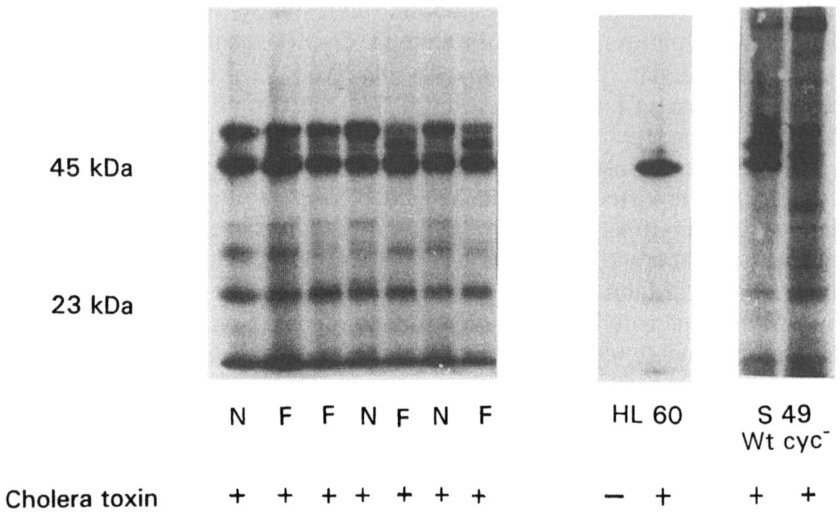

Fig. 3. Cholera toxin labeling of Gsα in membranes from the left ventricles of patients with heart failure (*F*; dilated cardiomyopathy) or from nonfailing hearts (*NF*). Cholera toxin substrates in HL60 cells and S49 wt mouse lymphoma cells in comparison to the cyc-variant. (Modified from [50])

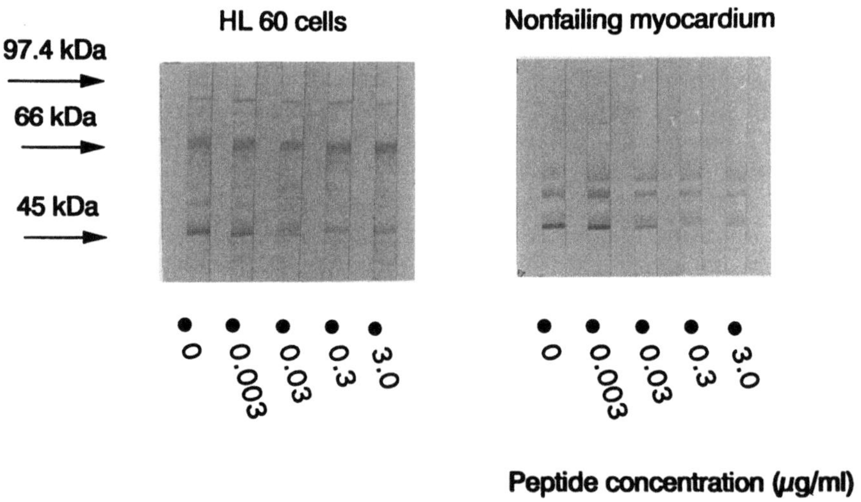

Fig. 4. Immunochemical detection of Gsα ($M_r \approx 45$ kDa) and inhibition of immunostaining by the synthetic C terminal decapeptide of Gsα RMHLRQYELL in membranes from HL60 cells and from human ventricular myocardium. Membrane proteins were separated by 10% SDS-PAGE before electrophoretic transfer to nitrocellulose membranes. Each lane contained 150 μg protein. (Unpublished experiment by H. Lensche and M. Böhm)

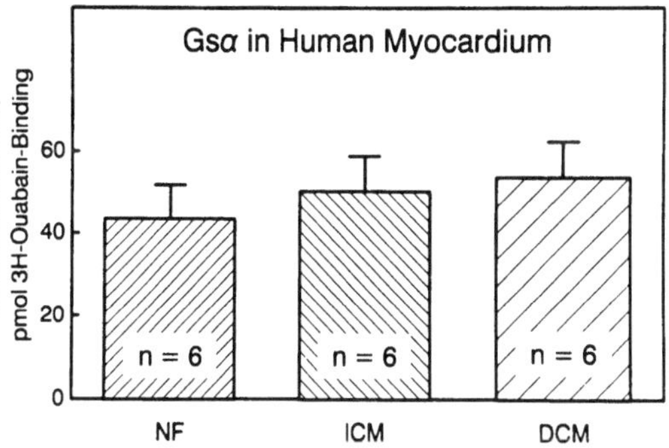

Fig. 5. Gsα content in membranes from the left ventricles of nonfailing hearts (*NF*) and from the left ventricles of patients with dilated (*DCM*) and ischemic (*IDC*) cardiomyopathy. *Ordinate*, Gsα content in densitometric units of Gsα (45 kDa) per pmol/[^{3}H]oubain binding as myocardial membrane marker. *Abscissa*, studied conditions. (Unpublished experiments by H. Lensche and M. Böhm)

failing myocardium into Gsα-deficient S49 cyc- mouse lymphoma cells and observed similar effects on adenylate cyclase activity in both groups. Taken together, hitherto, no change in Gsα has been observed using cholera toxin labeling, immunochemical quantification, or functional reconstitution experiments.

Inhibitory Guanine Nucleotide Binding Protein

Proteins resembling inhibitory guanine nucleotide protein (Giα) consist of one family of heterotrimeric G proteins with a molecular weight of approximately 39–42 kDa [24]. These proteins are subject to pertussis toxin catalyzed ADP ribosylation of their α subunits at a cysteine residue at the fourth position from their C terminus [57]. By using [^{32}P]NAD they can be identified and quantified on SDS-PAGE. The covalent modification by pertussis toxin induced ADP-ribosylation results in uncoupling of inhibitory receptors from the adenylate cyclase [24, 32, 41]. The mechanism of Giα-mediated inhibition of adenylate cyclase is not completely understood. Giα can interfere with coupling of agonist-occupied β-adrenoceptors [42], directly inhibit the catalyst of the adenylate cyclase, and produce a sink for Gsα by releasing βγ subunits [41]. In human dilated cardiomyopathy, an increase of 40%–100% in pertussis toxin substrates was reported in a number of investigations [7, 16, 29, 37, 45]. The data on failing myocardium due to ischemic heart disease are controversial. Bristow et al. [16] reported an increase of 100%, Neumann et al. [45] reported a decrease, while our group observed no change [7]. Since pertussis toxin labeling is potentially hampered by a limited accessability of the C terminus [24], a novel immunochemical method to quantify Giα proteins was developed [8]. Figure 6 shows inhibition of immunostaining of Giα by the synthetic C terminus of Giα. Quantification of Giα with the radioimmunoassay using the ^{125}I-labeled C terminus KENLKDCGLF is shown in Fig. 7. In dilated cardiomyopathy, an

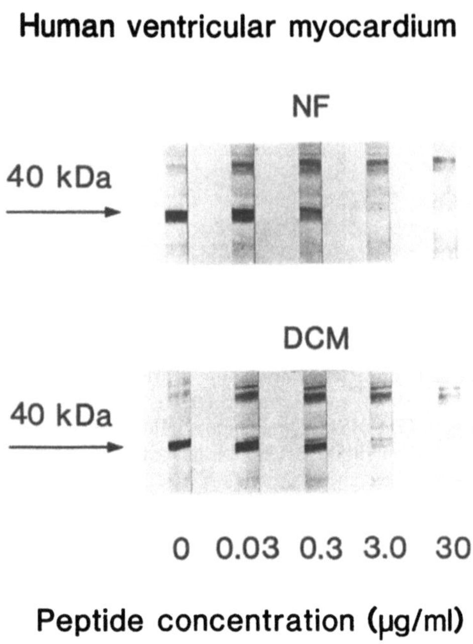

Fig. 6. Immunochemical detection and inhibition of immunostaining of Giα ($M_r \approx 40$ kDa) by the synthetic C-terminal decapeptide of retinal transducin α KENLKDCGLF in membranes from the left ventricles of a nonfailing heart (*NF*; *upper panel*) and from the left ventricles of patients with dilated cardiomyopathy (*DCM*; *lower panel*). Membrane proteins were separated by 10% SDS-PAGE before electrophoretic transfer to nitrocellulose membranes. Each lane contained 100 µg membrane protein

increase in Giα by 116% compared to nonfailing myocardium was observed. In ischemic cardiomyopathy, the increase was less pronounced (47%) than in dilated cardiomyopathy, but still significant. A similar increase in Giα was observed in one heart with aortic stenosis and myocarditis, indicating that Giα changes are not specific for any underlying cardiac disease. Feldman et al. [29] observed a similar basal adenylate cyclase activity in pertussis toxin treated membranes from failing and nonfailing hearts. The unchanged adenylate cyclase activity in these conditions, in which Giα is inactivated, suggests that the increase in Giα is functionally relevant in the reduced basal adenylate cyclase inhibition. This is supported by the close inverse correlation of Giα content to the Gpp(NH)p-stimulated adenylate cyclase activity (Fig. 8B). The hyperbolic relation between Giα-content and isoprenaline-stimulated adenylate cyclase reflects the influence of the decline in β-adrenoceptors on isoprenaline-stimulated enzyme activity. Surprisingly, the density, affinity states of agonist binding [7], inhibition of adenylate cyclase [37], and antiadrenergic effects on myocardial force of contraction [7] of Giα-coupled A_1-adenosine receptors and m-cholinoceptors were observed to be unchanged. Thus, different Giα subtypes could be involved in direct inhibition of adenylate cyclase and coupling of receptors. Taken

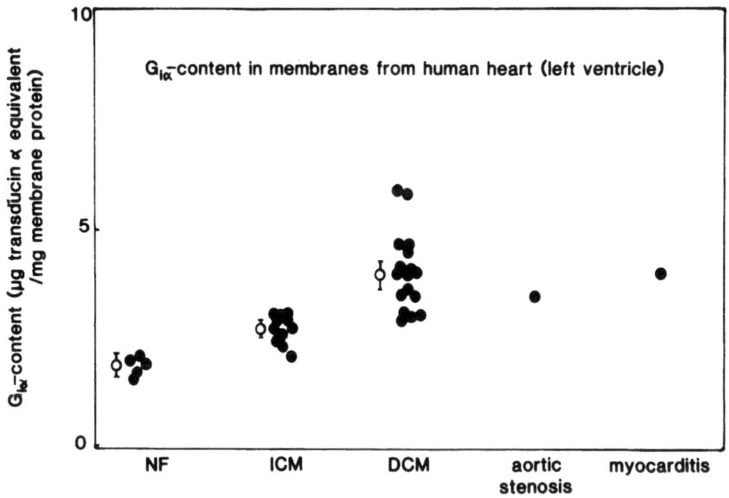

Fig. 7. Giα content in detergent extracts of membranes from the left ventricles of human heart with various cardiac diseases and nonfailing hearts. Giα was quantified by competition of membrane extracts to antiserum binding to the ^{125}I-iodinated C terminus of retinal transducin α as described elsewhere [8]. *Ordinate*, Giα content in micrograms of transducin α equivalents/ milligram membrane protein. *Abscissa*, studied conditions. (From [10])

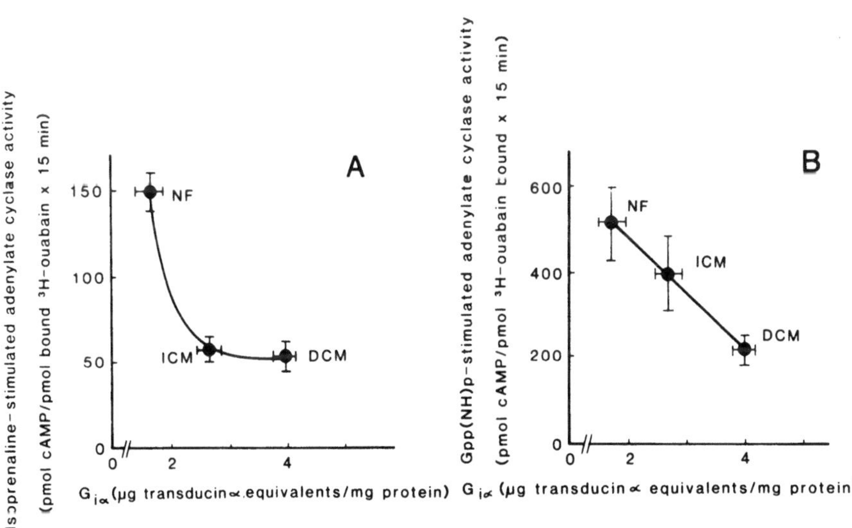

Fig. 8. Relation of immunochemically quantified Giα content in myocardial membrane extracts to the maximal stimulation of adenylate cyclase activity by isoprenaline (**A**; 10 μmol/l) and Gpp(NH)p (**B**; 100 μmol/l) in myocardial membranes of nonfailing hearts (*NF*) and hearts from patients with dilated (*DCM*) or ischemic (*ICM*) cardiomyopathy set.

together, the majority of investigations provided evidence for an increase in Giα and a functional role thereof in the diminished cAMP formation in heart failure.

Clinical Consequences

The experimental observations on reduced cAMP formation and diminished cAMP-dependent positive inotropic responses are in agreement with clinical observations. Following intravenous or intracoronary application of dobutamine or cAMP phosphodiesterase inhibitors [27, 30] the increase in myocardial contractility was reduced in patients with heart failure compared to controls. In addition, continuous intravenous infusion of dobutamine for 72 h or longer leads to the development of tolerance [54]. These clinical observations provide evidence that the reduction in β-adrenoceptors and the diminished cAMP formation are of clinical relevance in patients with heart failure and can be further aggravated by catecholamine stimulation of the myocardium. The observation that β-adrenoceptor blockade can upregulate β-adrenoceptors and can improve the functional status in heart failure patients [36, 56] suggests a pathogenetic role of the sympathetic stimulation in the development of the diminished cAMP formation in heart failure.

Acknowledgements. Experimental work was supported by the Deutsche Forschungsgemeinschaft (M.B.). We thank our colleaques from the Herzchirurgische Klinik for providing human myocardial tissue. The excellent technical assistance of Judith Sabo and Elisabeth Ronft is gratefully acknowledged. Experimental work was supported by the Deutsche Forschungsgemeinschaft (Bo 896).

References

1. Baumann G, Schrader J, Gerlach E (1981) Inhibitory action of adenosine on histamine- and dopamine-stimulated cardiac contractility and adenylate cyclase in guinea pigs. Circ Res 48:258–266
2. Böhm M, Brückner R, Neumann J, Schmitz W, Scholz H, Starbatty J (1986) Role of guanine nucleotide binding protein in the regulation by adenosine of cardiac potassium conductance and force of contraction. Naunyn Schmiedebergs Arch Pharmacol 332:403–405
3. Böhm M, Beuckelmann D, Brown L, Feiler G, Lorenz B, Näbauer M, Kemkes B, Erdmann E (1988) Reduction of beta-adrenoceptor density and evaluation of positive inotropic responses in isolated, diseased human myocardium. Eur Heart J 9:844–852
4. Böhm M, Diet F, Kemkes B, Erdmann E (1988) Enhancement of the effectiveness of milrinone to increase force of contraction by stimulation of cardiac beta-adrenoceptors in the failing human heart. Klin Wochenschr 66:957–962
5. Böhm M, Diet F, Feiler G, Kemkes B, Kreuzer E, Weinhold C, Erdmann E (1988) Subsensitivity of the failing human heart to isoprenaline and milrinone is related to beta-adrenoceptor downregulation. J Cardiovasc Pharmacol 12:726–732
6. Böhm M, Pieske B, Schnabel P, Schwinger RHG, Kemkes B, Klövekorn W-P, Erdmann E (1989) Reduced effects of dopexamine on force of contraction in the failing human heart despite preserved β_2-adrenoceptor subpopulation. J Cardiovasc Pharmacol 14:549–559

7. Böhm M, Gierschik P, Jakobs KH, Pieske B, Schnabel P, Ungerer M, Erdmann E (1990) Increase of Giα in human hearts with dilated but not ischemic cardiomyopathy. Circulation 82:1249–1265
8. Böhm M, Larisch K, Erdmann E, Camps M, Jakobs KH, Gierschik P (1991) Failure of (^{32}P)-ADP-ribosylation by pertussis toxin to determine Giα content in membranes from various human tissues. Improved radioimmunological quantification using the ^{125}I-labeled C-terminal decapeptide of retinal transducin. Biochem J 277:223–229
9. Böhm M, Morano I, Pieske B, Rüegg JC, Wankerl M, Zimmermann R, Erdmann E (1991) Contribution of cAMP-phosphodiesterase inhibition and sensitization of the contractile proteins for calcium to the inotropic effect of pimobendan in the failing human myocardium. Circ Res 68:689–701
10. Böhm M, Gierschik P, Erdmann E (1992) Quantification of Giα-proteins in the failing and nonfailing human myocardium. In: Hasenfuss G, Holubarsch Ch, Just H, Alpert NR (eds) Cellular and molecular alterations in the failing human heart. Steinkopff, Darmstadt; Springer, New York, pp 37–50
11. Bray P, Carter A, Simons C, Guo V, Puckett C, Kamholz J, Spiegel A, Nirenberg M (1986) Human cDNA clones for four species of Gαs signal transduction protein. Proc Natl Acad Sci USA 83:8893–9987
12. Bristow MR, Ginsburg R, Minobe W, Cubicciotti RS, Sageman WS, Lurie K, Billingham ME, Harrison DC, Stinson EB (1982) Decreased catecholamine sensitivity and β-adrenergic-receptor density in failing human hearts. N Engl J Med 307:205–211
13. Bristow MR, Ginsburg R, Strosberg A, Montgomery W, Minobe W (1984) Pharmacology and inotropic potential of forskolin in the human heart. J Clin Invest 74:212–223
14. Bristow MR, Ginsburg R, Umans V, Fowler MB, Minobe W, Rasmussen R, Zera P, Menlove R, Shah P, Jamieson S, Stinson EB (1986) β_1- and β_2-Adrenergic-receptor subpopulations in nonfailing and failing human ventricular myocardium: coupling of both receptor subtypes to muscle contraction and selective β_1-receptor down-regulation in heart failure. Circ Res 59:297–309
15. Bristow MR, Hershberger RE, Port JD, Minobe W, Rasmussen R (1989) β_1- and β_2-adrenergic receptor-mediated adenylate cyclase stimulation in nonfailing and failing human ventricular myocardium. Mol Pharmacol 35:295–303
16. Bristow MR, Anderson FL, Port JD, Skerl L, Hershberger RE, Larrabee P, O'Connell JB, Renlund DG, Volkamn K, Murray J, Feldman AM (1991) Differences in β-adrenergic neuroeffector mechanisms in ischemic versus idiopathic dilated cardiomyopathy. Circulation 84:1024–1039
17. Brodde OE, O'Hara N, Zerkowski HR, Rohm N (1984) Human cardiac β-adrenoceptors: both β_1- and β_2-adrenoceptors are functionally coupled to the adenylate cyclase in right atrium. J Cardiovasc Pharmacol 6:1184–1191
18. Brodde OE, Schüler S, Kretsch R, Brinkmann M, Borst HG, Hetzer R, Reidemeister JC, Warnecke H, Zerkowski HR (1986) Regional distribution of β-adrenoceptors in the human heart: coexistence of functional β_1- and β_2-adrenoceptors in both atria and ventricles in severe congestive cardiomyopathy. J Cardiovasc Pharmacol 8:1235–1242
19. Brodde O-E (1991) β_1- and β_2-Adrenoceptors in the human heart: properties, function, and alterations in chronic heart failure. Pharmacol Rev 43:203–242
20. Brodde O-E, Karad K, Zerkowski HR, Rohm N, Reidemeister JC (1983) Coexistence of β_1- and β_2-adrenoceptors in human right atrium. Direct identification by ($\pm$)-(^{125}I)-iodocyanopindolol binding. Circ Res 53:752–758
21. Brodde O-E, Zerkowski H-R, Doetsch N, Motomura S, Khamssi S, Michel MC (1989) Myocardial beta-adrenoceptor changes in heart failure: concomitant reduction in beta$_1$- and beta$_2$-adrenoceptor function related to the degree of heart failure in patients with mitral valve disease. J Am Coll Cardiol 14:323–331
22. Cassel D, Selinger Z (1977) Mechanism of adenylate cyclase activation by cholera toxin: inhibition of GTP hydrolysis at the regulatory site. Proc Natl Acad Sci USA 74:3307–3311
23. Chidsey CA, Harrison DC, Braunwald E (1962) Augmentation of the plasma noradrenaline response to exercise in patients with congestive heart failure. N Engl J Med 267:650–654

24. Codina J, Yatani A, VanDongen AMJV, Padrell E, Carty D, Mattera R, Brown AM, Iyengar R, Birnbaumer L (1990) Receptor-effector coupling by pertussis toxin substrates: studies with recombinant and native G-protein α subunits. In: Iyengar R, Birnbaumer L (eds) G proteins. Academic, New York, pp 267–294
25. Cohn JN, Levine TB, Olivari MT, Garberg V, Lura D, Francis GS, Simon AB, Rector T (1984) Plasma noreponephrine as a guide to prognosis in patients with chronic congestive heart failure. N Engl J Med 311:819–823
26. Cohn JN (1990) Abnormalities of peripheral nervous system control in congestive heart failure. Circulation 82 [Suppl 1]:159–167
27. Colucci WS, Leatherman GF, Ludmer PL, Gauthier DF (1987) β-Adrenergic inotropic responsiveness of patients with heart failure: studies with intracoronary dobutamine infusion. Circ Res 61 [Suppl 1]:82–86
28. Feldman MD, Copelas L, Gwathmey JK, Philips P, Warren SE, Schoen FJ, Grossman W, Morgan JP (1987) Deficient production of cyclic AMP: pharmacologic evidence of an important cause of contractile dysfunction in patients with end-stage heart failure. Circulation 75:331–339
29. Feldman AM, Cates AE, Veazey WB, Hershberger RE, Bristow MR, Baughman KL, Baumgartner WA, Van Dop C (1988) Increase of the 40000-mol wt pertussis toxin substrate (G protein) in the failing human heart. J Clin Invest 82:189–197
30. Gage J, Rutman H, Lucido D, Le Jemtel TH (1986) Additive effects of dobutamine and amrinone on myocardial contractility and ventricular performance in patients with severe heart failure. Circulation 74:367–373
31. Gille E, Lemoine H, Ehle B, Kaumann AJ (1985) The affinity of (−)-propranolol for β_1- and β_2-receptors of human heart. Differential antagonism of the positive inotropic effects and adenylate cyclase stimulation by (−)-noradrenaline and (−)-adrenaline. Naunyn Schmiedebergs Arch Pharmacol 331:60–70
32. Gilman AG (1984) G proteins and dual control of adenylate cyclase. Cell 36:577–579
33. Goldsmith SR, Francis GS, Cowley AW, Levine TB, Cohn JN (1983) Increased plasma arginine vasopressin levels in patients with congestive heart failure. J Am Coll Cardiol 1:1385–1390
34. Hausdorfer WP, Caron MG, Lefkowitz RJ (1990) Turning off the signal: desensitization of β-adrenergic receptor function. FASEB J 4:2881–2889
35. Hazeki O, Ui M (1981) Modification by islet-activating protein of receptor-mediated regulation of cyclic AMP accumulation in isolated rat heart cells. J Biol Chem 256:2856–2862
36. Heilbrunn SM, Shah P, Bristow MR, Valantine HA, Ginsburg R, Fowler MG (1989) Increased β-receptor density and improved hemodynamic response to catecholamine stimulation during long-term metoprolol therapy in heart failure from dilated cardiomyopathy. Circulation 79:483–490
37. Hershberger RE, Feldman AM, Bristow MR (1991) A_1-Adenosine receptor inhibition of adenylate cyclase in failing and nonfailing human ventricular myocardium. Circulation 83:1343–1351
38. Insel PA, Ransnäs LA (1988) G proteins and cardiovascular disease. Circulation 78:1511–1513
39. Kaumann AJ, Lemoine H (1987) β_2-Adrenoceptor-mediated positive inotropic effect of adrenaline in human ventricular myocardium. Quantitative discrepancies with binding and adenylate cyclase stimulation. Naunyn Schmiedebergs Arch Pharmacol 335:403–411
40. Lefkowitz RJ, Caron MG, Stiles GL (1984) Mechanisms of membrane-receptor-regulation. N Engl J Med 310:1570–1579
41. Levitzki A (1987) Regulation of adenylate cyclase by hormones and G proteins. FEBS Lett 211:113–118
42. Marbach I, Shiloach J, Levitzki A (1988) Gi affects the agonist-binding properties of β-adrenoceptors in the presence of Gs. Eur J Biochem 172:239–246
43. Moss J, Vaughan M (1988) ADP-ribosylation of guanyl nucleotide-binding regulatory proteins by bacterial toxins. Adv Enzymol 61:13303–13379

44. Movsesian MA, Smith CJ, Krall J, Bristow MR, Manganiello VC (1991) Sarcoplasmatic reticulum-associated cyclic adenosine 5′-monophosphate phosphodiesterase activity in normal and failing human hearts. J Clin Invest 88:15–19
45. Neumann J, Schmitz W, Scholz H, v Meyerinck L, Döring V, Kalmar P (1988) Increase in myocardial Gi-proteins in heart failure. Lancet 11:936–937
46. Packer M (1988) Neurohormonal interactions and adaptations in congestive heart failure. Circulation 77:721–730
47. Pierpont GL, Francis GS, DeMaster EG, Olivari MT, Ring WS, Goldberg IF, Reynolds S, Cohn JN (1987) Heterogeneous myocardial catecholamine concentrations in patients with congestive heart failure. Am J Cardiol 60:316–321
48. Ransnäs LA, Insel PA (1988) Quantification of the stimulatory guanine nucleotide binding protein Gs in S49 cell membranes, using anti-peptide antibodies to α_s. J Biol Chem 263:9482–9485
49. Schmitz W, von der Leyen H, Meyer W, Neumann J, Scholz H (1989) Phosphodiesterase inhibition and positive inotropic effects. J Cardiovasc Pharmacol 14 [Suppl 3]:S11–S14
50. Schnabel P, Böhm M, Gierschik P, Jakobs KH, Erdmann E (1990) Improvement of cholera toxin-catalyzed ADP-ribosylation by endogenous ADP-ribosylation factor from bovine brain provides evidence for an unchanged amount of Gsα in failing human myocardium. J Mol Cell Cardiol 22:73–82
51. Schwinger RHG, Böhm M, Erdmann E (1990) Evidence against spare or uncoupled β-adrenoceptors in the human heart. Am Heart J 119:899–904
52. Steinfath M, Geertz B, Schmitz W, Scholz H, Haverich A, Breil I, Hanrath P, Reupcke C, Sigmund M, Lo H-B (1991) Distinct down-regulation of cardiac β_1- and β_2-adrenoceptors in different human heart disease. Naunyn Schmiedebergs Arch Pharmacol 343:217–220
53. Stiles GL, Taylor S, Lefkowitz RJ (1983) Human cardiac beta-adrenergic receptors: subtype heterogeneity delineated by direct radioligand binding. Life Sci 33:467–473
54. Unverferth DV, Blanford M, Kates RE, Leier CV (1980) Tolerance to dobutamine after a 72-hour continuous infusion. Am J Med 69:262
55. von der Leyen H, Meyer W, Scholz H (1988) Similarly selective phosphodiesterase inhibition in nonfailing and failing human heart. Circulation 78 [Suppl II]:11–360
56. Waagstein F, Caidahl K, Wallentin I, Bergh C-H, Hjalmarson A (1989) Long-term β-blockade in dilated cardiomyopathy: effects of short- and long-term metoprolol treatment followed by withdrawal and readministration of metoprolol. Circulation 80:551–563
57. West REJr, Moss J, Vaughan M, Liu T, Liu TY (1985) Pertussis toxin-catalyzed ADP-ribosylation of transducin. Cysteine 347 is the ADP-ribose acceptor site. J Biol Chem 260:14428–14430

Abnormal Calcium Metabolism in Heart Muscle Dysfunction

J.K. Gwathmey and R.J. Hajjar

Introduction

The clinical demonstration that contractile function is depressed in failing human hearts resulted in several logical corollaries to explain the pathophysiology of heart failure. Only recently have investigators been able to study human myocardium as opposed to making extrapolations from animal studies. As a result of these studies several hypotheses have been dispelled and the inappropriateness of certain animal models eluded to. However, there are limitations to human studies. When studying human myocardium one is often left with studying only one point in time. Further, one is left with the question of whether one is studying a mechanism of the disease, an adaptive response to the disease, or changes in response to the varied pharmacological treatments. Therefore mechanisms involved in the pathogenesis or natural history of the disease process cannot be easily investigated. Furthermore, one is often left studying an end-stage process that may or may not be reflective of the disease process. Nevertheless, despite these limitations studies involving human myocardium have yielded important information regarding pathophysiology as well as markers of the disease state.

Excitation-Contraction Coupling

Excitation-contraction coupling (E-C) is initiated with depolarization of the sarcolemma. The depolarization results in an inward calcium current that induces the release of a much larger amount of calcium, called calcium-induced calcium release, from an intracellular store, the sarcoplasmic reticulum (SR), which then raises intracellular calcium levels to micromolar range. This much larger amount of calcium activates the myofilaments and results in cross-bridge formation and contraction. Calcium binds to troponin C (TnC), the calcium regulatory protein on the thin myofilaments, releasing the inhibition imposed by the regulatory subunit troponin I (TnI) by its

Cardiovascular Diseases and Muscle Research Laboratories, Charles A. Dana Research Institute and Harvard-Thorndike Laboratory of Beth Israel Hospital Department of Medicine (Cardiovascular Division), Harvard Medical School, the Department of Cellular and Molecular Physiology, Harvard University and the Massachusetts General Hospital Medical Service

Figulla et al. (Eds.)
Idiopathic Dilated Cardiomyopathy

blocking position, allowing tropomyosin to slide deeper into the groove between actin and myosin for actin and myosin to bind. Relaxation occurs when calcium dissociates from TnC and is resequestered by the SR. There are also extrusion mechanisms located in the sarcolemma, i.e., calcium ATPase and sodium-calcium exchange that assist in lowering the intracellular calcium concentration ($[Ca^{2+}]_i$) to resting nanomolar range. Although muscle contraction and relaxation would appear to follow the ebb and tide of intracellular calcium concentration, there are other important factors that affect this seemingly apparent relationship. Multiple factors affect calcium concentration, availability, mobilization, buffering, and binding. There are kinetic parameters, energy and mechanochemical transduction parameters that must be factored into E-C coupling. Although, as presented and as often perceived, E-C coupling is simplistic compared to many other biological processes, it is the complexity of simplicity that has resulted in decades of scientific research and investigation. It is this very complexity that must be forever in one's mind when investigating normal myocardium, but in particular pathophysiological states.

Contractile Dysfunction in Human Heart Failure

One of the first observations made in isolated muscles from patients with end-stage heart failure was a prolonged isometric twitch duration [10]. It was reported that the inotropic responsiveness was maintained to a digitalis such as glycoside acetylstrophanthidin, unlike the blunted inotropic response to isoproterenol. It was demonstrated that muscles from patients with end-stage heart failure failed to respond to increasing concentrations of milrinone, a phosphodiesterase inhibitor that depends on the presence of cyclic AMP for its inotropic action. This observation was also true for other phosphodiesterase inhibitors as well [10]. To date the only agents that demonstrate inotropic efficacy in isolated muscle preparations from patients with end-stage heart failure are agents that act via mechanisms other than β_1 receptor stimulation or phosphodiesterase inhibition. They are as follows: (a) DPI 201-106, a sodium channel agonist with calcium sensitizing properties at the level of the myofilaments, (b) MCI 154, a phosphodiesterase inhibitor with calcium sensitizing effects, (c) increased extracellular calcium concentrations ($[Ca^{2+}]_o$), and (d) digitalis, a Na-K ATPase inhibitor [10, 37, 38]. These data were interpreted to suggest that diminished cyclic AMP concentrations may play a role in the pathophysiology of muscles from heart failure patients [8, 10]. The suggested diminished intracellular cyclic AMP concentrations from these pharmacological experiments have recently been confirmed [2–4, 11].

Systolic Dysfunction

Systolic dysfunction has been difficult to demonstrate in isolated muscles from patients with end-stage heart failure unless stressed by higher contraction rates or exposure to toxic amounts of inotropic agents [12, 21, 31]. Muscles stimulated at relatively slow rates generate similar if not greater peak force compared to normal myocardium obtained from patients without known cardiac disease [5, 10, 14, 21]. Inotropic responsiveness to $[Ca^{2+}]_o$ has been demonstrated to be similar for normal and diseased myocardium [5, 10, 18]. This has been confirmed in other laboratories [31]. Higher rates of stimulation result in a negative-force interval relationship [12, 20, 31]. Addition of agents that increase $[Ca^{2+}]_i$ as the primary mechanism of action results in exacerbation of the contractile dysfunction [12, 21, 29, 31]. On the other hand, addition of inotropic agents that improve intracellular calcium mobilization by the SR appear to diminish the detrimental effects of increasing intracellular calcium in the presence of frequency perturbation [12, 31].

Myofibrillar ATPase activity has been reported to be significantly reduced in muscle from patients with ischemic and idiopathic cardiomyopathy [30]. Further myofibrillar protein content was significantly reduced by 30%. These findings would indicate that peak contractile performance should be reduced in the setting of heart failure.

Diastolic Dysfunction

Slowed relaxation has been a hallmark finding in muscles from patients with heart failure and from those with hypertrophic cardiomyopathy consistent with diastolic dysfunction. Increasing $[Ca^{2+}]_i$ with acetylstrophantidin or elevated $[Ca^{2+}]_o$ results in an increase in end-diastolic tension and further prolongation of the relaxation phase of the isometric twitch [21]. However, addition of an agent that directly activates adenylate cyclase i.e., forskolin or β_1 receptor stimulation with resultant activation of adenylate cyclase both result in abbreviation in the time course of the isometric twitch [12, 20, 21, 31]. Cyclic AMP acting in concert with protein kinases results in phosphorylation of the regulatory subunit TnI with improved relaxation and release of calcium from TnC and phosphorylation of phospholamban, a receptor protein that acts synergistically with the calcium ATPase pump in the SR thereby improving SR calcium sequestration. Both of the mentioned phosphorylation sites should result in improved muscle relaxation.

Membrane Receptors and Excitation-Contraction Coupling

One of the first recognized abnormalities in E-C coupling in failing human myocardium was a decrease in β_1 receptor number and coupling [9]. This led to the logical corollary that the contractile failure seen in the setting of heart disease was due to diminished inotropic responsiveness to β_1 receptor stimulation. Further, it has been reported that myocardium from patients with idiopathic dilated cardiomyopathy has a reduced number of calcium channels as assessed by radioligand binding studies [29]. Taken together these two observations suggest that failing human myocardium is under-stimulated thereby resulting in poor inotropy. We have found that the calcium channel number is increased in patients with ischemic cardiomyopathy by 20% using radioligand binding studies. Furthermore, in atrial tissue from patients with hypertrophic cardiomyopathy an increased number of calcium channels has been reported [36]. Further, calcium channel number in ventricular tissue from a patient with hypertrophic cardiomyopathy demonstrated an increase in calcium channel number [13]. Therefore the controversy remains regarding the potential for diseased human myocardium to generate normal levels of force and the role of calcium channel number. Key to this debate is the level of intracellular calcium and contractile activation of the myofilaments.

Electrical Properties

It was reported that action potential duration (APD) was prolonged in tissue from patients with end-stage heart failure due to either idiopathic cardiomyopathy or hypertrophic cardiomyopathy [18, 19, 21]. APD can reflect changes in the duration of activation of the slow inward voltage dependent calcium channels, delayed repolarization by potassium channels, and/or a large inward current generated by the sodium-calcium exchange [19]. In hypertrophied human myocardium it has been reported that a significant portion of the plateau phase is due to slow inward calcium current, but that the myocardium also demonstrated a significant sodium-calcium exchange current [33]. Interestingly, increased $[Ca^{2+}]_o$ results in a significant abbreviation of the action potential [20]. This abbreviation of the repolarization phase could result from an increase in resting/end-diastolic calcium concentration in myopathic myocardium that inactivates calcium channels or activation of calcium-dependent potassium channels to produce a shortening of the action potential [20]. Another potential mechanism for the observed abbreviation of APD is activation of ATP-inhibited potassium currents. It has been reported that ATP-dependent potassium channels in hypertrophied ventricular myocytes require a smaller decrease in ATP concentration for release of inhibition [6]. Therefore a small reduction in high-energy phosphate concentration such as might occur with lowered creatine kinase activity [26] could result in enhanced repolarization. Similarly, faster rates of stimulation result in a progressive abbreviation of the action potential (Fig. 1) which might result in an unstable trigger for calcium-induced calcium release from the SR and thereby graded SR calcium release [20]. Lederer et al. [27] reported in isolated myocytes from patients with heart disease that the peak inward calcium current decreased as the contraction rate increased, which supports our original observations regarding APD and hypotheses of an altered calcium trigger affecting SR calcium release.

Recent studies in voltage-clamped isolated myocytes indicate that the amount of calcium current produced by a depolarization is a determinant of the extent of calcium release caused by that depolarization. Further, after a phasic current, if depolarization persists, a noninactivating calcium current persists which leads to sustained SR calcium release (Fig. 2) [34]. These data

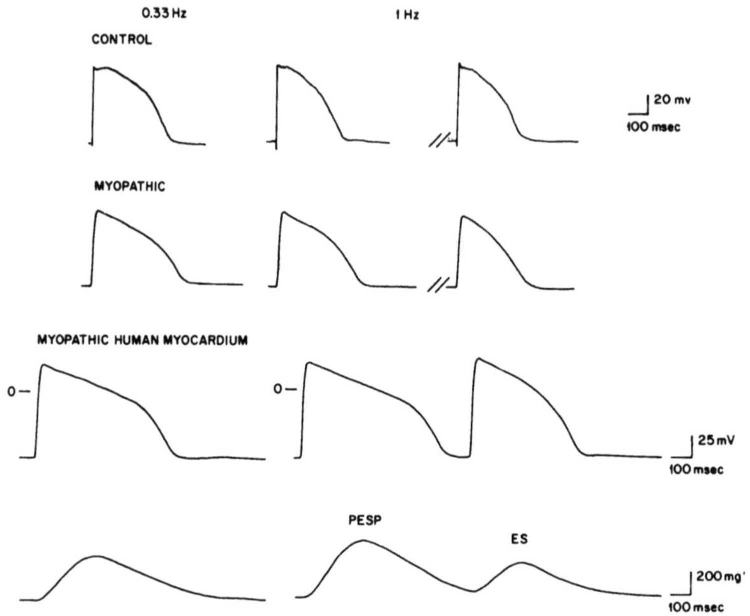

Fig. 1. *Above*, action potentials from control and myopathic human myocardium. Notice the longer APD seen in myopathic myocardium. With higher rates of stimulation the APD abbreviates in the control tissue and rapidly stabilizes at a new steady state. In myopathic myocardium there is progressive abbreviation of the action potential duration. *Below*, action potential recordings (*upper traces*) from a trabeculae carneae from a patient with dilated cardiomyopathy; *lower traces* are associated isometric contractions. *Right*, effect of inserting an extrasystole (*ES*) with resultant abbreviation of the associated action potential and diminished amplitude of the extrasystole. Note, however, that the action potential associated with the postextrasystole (*PESP*) is not abbreviated (slightly longer than ADP associated with conditioned contraction), and peak isometric twitch force is increased

fore support our original findings that a second slower component of the calcium transient found in myocardium from patients with end-stage heart failure reflects both enhanced transsarcolemmal calcium flux and release by the SR.

Therefore electrical properties of diseased myocardium may affect contractile performance and intracellular calcium mobilization. For example, although frequency potentiation is impaired in diseased myocardium, postextrasystolic potentiation (PESP) is maintained [20, 31]. This can be explained in part by restitution processes but also in light of the observation that the APD is not abbreviated with PESP and is if anything prolonged (Fig. 1), as opposed to frequency potentiation which is associated with a marked abbreviation in APD. These data therefore suggest a role for APD as a mechanism. Mechanical restitution has been reported to differ in human myocardium from that seen in other species [7].

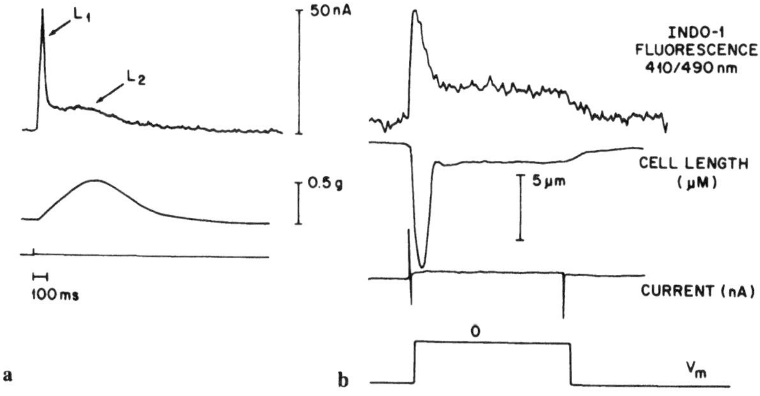

Fig. 2. a Aequorin signal from a patient with dilated cardiomyopathy. Notice the two components (L_1 and L_2). *Lower smooth trace* is associated isometric twitch. **b** Simultaneous recordings of changes in Ca_i (Indo fluorescence), cell length (μm) and membrane current (nA) in a voltage-clamped rat ventricular myocyte. (With permission from Dr. Edward G. Lakatta)

Intracellular Calcium Metabolism: Sarcoplasmic Reticulum

Calcium activating the myofilaments is derived primarily from calcium released by the SR, although there are data to suggest that in human myocardium calcium influx across the sarcolemma can also directly activate the myofilaments [19, 20]. In muscles from patients with end-stage heart failure the calcium transient consists of two components (Fig. 2). The first component, similar to that seen in normal human myocardium, reflects SR calcium release and is abolished upon addition of ryanodine, an agent that induces SR calcium leak. The second, slower component, seen only in hypertrophied and failing human myocardium, reflects in the case of failing myocardium SR calcium release and transsarcolemmal calcium flux. In hypertrophied myocardium the second, slower component reflects solely SR calcium release as is seen for the first, fast component in normal and diseased myocardium. In failing myocardium the origin of the second component has been demonstrated by addition of ryanodine, which diminished the amplitude of the second component but did not completely abolish it. Subsequent addition of verapamil abolished the remaining component, suggesting that a portion of the second, slower component arises from transsarcolemmal calcium flux.

Twitch force is determined not only by intracellular calcium availability but also the duration of TnC-calcium interaction. Further, it has been modeled that a significant portion of the free ionized calcium is quickly bound to the myofilaments and not reflected in the peak calcium transient as detected by intracellular calcium indicators [15]. Taken together the debate regarding peak SR calcium release would appear moot. It has nevertheless been reported that peak $[Ca^{2+}]_i$ detected with Indo-1 in myocytes isolated from failing human hearts was $1 \mu M$ at 0.4 Hz stimulation rates [27]. With repetitive stimulation there was a rise in resting calcium from 240 to 375 nM. It was noted that the increase in contractions was more dramatic than the increase in the amplitude of the intracellular calcium transients [27]. The authors concluded that although there was progressive loading of the SR, the rising diastolic $[Ca^{2+}]_i$ plays a major role in the contraction staircase. Using the calcium indicator aequorin, we found end-diastolic $[Ca^{2+}]$ to be 250 ± 20 and 370 ± 10 nM for control and myopathic tissue, respectively, at 0.33 Hz stimulation rates. Peak SR calcium release was similar for control and myopathic myocardium (1.2 ± 0.4 versus $1.26 \pm 0.29 \mu M$). At higher stimulation rates where a negative or blunted treppe was often noted the end-diastolic $[Ca^{2+}]$ increased in failing myocardium to 690 ± 20 nM whereas there was no significant difference in peak SR calcium release (2.26 ± 0.4 versus $2.43 \pm 0.28 \mu M$ for control and myopathic, respectively).

Calcium-force Relationships in Normal and Diseased Myocardium

Another proposed hypothesis to explain poor contractile performance in the setting of heart disease is a change in myofilament calcium responsiveness. A change in the calcium concentration required for 50% activation of the myofilaments ($[Ca^{2+}]$ 50%) or slope of the calcium-force relationship (Hill coefficient reflecting cooperativity between thick and thin myofilaments) could result in altered contractile performance. To bypass E-C coupling chemically skinned fibers can be studied in which the sarcolemma and SR have been removed and the $[Ca^{2+}]_i$ precisely controlled. In most animals models of hypertrophy and failure there have been no changes detected in the $[Ca^{2+}]$ 50% value, Hill coefficient, or maximal-calcium activated force [35]. However, in the myopathic Syrian hamster there is a shift to lower $[Ca^{2+}]_i$ in skinned fiber preparations as well as a decrease in maximal calcium-activated force.

Skinned Fiber

Myopathic hearts have different metabolic and structural properties and function under load conditions that are different from that in normal hearts. When compared to normal hearts myopathic hearts demonstrate different rates of tension development [10], intracellular calcium mobilization [18], creatine kinase activity [26] and sensitivity to inotropic interventions [10, 14, 16, 17, 25]. The differences in the activity of myopathic hearts have been attributed to alterations in E-C coupling, but they may also be due to differences in activity and regulation of the myofilaments. In trabeculae carneae obtained from human hearts without known cardiac disease and patients with end-stage heart failure the calcium-force relationships determined in skinned fiber preparations yielded $[Ca^{2+}]$ 50% values of 1.53 ± 0.32 versus $1.33 \pm 0.19 \mu M$ for control and myopathic muscles respectively

[25]. The Hill coefficients were 2.05 ± 0.19 versus 2.32 ± 0.21 respectively. These data indicate that there is no difference in the calcium-force relationship for normal and diseased human myocardium. Further, mean maximal calcium-activated force was not significantly different between the two groups [5, 25]. These data indicate that failing human myocardium has the potential to generate normal levels of force.

Tetanization

To compare maximal force in control and myopathic human myocardium steady-state calcium force relationships were generated in the presence of increasing $[Ca^{2+}]_o$ [16]. By utilizing steady-state measurements one avoids the issue of inhomogeneities of myoplasmic $[Ca^{2+}]$ and the effect on force production. Further utilization of this technique allows comparison of myofilament calcium responsiveness in intact preparations and thereby avoids the issue of potential errors in buffer composition that exists with skinned fiber preparations, particularly in the case of diseased myocardium where constituents are not known. In myopathic myocardium the Hill parameters were as follows: 0.54 ± 0.09 and 0.98 ± 0.03 for $[Ca^{2+}]$ 50% and 95%, respectively, with a Hill coefficient of 5.61 ± 0.6 versus for control $[Ca^{2+}]$ 50% and 95% 0.56 ± 0.05 and 0.97 ± 0.05 and Hill coefficient of 5.21 ± 0.2. Maximal calcium-activated force was 1.23 ± 0.49 versus 1.84 ± 0.74 g/mm^2 for control and myopathic tissues. Furthermore, increasing $[Ca^{2+}]_o$ resulted in similar intracellular calcium levels during tetanization for control and myopathic myocardium [16]. These data indicate that altered contractile function does not reside at the level of the thin myofilaments with regard to myofilament calcium responsiveness.

Twitch Force

Problems with utilization of peak force-peak calcium relationships as an indicator of myofilament calcium sensitivity have previously been discussed [16, 39]. The utilization of such a relationship if one is studying diseased states, which often involve time course changes in the intracellular calcium transient and/or twitch duration, can be particularly erroneous [15]. A clear example is seen when one plots peak force versus peak $[Ca^{2+}]$ in normal and diseased human myocardium. Because of the slow time course of the calcium transient the calcium-force relationship is shifted leftward in the case of myopathic myocardium [15, 16]. This means that despite no changes in myofilament calcium sensitivity the "apparent calcium" sensitivity is increased in failing human myocardium. Furthermore, peak SR calcium release in response to increasing $[Ca^{2+}]_o$ was not different between control and myopathic myocardium for twitch force [22].

Effect of Agents Targeted at the Level of the Myofilaments on Calcium-Force Relationships

Traditionally, force-pCa relationships derived from skinned fiber preparations have been thought to reflect changes at the level of the thin myofilaments. No change in the force-pCa relationship was interpreted to mean that there were no changes in thin myofilament composition. This was first demonstrated not to be the case by Hajjar et al. [25] and has since been confirmed in animal models of hypertrophy [1, 24]. Hajjar et al. [25] found that addition of the calcium sensitizing agent DPI 201-106 to skinned fiber preparations from control and myopathic hearts had differential effects. In control myocardium no calcium sensitization was noted. However, in myocardium from patients with end-stage heart failure there was a significant shift to lower $[Ca^{2+}]_i$ (0.2 pCa) in the presence of DPI. Supplemental experiments indicated that the site of the calcium sensitization most likely did not reside at the level of TnC or I. Troponin T (TnT) isoform composition was therefore implicated as playing a role in determining the calcium-force relationship. These experiments were subsequently followed by experiments which attempted to further address the issue of isoform composition in failing human myocardium and the effects on myofilament calcium activation. Protein kinase C activation results in phosphorylation of TnI and TnT. Exposure of human muscles to 12-deoxyphorbol-13-isobutyrate-20-acetate (DPBA), a protein kinase C activator, resulted in a shift to higher $[Ca^{2+}]_i$ in both control and myopathic myocardium [14, 17]. This observation most likely reflects a decrease in myofilament calcium sensitivity in response to phosphorylation of TnI. In control myocardium there was a corresponding decrease in the Hill coefficient reflecting reduced cooperativity. However, in myopathic myocardium the Hill coefficient was significantly higher after addition of DPBA, indicating increased cooperativity between the thick and thin myofilaments. It was proposed that the differential effect on the Hill coefficient between control and myopathic myocardium reflects different states of phosphorylation of TnT and/or the interaction of two TnT isoforms [16, 17]. TnT isoforms have been identified in failing human myocardium. Further, there appears to be an inverse relation between the occurrence of TnT_2 and myofibrillar ATPase activity. It was therefore suggested that TnT may play a role in myofilament calcium sensitivity.

These data indicate that there are isoform shifts at the level of the thin myofilaments that can potentially affect myofilament calcium-force activation in vivo with resultant changes in the "apparent" calcium responsiveness. One must now reconsider the interaction of thick and thin myofilaments as playing a role in the pathophysiology of cardiomyopathy. Altered thermodynamic interactions within the regulatory subunits of the thin myofilaments with resultant affects on myofibrillar ATPase activity and conformational changes in both the thin myofilament regulatory proteins and thick myofilaments associated with contraction now warrant investigation.

Cross-bridge Dynamics

To investigate mechanisms of systolic and diastolic abnormalities of heart failure the mechanical properties of cardiac muscles from normal and myopathic hearts were studied [23]. To avoid the complications of E-C coupling and the affect of calcium on cross-bridge dynamics fibers were saponin-skinned and maximally activated at pCa 4.0. Unloaded shortening velocity determined using Edman's slack test was 0.72 ± 0.09 versus 0.41 ± 0.06 L_0/s in normal and myopathic muscles, respectively. Stiffness was calculated from the ratio of force response to length oscillation amplitudes by sinusoidally oscillating the muscle at varying frequencies. At frequencies below 0.2 Hz stiffness was constant, but it was larger in myopathic muscles. Frequency of minimum stiffness was 0.78 ± 0.06 versus 0.42 ± 0.07 Hz for control and myopathic muscles, respectively. At higher frequencies the dynamic stiffness increased and reached a similar plateau for both groups of muscles. It is concluded that the number of force-generating cross-bridges are similar for normal and myopathic human myocardium. Furthermore, the slowed cross-bridge turnover rate might explain in part the diastolic abnormalities reported in failing human myocardium. Slower cross-bridge

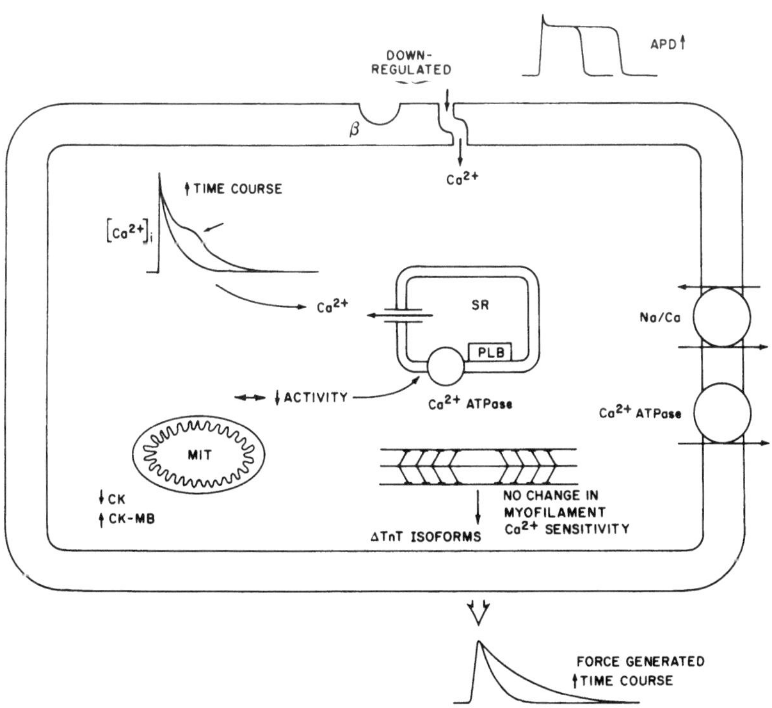

Fig. 3. Changes found to occur at the various steps of excitation contraction-coupling from human heart failure experiments. *PLB*, Phospholamban; *APD*, action potential duration

kinetics is often associated with myosin isoform shifts. We found that myosin ATPase activity is reduced in myopathic muscles from patients with both ischemic and idiopathic cardiomyopathy (Liao and Gwathmey, unpublished).

Conclusions

Over the past 10 years, the availability of normal and diseased human myocardium has provided great insight into mechanisms involving heart failure (Fig. 3). Certain findings have been surprising. These include the similarity of force generated between normal and myopathic muscles, the lack of differences in the amount of calcium released between normal and myopathic hearts during excitation, and maintenance of the sensitivity of the myofilaments to Ca^{2+} between normal and myopathic muscles. Other results were direct corollaries of established hypotheses concerning heart failure. These include the down-regulation of β_1 receptors and Ca^{2+} channels, the prolongation of force and $[Ca^{2+}]_i$ during contraction, and the reduction in energy reserve. The mismatch between energy supply and demand has proven to be the detrimental factor leading dilated but compensated hearts into contractile failure. Future research in the remodeling of energy-producing and -utilizing proteins will help us in understanding contactile dysfunction in heart failure.

Acknowledgements. This work was supported in part by CVD Institute, National Institutes of Health HL-39091, HL-36797, Glaxo Inc., Upjohn Co., a Beth Israel Hospital Biomedical Research Grant, and the John H. and Robnette T. Gwathmey Estate. J.K.G. is an Established Investigator of the American Heart Association.

References

1. Baudet S, Ventura-Clapier R (1990) Differential effects of caffeine on skinned fibers from control and hypertrophied ferret hearts. Am J Physiol 259:H1803–H1808
2. Böhm M, Beuckelmann D, Brown L, Feiler G, Lorenz B, Nabauer M, Kemkes B, Erdmann E (1988) Reduction of beta-adrenoceptor density and evaluation of positive inotropic responses in isolated, diseased human myocardium. Eur Heart J J9:844–852
3. Böhm M, Diet F, Feiler G, Kemkes B, Kreuzer E, Weinhold C, Erdmann E (1988) Subsensitivity of the failing human heart to isoprenaline and milrinone is related to beta-adrenoceptor downregulation. J Cardiovasc Pharmacol 12:726–732
4. Böhm M, Diet F, Kemkes B, Erdmann E (1988) Enhancement of the effectiveness of milrinone to increase force of contraction by stimulation of cardiac beta-adrenoceptors in the failing human heart. Klin Wochenschr 66:957–962
5. Böhm M, Morano I, Pieske B, Ruegg JC, Wankerl M, Zimmermann R, Erdmann E (1991) Contribution of cAMP-phosphodiesterase inhibition and sensitization of the contractile proteins for calcium to the inotropic effect of pimobendan in the failing human myocardium. Circ Res 68:689–701
6. Cameron JS, Kimura S, Jackson-Burns DA, Smith DB, Bassett AL (1988) ATP-sensitive K^+ channels are altered in hypertrophied ventricular myocytes. Am J Physiol 255:H1254–1258

7. Cooper IC, Fry (1990) Mechanical restitution in isolated mammalian myocardium:species differences and underlying mechanisms. J Mol Cell Cardiol 22:439–452
8. Danielsen W, von der Leyen H, Meyer W, Neumann J, Schmitz W, Scholz H, Starbatty J, Stein B, Doring V, Kalmar P (1989) Basal and isoprenaline-stimulated cAMP content in failing versus nonfailing human cardiac preparations. J Cardiovasc Pharmacol 14:171–173
9. Denniss AR, Colucci WS, Allen PD, Marsh JD (1989) Distribution and function of human ventricular beta adrenergic receptors in congestive heart failure. J Mol Cell Cardiol 21:651–660
10. Feldman MD, Copelas L, Gwathmey JK, Phillips P, Warren S, Schoen FJ, Grossman W, Morgan JP (1987) Deficient production of cyclic AMP: pharmacologic evidence of an important cause of contractile dysfunction in patients with end-stage heart failure. Circulation 75 2:331–339
11. Feldman AM, Cates AE, Veazey WB, Hershberger RE, Bristow MR, Baughman KL, Baumgartner WA, Van Dop C (1988) Increase in the 40,000-mol wt pertussis toxin substrate (G protein) in the failing human heart. J Clin Invest 82:189–197
12. Feldman MD, Gwathmey JK, Phillips P, Schoen F, Morgan JP (1988) Reversal of the force-frequency relationship in working myocardium from patients with end-stage heart failure. J App Cardiol 3:273–283
13. Finkel MS, Patterson RE, Roberts WC, Smith TD, Keiser HR (1988) Calcium channel binding characteristics in the human heart. Am J Cardiol 62:1281–1284
14. Gwathmey JK, Hajjar RJ (1990) Effect of protein kinase C activation on sarcoplasmic reticulum function and apparent myofibrillar Ca^{2+} sensitivity in intact and skinned muscles from normal and diseased human myocardium. Circ Res 67 3:744–752
15. Gwathmey JK, Hajjar RJ (1990) Intracellular calcium related to force development in twitch contraction of mammalian myocardium. Cell Calcium 11:531–538
16. Gwathmey JK, Hajjar RJ (1990) Relation between steady-state force and intracellular $[Ca^{2+}]$ in intact human myocardium. Index of myofibrillar responsiveness to Ca^{2+}. Circulation 82 4:1266–1278
17. Gwathmey JK, Hajjar RJ (1991) Protein kinase C activation in human ventricular myocardium. Biorheology 28:151–160
18. Gwathmey JK, Copelas L, Mackinnon R, Schoen FJ, Feldman MD, Grossman W, Morgan JP (1987) Abnormal intracellular calcium handling in myocardium from patients with end-stage heart failure. Circ Res 61:70–76
19. Gwathmey JK, Slawsky MT, Briggs GM, Morgan JP (1988) Role of intracellular sodium in the regulation of intracellular calcium and contractility. Effects of DPI 201-106 on excitation-contraction coupling in human ventricular myocardium. J Clin Invest 82:1592–1605
20. Gwathmey JK, Slawsky MT, Hajjar RJ, Briggs GM, Morgan JP (1990) Role of intracellular calcium handling in force-interval relationships of human myocardium. J Clin Invest 85:1599–1613
21. Gwathmey JK, Warren S, Briggs GM, Copelas L, Feldman MD, Phillips PJ, Callahan M, Schoen FJ, Grossman W, Morgan JP (1991) Diastolic dysfunction in hypertrophic cardiomyopathy. Effect on active force generation during systole. J Clin Invest 87:1023–1031
22. Gwathmey JK, Hajjar RJ, Allen PD (1992) Analysis of excitation-contraction coupling in human ventricular myocardium: relationship to the pathophysiology of heart failure. In: Gwathmey JK, Briggs GM, Allen PD (eds) Heart failure: basic science and clinical aspects. Dekker, New York (in press)
23. Hajjar RJ, Gwathmey JK (1991) Cross-bridge dynamics in human ventricular myocardium: regulatory properties of contractility in the Failing Heart. Circulation 86:1819–1826
24. Hajjar RJ, Gwathmey JK (1991) Modulation of calcium activation in control and pressure-overload hypertrophied ferret hearts: effect of DPI 201-106 on myofilament calcium responsiveness. J Mol Cell Cardiol 23:65–75
25. Hajjar RJ, Gwathmey JK, Briggs GM, Morgan JP (1988) Differential effects of DPI 201-106 on the sensitivity of the myofilaments to Ca^{2+} in intact and skinned trabeculae from control and myopathic human hearts. J Clin Invest 82:1578–1584

26. Ingwall JS, Kramer MF, Fifer MA, Lorell BH, Shemin R, Grossman W, Allen PD (1985) The creatine kinase system in normal and diseased human myocardium. N Engl J Med 313:1050–1054
27. Lederer WJ, Berlin JR, Cohen NM, Hadley RW, Bers DM, Cannell MB (1990) Excitation-contraction coupling in heart cells. Roles of the sodium-calcium exchange, the calcium current, and the sarcoplasmic reticulum. Ann NY Acad Sci 588:190–206
28. Marsh JD, Allen PD, Denniss AR (1987) Decreased calcium channel expression in sarcolemmal membranes from human hearts with congestive heart failure. J Mol Cell Cardiol 19 IV:533
29. Mulieri LA, Hasenfuss G, Leavitt B, Allen PD, Alpert NR (1992) Altered myocardial force-frequency relation in human heart failure. Circulation 85:1743–1750
30. Pagani ED, Alousi AA, Grant AM, Older TM, Dziuban SW, Allen PD (1988) Changes in myofibrillar content and Mg-ATPase activity in ventricular tissue from patients with heart failure caused by coronary artery disease, cardiomyopathy, or mitral valve insufficiency. Circ Res 63:380–385
31. Phillips PJ, Gwathmey JK, Feldman MD, Schoen FJ, Grossman W, Morgan JP (1990) Post-extrasystolic potentiation and the force-frequency relationship: differential augmentation of myocardial contractility in working myocardium from patients with end-stage heart failure. J Mol Cell Cardiol 22:99–110
32. Rasmussen R, Minobe M, Narita H, Ginsburg R, Bristow MR (1987) Are calcium channels altered in the failing human heart? Circulation 76 IV:535
33. Schouten VJA, ter Keurs HEDJ, Quaegebeur JM (1990) Influence of electrogenic Na/Ca exchange on the action potential in human heart muscle. Cardiovasc Res XXIV 9:758–767
34. Talo A, Stern MD, Spurgeon HA, Isenberg G, Lakatta EG (1990) Sustained subthreshold-for-twitch depolarization in rat single ventricular myocytes causes sustained calcium channel activation and sarcoplasmic reticulum calcium release. J Gen Physiol 96:1085–1103
35. Ventura-Clapier R (1990) Skinned fibres in cardiac overload. In: Swynghedauw B (ed) Research in cardiac hypertrophy and failure. INSERM/Libbey, London, pp 161–168
36. Wagner JA, Sax FL, Weisman HF, Potterfield J, McIntosh C, Weisfedlt ML, Snyder SH (1989) Calcium antagonist receptors in the atrial tissue of patients with hypertrophic cardiomyopathy. N Engl J Med 320:755–761
37. Warren SE, Gwathmey JK, Feldman MD, Phillips P, Grossman W, Morgan JP (1989) Inotropic and lusitropic effects of DPI 201-106 on human myocardium. J Appl Cardiol 4:177–188
38. Warren SE, Kihara Y, Pesaturo J, Gwathmey JK, Phillips P, Morgan JP (1989) Inotropic and lusitropic effects of MCI-154 (6-[4-(4-pyridyl) aminophenyl]-4,5-dihydro-3(2H)-pyridazinone) on human myocardium. J Mol Cell Cardiol 21:1037–1045
39. Yue DT, Marban E, Wier WG (1986) Relationship between force and intracellular calcium in tetanized mammalian heart muscle. J Gen Physiol 87:223–242

Energy Requirements of Contraction and Relaxation of the Heart

A.M. Katz

Introduction

While our knowledge of the myocardial abnormalities in patients with heart failure remains incomplete, there is substantial evidence that the failing heart is an energy-starved organ [7]. For this reason, the causes and consequences of an imbalance between energy production and energy utilization are of central importance in developing therapeutic strategies to deal with this increasingly common condition [8]. Among the many implications of a potential state of myocardial energy starvation for management of patients with congestive heart failure are the long-term consequences of interventions that increase myocardial energy demand [6]. For example, it is possible that lack of energy contributes to the detrimental effects of inotropic stimulation on the failing heart [14]. Conversely, both vasodilators and negative inotropic drugs, which can improve the balance between energy production and energy expenditure in the failing heart, might improve prognosis in patients with chronic heart failure [9].

Energy starvation in the failing heart can arise from either or both of two abnormalities: increased energy demands and decreased energy supply. As this chapter points out, both appear likely to occur in the majority of patients with heart failure.

Increased Myocardial Energy Demands in the Patient with Heart Failure

As heart failure can be caused by a variety of pathophysiological abnormalities, the energetic state of the failing heart varies among different patients. In patients with aggressive cardiomyopathies such as viral myocarditis, rapid destruction of heart muscle is far more important for prognosis that any cellular energy deficit. However, in most patients with heart failure, excessive energy expenditure represents a major component of the pathophysiological processes in the failing heart.

In patients whose heart failure is due to a chronic systolic overload, as with inadequately treated hypertension and aortic stenosis, the increased

Cardiology Division, Department of Medicine, University of Connecticut, Farmington, CT 06032, USA

Figulla et al. (Eds.)
Idiopathic Dilated Cardiomyopathy

work of left ventricular ejection against the increased afterload is a major cause of excessive energy demands. An increase in wall tension also occurs in patients with diastolic overload, due, for example, to regurgitant valve abnormalities and dilated cardiomyopathy. Myocardial energy demands are also increased in the noninfarcted myocardium of patients with regional wall motion abnormalities following acute myocardial infarction. While hypertrophy initially tends to normalize the tension on each myofiber in the walls of the ventricles [4], the enlarged muscle mass itself contributes to increased energy consumption (see below).

Decreased Myocardial Energy Supply in the Patient with Heart Failure

A number of abnormalities can impair the heart's ability to generate sufficient chemical energy needed to meet a chronically increased overload (for review see [9]). Several reflect alterations in the structure of the hypertrophied heart. Over 50 years ago, the growth of capillaries supplying the hypertrophied myocardium was observed not to keep up with the increased muscle mass. This observation, which has been confirmed in more recent morphological analyses [1], indicates that increased intracapillary distance and a deficit in arterial blood supply can deprive the chronically overloaded heart of an adequate supply of oxygen. Impaired perfusion of the failing heart has also been observed; most striking is a decreased vasodilator reserve that is especially severe in the relatively underperfused subendocardial regions of the ventricle [5].

Altered cell composition in the hypertrophied heart also contributes to a state of energy starvation in patients with heart failure. Important architectural alterations seen in long-standing pressure overload include an increased content of myofibrils, which consume energy, and a decreased number of ATP-generating mitochondria. The resulting increase in the number of myofibrils that must depend on each mitochondrion for an adequate supply of ATP can contribute to a chemical energy deficit in the failing heart [9].

Direct measurements of ATP and phosphocreatine contents have confirmed a decrease in myocardial high-energy phosphate content in animal models of heart failure caused by chronic overloading of the left and right ventricles. While similar data are much more difficult to obtain in clinical heart failure, endomyocardial biopsies from failing human hearts have also demonstrated a decreased ATP content that correlates with the extent of the impaired function [2].

Functional Significance of an Energy Deficit in the Failing Heart

It is now apparent that ATP has a dual effect in many aspects of myocardial cellular regulation [10]. This pattern is reflected in a large number of metabolic and regulatory processes, where low concentrations of ATP provide the substrate for energy-consuming reactions, and higher levels of this nucleotide serve a regulatory function. For example, micromolar ATP concentrations saturate the catalytic sites that allow the myosin cross-bridges to interact with actin, and supply energy to such ion pumps as the calcium pump of the sarcoplasmic reticulum. At higher concentrations, ATP influences a variety of systems through allosteric effects in which ATP is not hydrolyzed. These regulatory effects of ATP are generally stimulatory. In the case of the ion channels and ion pumps, where ATP can be viewed as a "lubricator," ATP binding to low-affinity sites accelerates a number of active and passive ion fluxes [10].

The two fundamentally different effects of ATP described above reflect the presence of two types of ATP binding site: high-affinity catalytic sites, which are half-saturated at submicromolar ATP concentrations, and low-affinity regulatory sites, which are half-saturated at ATP concentrations above $100\,\mu M$. Except in dying cells, it is unlikely that ATP levels fall to levels sufficiently low to impair energy consumption by depriving high-affinity sites of substrate. However, attenuation of the allosteric effects of ATP in the energy-starved heart probably contributes to important electrophysiological and mechanical abnormalities in patients with heart failure.

Effects of Energy Starvation in the Heart

Electrophysiological Effects of ATP Depletion

The ionic currents responsible for the cardiac action potential are generated by the passive, downhill fluxes of sodium, calcium, potassium, and possibly chloride ions. Depolarization is generated by inward sodium and calcium currents, while repolarization involves mainly outward potassium currents. Inhibition of depolarizing sodium and calcium currents, by slowing the action potential upstroke and reducing action potential amplitude, slow conduction and so contribute to the genesis of reentrant arrhythmias. Inhibition of repolarizing currents prolong the cardiac action potential, and are also arrhythmogenic. These effects, which can be caused when the stimulatory effects of high ATP concentrations are attenuated, likely contribute to the high incidence of sudden death in patients with heart failure.

Inhibition of the sodium pump, as occurs when the allosteric effect of ATP is attenuated in an energy-starved heart, reduces the potassium ion gradient and so depolarizes the resting myocardium. Depolarization in turn inactivates the sodium channels responsible for rapid conduction in the atria, ventricles, and His-Purkinje system, thereby increasing the propensity for reentrant arrhythmias.

A rise in cytosolic calcium concentration, as occurs in the energy-starved myocardium (see below), promotes oscillatory calcium release from the sarcoplasmic reticulum. The latter induces arrhythmogenic transient depolarizations that probably play an important role in generating the lethal arrhythmias that contribute to the poor long-term prognosis in patients with heart failure [10].

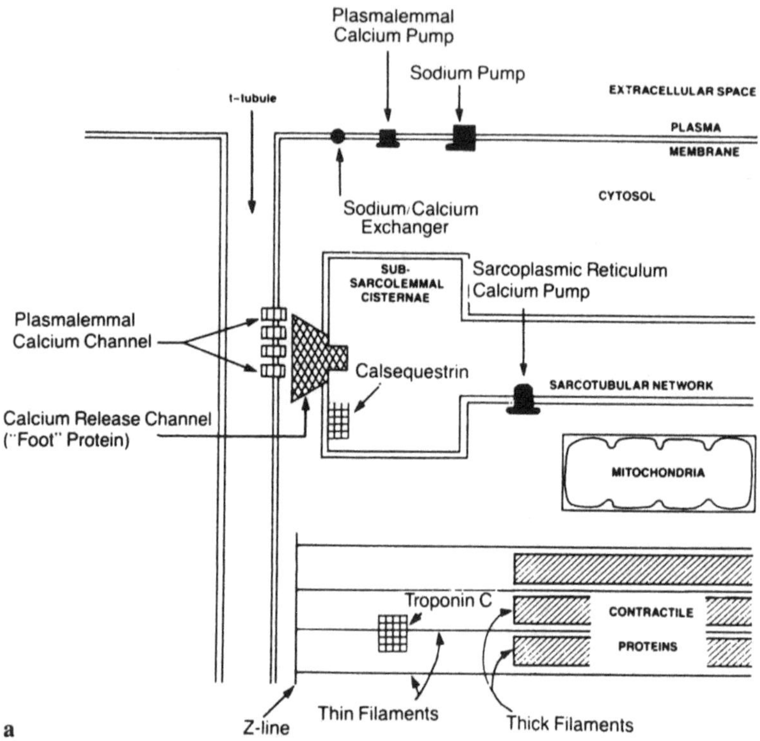

Plasmalemmal
Calcium Pump
Sodium Pump
t-tubule
EXTRACELLULAR SPACE
PLASMA
MEMBRANE
CYTOSOL
Sodium/Calcium
Exchanger
SUB-
SARCOLEMMAL
CISTERNAE
Sarcoplasmic Reticulum
Calcium Pump
Plasmalemmal
Calcium Channel
Calsequestrin
SARCOTUBULAR NETWORK
Calcium Release Channel
("Foot" Protein)
MITOCHONDRIA
Troponin C
CONTRACTILE
PROTEINS
Z-line
Thin Filaments
Thick Filaments

a

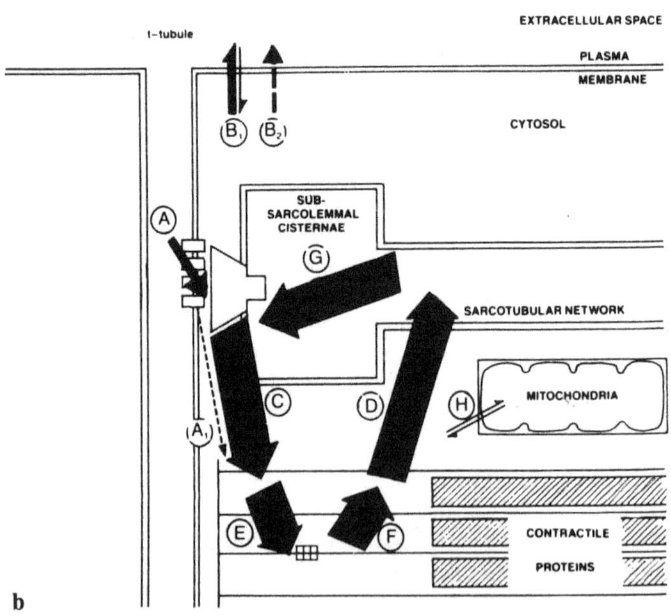

t-tubule
EXTRACELLULAR SPACE
PLASMA
MEMBRANE
CYTOSOL
B1
B2
SUB-
SARCOLEMMAL
CISTERNAE
A
G
SARCOTUBULAR NETWORK
C
D
H
MITOCHONDRIA
A1
E
F
CONTRACTILE
PROTEINS

b

Effects of ATP Depletion on Contraction and Relaxation

Contraction and relaxation both require the expenditure of energy, but energy is used quite differently in systole and diastole. During systole, the interactions between the contractile proteins that cause the heart to contract transduce chemical energy derived from ATP hydrolysis for the performance of mechanical work. The processes that activate the contractile proteins, however, do not require the expenditure of energy; instead, activator calcium is delivered to the cytosol by the passive diffusion of this ion through channels in the sarcolemmal and sarcoplasmic reticulum membranes (Fig. 1). This reflects that fact that cytosolic calcium concentration in the resting heart is approximately 0.1 μM, which is much lower than that in the extracellular space and within the sarcoplasmic reticulum, where calcium concentration is in the millimolar range. Thus, the signal that initiates cardiac contraction is mediated by the movement of calcium down an electrochemical gradient, a process that does not require direct expenditure of energy. As noted above, however, attenuation of the allosteric effects of ATP, which promote downhill calcium fluxes through calcium channels, can impair contractility in an energy-starved heart.

In contrast to the downhill calcium fluxes that activate contraction, the calcium fluxes that relax the heart require the expenditure of energy (Fig. 1). In order for the heart to relax, activator calcium must be pumped out of the cytosol into the sarcoplasmic reticulum and extracellular space to dissociate calcium from its binding site on troponin in the myofilaments. This requires that energy be supplied to membrane ion pumps. Both plasma membrane and sarcoplasmic reticulum calcium pumps are stimulated by allosteric effects of high ATP concentrations. The sodium/calcium exchanger, another important system that transports calcium out of the cell during diastole, uses the energy of the transsarcolemmal sodium gradient,

←

Fig. 1. Schematic diagram showing key structures (**a**) and major calcium fluxes (**b**) involved in cardiac excitation-contraction coupling. *Arrow thickness*, magnitude of the calcium fluxes; *arrow direction*, "energetics" of the calcium fluxes; *downward arrows*, passive calcium fluxes; *upward arrows*, energy-dependent calcium transport. Calcium enters the cell from the extracellular fluid via plasmalemmal calcium channels (A); while most of this calcium triggers calcium release from the sarcoplasmic reticulum, a small portion directly activates the contractile proteins (A_1). Calcium transport back into the extracellular fluid involves two plasma membrane systems: sodium/calcium exchange (B_1) and the plasmalemmal calcium pump (B_2). The sarcoplasmic reticulum membrane regulates two calcium fluxes: calcium release from the subsarcolemmal cisternae (C) and active calcium uptake by the calcium pump of the sarcotubular network (D). Calcium diffuses within the sarcoplasmic reticulum in a third calcium flux (G), returning to the subsarcolemmal cisternae where it is stored in complex with calsequestrin and other calcium-binding proteins. Binding (E) and dissociation (F) of calcium with the high-affinity calcium-binding sites of troponin C define the its affinity for calcium (ratio E/F). Movements of calcium into and out of mitochondria (H) buffer cytosolic Ca^{2+} concentration. (From [10]) rather than ATP, to effect the uphill movement of calcium out of the cell during diastole. Because turnover of the sodium/calcium exchanger is stimulated by an allosteric effect of high ATP concentrations, an energy deficit could slow this ion transport system as well, thereby impairing calcium removal from the cytosol. The ATP-dependent sodium pump that generates the sodium gradient which allows the sodium/calcium exchanger to transport calcium from the cytosol to the extracellular fluid, like the calcium pumps, is accelerated by an allosteric effect of the nucleotide.

In order for the contractile proteins to relax during diastole, ATP must bind to the hydrolytic site on the myosin cross-bridge [10]. Loss of this effect in skeletal muscle leads to rigor mortis. This action of ATP to dissociate rigor bonds linking actin and myosin at the end of the cross-bridge cycle, which plasticizes the contractile proteins, does not require that the nucleotide be hydrolyzed and so is analogous to the allosteric effects of ATP described above. Even a minor fall in ATP concentration can reduce this plasticizing effect of ATP, and so slows relaxation. The severe drop in ATP levels in the dying heart leads to the low energy state called ischemic contracture, or the "stone heart."

Energetics of the Calcium Fluxes During Contraction and Relaxation

As noted in Fig. 1, ionized calcium concentrations are much higher in the extracellular fluid and within the sarcoplasmic reticulum than in the cytosol. Thus, the calcium fluxes responsible for excitation are passive (downhill), while the transport of this activator out of the cytosol is an active (uphill) process that requires the expenditure of energy. In both failing and ischemic hearts, attenuation of the many allosteric effects of ATP described above

Table 1. Balances between calcium fluxes during activation and relaxation in the myocardium

Activation	
Total functional calcium channels/cell	14000
Functional plasma membrane calcium channels/cell	2500
Sarcoplasmic reticulum calcium release channels/cell	11500
Rate of calcium flux through each calcium channel	3×10^6 ions/s
Maximal calcium entry/cell during systole	4×10^{10} ions/s
Relaxation	
Sarcoplasmic reticulum calcium pump sites/cell	166000000
Rate of calcium transport during relaxation	30 ions/s
Maximal calcium removal/cell during diastole	5×10^9 ions/s
Balances	
Maximal rate of calcium entry during systole/maximal rate of calcium removal during diastole	6–8
Maximal rate of calcium entry during systole/average rate of calcium removal during diastole	13–16

Assumptions and sources of data are detailed in Chap. 11 of [10].

probably represents a major cause of the lusitropic (relaxation) abnormalities so prominent in these conditions. In this context, it is important to appreciate the precarious state of the calcium fluxes involved in relaxation, compared to those that deliver activator calcium to the contractile proteins (Table 1).

A fundamental difference between the calcium fluxes responsible for contraction and those that relax the heart lies in their intrinsic rates [10]. Passive diffusion, which delivers calcium to the contractile proteins, is extremely rapid; calcium entry via a single plasma membrane calcium channel has been estimated to be approximately 3 000 000 ions per second. In contrast, calcium transport by the ATP-dependent calcium pumps that relax the heart is much slower. The maximal calcium flux by a single calcium pump site of the sarcoplasmic reticulum at 37°C is approximately 30 ions per second, which is only 1/100 000 the rate of calcium entry through a calcium channel. Compensation for this enormous difference in rates is provided by a much greater number of calcium pump molecules than calcium channels, but, as discussed below, even this fails to provide a large reserve for the calcium fluxes needed to relax the heart.

Comparison of the relative rates of the calcium fluxes that activate and relax the heart (Table 1) reveals a precarious situation that helps explain why the energy-starved heart is prone to calcium overload and so to impaired relaxation. Even the maximal rate of calcium removal from the cytosol during diastole, which occurs when the calcium pump is operating at a saturating concentration of calcium, is much slower than that of calcium diffusion into the cytosol through calcium channels (Table 1). A more realistic calculation, which assumes that the calcium pump operates at about half its maximal rate throughout diastole, takes into account the fall in cytosolic calcium concentration during diastole; this rate is less than 10% that of the calcium fluxes which activate the heart (Table 1).

Examination of the calcium flux rates in Table 1 makes clear just how precarious relaxation can become in an energy-starved heart. Even under normal conditions, calcium removal from the heart's cytosol is about an order of magnitude slower than that of the calcium delivery during activation. Because active transport is so much slower than diffusion, even the densely packed calcium pump proteins in the sarcoplasmic reticulum membrane, which resemble an array of soldiers standing at attention on parade, cannot compensate for the approximately 100 000-fold greater rate of diffusion through the limited number of channels that allow calcium to enter the cytosol. That the heart is able to relax in spite of this difference is simply because calcium release at the onset of systole occurs over a much shorter time than does calcium uptake during diastole. However, if a deficit in ATP, by attenuating the allosteric effect to accelerate the calcium pump, slows calcium removal from the cytosol, the sarcoplasmic reticulum will be unable to lower cytosolic calcium to allow for complete relaxation, even during a normal diastole. This susceptibility of the failing heart to calcium overload also explains why tachycardia can be so poorly tolerated in patients with heart failure.

Effects of the Cardiomyopathy of Overload

There is now abundant evidence that the density of calcium pump molecules is reduced in the sarcoplasmic of hypertrophied hearts [3, 11, 13]. This abnormality, by further slowing calcium removal from the cytosol of the overloaded failing heart, would exacerbate the consequences of the energy deficit described above. It is possible that other abnormalities in the composition of the hypertrophied heart [12] also contribute to the prominent relaxation abnormalities in patients with heart failure.

Acknowledgement. Supported in part by a Program Project HL-33026 from the National Heart Lung and Blood Institute.

References

1. Anversa P, Olivetti G, Melissari M, Loud AV (1980) Stereological measurement of cellular and subcellular hypertrophy and hyperplasia in the papillary muscle of adult rat. J Mol Cell Cardiol 12:781–795
2. Bashore TM, Magorien DJ, Letterio J, Shaffer P, Unverferth DV (1978) Histologic and biochemical correlates of left ventricular chamber dynamics in man. J Am Coll Cardiol 9:734–742
3. De la Bastie D, Levitsky D, Rappaport L, Mercadier J-J, Marotte F, Wisnewsky C, Brokovich V, Schwartz K, Lompre A-M (1990) Function of the sarcoplasmic reticulum and expression of its Ca^{2+} ATPase gene in pressure-overloaded cardiac hypertrophy in the rat. Circ Res 66:554–564
4. Grossman W, Jones D, McLaurin LP (1975) Wall stress and patterns of hypertrophy in the human left ventricle. J Clin Invest 56:56–64
5. Hoffman JEI (1987) Transmural myocardial perfusion. Prog Cardiovasc Dis 29:429–464
6. Katz AM (1986) Potential deleterious effects of inotropic agents in the therapy of chronic heart failure. Circulation 73 [Suppl III]:184–188
7. Katz AM (1989) The myocardium in congestive heart failure. Am J Cardiol 63:2A–16A
8. Katz AM (1989) Changing strategies in the management of congestive heart failure. J Am Coll Cardiol 13:512–523
9. Katz AM (1990) Cardiomyopathy of overload. A major determinant of prognosis in congestive heart failure. N Engl J Med 322:100–110
10. Katz AM (1992) Physiology of the heart, 2nd edn. Raven, New York
11. Komuro I, Kurabayashi M, Shibazaki Y, Takaku F, Yazaki Y (1989) Molecular cloning and characterization of a Ca^{2+} + Mg^{2+}-dependent adenosine triphsophatase from rat cardiac sarcoplasmic reticulum. Regulation of its expression by pressure overload and developmental stage. J Clin Invest 83:1102–1108
12. Legere J, Dechesne C, Bouvagnet P, Leger J-J (1990) Contractile proteins in the normal and hypertrophied human heart. In: Swynghedauw B (ed) Cardiac hypertrophy and failure. Libbey, London, pp 237–250
13. Nagai R, Zarain-Herzberg A, Brandl CJ, Fujii J, Tada M, MacLennan DH, Alpert NR, Periasamy M (1989) Regulation of myocardial Ca^{2+} ATPase and phospholamban mRNA expression in response to pressure overload and thyroid hormone. Proc Natl Acad Sci USA 86:2966–2970
14. Packer M, Carver, JR, Rodeheffer RJ et al. (1991) Effect of oral milrinone on mortality in severe heart failure. N Engl J Med 325:1468–1475

VI. Microcirculation and Oxygen Supply in Cardiac Hypertrophy and IDC

Capillary Density, Distribution, and Length Parameters Related to Oxygen Supply in Myocardial Hypertrophy and Atrophy

K. Rakusan[1], S. Batra[1], P. Veprek[2], and Z. Turek[3]

Introduction

Cardiac muscle is well known to be vulnerable to anoxia, even for a very short duration, due to its almost entirely oxidative metabolism. For proper cardiac function the balance between oxygen supply and demand must be met not only at the level of the whole organ but also at that of the individual working units, i.e. cardiac myocytes. Previous attempts to measure the adequacy of the oxygen supply-to-demand ratio were limited by an emphasis on measuring the basic parameters only at the level of the whole organ. This was due mainly to lack of data on the local distribution of oxygen sources and sinks, i.e. basic supply and demand elements. Recent developments in the quantitative analysis of cardiac tissue should permit such an investigation at the tissue level as well.

The objectives of this presentation are the following: (a) to review advanced morphometric methods available for evaluating the coronary capillaries; (b) to present some new data concerning changes in capillary net associated with normal and pathological growth as well as in cardiac atrophy (most of these data originate from our own laboratory); and (c) to discuss some of the functional consequences of the above changes.

Search for a Better Approach for Evaluating Geometry of the Capillary Bed

Traditional Parameters

The capillary supply of the heart is traditionally characterized by the capillary diameter, capillary density and fiber-capillary ratio. Capillary diameter is a quantitative parameter which is quite variable in vivo depending on the presence or absence of the red blood cells within the vessel as well as on the extravascular pressure. Moreover, its size measured on histological section is obviously influenced by the fixation and processing of the tissue. Capillary

[1] Department of Physiology, University of Ottawa, Ottawa, Canada
[2] Kardiocentrum, Prague, Czechoslovakia
[3] Department of Physiology, University of Nijmegen, Nijmegen, The Netherlands

Figulla et al. (Eds.)
Idiopathic Dilated Cardiomyopathy

density simply means the number of capillaries per unit of tissue cross-section. Consequently, it is possible to estimate the average intercapillary distance, assuming a certain geometrical arrangement of capillaries. Finally, the fiber/capillary ratio compares the number of muscle cells supplied by a single capillary. This index, however, ignores the size of the individual muscle cells.

Global Indicators of Capillary Supply

Capillary density as described above means simply the numerical density of capillaries, and as such may serve as a global indicator of capillary supply. Capillary length density, expressed as length per tissue cross-section (mm/mm^2) or length per volume tissue (mm/mm^3), is a more sophisticated index derived along similar lines as numerical density. Finally, one may also estimate the volume density of coronary capillaries (mm^3/mm^3). This index, however, is very sensitive to the values of capillary diameters, with all the limitations of their measurements, as described above.

Advanced Morphometric Parameters

A basic deficiency inherent in a calculation of intercapillary distance (ICD) from capillary density as described above is the assumption of uniform distribution of capillaries. However, myocardial oxygenation is influenced not only by the average ICD as obtained from the capillary density but also by the variability of the ICDs, which is an expression of the heterogeneity of capillary spacing. Evaluation of the local capillary supply region and subsequent estimates of capillary spacing may be derived from so-called capillary domains. These are polygonal regions which are closer to a given capillary than to any other. From each domain area, the equivalent radius of the Krogh cylinder with the same area may be calculated. As the distribution of these radii is log-normal, their variability is best characterized by the standard deviation of the logarithmic distribution (SD log).

Heterogeneity of capillary spacing is the result of two components. The first source of heterogeneity is simply an uneven distribution of capillaries in space, with all units running in parallel along a preferred axis, i.e. anisotropy. This probably reflects variability in diameter of the muscle cells supplied by these capillaries. The second source is the deviation from ideal alignment with the axis of anisotropy, i.e., a less uniform arrangement around a preferred axis. The degree of anisotropy may be obtained by applying the short-cut estimation procedure proposed by Mathieu and coworkers [9]. For a typical printout of the capillary domain and a graphic representation of two possible sources of heterogeneity, see Fig. 1.

Until recently, data concerning longitudinal parameters of coronary capillaries were rare and at best semiquantitative. The task of following capillaries within the longitudinal sections is rather difficult. Part of the restriction lies in the procurement of tissue sections that capture a significant part of the capillary network. Furthermore, it is difficult to establish valid

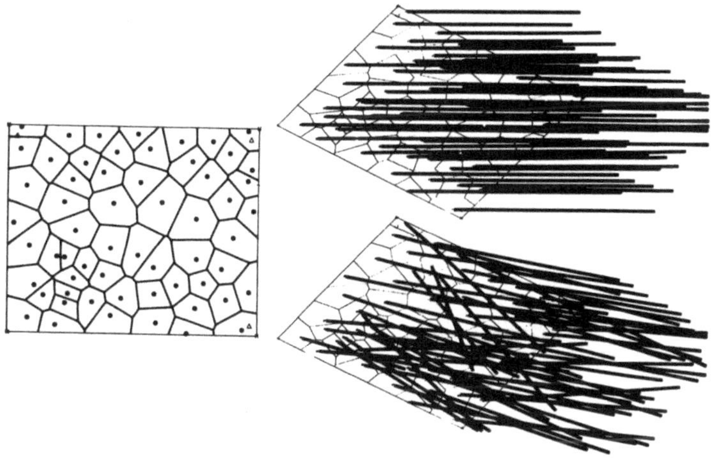

Fig. 1. Capillary domain areas and two possible sources fo their variability

standards for histologically demarcating capillaries from arterioles to venules. In an effort to overcome some of these hindrances, we have employed a combination of histochemical methods that distinguish the arteriolar and venular portions of capillaries by colour [7]. We have used this approach in analysing changes of the capillary net during normal and pathological geometry, which is published in a series of recent articles [1–3]. The most important parameters derived in this fashion are minimal capillary length, defined as the shortest contiguous pathway from terminal arteriole to collecting venule, capillary set length, which includes all the capillaries connected to the same arteriole and venule and the length of capillary segments (the distance between two ensuing branch points).

Geometry of Coronary Capillary Networks During Normal Growth, Cardiac Hypertrophy and Atrophy

Heterogeneity of Capillary Spacing

To demonstrate the application of advanced morphometric parameters as described above, we have reviewed known data on the heterogeneity of coronary capillary spacing as well as parameters based on double-staining of coronary capillaries. Data refer to various types of normal and pathological growth of the heart, both in man and various experimental animals. They are based predominantly (SD log) or entirely (remaining parameters) on measurements from our own laboratory.

Changes in the heterogeneity of capillary spacing (SD log) are summarized in Table 1. Normal cardiac growth and development in rats is characterized by a brief drop in SD log when the hearts from rats killed

Table 1. Heterogeneity of coronary capillary spacing (SD log)

Normal development		
Rat	↓	25–40 days [11]
		21–240 days [2]
	↑	Arteriolar regions
	(NS)	venular regions
Man	(NS)	infant–young adults [13]
Pressure overload hypertrophy		
Rat	↑	Senescent SHR [14]
	↑	Renal hypertension [15]
	↑	Aortic constriction 5 days–5 months [6]
		Aortic constriction 5–28 days [12]
	(↑)	Arteriolar regions
	(NS)	venular regions
Dog	(NS)	renal hypertension [18]
Man	(↑)	Inborn aortic valve stenosis
	↑	Acquired aortic valve stenosis [13]
Volume overload hypertrophy		
Rat	↓	AV fistula, both arteriolar and venular regions [1]
Dog	(NS)	AV fistula [6]
Atrophy		
Rat	↑	Both arteriolar and venular regions [22]

↑, Significant increase; ↓, significant decrease; (↑), borderline increase; NS, no significant changes.

during the weaning period were compared to hearts from rats killed 2 weeks later [11]. Extended studies of the postnatal developmental period revealed an increase in heterogeneity of capillary spacing in portions of the capillaries close to arterioles, while in the venular regions no changes were detected [2]. The heterogeneity of all capillaries does not significantly change in human hearts during normal development [13].

Pressure overload hypertrophy, usually characterized by a decreased capillary density, is accompanied by increased heterogeneity of capillary spacing. This increase was most pronounced in rat hearts with spontaneous hypertension [14] or with long-term aortic constriction [12]. It was also present in human hearts with acquired aortic valve stenosis [13]. In some cases of pressure overload hypertrophy, the increases were only of borderline significance [3]. On the other hand, volume overload hypertrophy was characterized either by a decrease or no change in the heterogeneity of capillary spacing [1, 18]. Finally, significant increases in SD log were found in both arteriolar and venular regions of rat hearts that had atrophied due to heterotopic transplantation [22].

Capillary Domains

Table 2 summarizes the data obtained using double-staining method, which enables us to distinguish between the arteriolar and venular portions of

Table 2. Morphometry of coronary capillary networks

Arteriolar capillary domain area	↑↑NG, ↑POH, (NS) VOH, AT mid-, ↓ AT endo-
Venular capillary domain area	↑↑NG, (NS) POH, AT mid-, ↓VOH, AT endo-
Capillary set length	↑VOH, ↓NG, POH, ↓↓AT mid-
AC segment length	↑NG, VOH, (NS) POH, ↓AT mid-
VC segment length	↑NG, (NS) POH, VOH, ↓AT mid-
Capillary set area	↑↑NG, ↑POH, VOH ↓AT
Length/area	↓↓NG, (↓) POH, (NS) VOH, AT mid-

↑↑, Massive; ↑, significant; (↑), borderline increase (or decrease, ↓).
All experiments on rats; NG, normal growth 21–240 days [2]; POH, pressure overload hypertrophy, aortic constriction 5–28 days [3]; VOH, volume overload hypertrophy, AV fistula [1]; AT, atrophy, heterotopic transplantation [22]; mid-, midsection; endo-, endomyocardium; NS, no significant changes.

capillaries and consequently facilitates the procurement of longitudinal data. For comprehensive reports, see the original publications referred to in Table 1.

The most prominent features of the geometry of the coronary capillary net are the following. (a) Capillary domain area, which reflects capillary density, is in all situations significantly larger in the arteriolar region than in the venular region of coronary capillaries. (For a possible significance of this particular geometrical arrangement, see the subsequent section). (b) Both domain areas increase during the normal growth of the rat heart, reflecting the postnatal decrease in capillary density. (c) Pressure overload hypertrophy is characterized by an increased capillary domain area on the arteriolar region while no changes are detected on the venular side. In contrast, capillary domain area decreases on the venular side in the case of volume overload hypertrophy while the arteriolar region remains unchanged. (d) Finally, both regions decrease in the endomyocardium of atrophic hearts even though in the midmyocardium they remain essentially the same.

Longitudinal Parameters

In the same table, the longitudinal parameters derived from the double-staining method are also included. Capillary set length decreases during normal growth of the heart as well as in pressure overload hypertrophy. These decreases are accompanied by increases in the individual capillary segment length. Thus, the total number of capillary segments per capillary set decreases. The most probable explanation for this finding is the appearance of new arterioles and new capillary sets.

On the other hand, in volume overload hypertrophy a significant increase is found in the length of combined capillaries connecting the same arteriole and venule. This is accompanied by a concomitant increase in the length of arteriolar segments. Finally, a decrease in capillary set length together with decreases in both arteriolar and venular segment length is present in cardiac atrophy.

Capillary set area is significantly increased during normal growth and in both types of pathological growth. The opposite is true in case of cardiac atrophy. Capillary length density (length/area) decreases significantly during normal growth. The decrease is less pronounced in pressure overload hypertrophy, and no significant changes are detected in either volume overload hypertrophy or in the midsection of atrophic transplants.

Functional Consequences of the Changes in Capillary Geometry

In the previous sections we introduced several new concepts. Foremost was the concept of the capillary domain, which is illustrated in Fig. 1. In this case, the locations of all capillaries on the photomicrograph are recorded by an image analyser as pairs of coordinates. Subsequently, a computer program finds equidistant border lines between each capillary and all its neighbours. These lines intersect and thus demarcate a polygonal region of tissue which we call the capillary domain. This approach enables us to evaluate the capillary supply at the local level. At the same time, variability of domain area and radius yields SD log, an index of heterogeneity of capillary spacing, as described above. The double-staining method helps us to distinguish between the earlier (arteriolar) and latter (venular) regions of coronary capillaries as well as their respective capillary domains. Finally, the double-staining method facilitates the collection of longitudinal parameters.

The heterogeneity of capillary spacing directly influences the local myocardial tissue PO_2; this effect is independent of the average intercapillary distance. This was found by applying our multicylindrical model of oxygen transport that represents myocardial tissue as a set of overlapping Kroghian tissue cylinders [20]. This model allows calculation of tissue PO_2 histograms under varying conditions. Calculated PO_2 values are strongly dependent on heterogeneity in capillary spacing and capillary blood flow. This dependency, however, decreases with the inclusion of an additional capillary barrier combined with Michaelis-Menten type kinetics of O_2 consumption.

Figure 2 represents the application of more realistic geometry of capillary networks to the calculation of tissue PO_2 in hypertrophied rat heart (pressure overload hypertrophy). Tissue PO_2 histograms based on the classical Krogh model cylinder, as used previously, are compared with PO_2 histograms based on our finding of stepwise decreases in intercapillary distance with corresponding decreases in heterogeneity of capillary spacing [3]. In this case, a truncated cone would be a better approximation of the tissue arrangement around each capillary: arteriolar domains are always larger than the venular domains. When the more realistic values of the truncated cone model are used, the calculated PO_2 histograms become less heterogenous. Furthermore, both the percentage of PO_2 lower than 10 mmHg and the percentage of anoxic tissue are decreased. These results are consistent with the view that a shorter intercapillary distance and a less pronounced heterogeneity are more effective when the capillary PO_2 is already low at the onset [19].

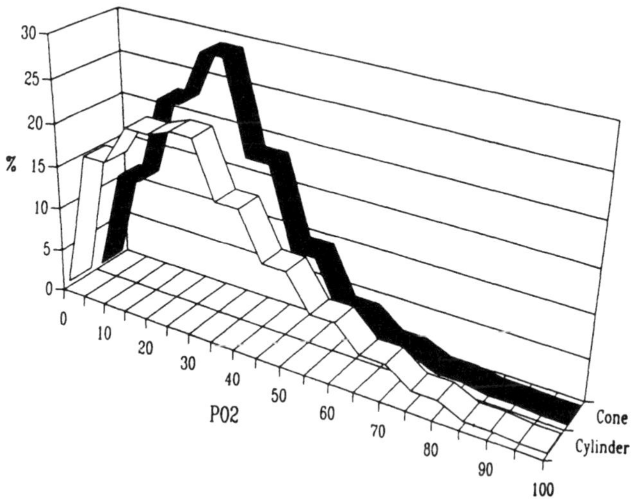

Fig. 2. Calculated myocardial tissue PO_2 histograms based on the same data obtained from pressure-overloaded hypertrophy, arranged as cylinder and truncated cone models. (From [19])

The double-staining of coronary capillaries also enables measurements of longitudinal parameters. At present, we do not have a comprehensive tissue PO_2 model which also incorporates the longitudinal parameters. It is clear, however, that capillary length and its branching pattern are important determinants of the red blood cell transit time and, indirectly, of tissue PO_2 as well.

An overview of our morphometric data presented in a previous section demonstrates that an increase in capillary domain areas, signalling decreasing capillary density during normal cardiac growth and in pressure overload hypertrophy, is usually associated with increasing heterogeneity of capillary spacing. Thus, morphological conditions for oxygen supply are altered detrimentally due to expanding intercapillary distances and their increasing variability. The opposite is true in the case of volume overload hypertrophy. Finally, the main feature of cardiac atrophy is an increased heterogeneity of capillary spacing which signals worsening geometrical conditions for the oxygen supply.

It is of interest to note that geometrical conditions along the venular portions of capillaries appear to be more advantageous for oxygen transfer: smaller capillary domains and often less heterogeneity of capillary spacing. These conditions are usually also less affected in various experimental conditions in an attempt to compensate for a decreasing PO_2 in the venular portions of capillaries.

Implication of the Morphometric Findings for the Situation of Idiopathic Dilated Cardiomyopathy

A correlation between functional changes and pathological findings in idiopathic dilated cardiomyopathy has been described by several authors. Mall and coworkers [8] and Schwarz and coworkers [17] reported that decreasing ejection fraction is associated with increasing cell size and increasing degree of fibrosis as well as with the decreasing volume fraction of myofibrils. Pathological findings such as interstitial fibrosis, cellular oedema, increased numbers of lipid droplets, lysosome and lipofuscin granules as well as various mitochondrial alterations are predominantly located in the subendocardial regions [4, 5, 16, 21]. The most probable explanation for this is a decrease in the local oxygen supply-to-demand ratio [4]. In this case, altered geometry of the capillary bed may play an important role.

We are not aware of any comprehensive study dealing with morphometry of coronary capillaries in idiopathic dilated cardiomyopathy or in any other class of cardiomyopathy. Nevertheless, one can speculate that the geometrical conditions for oxygen transfer are impaired in this pathological unit. Recently, Mosseri and coworkers [10] reported an unchanged capillary-to-myocytes ratio in endomyocardial biopsy samples from patients with idiopathic dilated cardiomyopathy. In conjunction with the reports of increased cell size in this pathological situation, this may be interpreted as a sign of decreased capillary supply per tissue unit. The increased intercapillary distances are probably also associated with an increased heterogeneity of capillary spacing, judging from the reported disarray in the orientation of muscle fibres.

Summary and Conclusions

This report starts with a brief overview of newer and more discriminating methods used for evaluation of the geometry of the capillary bed. Special attention was focussed on the concept of capillary domain which enables us to evaluate the local capillary supply and, subsequently the heterogeneity of capillary spacing. The recent application of double-staining histochemical methods has made it possible to distinguish between the arteriolar and venular portions of capillaries and their respective domain areas. This method also facilitates the collection of longitudinal data.

A summary of quantitative results dealing with the geometry of coronary capillary networks based on the above methods was presented. These results were obtained during conditions of normal growth, in various types of cardiac hypertrophy and in cardiac atrophy. The results clearly demonstrate marked plasticity of coronary capillaries in the face of physiological and pathological challenge.

Final sections deal with the functional consequences of the described changes in capillary geometry, especially as related to oxygen transport and myocardial tissue PO_2. Possible implications of these findings for the situation of idiopathic dilated cardiomyopathy are also discussed. In this case a decrease in the local O_2 supply-to-demand ratio may play an important role.

References

1. Batra S, Rakusan K (1991) Geometry of capillary networks in volume overloaded rat heart. Microvasc Res 49:39–50
2. Batra S, Rakusan K (1992) Capillary network geometry during postnatal growth in the heart. Am J Physiol 262:H635–H640
3. Batra S, Rakusan K, Campbell SE (1991) Geometry of capillary networks in hypertrophied rat heart. Microvasc Res 41:29–40
4. Dick MR, Unverferth DV, Baba N (1982) The pattern of myocardial degeneration in nonischemic congestive cardiomyopathy. Hum Pathol 13:740–744
5. Figulla HR, Rahlf G, Nieger M, Luig H, Kreuzer H (1985) Spontaneous hemodynamic improvement or stabilization and associated biopsy findings in patients with congestive cardiomyopathy. Circulation 71:1095–1104
6. Legault F, Rouleau JL, Juneau C, Rose C, Rakusan K (1990) Functional and morphological characteristics of compensated and decompensated cardiac hypertrophy in dogs with chronic infrarenal aorto-caval fistulas. Circ Res 66:846–859
7. Lojda Z (1979) Studies on dipeptidyl (amino) peptidase IV (glycyl-proline naphtylamidase). Histochemistry 59:153–166
8. Mall G, Schwarz F, Derks H (1982) Clinicopathologic correlations in congestive cardiomyopathy. Virchows Arch 397:67–82
9. Mathieu O, Cruz-Orive LM, Hoppeler H, Weibel ER (1983) Estimating length density and quantifying anisotropy in skeletal muscle capillaries. J Microsc 131:131–146
10. Mosseri M, Schaper J, Admon D, Hasin Y, Gotsman MS, Sapoznikov D, Pickering JG, Yarom R (1991) Coronary capillaries in patients with congestive cardiomyopathy or angina pectoris with patent main coronary arteries. Circulation 84:203–210
11. Rakusan K, Turek Z (1985) Protamine inhibits capillary formation in growing rat hearts. Circ Res 37:393–399
12. Rakusan K, Turek Z (1987) Oxygen sources and oxygen sinks in myocardial hypertrophy. In: Dhalla NS, Singal PK, Beamish RE (eds) Pathophysiology of heart disease. Nijhoff, Boston, pp 83–92
13. Rakusan K, Flanagan MF, Geva T, Southern J, Van Praagh R (1992) Morphometry of human coronary capillaries during normal growth and the effect of age in left ventricular pressure-overload hypertrophy. Circulation 86:38–46
14. Rakusan K, Hrdina PW, Turek Z, Lakatta EG, Spurgeon HA, Wolford GD (1984) Cell size and capillary supply of the hypertensive rat heart. Basic Res Cardiol 79:389–395
15. Rakusan K, Wicker P, Abdul-Samad M, Healy B, Turek Z (1987) Failure of swimming exercise to improve capillarization in cardiac hypertrophy of renal hypertensive rat. Circ Res 61:641–647
16. Roberts WC, Ferrans VJ (1975) Pathologic anatomy of the cardiomyopathies. Hum Pathol 6:287–342
17. Schwarz F, Mall G, Zebe H, Blickle J, Derks H, Manthey J, Kübler W (1983) Quantitative morphologic findings of the myocardium in idiopathic dilated cardiomyopathy. Am J Cardiol 51:501–506
18. Tomanek RJ, Wessel T, Harrison DG (1991) Capillary growth and geometry during long-term hypertension and myocardial hypertrophy in dogs. Am J Physiol 261:H1011–H1018
19. Turek Z, Hoofd L, Batra S, Rakusan K (1992) The effect of realistic geometry of capillary networks on tissue PO_2 in hypertrophied rat heart. In: Oxygen supply to tissue XIV Plenum, New York (in press)

20. Turek Z, Rakusan K, Olders J, Hoofd L, Kreuzer F (1991) Computed myocardial PO_2 histograms: effects of various geometrical and functional conditions. J Appl Physiol 70:1845–1853
21. Unverferth DV, Baker PB, Swift SE, Chaffee R, Fetters JK, Uretsky BF, Thompson ME, Leier CV (1986) Extent of myocardial fibrosis and cellular hypertrophy in dilated cardiomyopathy. Am J Cardiol 57:816–820
22. Veprek P, Batra S, Korecky B, Rakusan K (1992) Geometry of capillary networks in atrophied rat heart (in preparation)

Microcirculation in the Cardiomyopathic Syrian Hamster and in Human Congestive Cardiomyopathy

S.M. Factor

Introduction

The inbred cardiomyopathic Syrian hamster (CSH) is an excellent model of human congestive cardiomyopathy, with well-characterized pathological features. Since its original description by Homburger et al. [19], the hamster has provided insight into the pathogenesis of myocardial injury common to many other animal models of cardiomyopathy, as well as that characteristic of human disease. The CSH develops cardiomyopathy and skeletal myopathy as a result of an autosomal recessive trait that is phenotypically expressed in 100% of affected animals [33]. Although the specific genetic defect remains undefined (and it is generally assumed that most human congestive cardiomyopathy is not genetically determined), the fact that the CSH develops cardiac damage in a predictable sequence has permitted extrapolation to human disease.

The CSH dies of congestive heart failure or arrhythmia at approximately 200–400 days of life, or one-third to one-half the normal life expectancy. Yet myocardial damage begins insidiously at 40–75 days of age, progresses actively for another 3–4 months, and then wanes [9]. Concurrently, the remaining viable myocardium undergoes compensatory hypertrophy so that the total cardiac mass increases despite the loss of 5%–12% of the left ventricular myocardium [14]. After about 250 days of age the compensatory hypertrophy stage abates, the scarred left ventricle dilates with mural thinning, and the animal begins to manifest signs of congestive heart failure (peripheral and pulmonary edema, and ascites). Thus, the animal progresses from a cardiac damage phase to one of compensation before overt congestive heart failure ensues. This mimics the sequence of events in many human patients in which stages of subclinical injury and compensation may precede the clinical manifestations of congestive heart failure. Human cardiomyopathic changes such as scarring and myocellular hypertrophy may be identified at postmortem examination following sudden unexpected death or accidental demise, or by a fortuitous endomyocardial biopsy for the diagnosis of unexplained cardiac complaints (e.g., arrhythmia or chest pain), well before there are any signs or symptoms of significant heart failure.

Department of Pathology, Albert Einstein College of Medicine, 1300 Morris Park Avenue, Bronx, NY 10461, USA

Figulla et al. (Eds.)
Idiopathic Dilated Cardiomyopathy

Thus, one can often separate the events leading to myocardial injury from the clinical manifestations of congestive heart failure, wherein it is recognized that the latter represents the systemic response to ventricular failure [11]. Such insights, applicable to clinical cardiomyopathy, have been derived to a great extent from analysis of the disease in the CSH.

Pathogenesis of Hamster Cardiomyopathy

The end-stage of hamster cardiomyopathy has many features in common with human idiopathic dilated cardiomyopathy: multifocal scarring (although in the hamster, as in many small species the scarred foci tend to calcify) with localization predominantly in the mid-ventricular wall and subepicardium, myocyte hypertrophy, and ventricular mural thinning with a dilated cavity [9]. The multifocal scarring results from discrete areas of myocytolytic necrosis, a lesion that begins as contraction band injury and progresses to a stage of myocyte destruction that elicits a mononuclear inflammatory response, fibroblastic proliferation, and connective tissue deposition.

Role of Catecholamines

Myocytolytic necrosis is thought to be a lesion that derives from the influence of endogenous or exogenously administered catecholamines or other vasoactive mediators. The relationship of catecholamines to hamster cardiomyopathy has been well established. Bajusz et al. [1] originally suggested this association, based on the development of fulminant myocardial necrosis, congestive heart failure, and death in young hamsters by using epinephrine in doses that did not affect control animals. Lossnitzer et al. [24] used isoproterenol and showed an increased calcium uptake in the hearts of CSH, at a time prior to the onset of cardiomyopathy. Urine norepinephrine was found to be elevated even in juvenile CSH [20], and propranolol was shown to be effective in preventing damage from exogenously administered catecholamines [1], although it did not prevent the development of the spontaneous disease [20]. Sole and colleagues [31, 32] found that cardiac norepinephrine turnover was increased in the CSH, even in the earliest stages of the heart disease, and that dopamine was increased in the failing hamster heart. Lund et al. [25] have shown an increase of some of the enzymes required for norepinephrine synthesis in the hearts of prenecrotic hamsters. Other indirect evidence for the role of catecholamines in hamster cardiomyopathy comes from the work of Karliner and colleagues [22] who showed increased numbers of α_1- and β-adrenoreceptors in hearts of CSH as compared to controls.

Role of Calcium

Despite the strong circumstantial evidence that hamster cardiomyopathy is induced by catecholamines through direct effects of these substances on the heart, there is also equally compelling evidence that abnormalities of calcium metabolism may play a role. It is important to emphasize, however, that there may be an interplay between catecholamine effects and calcium metabolism, so that abnormalities of each are not mutually exclusive. By the same token, both catecholamines and calcium may interact at the myocyte level through the microcirculation, to eventuate in myocyte damage, as discussed below.

Lossnitzer and colleagues [24] initially suggested that disturbed myocardial calcium metabolism might be the cause of hamster cardiomyopathy. Wrogemann and Nylen [38] demonstrated mitochondrial calcium loading in the CSH, a finding that might be explained by the increased mitochondrial calcium seen in numerous species following coronary reperfusion-induced injury to myocytes. Evidence also exists to suggest that there is a generalized membrane defect in the CSH that may make myocytes susceptible to the effects of increased trans-sarcolemmal calcium flux [21, 28]. The defect may exist at the level of the calcium channel, as two independent groups have shown increases in the hearts of affected CSH [17, 36]. Other membrane abnormalities that may contribute to or, in fact, be the primary genetically determined defect, include observations of differences in sarcolemmal lipid [30]. Morphologically, several studies have noted differences in the freeze-fracture appearance of intramembrane particles in the cardiac sarcolemma of CSH and control animals [2, 18].

Role of the Microcirculation

Despite the strong evidence that this genetically determined cardiomyopathy is associated with abnormalities of catecholamine and calcium metabolism, one theoretical consideration and one observational fact cannot be explained directly by the aforementioned findings: (a) a genetic disease, whether inducing an increased sensitivity to catecholamines or a sarcolemmal membrane defect leading to increased permeability for calcium, should have the same phenotypic response in every cardiac cell; but (b) the pathology of myocardial injury in the CSH, like that in most human congestive cardiomyopathies, is that of *focal* necrosis and scarring, not the diffuse lesions anticipated with a generalized genetic defect. This paradox suggested to us a decade ago that we should focus on the microcirculation, where abnormalities could induce focal myocyte injury. Both direct and indirect evidence derived from work in our laboratory now support this concept, as described below. Furthermore, there is additional evidence from studies of other models of human cardiomyopathy that similar microcirculatory abnormalities may lead to focal myocyte necrosis, and, by inference, similar pathogenetic events may occur in human cardiomyopathies of diverse etiology, but characterized by focal pathology.

Microcirculation in the Cardiomyopathic Hamster. We have shown that the microvasculature of mammalian hearts is organized into an end-capillary network, with a discrete volume of myocytes supplied by nonanastomosing precapillary arterioles [7, 10, 27]. Thus, anatomical (e.g., thrombi or arteriolopathy) or functional (e.g., spasm) lesions of the terminal arterioles or capillaries could lead to damage or necrosis of myocytes supplied by these vessels. Obstruction of precapillary arterioles with microspheres 25 μm in diameter leads to necrosis of myocytes in the canine or rat heart; these lesions have a fairly constant diameter of about 200 μm in greatest dimension, although larger lesions can develop secondary to confluence of smaller foci [4]. Although we have not directly studied the anatomy of the hamster heart for the presence of an end-capillary network due to technical limitations, there appears to be no reason to assume that it is any different than other species which we have studied directly. Furthermore, the size of the myocytolytic foci that develop during the active necrotizing and healing stages of the disease in the CSH suggest comparability to other species such as dog or rat where we have specifically caused myocytolysis by embolizing the precapillary arterioles. Finally, there is no evidence that the microcirculation or larger vessels in the CSH has any pathoanatomic lesions [34], whether associated with myocytolytic foci or distant from them (Fig. 1). Thus, if the microcirculation plays a role in the development of myocyte necrosis in the CSH, functional abnormalities, including spasm, may be to blame.

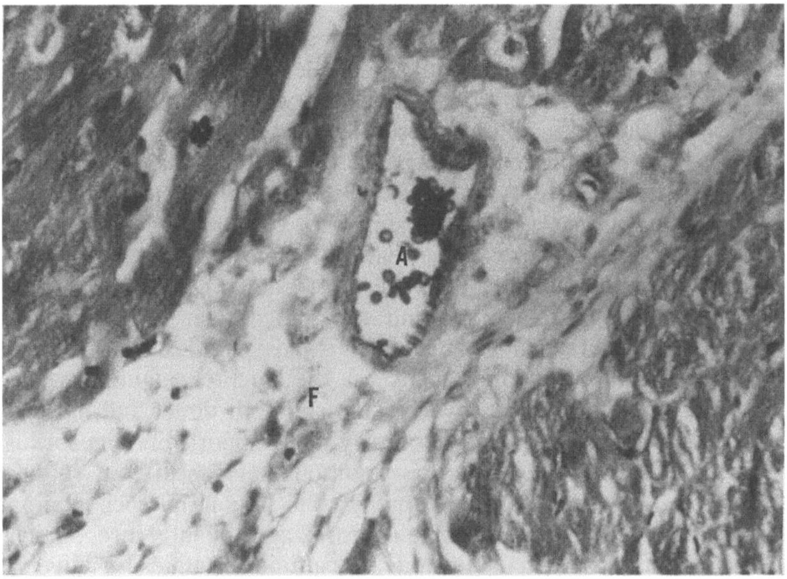

Fig. 1. A section of left ventricle from a cardiomyopathic Syrian hamster. There is a fibrotic scar (*F*) sharply demarcated from the surrounding morphologically normal myocardium. This scarred region represents a healed myocytolytic focus. The central small arterial blood vessel (*A*) within the affected area is completely normal, with no evidence of luminal occlusion or narrowing. Hematoxylin-eosin, original magnification ×80

Microvascular Spasm in the Hamster

To study the microcirculation in the CSH we carried out a series of investigations with in vivo coronary perfusion of beating hearts just prior to sacrifice. We employed silicone rubber solutions which completely fill the arterial side of the circulation down to capillaries and then enter the veins. The silicone rubber hardens in situ after about 15 min, producing an intravascular cast of the vessel structure at the time of death. After retrograde coronary perfusion of the silicone rubber through the root of the aorta and cessation of heart beat the heart is removed, fixed, and chemically "cleared." The latter leaves the anatomy intact but makes the optically opaque silicone rubber visible against a transparent background of tissue. Thus, in thick sections (several hundred micrometers) of myocardium, the vasculature can be studied comparable to a microangiogram, if it could be seen radiographically.

With this technique, we studied hamsters at all stages of disease, even in the prenecrotic phase, and we demonstrated significant abnormalities not seen in age-matched controls. Specifically, we identified changes in arteriolar-sized vessels consistent with patterns of microvascular spasm [9]. These patterns included segments of vascular constriction, both individual and multiple, areas of diffuse vessel narrowing, and luminal irregularity

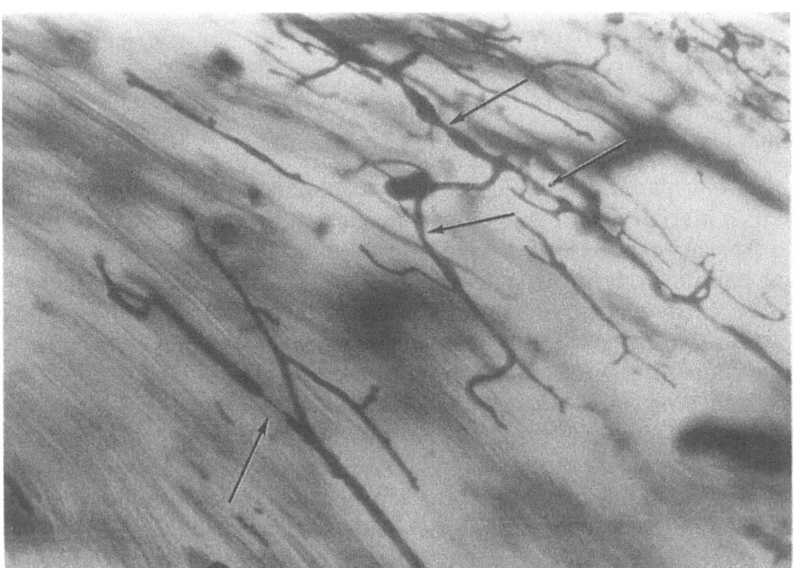

Fig. 2. A section of left ventricle from an untreated cardiomyopathic hamster after perfusion of the microcirculation with silicone rubber. The tissues have been cleared to allow visualization of the vasculature. Note the numerous vascular irregularities, spastic narrowings, and dilatations (*arrows*) consistent with a hyperreactive coronary microcirculation. Such lesions were abolished with appropriate therapy (see text), myocardial injury was ameliorated, and cardiomyopathy was prevented. Cleared, unstained tissue, original magnification ×30

(Fig. 2). Focal aneurysmal dilatations were also seen adjacent to constricted zones. Since this technique provides an equivalent of a "stop-action" view of the microcirculation at the time of death, it does not establish the temporal duration of the spastic lesions, nor does it permit any direct correlation with a vascular lesion and a focus of myocyte necrosis. Presumably, however, it does indicate that the microvasculature of the CSH is inordinately sensitive to stimulation and can undergo spastic changes that theoretically could lead to reperfusion injury of myocytes.

Of course it could be argued that the changes seen were artifacts of manipulation, silicone rubber perfusion, fixation, or clearing. Although these criticisms are cogent, they fail to account for the absence of such changes in control animals handled similarly. Despite this, we elected to study the microvasculature in vivo by using the well-characterized cremaster muscle preparation [3]. Cardiomyopathic hamsters were studied during the active stage of necrosis and during the healing, hypertrophy stage of the disease, and were compared to age-matched controls. With topically applied norepinephrine in the CSH, second-order arterioles constricted to 35%–50% of their resting internal diameter and at significantly lower concentrations of agonist than the controls. This indicated both a reduced threshold to norepinephrine and an enhanced response to the agonist. Of interest in the older animals in the absence of active myocyte necrosis in the skeletal muscle or heart, the cremaster microcirculation showed no differences in arteriolar responsiveness to applied norepinephrine. These results support a direct relationship of a hyperreactive microcirculation and active necrosis in this model, possibly related to catecholamines.

Prevention of Microvascular Spasm and Myocyte Injury

If the arteriolar pathology noted above is not simply an epiphenomenon but represents a marker of a real pathogenetic mechanism, prevention of microvascular spasm should lead to the prevention of myocyte necrosis and eventual cardiomyopathy despite the presence of the underlying genetic defect. We carried out two sets of studies in both the acute and chronic stages of disease. The calcium channel blocking agent verapamil given to young hamsters for 2 weeks at the onset of the disease completely prevented myocardial necrosis and ameliorated microvascular spasm [9]. To rule out the possible effects of verapamil on calcium flux across the myocyte sarcolemmal membrane rather than on the microvasculature, we elected to use the α_1-adrenergic antagonist prazosin because of its specificity at the vascular level. Treatment of CSH for 2 weeks, again during the active stage of necrosis, led to virtually complete absence of necrotic lesions (0.66% in treated versus 8.8% in untreated). This strongly supported the concept that the microvasculature was the primary site of abnormality leading to myocyte necrosis and the subsequent development of cardiomyopathy.

Although the acute studies indicated that short-term prevention of microvascular spasm could eliminate myocyte necrosis, it still could not eliminate the possibility that the genetically determined cardiomyopathy resulted from some cellular metabolic defect independent of myocyte loss caused by the vasculature. To address this possibility, CSH were treated for 7–8 months with oral verapamil therapy, to a point where they should have manifested early heart failure or death. Chronic treatment led to a highly significant difference in the percentage of ventricular necrosis (1.73% treated versus 12.45% untreated); additionally, contractile protein and myosin isoenzyme patterns were normalized by the verapamil treatment [14]. Thus, these observations indicate clearly that the phenotypic manifestations of this genetically determined disease can be prevented by inhibiting cardiac necrosis elicited through microvascular spasm.

Several other studies from different laboratories buttress these arguments. Rouleau and colleagues [29] showed that verapamil preserved cardiac contractility in the CSH. Wikman-Coffelt et al. [37] demonstrated that varapamil preserved the adenine nucleotide pool in the cardiomyopathic animals. Kobayashi et al. [23] also showed that verapamil could prevent long-term damage in the CSH, although they followed the animals only to 90 days of age. Most relevant to the question of the role of the microvasculature and the development of cardiomyopathy is the study of Figulla and associates [16], in which they showed the presence of inhomogeneous capillary flow in the affected hearts and the prevention of this change by verapamil and hydralazine. Inhomogeneous capillary flow is precisely what one would expect in the presence of microvascular spasm affecting focal areas of the microcirculation.

Microvascular Spasm in Other Models of Human Cardiomyopathy

As previously noted, the cardiomyopathic hamster is relatively unusual as a model of human congestive cardiomyopathy since the disease is caused by some as a yet undefined genetic abnormality. To establish that the alterations in the hamster are not unique and to relate abnormalities of the microvasculature to clinical cardiomyopathy, one must either study these vessels in vivo in human hearts or establish the generality of microvascular pathology in other examples of cardiomyopathy and extrapolate to the clinical situation. We have identified three models typified by focal myocardial pathology in which microvascular spasm appears to play a significant role and can be ameliorated or prevented with appropriate therapy.

As described in an earlier section, embolization of microspheres into the microcirculation of dogs or rats leads to the development of microscopic foci of myocardial necrosis [4]. Although we originally considered that these foci were the result of obstruction of the precapillary arterioles by microspheres, the morphology of the lesions suggested a reperfusion type of injury (e.g., contraction bands). Accordingly, drugs to prevent microvascular spasm were administered and were found to be effective in preventing necrosis, despite the continued presence of microspheres in the arterioles. Both prazosin and verapamil prevented lesion development, as they did in the hamster. Thus, we concluded that the microspheres elicited a vasoreactive response which then led to transient vessel occlusion and reperfusion injury to myocytes.

In a second model, the hypertensive and diabetic rat, the pathology is characterized by multifocal necrosis, interstitial and perivascular fibrosis, and cardiac hypertrophy [8]. Although in this model, as in human disease, there are morphological alterations of the microvasculature (e.g., basement membrane abnormalities), the determination of functional changes required a study with the silicone rubber perfusion technique. When we carried out such a study [12], we demonstrated the presence of numerous areas of microvascular tortuousity, focal constriction, and microaneurysm formation. Similar but quantitatively less frequent abnormalities were also present in hypertensive animals and in diabetic animals, but there was a significant difference when compared to the hypertensive and diabetic animals with severe cardiomyopathy. Of note, although we have not directly examined the microvasculature following calcium channel blocking agent therapy, we recently showed a marked reduction of mortality in cardiomyopathic animals treated with diltiazem at levels of therapy that did not affect blood pressure levels [15].

Finally, in the mouse infected with *Trypanosoma cruzi*, a model of human Chagas' disease, we have shown the presence of coronary microcirculatory abnormalities in acute disease [13]. Here too, in this model typified by focal pathology, the microvasculature shows the presence of spasm and other lesions similar to those seen in the hamster. Again, calcium channel blockade with verapamil significantly ameliorates the development of pathology, along with biochemical and clinical manifestations of disease [26, 35].

Thus, it appears that in several models of acquired cardiac damage and cardiomyopathy, the microcirculation shows evidence of hyperreactivity with features of spasm similar to that which we initially identified in the hamster model. Furthermore, in all of these acquired models comparable to the genetically determined hamster cardiomyopathy, calcium channel blockade and/or α_1-adrenergic blockade either prevents or lessens the degree of myocardial pathology.

Microvascular Spasm in Human Cardiomyopathy

There is little direct evidence for the existence of microvascular spasm or other microvascular lesions in human congestive cardiomyopathy. In postmortem perfusions of silicone rubber into diabetic hearts we have shown the presence of microaneurysms and other lesions [6]; however, whether these lesions are secondary to spasm cannot be determined. As noted above, several of the models of human cardiomyopathy, including hypertension and diabetes, and Chagas' disease with features very similar to human disease do have spasm, but this is only indirect evidence. Patients with congestive cardiomyopathy may have atypical chest pain consistent with ischemia in the absence of epicardial coronary lesions, suggesting the presence of intramyocardial vascular disease. Such patients may also have atypical thallium perfusion scans with features of focal ischemia. Patients with alcoholic cardiomyopathy may have morphologically abnormal small vessels [5], but these lesions cannot be extrapolated to indicate the existence of functional microvascular disease. Thus, until a more specific test can be developed to establish with certainty the presence of microvascular abnormalities in human cardiomyopathy, such lesions only can be considered possible, although highly probable based on the evidence developed to date.

Conclusions

There is now a large body of data to support the conclusion that cardiomyopathy in the Syrian hamster is pathogenetically related to microvascular spasm. Although the underlying genetic defect that causes the spasm is unknown, it is most likely some abnormality in cell membranes that could make vascular smooth muscle increasingly sensitive to catecholamines and/or calcium, and could make groups of myocytes supplied by small intramyocardial vessels increasingly sensitive to the effects of transient ischemia. Although we cannot absolutely disregard the possibility that myocytes may be adversely affected directly by calcium or catecholamines, the fact that abolition of microvascular spasm eliminates or lessens myocardial damage strongly suggests a major role for this pathogenetic sequence of events. Further, the similarity of so many models devoid of a genetic predisposition to that of hamster cardiomyopathy, with the presence of focal myocardial pathology and an association with preventable microvascular spasm, supports a role of the microvasculature in the genesis of myocardial injury. A common denominator of all of these models is the focal nature of the myocardial pathology. It awaits to be seen whether all cardiac disease typified by small regions of focal myocardial injury, including that seen in human congestive cardiomyopathy, are manifestations of microvascular abnormalities.

References

1. Bajusz E, Homburger F, Baker JR, Bogdonoff P (1969) Dissociation of factors influencing myocardial degeneration and circulatory failure. Ann NY Acad Sci 156:397–420
2. Berry B, Poulsen R, Yunge L et al. (1983) Numerical densities of intramembrane particles in the cardiac sarcolemma of normal and myopathic Syrian hamsters. J Mol Cell Cardiol 15:503–513
3. Conway RS, Factor SM, Sonnenblick EH, Baez S (1987) Microvascular reactivity of the myopathic Syrian hamster cremaster muscle. Cardiovasc Res 11:796–803
4. Eng C, Cho S, Factor SM, Sonnenblick EH, Kirk ES (1984) Myocardial micronecrosis produced by microsphere embolization. Role of alpha-adrenergic tonic influence on the coronary microcirculation. Circ Res 54:74–82
5. Factor SM (1976) Intramyocardial small vessel disease in chronic alcoholism. Am Heart J 92:561–575
6. Factor SM, Okun EM, Minase T (1980) Capillary microaneurysms in the human diabetic heart. N Engl J Med 302:384–388
7. Factor SM, Okun EM, Kirk ES (1981) The histological lateral border of acute canine myocardial infarction: a function of microcirculation. Circ Res 48:640–649

8. Factor SM, Bhan R, Minase T, Wolinsky H, Sonnenblick EH (1981b) Hypertensive-diabetic cardiomyopathy in the rat: an experimental model of human disease. Am J Pathol 102:219–228
9. Factor SM, Minase T, Cho S, Dominitz R, Sonnenblick EH (1982) Microvascular spasm in the cardiomyopathic Syrian hamster: a preventable cause of focal myocardial pathology. Circulation 66:342–354
10. Factor SM, Okun EM, Minase T, Kirk ES (1982) The microcirculation of the human heart: end-capillary loops with discrete perfusion fields. Circulation 66:1241–1248
11. Factor SM, Sonnenblick EH (1983) Microvascular spasm as a cause of cardiomyopathies. Cardiovasc Rev Rep 4:1177–1182
12. Factor SM, Minase T, Cho S, Fein F, Capasso JM, Sonnenblick EH (1984) Coronary microvascular abnormalities in the hypertensive-diabetic rat. A cause of cardiomyopathy? Am J Pathol 116:9–20
13. Factor SM, Cho S, Wittner M, Tanowitz H (1985) Abnormalities of the coronary microcirculation in acute murine Chagas' disease. Am J Trop Med Hyg 34:246–253
14. Factor SM, Malhotra A, Cho S, Scheuer J, Sonnenblick EH, Capasso JM (1988) Prevention of hereditary cardiomyopathy in the Syrian hamster with chronic verapamil therapy. J Am Coll Cardiol 12:1599–1604
15. Fein FS, Cho S, Malhotra A, Akella J, van Hoeven KH, Sonnenblick EH, Factor SM (1991) Beneficial effects of diltiazem on the natural history of hypertensive diabetic cardiomyopathy in rats. J Am Coll Cardiol 18:1406–1417
16. Figulla HR, Vetterlein F, Glaubitz M, Kreuzer H (1987) Inhomogeneous capillary flow and its prevention by verapamil and hydralazine in the cardiomyopathic Syrian hamster. Circulation 76:208–216
17. Finkel MS, Marks ES, Patterson RE, Speir EH, Steadman K, Keiser HR (1986) Increased cardiac calcium channels in hamster cardiomyopathy. Am J Cardiol 57:1205–1206
18. Graham KA, Shivers RR, Atkinson BG (1984) A freeze-fracture analysis of intramembrane particle densities on dystrophic hamster heart sarcolemma. Muscle Nerve 7:513–523
19. Homburger F, Baker JR, Nixon CW, Whitney R (1962) Primary generalized polymyopathy and cardiac necrosis in an inbred line of Syrian hamsters. Med Exp 6:339–345
20. Jasmin G, Solymoss B, Proschek L (1979) Therapeutic trials in hamster dystrophy. Ann NY Acad Sci 317:338–348
21. Jasmin G, Proschek L (1984) Calcium and myocardial injury: an appraisal in the cardiomyopathic hamster. Can J Physiol Pharmacol 18:567–572
22. Karliner JS, Alabaster C, Stephens H, Barnes P, Dollery C (1981) Enhanced noradrenaline response in cardiomyopathic hamsters: possible relation to studies in adrenoceptors studied by radioligand binding. Cardiovasc Res 15:296–304
23. Kobayashi A, Yamashita T, Kaneko M, Nishiyama T, Hayashi H, Yamazaki N (1987) Effects of verapamil on experimental cardiomyopathy in the BIO 14.6 Syrian hamster. J Am Coll Cardiol 10:1128–1134
24. Lossnitzer K, Janke J, Hein B, Stauch M, Fleckenstein A (1975) Disturbed myocardial calcium metabolism: a possible pathogenetic factor in the hereditary cardiomyopathy of the Syrian hamster. In: Fleckenstein A, Rona G (eds) Recent advances in the study of cardiac structure and metabolism. University Park Press, Baltimore, pp 207–217
25. Lund DD, Schmid PG, Bhatnager RK, Roskoski R (1982) Changes in parasympathetic and sympathetic neurochemical indices in hearts of myopathic hamsters. J Auton Nerv Sys 5:237–246
26. Morris SA, Weiss L, Factor SM, Bilezikian JP, Tanowitz H, Wittner M (1989) Verapamil ameliorates clinical, pathological and biochemical manifestation of experimental Chagasic cardiomyopathy in mice. J Am Coll Cardiol 14:782–789
27. Okun EM, Factor SM, Kirk ES (1979) End-capillary loops in the heart: an explanation for discrete myocardial infarctions without border zones. Science 206:565–567
28. Panagia V, Singh JN, Anand-Srivastava MB, Pierce GN, Jasmin G, Dhalla NS (1984) Sarcolemmal alterations during the development of genetically determined cardiomyopathy. Cardiovasc Res 18:567–572

29. Rouleau JL, Chuck LHS, Hollosi G et al. (1982) Verapamil preserves myocardial contractility in the hereditary cardiomyopathy of the Syrian hamster. Circ Res 50:405–412
30. Slack BE, Boegman RJ, Downie JW, Jasmin G (1980) Cardiac membrane cholesterol in dystrophic and verapamil-treated hamsters. J Mol Cell Cardiol 12:179–185
31. Sole MJ, Lo C-M, Laird CW, Sonnenblick EH, Wurtman RJ (1975) Norepinephrine turnover in the heart and spleen of the cardiomyopathic Syrian hamster. Circ Res 37:855–862
32. Sole MJ, Helke CJ, Jackowbowicz DM (1982) Increased dopamine in the failing hamster heart: transvesicular transport of dopamine limits the rate of norepinephrine synthesis. Am J Cardiol 49:1682–1690
33. Sole MJ, Factor SM (1985) Hamster cardiomyopathy: a genetically-transmitted sympathetic dystrophy? In: Beamish RE, Panagia V, Dhalla NS (eds) Pathogenesis of stress-induced heart disease. Nijhoff, Boston, pp 34–43
34. Strobeck JE, Factor SM, Bhan A et al. (1979) Hereditary and acquired cardiomyopathies in experimental animals: mechanical, biochemical, and structural features. Ann NY Acad Sci 317:59–88
35. Tanowitz HB, Morris SA, Weiss LM, Bilezikian JP, Factor SM, Wittner M (1989) Effect of verapamil on the development of chronic experimental Chagas' disease. Am J Trop Med Hyg 41:651–657
36. Wagner JA, Reynolds IJ, Weisman HF, Dudeck P, Weisfeldt ML, Snyder SH (1986) Calcium antagonist receptors in cardiomyopathic hamster: selective increases in heart, muscle, brain. Science 232:515–518
37. Wikman-Coffelt J, Sievers R, Parmley WW, Jasmin G (1986) Verapamil preserves adenine nucleotide pool in cardiomyopathic Syrian hamster. Am J Physiol 250:H22–H28
38. Wrogemann K, Nylen EG (1978) Mitochondrial calcium overloading in cardiomyopathic hamsters. J Mol Cell Cardiol 10:185–195

The Microcirculation in Cardiac Hypertrophy

R.O. Cannon, III

Left ventricular hypertrophy has long been associated with increased risk of fatal and nonfatal cardiovascular events. In the Framingham experience, individuals with EKG evidence of left ventricular hypertrophy had an approximately five fold increased risk of death, new onset of angina pectoris, or nonfatal myocardial infarction [23]. Recent studies have shown that over one-third of hypertensive patients have measurable increases in left ventricular mass, even if the EKG is normal [13, 15]: these patients also represent a high-risk group for cardiovascular events [11, 26]. Several observations in humans with left ventricular hypertrophy, whether primary (e.g., hypertrophic cardiomyopathy) or secondary (e.g., hypertension, aortic stenosis) suggest that myocardial ischemia may mediate many of these events, such as complaints of anginalike chest pain, often with abnormal noninvasive testing similar to that seen in patients with coronary artery disease [9, 19, 28, 53]. Because epicardial coronary arteries commonly appear normal during angiography, consideration has been made that the coronary microcirculation may be abnormal in structure or function in patients with left ventricular hypertrophy.

Animal Models of Left Ventricular Hypertrophy

Several studies have shown coronary flow reserve of hypertrophied hearts in response to a potent arteriolar dilator such as adenosine to be less than nonhypertrophied hearts [32, 34, 38]. Limited coronary flow reserve could be a consequence of elevated resting absolute flow requirements due to increased oxygen demands of pressure-loaded hypertrophy, with partial exhaustion of arteriolar vasodilator capacity, and/or limitation in either absolute or relative (to the mass of the myocardium) maximum vasodilator capacity. Indeed, several studies suggest that impairment in peak flow capacity may be particularly important in left ventricular hypertrophy. Morphometric and microsphere studies suggest absent or insufficient neovascularization in response to the hypertrophic response to pressure load [37, 46, 54]. This could lead to insufficient oxygen delivery during stress as

Cardiovascular Diagnosis Section Cardiology Branch, NHLBI National Institutes of Health
Bethesda, MD 20892, USA

Figulla et al. (Eds.)
Idiopathic Dilated Cardiomyopathy

a consequence of a relative paucity of capillaries, with attendant increases in diffusion distances [17]. Intrinsic vascular and perivascular alterations accompanying left ventricular hypertrophy have also been described, including perivascular and intravascular fibrosis, and medial hypertrophy and hyperplasia [2, 3, 24, 55]. Consequences of these morphologic alterations might include compromise of the vascular lumen, limitations in vasodilator capacity, and heightened sensitivity to neurohumoral constrictor influences. Experimental imposition of left ventricular pressure overload in older animals may produce more coronary microvascular changes than hypertrophy in contrast to younger animals [21, 54]. Dogs with long-standing pressure-induced left ventricular hypertrophy have also been shown to have abnormal transmural distribution of coronary flow with relative deficiency in subendocardial flow [18]. Impaired autoregulation, particularly at lower systemic pressures, has been demonstrated in experimental left ventricular hypertrophy [16]. Finally, coronary occlusion in dogs with experimental left ventricular hypertrophy results in larger infarcts for a given risk area and higher mortality than coronary occlusion in dogs without left ventricular hypertrophy [25].

Hypertrophic Cardiomyopathy

However elegant the experimental design of animal models of left ventricular hypertrophy, extrapolation of these results to the awake, ambulatory, noninstrumented human must obviously be made with caution. Indeed, with hypertrophic cardiomyopathy , a familial (in most cases) disease of unexplained hypertrophy with often massive thickening of various wall segments (especially the septum) [30], no animal model exists. Myocardial ischemia is common in this disease, even in asymptomatic individuals, and may be demonstrated by myocardial lactate production during pacing or isoproterenol stress and reversible thallium perfusion defects during exercise stress testing [6, 7, 9, 33]. Thallium perfusion defects may be regional and subendocardial (manifest as an apparent cavity dilatation postexercise in the absence of actual chamber enlargement), the latter commonly associated with elevated left ventricular filling pressure during stress and increased lung uptake of thallium, suggesting pulmonary congestion [9].

Myocardial ischemia in hypertrophic cardiomyopathy undoubtedly has multiple mechanisms, given the complex pathophysiology of this disease, including elevated resting and stress-related coronary flow requirements, especially in those patients with elevated left ventricular systolic pressures as a consequence of outflow obstruction and limitation in appropriate flow delivery [6, 7]. Delayed and incomplete diastolic relaxation is the pathophysiologic hallmark of this disease and could limit the rate and magnitude of coronary flow into the myocardium, particularly the subendocardium, during diastole. Abnormal myocellular-capillary relationships, the histologic hallmark of this disease, may limit appropriate oxygen delivery during stress [29]. Abnormal intramural coronary arteries, often with striking medial and intimal thickening, may limit appropriate coronary flow during stress. As a consequence of these abnormalities affecting both myocardial oxygen demand and coronary flow capacity, coronary flow reserve is often severely compromised in both hypertrophied and nonhypertrophied regions [5]. Studies of myocardial metabolism during pacing stress at our institution have shown that despite myocardial ischemia during pacing stress, myocardial oxygen extraction often paradoxically decreases in this disease [7], in contrast to increased myocardial oxygen extraction during pacing stress in patients with coronary artery disease. The cause of this phenomenon is unclear, but potential mechanisms include shunting of oxygen-rich blood directly into the coronary venous drainage because of compressive effects of elevated filling pressures on the subendocardial microcirculation with opening of potential arterial-venous shunts, or vasodilatation of epicardial vessels with flow in excess of epicardial myocardial demands despite subendocardial ischemia.

Verapamil has been shown to improve symptom status, prolong effort tolerance, and improve thallium perfusion abnormalities of patients with hypertrophic cardiomyopathy [1, 49], with multiple pharmacologic actions that could decrease ischemia, including a reduction in left ventricular outflow obstruction, improvement in diastolic relaxation, and small vessel vasodilatation. Operative relief of left ventricular outflow obstruction results in a reduction in rest and stress-related coronary flow requirements, in association with a reduction in left ventricular systolic pressure, and reduction in the magnitude and prevalence of ischemia during stress [8]. More recently we have found that regional and subendocardial thallium abnormalities, so common in this disease, are substantially reduced or eliminated following operative relief of left ventricular outflow tract obstruction [10]. The impact of a newer technique for reducing left ventricular outflow tract obstruction, dual chamber pacing, on metabolic and thallium scintigraphic markers of myocardial ischemia is unknown at present.

Pressure-Induced Left Ventricular Hypertrophy in Humans

Studies investigating the impact of left ventricular hypertrophy in animal models may provide more insight as to the consequences of pressure-induced left ventricular hypertrophy in humans. Unfortunately, techniques for measuring the magnitude of left ventricular hypertrophy and coronary blood flow in humans are less precise than those used in animal models. For example, thermodilution measurements of coronary venous outflow can be affected by right atrial reflux and catheter movement and may underestimate the magnitude of high flows after pharmacologic stimulation. Inert gas saturation or washout techniques require multiple blood samples over a period of time, which make measurement of peak flows difficult after a drug such as dipyridamole that has variable time of onset, duration of maximum effect, and relative potency in humans. Also, patients are commonly on therapy prior to study, with potential residual effects on coronary flow dynamics despite cessation of drugs at several pharmacologic half-lives before performance of the study. Coronary venous sampling for metabolic studies gives only an average of what are undoubtedly transmural and possibly regional differences in myocardial energetics in pressure-induced left ventricular hypertrophy. Finally, assessment of coronary microvascular morphology in living humans is limited to endocardial biopsies, which sample only the smallest intramyocardial vessels.

Despite these caveats, several studies suggest that although resting coronary flow per unit mass may be normal in hypertensive patients with left ventricular hypertrophy and normal systolic function, coronary flow reserve may be limited. Strauer performed studies of coronary flow (inert gas technique) in 114 patients with hypertension, and found significant reduction in coronary flow reserve in response to dipyridamole 0.5 mg/kg, compared to normotensive controls [43]. Strauer has emphasized the importance of wall stress in determining not only basal coronary flow requirements but flow reserve as well [44]. Thus the greater the wall stress, the lower is the flow reserve, presumably because of partial exhaustion of vasodilator capacity due to elevated coronary flow requirements. Opherk et al. reported that 16 hypertensive patients with chest pain, normal coronary angiograms, and EKG evidence of left ventricular hypertrophy had a significantly higher minimum coronary resistance following dipyridamole 0.5 mg/kg (flow measured by inert gas method) compared to normotensive controls [35]. Left ventricular biopsies performed in seven patients showed myocyte hypertrophy, but no vascular abnormalities or interstitial fibrosis.

Vogt et al. measured coronary flow reserve (inert gas method), following dipyridamole 0.5 mg/kg in 54 hypertensive patients with chest pain and normal coronary angiograms [50]. They were found to have resting coronary flow and myocardial oxygen consumption per unit mass similar to 12 normotensive controls. However, there was a significant reduction in the flow response to dipyridamole, with elevation of minimum coronary resistance compared to the normotensive controls. The authors found no significant correlation between minimum coronary resistance or coronary reserve and the left ventricular muscle mass or diastolic wall stress, concluding that coronary microvascular abnormalities might contribute to impaired coronary flow reserve independent of left ventricular hypertrophy. Houghton et al. performed intracoronary Doppler flow velocity measurements in 40 hypertensive patients with chest pain and no or minor coronary artery disease, some of whom had thallium scintigraphic abnormalities following exercise stress [19]. Flow reserve was assessed following dipyridamole (average dose 0.59 mg/kg). The authors found that patients with thallium perfusion abnormalities were more likely to have a limited flow velocity response to dipyridamole than those with normal thallium perfusion scans, and that patients with thallium defects were more likely to have a greater degree of left ventricular hypertrophy than those with normal thallium scans. However, the decrement in flow reserve was not linearly related to the magnitude of left ventricular hypertrophy.

Our group reported that 12 hypertensive patients with chest pain and normal coronary angiograms had a coronary microvascular constrictor response to ergonovine 0.15 mg intravenously; all had no or minimal left ventricular hypertrophy by echo [4]. In contrast, 13 normotensive subjects without pacing-induced chest pain responses had a dilator response to ergonovine. More recently, Treasure et al. presented preliminary data to indicate that endothelium-mediated relaxation to acetylcholine is impaired in the microcirculation of hypertensive patients with left ventricular hypertrophy, with flow measured by the intracoronary Doppler catheter [48]. Thus, patients with hypertension and left ventricular hypertrophy commonly have abnormal flow reserve, with several studies suggesting an independent contribution of the coronary microcirculation, possibly related to endothelial dysfunction.

Morphologic abnormalities in microvascular density and structure may also limit appropriate coronary flow in humans with pressure-induced left ventricular hypertrophy. Anatomic studies by Rakusan showed substantial reduction in capillary density in hypertrophied human hearts [37]. Tanaka et al. reported luminal narrowing of intramyocardial small arteries in hearts of patients with left ventricular hypertrophy at necropsy [45]. In a preliminary report, Schwartzkopff et al. found luminal narrowing of intramural arteries from right ventricular septal biopsies in patients with chest pain, hypertension and normal coronary angiograms [41]. They found a correlation between the minimum coronary resistance following dipyridamole 0.5 mg/kg (inert gas method) and the arterial media wall area. No correlation was found between left ventricular mass index by echocardiography and the minimum coronary resistance. They concluded that the wall thickening was causally related to the diminished coronary reserve, in part independent of myocardial hypertrophy. Abnormal structure and function of the coronary microcirculation in left ventricular hypertrophy may contribute to impaired autoregulation of coronary flow with reductions in systemic pressure, as reported by Polese et al. [36].

Regression of Left Ventricular Hypertrophy and Coronary Flow

Therapeutic studies in animal models of left ventricular hypertrophy provide evidence that many of the abnormalities resulting in impaired myocardial perfusion can be reversed. Antihypertensive therapy resulting in regression of left ventricular hypertrophy can increase coronary flow reserve and reduce pharmacologic minimum coronary resistance in animals [2, 3, 27, 39, 42, 52]. Further, myocardial collagen content is reduced along with left ventricular hypertrophy regression in treated animals which could improve myocardial perfusion by improving diastolic affects on intramyocardial coronary flow [3, 12, 31]. Non-pharmacologic reversal of pressure overload has been found to regress left ventricular hypertrophy and improve coronary flow dynamics in rats following 4 weeks of aortic banding along with improved autoregulatory control of myocardial blood flow [20, 40], with less improvement in coronary flow dynamics in animals with 10 weeks of banding [22]. Thus, at least in animal models, the impact of reversibility of hypertrophy on the coronary circulation may in part depend on the duration of pressure load hypertrophy.

Therapeutic Studies in Humans

Fewer studies have been reported in humans regarding the impact of therapy on coronary flow dynamics in the setting of pressure-induced left ventricular hypertrophy. Aortic stenosis was reported by Marcus et al. to be associated with limited coronary flow response (Doppler technique) to brief intraoperative coronary artery occlusion [28]. Eberli et al. [14] found the coronary flow response to dipyridamole 0.5 mg/kg (thermodilution measurement) significantly greater in patients following aortic valve replacement compared to unoperated patients. Although regression of left ventricular hypertrophy was noted in most patients, some degree of hypertrophy persisted even following successful valve replacement. Of interest, the primary effect of early valve replacement on coronary flow dynamics appeared to be a reduction in basal flow, with maximum flow to dipyridamole being similar in the preoperative and postoperative patients. Operative relief of left ventricular outflow obstruction in hypertrophic cardiomyopathy also reduces basal, as well as stress-induced coronary flow and myocardial oxygen consumption [8].

Although several antihypertensive agents have been shown to regress left ventricular hypertrophy in humans, there are few data on the impact of left ventricular hypertrophy regression on coronary dynamics. Vogt et al. in a preliminary report found that 12 months of therapy with the angiotensin-converting enzyme inhibitor enalapril increased the maximum coronary flow response (inert gas technique) to dipyridamole 0.5 mg/kg and reduced the minimum coronary resistance compared to pretreatment study [51]. There was only a mild (−8%) reduction in echocardiographic left ventricular mass, suggesting that the effect of enalapril was more a consequence of vascular effects of the drug rather than left ventricular hypertrophy regression. Further studies in humans are necessary to assess the impact of antihypertensive therapy and left ventricular hypertrophy regression on coronary vascular morphology, endothelial function, autoregulation, and coronary flow reserve with hopes of reversal of the morbid and mortal risks associated with left ventricular hypertrophy.

References

1. Bonow RO, Rosing DR, Bacharach SL, Green MV, Kent KM, Lipson LC, Maron BJ, Leon MB, Epstein SE (1981) Effects of verapamil on left ventricular systolic function and diastolic filling in patients with hypertrophic cardiomyopathy. Circulation 64:787–796
2. Bostrom SL, Fryklund J (1979) Protein synthesis and composition of cardiac and vascular tissue in spontaneously hypertensive rats and the effects of beta-adrenergic antagonist treatment. Clin Sci 57:39–41

3. Brilla CG, Janicki JS, Weber KT (1991) Diastolic function and coronary reserve in genetic hypertension. Role of interstitial fibrosis and medial thickening of intramyocardial coronary arteries. Circ Res 69:107–115
4. Brush JE, Cannon RO, Schenke WH, Bonow RO, Leon MB, Maron BJ, Epstein SE (1988) Angina due to coronary microvascular disease in hypertensive patients without left ventricular hypertrophy. N Engl J Med 319:1302–1307
5. Camici P, Chiriatti G, Lorenzoni R, Bellina RC, Gistri R, Italiani G, Parodi O, Salvadori PA, Nista N, Papi L, L'Abbate A (1991) Coronary vasodilation is impaired in both hypertrophied and nonhypertrophied myocardium of patients with hypertrophic cardiomyopathy: a study with nitrogen-12 ammonia and positron emission tomography. J Am Coll Cardiol 17:879–886
6. Cannon RO, Rosing DR, Maron BJ, Leon MB, Bonow RO, Watson RM, Epstein SE (1985) Myocardial ischemia in patients with hypertrophic cardiomyopathy: contribution of inadequate vasodilator reserve and elevated left ventricular filling pressures. Circulation 71:234–243
7. Cannon RO, Schenke WH, Maron BJ, Tracy CM, Leon MB, Brush JE, Rosing DR, Epstein SE (1987) Differences in coronary flow and myocardial metabolism at rest and during pacing between patients with obstructive and patients with nonobstructive cardiomyopathy. J Am Coll Cardiol 10:53–62
8. Cannon RO, McIntosh CL, Schenke WH, Maron BJ, Bonow RO, Epstein SE (1989) Effects of surgical reduction of left ventricular outflow obstruction on hemodynamic, coronary flow and myocardial metabolism in hypertrophic cardiomyopathy. Circulation 79:766–775
9. Cannon RO, Dilsizian V, O'Gara PT, Udelson JE, Schenke WH, Quyyumi A, Fananapazir L, Bonow RO (1991) Myocardial metabolic, hemodynamic and electrocardiographic significance of reversible thallium-201 abnormalities in hypertrophic cardiomyopathy. Circulation 83:1660–1667
10. Cannon RO, Dilsizian V, O'Gara PT, Udelson JE, Tucker E, Panza JA, Fananapazir L, McIntosh CL, Wallace RB, Bonow RO (1992) Impact of operative relief of outflow obstruction on thallium perfusion abnormalities in hypertrophic cardiomyopathy. Circulation 85 (in press)
11. Casale PN, Devereux RB, Milner M, Zullo G, Harshfield GA, Pickering TG, Laragh JH (1986) Value of echocardiographic measurement of left ventricular mass in predicting cardiovascular morbid events in hypertensive men. Ann Intern Med 105:173–178
12. Cutilletta AF, Dowell RT, Rudnik M, Arcilla RA, Zak R (1975) Regression of myocardial hypertrophy. I. Experimental model, changes in heart weight, nucleic acids and collagen. J Mol Cell Cardiol 7:767–781
13. Devereux RB, Savage DD, Sachs I, Laragh JH (1983) Relation of hemodynamic load to left ventricular hypertrophy and performance in hypertension. Am J Cardiol 51:171–176
14. Eberli FR, Ritter M, Schwitter M, Bortone A, Schneider J, Hess OM, Krayenbuehl HP (1991) Coronary reserve in patients with arotic valve disease before and after successful aortic valve replacement. Eur Heart J 12:127–138
15. Hammond IW, Devereux RB, Aldeman MH, Lutas EM, Spitzer MC, Crowley JS, Laragh JH (1986) The prevalence and correlates of echocardiographic left ventricular hypertrophy among employed patients with uncomplicated hypertension. J Am Coll Cardiol 7:639–650
16. Harrison DG, Florentine MS, Brooks LA, Cooper SM, Marcus ML (1988) The effect of hypertension and left ventricular hypertrophy on the lower range of coronary autoregulation. Circulation 77:1108–1115
17. Henquell L, Odoroff CC, Honig CR (1977) Intercapillary distance and capillary reserve in hypertrophied rat hearts beating in situ. Circ Res 41:400–408
18. Holtz J, Restorff WV, Bard P, Bassenge E (1977) Transmural distribution of myocardial blood flow of coronary reserve in canine left ventricular hypertrophy. Basic Res Cardiol 72:286–292
19. Houghton JL, Frank MJ, Carr AA, Von Dohlen TW, Prisant LM (1990) Relations among impaired coronary flow reserve, left ventricular hypertrophy, and thallium perfusion defects in hypertensive patients without obstructive coronary artery disease. J Am Coll Cardiol 15:43–51

20. Isoyama S, Ito N, Kuroha M, Takishima T (1989) Complete reversibility of physiological coronary vascular abnormalities in hypertrophied hearts produced by pressure overload in the rat. J Clin Invest 84:288–294
21. Isoyama S, Sato F, Takishima T (1991) Effect of age on coronary circulation after imposition of pressure-overload in rats. Hypertension 17:369–377
22. Ito N, Isoyama S, Kuroha M, Takishima T (1990) Duration of pressure overload alters regression of coronary circulation abnormalities. Am J Phsiol 258:H1753–H1760
23. Kannel WB, Gordon T, Offutt D (1969) Left ventricular hypertrophy by electrocardiogram: prevalence, incidence, and mortality in the Framingham Study. Ann Intern Med 71:89–105
24. Klepzig M, Eisenlohr H, Steindl J, Schmiebusch H, Strauer BE (1987) Media hypertrophy in hypertensive coronary resistance vessels. J Cardiovasc Pharmacol 10 [Suppl 6]:S97–S102
25. Koyanagi S, Eastham CL, Harrison DG, Marcus ML (1982) Increased size of myocardial infarction in dogs with chronic hypertension and left ventricular hypertrophy. Circ Res 50:55–62
26. Levy D, Garrison RJ, Savage DD, Kannel WB, Castelli WP (1990) Prognostic implications of echocardiographically determined left ventricular mass in the Framingham Heart Study. N Engl J Med 322:1561–1566
27. Lundin SA, Hallback-Nordlander M (1984) Regression of structural cardiovascular changes by antihypertensive therapy in spontaneously hypertensive rats. J Hypertension 2:11–18
28. Marcus ML, Doty DB, Hiratzka LF, Wright CB, Eastham CL (1982) Decreased coronary reserve: a mechanism for angina pectoris in patients with aortic stenosis and normal coronary arteries. N Engl J Med 307:1362–1366
29. Maron BJ, Wolfson JK, Epstein SE, Roberts WC (1986) Intramural ("small vessel") coronary artery disease in hypertrophic cardiomyopathy. J Am Coll Cardiol 8:545–557
30. Maron BJ, Bonow RO, Cannon RO, Leon MB, Epstein SE (1987) Hypertrophic cardiomyopathy. Interrelation of clinical manifestations, pathophysiology and therapy. N Engl J Med 316:780–789
31. Motz W, Strauer BE (1989) Left ventricular function and collagen content after regression of hypertensive hypertrophy. Hypertension 13:43–50
32. Mueller TM, Marcus ML, Kerber RE, Young JA, Barnes RW, Abboud FM (1978) Effect of renal hypertension and left ventricular hypertrophy on the coronary circulation in dogs. Circ Res 42:543–549
33. O'Gara PT, Bonow RO, Maron BJ, Damske BA, Van Lingen A, Bacharach SL, Larson SM, Epstein SE (1987) Myocardial perfusion abnormalities in patients with hypertrophic cardiomyopathy; assessment with thallium-201 computed tomography. Circulation 76: 1214–1223
34. O'Keefe DD, Hoffman JE, Chietlin M, O'Neill JO, Allard JR, Shapkin E (1978) Coronary blood flow in experimental canine left ventricular hypertrophy. Circ Res 43:43–51
35. Opherk D, Mall G, Zebe H, Schwarz F, Weihe E, Manthey J, Kubler W (1984) Reduction of coronary reserve: a mechanism for angina pectoris and normal coronary arteries. Circulation 69:1–7
36. Polese A, De Cesare N, Montorsi P, Fabbiocchi F, Guazzi M, Loaldi A, Guazzi MD (1991) Upward shift of the lower range of coronary flow autoregulation in hypertensive patients with hypertrophy of the left ventricle. Circulation 83:845–853
37. Rakusan K (1971) Quantitative morphology of capillaries of the heart. Number of capillaries in animal and human hearts under normal and pathological conditions. Methods Achiev Exp Pathol 5:272–286
38. Rembert JC, Kleinman LH, Fedor JM, Wechsler AS, Greenfield JC (1978) Myocardial blood flow distribution in concentric left ventricular hypertrophy. J Clin Invest 62:379–386
39. Sano T, Tarazi RC (1987) Differential structural responses of small resistance vessels to antihypertensive therapy. Circulation 75:618–626
40. Sato F, Isoyama S, Takishima T (1990) Normalization of impaired coronary circulation in hypertrophied rat hearts. Hypertension 16:26–34
41. Schwartzkopff B, Vogt M, Knauer S, Motz W, Strauer BE (1991) Medial hypertrophy of intramural coronary arteries in patients with reduced coronary reserve in hypertensive heart disease. Circulation 84:II–479 (abstr)

42. Sen S, Tarazi RC, Bumpus FM (1981) Reversal of cardiac hypertrophy in renal hypertensive rats: medical versus surgical therapy. Am J Physiol 240:H408–H412
43. Strauer BE (1979) Ventricular function and coronary hemodynamics in hypertensive heart disease. Am J Cardiol 44:999–1002
44. Strauer BE (1979) Myocardial oxygen consumption in chronic heart disease: role of wall stress, hypertrophy, and coronary reserve. Am J Cardiol 44:730–740
45. Tanaka M, Fujiwara H, Onodera T, Wu D-J, Matsuda M, Hamashima Y, Kawai C (1987) Quantitative analysis of narrowings of intramyocardial small arteries in normal hearts, hypertensive hearts, and hearts with hypertrophic cardiomyopathy. Circulation 6:1130–1139
46. Tomanek RJ, Searls JC, Lachenbruch PA (1982) Quantitative changes in the capillary bed during developing, peak, and stabilized cardiac hypertrophy in the spontaneously hypertensive rat. Circ Res 51:295–304
47. Tomanek RJ, Aydelotte MR, Butters CA (1990) Late-onset renal hypertension in old rats alters myocardial microvessels. Am J Physiol 259:H1681–H1687
48. Treasure CB, Lein JL, Vita JA, Selwyn AP, Alexander RW, Ganz P (1991) Left ventricular hypertrophy secondary to hypertension is associated with impaired endothelium-mediated relaxation in the coronary microvessels. J Am Coll Cardiol 17:237A (abstr)
49. Udelson JE, Bonow RO, O'Gara PT, Maron BJ, Van Lingen A, Bacharach SL, Epstein SE (1989) Verapamil prevents silent myocardial perfusion abnormalities during exercise in asymptomatic patients with hypertrophic cardiomyopathy. Circulation 79:1052–1060
50. Vogt M, Motz W, Schwartzkopff B, Strauer BE (1990) Coronary microangiopathy and cardiac hypertrophy. Eur Heart J 11 [Suppl B]:133–138
51. Vogt M, Motz W, Politz B, Scheler S, Strauer BE (1991) Improvement of coronary reserve by chronic treatment with ACE-inhibitors. Circulation 84:II–136 (abstr)
52. Warshaw DM, Root DT, Halpern W (1980) Effects of antihypertensive drug therapy on the morphology and mechanics of resistance arteries from spontaneously hypertensive rats. Blood Vessels 17:257–270
53. Wasserman AD, Katz RJ, Varghese PJ, Leiboff RH, Bren GG, Schlesselman S, Varma VM, Reba RC, Ross AM (1984) Exercise radionuclide ventriculographic responses in hypertensive patients with chest pain. N Engl J Med 311:1276–1280
54. Wearn JT (1928) The extent of the capillary bed of the heart. J Exp Med 47:273–292
55. Yamori Y, Mori C, Nishio T, Ooshima A, Horie R, Ohtaka M, Soeda T, Saito M, Abe K, Nara Y, Nakao Y, Kihara M (1979) Cardiac hypertrophy in early hypertension. Am J Cardiol 44:964–969

Myocardial Oxygen Consumption in Patients with Idiopathic Dilated Cardiomyopathy: Effects of Vasodilating and Inotropic Agents

Ch. Holubarsch, G. Hasenfuss, and H. Just

Introduction

Idiopathic dilated cardiomyopathy is a myocardial disease of unknown origin. Several causal factors are discussed. One hypothesis involves energy depriviation in the failing myocardium, and a number of arguments exist for such an energy-starving situation [22, 23]. These include the following: (a) In some studies investigating the energy state of hypertrophied and failing myocardium, a reduced content of high energy phosphates was reported [5, 39]. However, these findings were not confirmed by others [33, 34]. (b) In the failing human myocardium, a reduction in myofibrillar ATPase activity [1, 25, 30] and a shift of the isoenzyme pattern of lactate dehydrogenase [36] has been observed. Both changes may be interpreted in terms of adaptation to energy depletion; reduced myofibrillar ATPase activity is associated with more economical tension development, as shown in mammalian myocardium [2, 15, 16] and in human myocardium [14], and the shifts in the isoforms of the lactate dehydrogenase are indicators of anaerobic glycolysis [36]. (c) A reduction in capillary density may lead to increased diffusion distances [4]. (d) A reduced ratio of mitochondria and myofibrillar volumes indicates a mismatch between the number of ATP synthesizing sites and the number of ATP consuming sites [4, 31]. (e) Reduced coronary reserve has been observed in patients with dilated cardiomyopathy using the dipyridamole test [28, 29, 32]. Although it is currently unknown whether this reduction in coronary reserve is due to elevated filling pressure, results from a mismatch between coronary vessel growth and cardiac hypertrophy, or represents impairment of functional vasodilating mechanisms, it certainly may limit the delivery of oxygen and other nutrients to the myocytes during exercise. However, some authors have postulated that coronary perfusion is even insufficient during rest [10, 32]. Because energy supply may be a limiting factor in cardiac hypertrophy and myocardial failure, it is important to know how pharmacological interventions may influence energy demands and myocardial energetics. We therefore investigated myocardial energetics of patients with dilated cardiomyopathy at rest and after administration of a variety of vasodilators and positive inotropic compounds.

Department of Cardiology, Internal Medicine, University of Freiburg, Hugstetter Strasse 55, W-7800 Freiburg, Fed. Rep. of Germany

Figulla et al. (Eds.)
Idiopathic Dilated Cardiomyopathy

In studying myocardial energetics it is not sufficient simply to measure myocardial oxygen consumption. This is best demonstrated by comparing some studies in which the influence of intravenous application of enoximone on myocardial oxygen consumption was investigated. Enoximone may either increase or decrease myocardial oxygen consumption: findings have been a decrease of 18% [3], increase of 7% [26], and increase of 31% [42]. Therefore it is necessary to relate myocardial oxygen consumption to the decisive hemodynamic or mechanical parameters. There are four major parameters that are thought to determine myocardial oxygen consumption: heart rate [6], peak wall stress or systolic stress-time integral [13, 17, 24, 38], maximum rise of pressure development [11], and pressure-volume work [35]. Heart rate was taken into account by normalizing myocardial oxygen consumption to one cardiac cycle. Therefore, in the present study, all parameters – energetic and mechanical – are calculated for one cardiac cycle. In previous studies [13, 17], we have shown that myocardial oxygen consumption is correlated best with systolic stress-time integral. Therefore, we analyzed the ratio between systolic stress-time integral and myocardial oxygen consumption per beat, which is a measure of myocardial economy of force generation. Furthermore, pressure-volume work was proposed to be an important determinant of myocardial oxgen consumption [35] and was shown to be positively and significantly related to myocardial oxygen consumption in patients with dilated cardiomyopathy [13, 21].

Both ratios, stress-time integral per myocardial oxygen consumption and pressure-volume work per myocardial oxygen consumption, represent basic physical terms. Myocardial economy of contraction indicates the force developed for a unit of energy, and myocardial external efficiency indicates mechanical work produced for a unit of energy consumed. Therefore, we used these two terms to analyze the influence of pharmacological interventions on myocardial energetics.

Methods

Myocardial energetics were studied in patients with normal ventricles and with dilated cardiomyopathy. Coronary artery disease was excluded by coronary angiography. Patients were excluded from the study, when valvular heart disease or arterial hypertension were diagnosed. Patients with dilated cardiomyopathy were clinically in NYHA stages II–III. In all of the studied patients myocardial blood flow was measured using the argon method. Myocardial oxygen consumption was calculated from myocardial blood flow and coronary artery/coronary sinus oxygen content difference. To define left ventricular dimensions and calculate left ventricular wall stress, left ventricular angiocardiography was performed using Millar microtip-catheter for simultaneous pressure measurements. By coordinating the pressure and volume data obtained from cine frames every 20 ms, a pressure-volume loop, which represents pressure-volume work, can be constructed. Assuming an ellipsoidal model of the left ventricle [27], left ventricular wall stress can be calculated from the instantaneous values of pressure, wall thickness, and major and minor semiaxis of the left ventricle using our own computer program [41]. The stress-time integral is obtained by integration of the stress-time function from end-diastole to end-systole. The oxygen supply-demand ratio is arterial oxygen saturation divided by arterial minus coronary sinus oxygen saturation [12].

Pharmacological interventions were performed in patients with dilated cardiomyopathy. The first group of patients received sodium nitroprusside intravenously at a concentration of 71 ± 58 μg/min. The phosphodiesterase inhibitor enoximone was given intravenously in an average dose of 1.42 ± 0.52 mg/kg body weight [13]. The dose of xamoterol given intravenously over a period of 5 min was 0.2 mg/kg body weight [40]. The angiotensin-converting enzyme inhibitor benazepril was injected at a dose of 3 mg [21].

Results and Discussion

Myocardial Energetics at Rest: Dilated Cardiomyopathy Versus Normal

Left ventricular muscle mass and LV end-diastolic volumes were significantly higher in patients with dilated cardiomyopathy compared to normal, and ejection fraction was substantially decreased (Table 1). The presented data indicate typical characteristics of dilated cardiomyopathy: left ventricular dilatation and left ventricular hypertrophy.

Neither myocardial blood flow, myocardial oxygen consumption, nor the oxygen supply-demand ratio were significantly different between the two groups of patients (Table 1). The oxygen supply-demand ratio is a very sensitive marker for oxygen delivery to the myocardial cells in relation to the tissue oxygen uptake. If there is insufficient myocardial blood flow, this ratio would decrease. However, such a decrease was not observed in dilated ventricles. Therefore, both coronary vessel growth and coronary vessel dilatation appear to meet perfectly the oxygen demand of the left ventricle in patients with dilated cardiomyopathy at rest. In contrast, decreased coronary reserve was clearly shown by the dipyridamole test in patients with dilated cardiomyopathy [28, 29], which could explain the pain of some of these patients during exercise [32]. In such a situation, any increase in

Table 1. Normal parameters versus those in idiopathic dilated cardiomyopathy (IDC)

	Normal	IDC	*p*
LV muscle mass (g)	194 ± 58	311 ± 66	<0.01
LVEDV (ml)	177 ± 55	307 ± 56	<0.01
EF (%)	76 ± 8	40 ± 10	<0.01
MBF (ml $100\,g^{-1}\,min^{-1}$)	83 ± 20	98 ± 34	NS
MVO_2 (ml $100\,g^{-1}\,min^{-1}$)	11.4 ± 2.7	12.3 ± 3.8	NS
Oxygen supply-demand ratio	1.49 ± 0.12	1.44 ± 0.10	NS

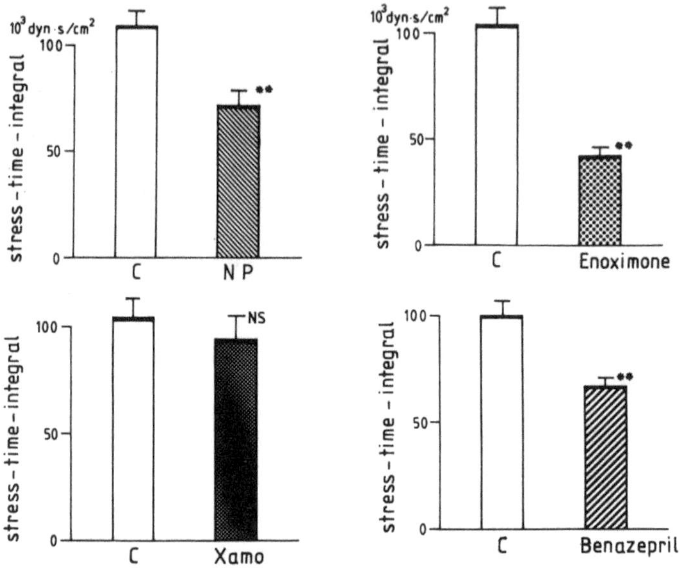

Fig. 1. Stress-time integral as calculated from left ventricular pressure, left ventricular volume, and left ventricular wall thickness from end-diastole to end-systole. Significant reduction in stress-time integral, i.e., the afterlaod of left ventricular myocardium, is observed with nitroprusside (*NP*; *upper left*), the phosphodiesterase inhibitor enoximone (*upper right*), and the angiotensin-converting enzyme inhibitor benazepril (*lower right*). The partial beta-1 agonist xamoterol (*Xamo*; *lower left*) has no significant effect on systolic stress-time integral

oxygen consumption may be harmful and is therefore to be avoided under all circumstances.

Influence of Vasodilators and Positive Inotropic Compounds on Myocardial Energetics

Mechanics. The stress-time integral as calculated from left ventricular dimensions, left ventricular wall thickness and left ventricular pressure represents the afterload of left ventricular myocardium. Sodium nitroprusside and benazepril induce a significant reduction in the stress-time integral (Fig. 1) due to their effects on preload and afterload. The reduction in the stress-time integral is even more pronounced with intravenous injection of the phosphodiesterase inhibitor enoximone, whereas the small decrease in afterlaod is not significant with the partial beta-1 agonist xamoterol.

Energetics. As discussed above, to study myocardial energetics it is not sufficient simply to measure myocardial oxygen consumption; rather, myocardial oxygen consumption and its alterations must be related to the important hemodynamic or mechanical parameters. We used the ratio of stress-time integral and myocardial oxygen consumption and that of pressure-volume work and myocardial oxygen consumption to analyze myo-

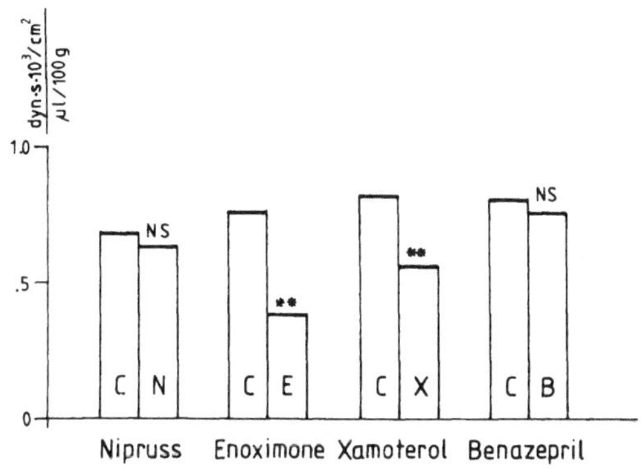

Fig. 2. Stress-time integral divided by myocardial oxygen consumption per beat: economy of myocardial force generation. With nitroprusside and benazepril, economy of myocardial force generation is unaltered, with the phosphodiesterase inhibitor enoximone and the partial beta-1 agonist xamoterol – both of which increase cyclic AMP – economy of myocardial force generation is significantly decreased

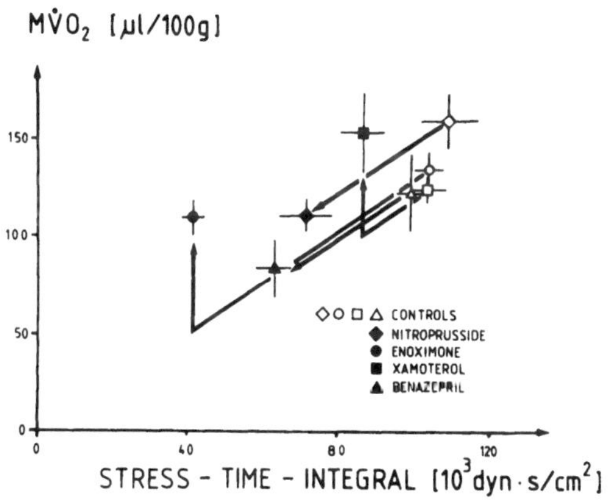

Fig. 3. Myocardial oxygen consumption per beat as a function of systolic stress-time integral. Both nitroprusside and benazepril decrease myocardial oxygen consumption per beat in proportion to the reduction in the stress-time integral (*parallel lines*, the established function between the two parameters; see [13, 21]. Xamoterol, the partial beta-1 agonist, has no significant effect on systolic stress-time integral but significantly increases myocardial oxygen consumption per beat. Enoximone, the phosphodiesterase inhibitor, has two effects: vasodilation (venous and arteriolar) decreases greatly the systolic stress-time integral; however, myocardial oxygen consumption is not proportionally reduced. The measured data point is far away from the linear relationship established for the function between myocardial oxygen consumption and stress-time integral. This deviation reflects additional energy consumption, which is attributed to the specific stimulation of the myocardium by phosphodiesterase inhibition

cardial energetics and its alterations due to pharmacological interventions in patients with dilated cardiomyopathy. Benazepril and nitroprusside did not alter the ratio between stress-time integral and myocardial oxygen consumption, whereas enoximone and the partial beta-1 agonist xamoterol decrease this index of economy of myocardial contraction (Fig. 2). The effect of these drugs on myocardial oxygen consumption and stress-time integral can be seen in Fig. 3. Benazepril and nitroprusside reduce myocardial oxygen consumption in proportion to the decrease in stress-time integral. Therefore, both drugs decrease myocardial oxygen consumption due to unloading of the left ventricle, i.e., reduction in preload and afterload.

In contrast, with xamoterol, the partial beta-1 agonist, the stress-time integral is little and nonsignificantly changed, whereas myocardial oxygen consumption is significantly increased. The direct stimulation of the myocardium via beta-1 receptors leads to an additional energy consumption which decreases myocardial economy of contraction (Fig. 2) and myocardial external efficiency (Fig. 4).

Enoximone, the phosphodiesterase inhibitor, has two effects: vasodilation greatly decreases the left ventricular stress-time integral; however, oxygen consumption is left almost unchanged, reflecting an additional positive inotropic stimulation with a relative increase in oxygen consumption. The energy-saving effect of enoximone due to venous and arteriolar vasodilation is offset by the energy-consuming effect induced by direct myocardial stimulation.

The results that substances such as beta-1 adrenoceptor agonists and phosphodiesterase inhibitors, both of which increase cyclic AMP, lead to an additional energy consumption of the myocardium, are also reflected in values of external efficiency. Whereas nitroprusside showed a tendency to increase myocardial efficiency, this index was decreased by 20% (NS) and 22% ($p < 0.05$) after intravenous administration of enoximone and xamoterol, respectively (Fig. 4).

CONTROL	14.6±6.2	19.8±5.3	17.6±7.8
TREATMENT	16.1±6.6	16.2±6.9	13.8±6.5
	NS	NS	P<0.05

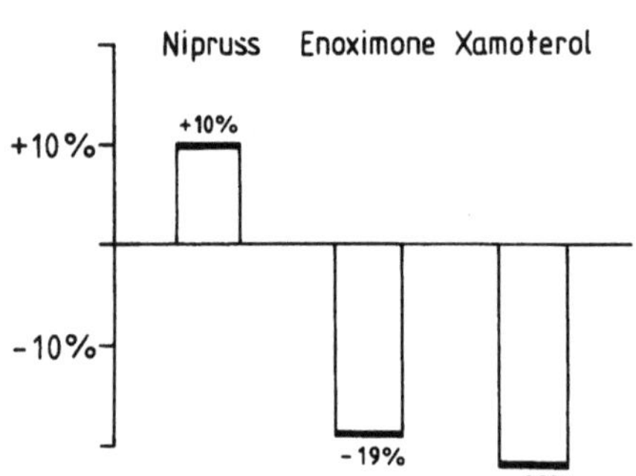

Fig. 4. External myocardial efficiency: pressure-volume work divided by myocardial oxygen consumption. In all patients with dilated cardiomyopathy, external efficiency is decreased substantially compared to the 30% value in normal hearts. Under conditions of nitroprusside, external efficiency showed a tendency to increase; with enoximone a reduction by 19% was observed. Xamoterol significantly decreased external myocardial efficiency by 22%

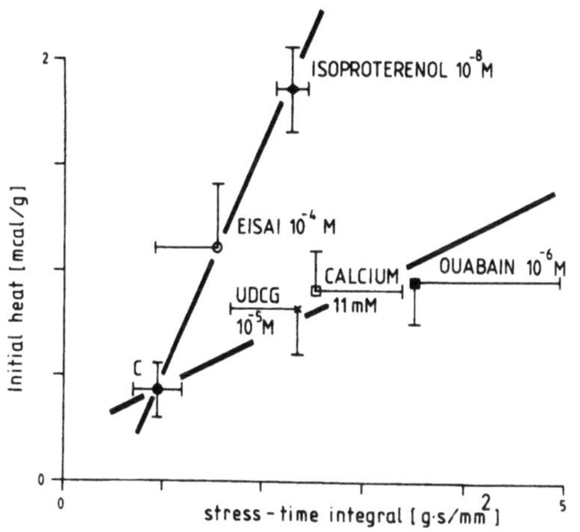

Fig. 5. Initial heat as a function of the tension-time integral. Measurements were performed in right ventricular guinea pig papillary muscles at 21°C and stimulation rates of 0.2 Hz using antimony-bismuth thermopiles and isometric force transducers [20]. Two groups of positive inotropic compounds were separated with respect to myocardial energetics: substances that increase cyclic AMP, i.e., isoproterenol and Eisai-1020, an phosphodiesterase inhibitor, induce overproportional increases in energy expenditure (*upper left line*); high calcium (11 m*M*), ouabain, and the calcium sensitizer UDCG-115 stimulate the myocardium in a more economical way (*lower right line*; see [18–20])

It seems from these data that additional energy consumption is unique for those substances that increase cyclic AMP. This view is supported by our experiments on isolated papillary muscles from giunea pigs: In these investigations, two groups of positive intropic compounds were found: (a) those substances that increase cyclic AMP and thereby decrease myocardial economy of contraction, and (b) digitalis and calcium sensitizers that leave economy of myocardial contraction unchanged (Fig. 5) [18–20].

Conclusions

Substantial reduction in myocardial oxygen consumption can be achieved only by sodium nitroprusside or angiotensin-converting enzyme inhibitors. Beta-1 agonists increase myocardial oxygen consumption. With phosphodiesterase inhibitors the energy-consuming and energy-saving effects are balanced. One is tempted to speculate that reduction of myocardial oxygen consumption may be the main effect that could explain the beneficial outcome of trials investigating the long-term effect of vasodilators and angiotensin-converting enzyme inhibitors in patients with heart failure [7, 8, 37]. In contrast, increases in intracellular cyclic AMP have unfavorable effects with respect to myocardial oxygen consumption and arrhythmogeneity and may explain the detrimental outcome in some trials investigating beta-1 adrenoceptor agonists [9, 43] and phosphodiestrase inhibitors.

References

1. Alpert NR, Gordon MS (1962) Myofibrillar adenosine triphosphatase activity in congestive heart failure. Am J Physiol 202:940–946
2. Alpert NR, Mulieri LA (1982) Increasing myothermal economy of isometric force generation in compensated cardiac hypertrophy induced by pulmonary artery constriction in rabbit. A characterization of heat liberation in normal and hypertrophied papillary muscle. Circ Res 50:491–500
3. Amin DK, Shah PK, Hulse S, Shellock FG, Swan HJC (1984) Myocardial metabolic and hemodynamic effects of intravenous MDL 17043, a new cardiotonic drug in patients with severe heart failure. Am Heart J 108:1285–1292
4. Anversa P, Olivetti C, Melissari M, Loud AV (1980) Stereological measurement of cellular and subcellular hypertrophy and hyperplasia in the papillary muscle of the adult rat. J Moll Cell Cardiol 12:781–795
5. Bashore TM, Magorien DJ, Letterio J, Shaffer P, Unverferth DV (1987) Histologic and biochemical correlates of the left ventricular chamber dynamics in man. J Am Coll Cardiol 9:734–742
6. Braunwald E (1971) Control of myocardial oxygen consumption: physiologic and clinical considerations. Am J Cardiol 27:416–432
7. Cohn JN, Archibald DG, Ziesche S et al. (1986) Effects of vasodilator therapy on mortality in chronic congestive heart failure. Results of a Veterans Administration Cooperative Study. N Engl J Med 314:1547–1552
8. CONSENSUS Trial Study Group (1987) Effects of enalapril on mortality in severe congestive heart failure: results of the Cooperative North Scandinavian Enalapril Survival Study (CONSENSUS). N Engl J Med 316:1429–1435
9. Dies F, Krell MJ, Whitlow P, Liang C-S, Goldenberg I, Applefeld MM, Gilbert EM (1986) Intermittent dobutamine in ambulatory outpatients with chronic cardiac failure. Circulation 74 [Suppl II]:152
10. Dubiel IS, Dubiel IP, Frendo J, Zmudka K, Horzela T (1986) Coronary circulation and myocardial metabolism in dilated cardiomyopathy. Cor Vasa 28:356–364
11. Graham TP, Covell JW, Sonnenblick EH, Ross J Jr, Braunwald E (1968) Control of myocardial oxygen consumption: relative influence of contractile state and tension development. J Clin Invest 47:375–385
12. Hasenfuss G, Holubarsch C, Heiss HW, Bonzel T, Funfack M, Revenaugh M, Meinertz T, Just H (1987) Hemodynamics and myocardial oxygen metabolism of pulsus alternans in patients with dilated cardiomyopathy. Clin Cardiol 10:223–327
13. Hasenfuss G, Holubarsch C, Heiss HW, Meinertz T, Bonzel T, Wais U, Lehmann M, Just H (1989) Myocardial energetics in patients with dilative cardiomyopathy. Influence of nitroprusside and enoximone. Circulation 80:51–64
14. Hasenfuss G, Mulieri LA, Blanchard EM, Holubarsch C, Leavitt BJ, Ittleman F, Alpert NR (1991) Energetics of isometric force development in control and volume-overload human myocardium. Comparison with animal species. Circ Res 68:836–846
15. Holubarsch C, Goulette RP, Mulieri LA, Alpert NR (1983) Heat liberation in experimentally induced tetanic contractions of myocardium from normal and Goldblatt rats. In: Jacob R (ed) Cardiac adaptation of hemodynamic overload, training and stress. Steinkopff, Darmstadt, pp 158–166
16. Holubarsch C, Litten RZ, Mulieri LA, Alpert NR (1985) Energetic changes of myocardium as an adaptation to chronic hemodynamic overload and thyroid gland activity. Basic Res Cardiol 80:582–593
17. Holubarsch C, Hasenfuss G, Heiss HW, Meinertz T, Just H (1987) Acute and chronic changes of myocardial energetics in the mammalian and human heart. Basic Res Cardiol 82 [Suppl II]:377–388
18. Holubarsch C, Hasenfuss G, Just H, Blanchard E, Mulieri LA, Alpert NR (1989) Influence of the positive inotropic substance pimobendane (UDCG 115 BS) on contractile economy of guinea pig papillary muscles. J Cardiovasc Pharmacol 14 [Suppl 2]:S13–S17

19. Holubarsch C, Hasenfuss G, Blanchard E, Mulieri LA, Alpert NR (1989) Influence of the phosphodiesterase inhibitor E-1020 on mechanics and energetics in guinea pig papillary muscles. Circulation 80 [Suppl II]:612
20. Holubarsch C, Hasenfuss G, Just H, Blanchard EM, Mulieri LA, Alpert NR (1990) Modulation of myothermal economy of isometric force generation by positive inotropic interventions in the guinea pig myocardium. Cardioscience 1:33–41
21. Holubarsch Ch, Hasenfuss G, Thiefelder L, Just H (1991) Energetic consequences of substances currently used or recommended for long-term treatment of chronic heart failure. Bas Res Cardiol 86, Suppl I:107–112
22. Katz AM (1986) Potential deleterious effects of inotropic agents in the therapy of chronic heart failure. Circulation 73 [Suppl III]:184–188
23. Katz AM (1989) Energy requirements of contraction and relaxation: implications for inotropic stimulation of the failing heart. Basic Res Cardiol 84 [Suppl I]:47–53
24. Laskey WK, Reichek N, Sutton MS, Untereker WJ, Hirshfeld JW (1983) Myocardial oxygen consumption in left ventricular hypertrophy and its relation to left ventricular mechanics. Am J Cardiol 52:852–858
25. Leclerque JF, Swynghedauw B (1976) Myofibrillar ATPase, DNA and hydroxyproline content of human hypertrophied heart. Eur J Clin Invest 6:27–33
26. Martin JL, Likoff MJ, Janicki JS, Laskey WK, Hirshfeld JW, Weber KT (1984) Myocardial energetics and clinical response to the cardiotonic agent MDL 17043 in advanced heart failure. J Am Coll Cardiol 4:875–883
27. Mirsky I (1979) Elastic properties of the myocardium: a quantitative approach with physiological and clinical applications. In: Berne RM (ed) Handbook of physiology. The cardiovascular system. American Physiological Society, Washington, pp 497–531
28. Nitenberg A, Foult IM, Blanchet F, Zonioneche S (1985) Multifactorial determinants of reduced coronary flow reserve after dipyridanole in dilated cardiomyopathy. Am J Cardiol 55:748–754
29. Opherk D, Schwarz F, Mall G, Manthey J, Baller D, Kuebler W (1983) Coronary dilator capacity in idiopathic dilated cardiomyopathy: analysis of 16 patients. Am J Cardiol 51:1657–1662
30. Pagani ED, Alousi AA, Grant AM, Order TM, Dziuban SW, Allen PD (1988) Changes in myofibrillar content and Mg-ATPase activity in ventricular tissue from patients with heart failure caused by coronary artery disease, cardiomyopathy, or mitral valve insufficiency. Circ Res 63:380–385
31. Page E, McCalister LP (1973) Quantitative electron microscopic description of heart muscle cells. Application to normal, hypertrophied, and thyroxin-stimulated heart. Am J Cardiol 31:172–181
32. Pasternac A, Noble J, Streulens Y, Elie R, Heuschke C, Bourassa MG (1982) Pathophysiology of chest pain in patients with cardiomyopathies and normal coronary arteries. Circulation 65:778–789
33. Regitz V, Rappolder M, Bossaller C, Strasser R, Fleck E (1990) Myocardial ATP, ADP and AMP in heart failure of different origins. Eur Heart J 11 [Suppl]:189
34. Regitz V, Fleck E (1992) Adenine nucleotide metabolism and contracile dysfunction in heart failure. Basic Res Cardiol 87, Suppl I:321–329
35. Rooke GA, Feigl EO (1982) Work as a correlate of canine left ventricular oxygen consumption, and the problem of catecholamine oxygen wasting. Circ Res 50:273–286
36. Schultheiss HP (1990) Effect on the myocardial energy metabolism of angiotensin-converting enzyme inhibition in chronic heart failure. Am J Cardiol 65:G74–G81
37. SOLVD investigators (1991) Effect of enalapril on survival in patients with reduced left ventricular ejection farctions and congestive heart failure. N Engl J Med 325:293–302
38. Strauer BE (1979) Ventricular function and coronary hemodynamics in hypertensive heart disease. Am J Cardiol 44:999–1006
39. Swain JL, Sabina RL, Peyton RB et al. (1979) Derangements in myocardial purin and pyrimidin metabolism in patients with coronary artery disease and left ventricular hypertrophy. Proc Natl Acad Sci USA 79:655–659

40. Thierfelder L, Holubarsch C, Hasenfuss G, Heiss HW, Just H (1991) The effect of the partial beta-1-adrenoceptor agonist xamoterol on hemodynamics and myocardial energetics in patients with idiopathic dilated cardiomyopathy. J Cardiovasc Pharmacol 17:593–599
41. von Herrath M, Hasenfuss G, Holubarsch C, Hofmann T, Heiss HW, Just H (1990) Repeated determination of left ventricular wall thickness from mass and volume during one cardiac cycle for the calculation of left ventricular wall stress parameters. Clin Cardiol 13:218–220
42. Viquerat CE, Kereiakes D, Morris L, Daly PA, Wexman M, Frank P, Parmley WW, Chatterjee K (1985) Alterations in left ventricular function, coronary hemodynamics and myocardial catecholamine balance with MDL 17043, a new inotropic vasodilator agent, in patients with severe heart failure. J Am Coll Cardiol 5:326–332
43. Xamoterol in Severe Heart Failure Study Group (1990) Xamoterol in severe heart failure. Lancet 336:1–6

Regression of Cardiac Hypertrophy with Pharmacotherapeutic Regimen

W. Motz, M. Vogt, S. Scheler, and B.E. Strauer

Epidemiology

Left ventricular hypertrophy represents the general structural mechanism of adaptation of the heart in response to a chronic pressure load of the left ventricle. This mechanism of adaptation enables the left ventricle to eject a normal stroke volume into the periphery in spite of an excessively elevated systolic pressure in the left ventricle. From this point of view, left ventricular hypertrophy should be considered as a desirable and favorable mechanism of adaptation of the heart to cope with an increased ventricular pressure burden [12, 16, 27, 44, 45].

However, former and recent epidemiologic data indicate that left ventricular hypertrophy, even in the absence of arterial hypertension, is a serious risk factor for cardiovascular events and is associated with later congestive heart failure. In 75% of all persons who developed congestive heart failure in the Framingham Study, arterial hypertension was the underlying disease or at least a side diagnosis. Individuals with hypertension and definite signs of left ventricular hypertrophy noted in an electrocardiogram had a tenfold greater risk of developing heart failure than hypertensive individuals without electrocardiographic signs of hypertrophy. Left ventricular hypertrophy on an echocardiogram has also been identified as a severe cardiovascular risk factor since it was found to be associated with death from cardiovascular disease [20, 21, 26].

Although no definite proof exists yet that reversing left ventricular hypertrophy by chronic antihypertensive treatment will lower cardiac mortality and improve prognosis, recent data indicate that regression or prevention of left ventricular hypertrophy along with medical blood pressure reduction significantly lowers the frequency of cardiovascular events [23].

Left Ventricular Hypertrophy, Myocardial Collagen, and the Renin-Angiotensin System

Left ventricular hypertrophy in arterial hypertension is characterized by cardiac myocyte growth accompanying an increased myocyte workload [2],

Medical Clinic B, Heinrich-Heine-University of Düsseldorf, W-4000 Düsseldorf 1, Fed. Rep. of Germany

Figulla et al. (Eds.)
Idiopathic Dilated Cardiomyopathy

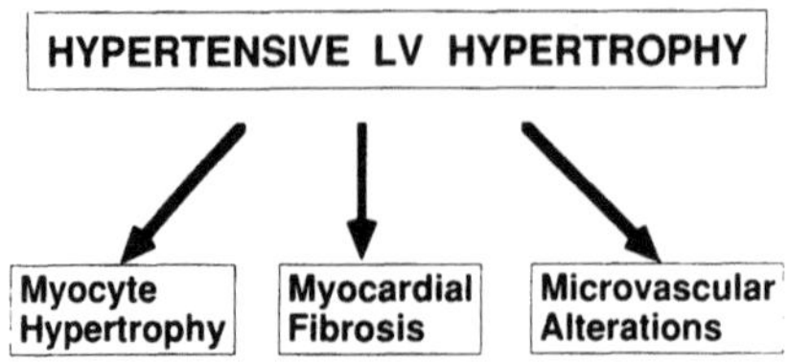

Fig. 1. Hypertensive hypertrophy is characterized by activation of the myocyte, interstitial, and vascular compartments

which is accompanied by an activation of nonmyocyte cells such as cardiac fibroblasts, vascular smooth muscle, and endothelial cells. The appearance and extent of altered growth and metabolism of nonmyocytic cells especially determines pathologic left ventricular hypertrophy in arterial hypertension. Accordingly, left ventricular hypertrophy in hypertension, which is characterized by myocardial fibrosis and structural alterations of the small intramural arteries, is clearly different from physiologic left ventricular hypertrophy due to athletic training, which reflects adaptive growth of myocytes [52–56] (Fig. 1).

The myocardial collagen is a network which connects the single myocytes with each other and the myocytes with the supporting coronary capillaries and vessels. Both the quantitative amount and the elastic material properties of the myocardial collagen (types I and III) determine the myocardial stiffness, i.e., the passive elastic material properties of the myocardial tissue. During the process of left ventricular hypertrophy, myocardial collagen content increases due to activation of interstitial fibroblasts. Two major types of fibrosis appear: replacement fibrosis and reactive fibrosis. Replacement fibrosis restores structural integrity in the case of parenchymal cell injury. Reactive fibrosis reflects an inappropriate enhancement of collagen synthesis leading to accumulation of fibrillar collagen. This in turn alters the passive elastic material properties of the myocardium, which leads to increased myocardial stiffness and reduces the diastolic distensibility of the left ventricle [52–56]. Recent studies indicate that myocardial fibrosis is the consequence of chronically elevated circulating mineralocorticoid levels such as aldosterone and deoxycorticosterone and the height of the angiotensin II levels, rather than the consequence of the prolonged pressure load alone [55].

The conversion of the inactive angiotensin I to the active angiotensin II is rate-limited by the local concentration of angiotensin-converting enzyme (ACE). In situ labeling of ACE by a radioligand binding method showed that local ACE is nonuniformly distributed within the heart, with the highest concentration in the leaflets and the lowest in the endocardium. High ACE concentrations were also found in the small intramyocardial arteries [58]. The role of the renin-angiotensin aldosterone system (RAAS) for cardiovascular remodeling is also emphasized by the fact that angiotensin II can augment the hypertrophy process of the left ventricle when exposed to a pressure load. However, angiotensin II alone without a concomitant pressure load has no trophic effect. Accordingly, angiotensin II only reinforces the growth response of the myocardium to a prolonged fiber stretch as a growth cofactor [38].

The impact of the RAAS on cardiovascular remodeling is also shown by the fact that ACE inhibitors can reverse hypertrophy with blood pressure normalization [29, 34]. ACE inhibitors were also found to lead to a regression in myocardial collagen and improve coronary flow reserve in experimental hypertension [3].

Left Ventricular Systolic and Diastolic Function

Left ventricular wall thickness increases due to myocardial cell hypertrophy and the parallel alignment of sarcomeres [2]. Consequently, the relation between left ventricular wall thickness and left ventricular radius, i.e., wall thickness-radius ratio, is augmented because left ventricular end-diastolic diameter remains unchanged or is even reduced. The process of wall thickening reduces the physical force, i.e., stress on the single cell in the myocardial texture exerted by the generated pressure in the systole of left ventricle [12, 16]. Accordingly, left ventricular hypertrophy in proportion to the increased left ventricular systolic pressure is considered to be adaptive in nature. In this type of concentric left ventricular hypertrophy, systolic wall stress remains constant. Since systolic wall stress, which reflects the left ventricular afterload, is the main determinant of left ventricular systolic function and myocardial oxygen consumption, systolic pump function and oxygen consumption per weight unit myocardium are normal in turn [44, 45]. Iso-volumetric contractility indices at this stage are normal as well as isometric and isotonic contractility indices of isolated papillary muscles in experimental models of slowly developing hypertension [5, 43].

Although in this early form of left ventricular hypertrophy systolic pump function, as measured by ejection fraction, is not impaired [16, 44, 45], diastolic filling of the left ventricle is already compromised by the process of left ventricular hypertrophy [14, 40, 42]. Even before left ventricular hypertrophy is detectable by clinical methods, diastolic dysfunction is present [17]. Mere wall thickening due to augmentation of myocardial sarcomeres reduces left ventricular distensibility or left ventricular compliance without any necessary change in the elastic material properties of the myocardium. The pressure-volume curve of the left ventricle is shifted upward and leftward. An increase in myocardial stiffness along with progressive interstitial fibrosis further moves the pressure-volume curve upwards and leftwards [15]. Due to this marked increase in left ventricular chamber stiffness caused by both left ventricular geometry and altered myocardial material properties, end-diastolic pressure and, in turn, pulmonary artery pressure rise [40, 42]. This pathophysiologic constellation explains that patients with arterial hypertension complain of exertional dyspnea and have a pathologic increase in pulmonary artery pressure under exercise inspite of a normal systolic pump function.

After a prolonged period of exposure to hypertensive pressure levels, left ventricular dilation occurs [27, 40, 44, 45]. The ultimate cause of structural dilation of the left ventricle, which is due to slippage and reorganization of the sarcomeres, in parallel is still unclear [18]. Due to alteration in left ventricular geometry such as wall thinning with simultaneous increase in left ventricular diameter, afterloading of the left ventricle rises due to the increment in systolic wall stress [12, 16]. Simultaneously, left ventricular ejection fraction drops with a parallel increment in myocardial oxygen demand [12, 16, 44, 45]. At this stage, isovolumetric contractility is also impaired as well as isometric and isotonic contractility indices of isolated papillary muscles in spontaneously hypertensive rats and other experimental models of slowly developing arterial hypertension [5, 43]. At this stage, diastolic function is characterized by a rightward shift and a steep course of the pressure-volume curve due to a marked increment in myocardial stiffness [15]. This enables the heart to handle large stroke volumes at low pressures over a wide volume scale. However, when a certain volume is exceeded, a sharp and tremendous rise in ventricular pressures occurs. The decrement in left ventricle pump function is similarly associated with an increase in systolic wall stress, for example, afterload due to structural ventricular dilation, a decline in intrinsic myocardial contractility, and a decrease in the elastic material properties of the myocardium due to progressive interstitial fibrosis. This demonstrates the close interrelationship between diastolic and systolic function [5]. The myocardium is an elastic network of myocytes enmashed in a collagen matrix, which connects the myocytes and the supporting coronary vasculature [53, 54]. Accordingly, extensive alterations in the functional and structural composition of this elastic network of contractile and noncontractile components is reflected by corresponding changes in systolic and diastolic function [7, 52, 54]. Dysfunction of the left ventricle in hypertension starts continuously at the level of the diastole and ends at the systole.

Involvement of Coronary Microcirculation

Patients with arterial hypertension often have angina pectoris and positive results on the exercise tolerance test even when their epicardial coronary arteries are of normal anatomical appearance [4, 6, 44, 45]. Determination of coronary flow reserve allows for the evaluation of the functional impairment of coronary regulatory capacity and, consequently, the evaluation of microvascular angina [6]. In hypertensive patients, the coronary vasodilator reserve is frequently impaired. Maximal coronary blood flow after dipyridamole administration is reduced by 30%–50% and, the minimal coronary resistance is accordingly elevated. Coronary reserve – the ratio between resting coronary resistance and that observed under maximal vasodilation (dipyridamole) – is likewise reduced relative to that seen in normotensive individuals [33, 44, 45, 50]. The impairment of coronary vasodilator reserve in arterial hypertension can be attributed primarily to structural alterations in coronary resistance vessels (vascular component) [11, 19, 39, 48, 49]. Myocardial hypertrophy (myocardial component), could also contribute due to extravascular compressive forces that elevate the extravascular component of coronary resistance. The myocardial component of coronary resistance reflects forces that appear during the myocardial contraction and relaxation processes. Interstitial and perivascular myocardial fibrosis may reduce reserve by increasing the myocardial component of coronary resistance due to alteration of the elastic material properties of the myocardium [55, 56].

In left ventricular hypertrophy of moderate severity, the reduction in coronary flow reserve is not proportional to the elevation in left ventricular muscle mass [50]. This indicates that the myocardial factor, expressed indirectly as echocardiographic left ventricular muscle mass, does not directly determine the reduction in vasodilator reserve. Moreover, an impaired vasodilator reserve was reported even in hypertensive patients without left ventricular hypertrophy [4]. Hypertensive remodeling of the coronary microcirculation includes alterations in the composition of the vessel wall leading to a decreased vasodilator capacity and a reduction in the number of parallel vascular pathways in the myocardium [49]. This can be provoked by a disproportionate growth response of the myocyte compartment without a corresponding growth adaptation of the coronary resistance vessels or capillaries or a disproportionate loss of resistance vessels without any necessary alteration in the myocyte compartment. Recently, medial thickening of intramural coronary arteries in vivo has been demonstrated in hypertensive patients by morphometric evaluations of transvenous endomyocardial biopsies [39].

According to Folkow's theory [11], structural changes of the resistance vessels such as vascular smooth muscle hypertrophy are the consequence of hypertensive tissue perfusion. Every contraction of vascular smooth muscle in the presence of a high vessel wall-lumen ratio leads to a more progressive luminal encroachment than the same smooth muscle contraction when the vessel wall-lumen ratio is low. Coronary involvement through small resistance vessel hypertrophy might be the key pathologic-anatomic background of the impaired coronary reserve and the syndrome of angina pectoris in hypertensive patients [39, 48]. In addition to structural factors that may contribute to the impairment in coronary reserve in left ventricular hypertrophy, attenuations in endothelial-mediated control of coronary resistance vessels may also exist. Impaired endothelial-dependent coronary vasodilation to acetylcholine has been observed in preliminary studies in patients with established essential hypertension and would be consistent with an alteration of the EDRF system in human hypertension [32]. An alternative mechanism may be enhanced activity of endothelium-derived vasoconstricting factors.

Coronary reserve can also be metabolically impaired by enhancing blood flow requirements under basal conditions without any necessary alterations on the level of the coronary resistance vessels or the myocardium [44]. At least in concentric left ventricular hypertrophy, for example, left

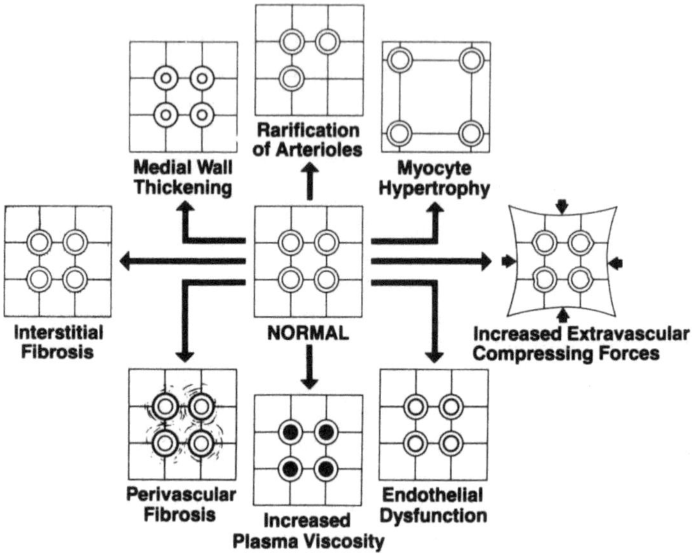

Fig. 2. Possible causes of disturbances in the coronary microcirculation which limit vasodilator reserve

ventricular wall thickening without chamber enlargement, both myocardial oxygen consumption and coronary blood flow per 100 g myocardium are unchanged due to normal systolic wall stress levels. Only in left ventricular hypertrophy with chamber enlargement, i.e., eccentric left ventricular hypertrophy, is coronary flow reserve also metabolically reduced due to increased myocardial oxygen consumption and coronary blood flow per 100 g myocardium as a consequence of the high afterload or wall stress [44, 45] (Fig. 2).

Therapeutic Consequences

Left Ventricular Hypertrophy

Since left ventricular hypertrophy and involvement of the coronary microcirculation are the main cardiac manifestations in arterial hypertension, medical treatment must aim at reversing or preventing left ventricular hypertrophy and improving coronary flow reserve [31]. Clinical studies have shown that not all forms of antihypertensive therapy leads to regression of myocardial hypertrophy. After chronic treatment with diuretics no reversal of cardiac hypertrophy was demonstrated although arterial blood pressure was reduced to normotensive levels [31, 57]. When α-methyldopa was given in addition to ongoing diuretic treatment, some of the patients examined by Fouad et al. [13] showed a significant reduction in blood pressure. During therapy with enalapril [29, 34], calcium-channel blockers [30], prazosin, and clonidin [47] the reduction in blood pressure was paralleled by regression of cardiac hypertrophy. After treatment with β-adrenergic antagonists, heterogenous results were obtained regarding reversal of left ventricular hypertrophy [9, 36, 37]. The intrinsic sympathomimetic activity appears to play an important role in the reversal of cardiac hypertrophy during treatment with β-adrenergic antagonists. Treatment with acebutolol, a β-adrenergic antagonist with intrinsic sympathomimetic activity, did not produce any regression of cardiac hypertrophy, but when therapy in the same patients was switched to atenolol, a β-adrenergic blocker without intrinsic sympathomimetic activity, there was regression of cardiac hypertrophy while blood pressure remained at comparable levels [36].

Recently, Schulman and coworkers [37] demonstrated that atenolol has failed to produce left ventricular hypertrophy regression in elderly patients in contrast to the calcium channel blocker verapamil. At the present time there is no clear and convincing explanation, with β-adrenergic antagonists control left ventricular hypertrophy in some patients but not in others [9, 37]. They probably act on the hypertrophy process only indirectly by reducing the systolic pressure load and peak systolic wall stress by shifting peak systolic pressure later in the auxotonic phase of contraction. A direct myocardial effect is unlikely because norenephrine-mediated growth of isolated myocytes can only be prevented by blockade of α_1-receptors and not by β-receptors [41].

Catecholamines and angiotensin II have been identified as growth cofactors for left ventricular hypertrophy, which reinforce the growth response of the left ventricle to an elevated fiber stretch as a consequence of an increased systolic pressure load [22, 24, 25, 35, 38]. Accordingly, antihypertensive substances, which stimulate the sympatho-adrenergic system on the RAAS, such as diuretics and arterial vasodilators, should not be

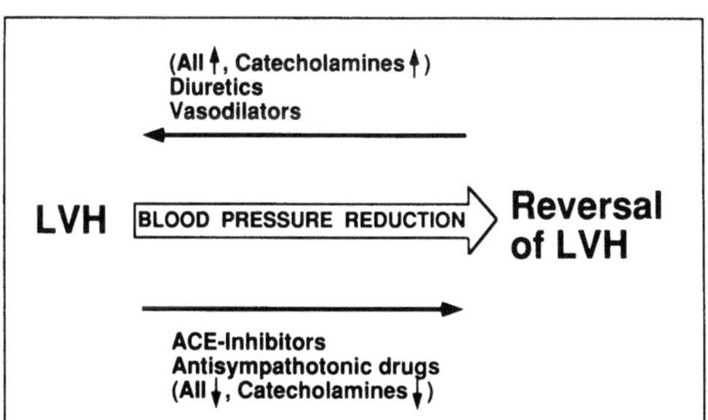

Fig. 3. Modulating effects of antihypertensive drugs on reversal of left ventricular hypertrophy (*LVH*) along with blood pressure reduction. Substances which increase catecholamines and/or angiotensin II attenuate the effect of systolic unloading; those which decrease catecholamines and/or angiotensin II potentiate the effect of systolic unloading in terms of LVH reversal

employed in treating patients with left ventricular hypertrophy because they attenuate or even compensate for their ventricular unloading effect due to blood pressure normalization (Fig. 3).

Calcium channel blockers seem to induce reversal of left ventricular hypertrophy primarily through the antihypertensive action and by the lack of undesired stimulating the sympathicoadrenergic system and the RAAS when given on the long term [46]. ACE inhibitors are very potent in reversing left ventricular hypertrophy because they further support the physical unloading of the left ventricle by virtue of inhibiting the growth cofactor angiotensin II. The spectrum of action of antihypertensive pharmacotherapy on myocardial hypertrophy thus ranges from complete absence of effect, through a largely pressure-related decrease in muscle mass, to an over-proportional reversal of myocardial hypertrophy.

Coronary Microcirculation

Whereas it has been shown that reversal of left ventricular hypertrophy can be achieved by suitable antihypertensive agents, information regarding the ability to repair coronary microcirculation is needed.

In spontaneously hypertensive rats, it is known that coronary reserve is enhanced after hydralazine, without a concomitant regression of left ventricular hypertrophy [1], while administration of the calcium antagonist felodipine led to a reversal of medial hypertrophy in coronary resistance vessels [10, 48]. Lisinopril was shown to improve coronary reserve along with reversal of both medial hypertrophy and myocardial fibrosis in spontaneously hypertensive rats [3]. A reduction in the volume density of myocardial volume and an increase in the length density of capillaries with either nifedipine or moxonidine was also found in experimental hypertension [28]. However, these experimental findings obtained in genetically determined hypertension in spontaneously hypertensive rats should be interpreted with caution in respect to extrapolations to man (Table 1). However, first preliminary clinical data indicate that after a prolonged antihypertensive treatment coronary flow reserve can be improved in hypertensive microvascular hypertension patients [51].

In nine hypertensive patients with normal coronary arteries and an impaired coronary reserve, systolic blood pressure was lowered by medical therapy for 9–12 months. The antihypertensive treatment consisted of β-

Table 1. Repair of coronary microcirculation: experimental data

	Structural reparation	Coronary flow reserve	Reference
Hydralazine	+	↑	[1]
Felodipine	+	↑	[48]
Lisinopril	+	↑	[3]
Nifedipine	+	?	[28]
Moxonidine	+	?	[28]

Table 2. Patient characteristics and antihypertensive treatment

Patient no.	Age (years)	Sex	Therapy: Ca^{2+} blockers	Therapy: Beta-blockers	Therapy: ACE inhibitors	Diuretics	Duration of therapy (months)
1	57	M	•			•	11
2	48	M	•	•	•	•	13
3	61	M		•		•	11
4	54	M	•			•	17
5	48	M	•	•		•	13
6	62	M	•	•		•	12
7	42	F	•				19
8	52	M		•			11
9	58	M		•			9

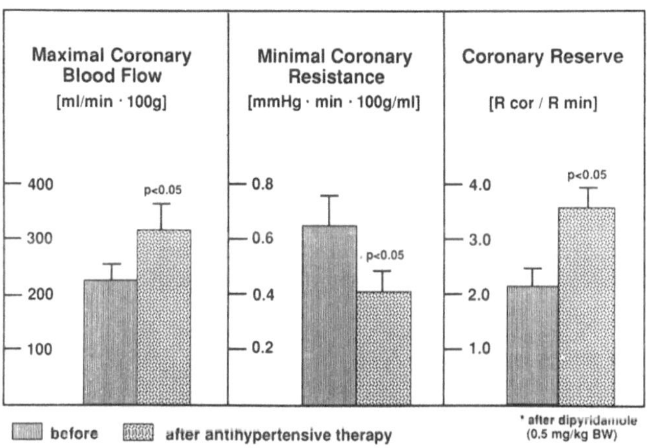

Fig. 4. Maximal coronary blood flow, minimal coronary resistance, and coronary reserve after long-term antihypertensive therapy

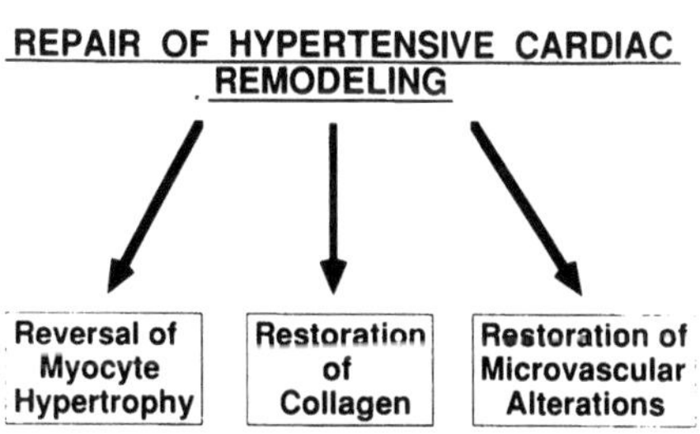

Fig. 5. Repair of hypertensive cardiac remodeling aims at reversal of myocyte hypertrophy, restoration of collagen, and improvement of coronary flow reserve by reversal of structural microvascular alterations

receptor blockers, calcium channel blockers, diuretics, and in one patient an ACE inhibitor (Table 2). Coronary reserve was measured (a) before antihypertensive therapy had been started and (b) after 9–12 months of treatment. To assess the long-term effect of antihypertensive therapy on coronary microcirculation apart from direct acute effects of the antihypertensive agents, treatment had been discontinued about 1 week before the second measurement of coronary reserve was performed. At this time all patients had similar elevated systolic and diastolic pressures as before the treatment had been started. After the prolonged phase of blood pressure reduction maximal coronary blood flow was significantly elevated, and minimal coronary resistance as the reciprocal index of overall coronary conductance was significantly lowered (Fig. 4). Accordingly, coronary reserve, i.e., the quotient of coronary resistance under resting conditions and after dipyridamole, was enhanced by about 30%.

This is the first clinical evidence that coronary flow reserve can be improved by long-term antihypertensive therapy. Further studies are needed to determine whether the improved coronary flow reserve after medical treatment of arterial hypertension is due to an influence of myocardial factors, such as left ventricular hypertrophy or myocardial fibrosis, or to a repair of the structurally remodeled microcirculation, such as restoration of structural abnormalities: decrease in medial thickening, increase in capillary density, or decrease in perivascular Fibrosis (Fig. 5).

References

1. Anderson PG, Bishop SP, Digerness SB (1989) Vascular remodeling and improvement of coronary reserve after hydralazine treatment in spontaneously hypertensive rats. Circ Res 64:1127–1136
2. Anversa P, Ricci R, Olivetti G (1986) Quantitative structural analysis of the myocardium during physiologic growth and induced cardiac hypertrophy: a review. J Am Coll Cardiol 7:1140–1149
3. Brilla CG, Janicki JS, Weber KT (1991) Cardioreparative effects of lisinopril in rats with genetic hypertension and left ventricular hypertrophy. Circulation 83:1771–1779
4. Brush JE, Cannon RO III, Schenke WH et al. (1988) Angina due to coronary microvascular disease in hypertensive patients without left ventricular hypertrophy. N Engl J Med 319:1302
5. Burger SB, Strauer BE (1981) Left ventricular hypertrophy in chronic pressure load due to spontaneous essential hypertension. II. Contractility of the isolated left ventricular myocardium and left ventricular stiffness. In: Strauer BE (ed) The heart in hypertension. Springer, Berlin Heidelberg New York, pp 37–52
6. Cannon RO III, Watson RM, Rosing DR, Epstein SE (1989) Angina caused by reduced vasodilator reserve of the small coronary arteries. J Am Coll Cardiol 1:1359–1373
7. Doering CW, Jalil JE, Janicki JS et al. (1988) Collagen network remodeling and diastolic stiffness of the rat left ventricle with pressure overload hypertrophy. Cardiovasc Res 22:686–695
8. Drayer JIM, Garding JM, Weber MS, Aronow WS (1983) Cardiac muscle mass during vasodilatation therapy of hypertension. Clin Pharmacol Ther 33:727–732
9. Dunn FG, Ventura HO, Messerli FH, Kobrin I, Frohlich E (1987) Time course of regression of left ventricular hypertrophy in hypertensive patients treated with atenolol. Circulation 76:254–258
10. Eisenlohr H, Schmiebusch H, Strauer BE (1988) Regression of mediahypertrophy in hypertensive coronary resistance vessels by antihypertensive therapy. Circulation 78 [Suppl II]:169 (abstr)
11. Folkow B (1982) Physiological aspects of primary hypertension. Physiol Rev 62:347–467
12. Ford EF (1985) Heart size. Circ Res 39:297–303

13. Fouad FM, Nakashima Y, Tarazi RC, Salcedo EE (1982) Reversal of left ventricular hypertrophy in hypertensive patients treated with methyldopa. Lack of association with blood pressure control. Am J Cardiol 49:795–801
14. Fouad FM, Slominski JM, Tarazi RC (1984) Left ventricular diastolic function in hypertension: relation to left ventricular mass and systolic function. J Am Coll Cardiol 3:1500–1506
15. Gaasch WH, Levine HJ, Quinones MA, Alexander JK (1976) Left ventricular compliance: mechanisms and clinical implications. Am J Cardiol 38:645–653
16. Grossmann W, Jones D, McLaurin LP (1975) Wall stress and patterns of hypertrophy in the human left ventricle. J Clin Invest 56:56–64
17. Hartford M, Wikstrand J, Wallentin I, Ljungman S, Wilhelmsen L, Berglund G (1984) Diastolic function of the heart in untreated primary hypertension. Hypertension 6:329–338
18. Jakob R, Brändle M, Dierberger B, Rupp H (1991) Functional consequences of cardiac hypertrophy and dilatation. In: Smits JFM (ed) Pharmacology of cardiac and vascular remodelling. Steinkopff, Darmstadt, pp 113–130
19. James TN (1990) Morphologic characteristics and functional significance of focal fibromuscular dysplasia of small coronary arteries. Am J Cardiol 65:126–226
20. Kannel WB, Dawber TR (1973) Hypertensive cardiovascular disease. The Framingham Study. In: Onesti G, Kim KE, Maya JH (eds) Hypertension: mechanisms and management. Grune and Stratton, New York, pp 93–99
21. Kannel WB, Daunenberg AL, Levy D (1987) Population implications of electrocardiographic left ventricular hypertrophy. Am J Cardiol 60:851–931
22. Kent RL, Mann DL, Cooper G IV (1992) Signals for cardiac muscle hypertrophy in hypertension. J Cardiovasc Pharmacol (in press)
23. Koren MJ, Devereux RB, Casale PN, Savage DD, Laragh JH (1991) Relation of left ventricular mass and geometry to morbidity and mortality in uncomplicated essential hypertension. Ann Intern Med 114:345–352
24. Kromer EP, Riegger GA (1988) Effects of long-term angiotensin converting enzyme inhibition on myocardial hypertrophy in experimental aortic stenosis in the rat. Am J Cardiol 62:161–163
25. Laks MN (1976) Norepinephrine – the myocardial hypertrophy hormone? Am Heart J 91:674–675
26. Levy D, Garrison RJ, Savage DD, Kannel WB, Castelli WP (1990) Prognostic implications of echocardiographically determined left ventricular mass in the Framingham Heart Study. N Engl J Med 322:1561–1566
27. Linzbach AJ (1981) Structural adaptation of the heart in hypertension and the physical consequences. In: Strauer BE (ed) The heart in hypertension. Springer, Berlin Heidelberg New York, pp 243–250
28. Mall G, Greber D, Gharebhaghi H, Wiest G, Mattfeldt T, Ganten E (1992) Myokardprotektion und Hypertrophieregression bei spontan hypertensiven Ratten durch Nifedipin und Moxonidin – Stereologische Untersuchungen. In: Herz-Kreislauf-Regulation, Organprotektion und Organschäden. Schattauer, Stuttgart (in press)
29. Motz W, Strauer BE (1988) Rückbildung der hypertensiven Herzhypertrophie durch chronische Angiotensin-Konversionsenzymhemmung. Z Kardiol 77:53–60
30. Motz W, Strauer BE (1989) Left ventricular function and collagen content after regression of hypertensive hypertrophy. Hypertension 13:43–50
31. Motz W, Strauer BE (1990) Differential therapy of hypertensive heart disease. Am J Cardiol 65:60G–64G
32. Motz W, Vogt M, Rabenau O, Scheler S, Lückhoff A, Strauer BE (1991) Evidence of endothelial dysfunction in coronary resistance vessels in patients with angina pectoris and normal coronary angiograms. Am J Cardiol 68:996–1003
33. Motz W, Vogt M, Scheler S, Schwartzkopff B, Strauer BE (1991) Coronary circulation in arterial hypertension. J Cardiovasc Pharmacol 17 [Suppl 2]:S35–S39
34. Nakashima Y, Fouad FM, Tarazi RC (1984) Regression of left ventricular hypertrophy from systemic hypertension by enalapril. Am J Cardiol 53:1044–1049
35. Östman-Smith I (1981) Cardiac sympathetic nerves as the final common pathway in the induction of adaptive cardiac hypertrophy. Clin Sci 61:265–272

36. Sau F, Seguro C, Merano G, Cherchi A (1986) Atenolol but not acebutolol reverses left ventricular hypertrophy secondary to arterial hypertension. J Am Coll Cardiol 7 [Suppl]:A186
37. Schulman SP, Weiss JL, Becker LC et al. (1990) The effects of antihypertensive therapy on left ventricular mass in elderly patients. N Engl J Med 322:1350–1356
38. Schunkert H, Dzau VJ, Tang SS, Hirsch AT, Apskin CS, Lorell BH (1990) Increased rat cardiac angiotensin converting enzyme activity and mRNA expression in pressure overload left ventricular hypertrophy. J Clin Invest 86:1913–1920
39. Schwartzkopff B, Motz W, Knauer S, Vogt M, Frenzel H, Strauer BE (1992) Thickening of the medial wall of intramyocardial arterioles in patients with arterial hypertension and reduced coronary vasodilator reserve (submitted)
40. Serizawa T, Mirsky J, Carabello BA, Grossman W (1982) Diastolic myocardial stiffness in gradually developing left ventricular hypertrophy in dog. Am J Physiol 242:H-633–H-637
41. Simpson P (1983) Norepinephrine-stimulated hypertrophy of cultured rat myocardial cell is an alpha-1-adrenergic response. J Clin Invest 72:732–738
42. Smith VE, Schulman P, Karimeddini MK, White WB, Meeran MK, Katz AM (1985) Rapid ventricular filling in left ventricular hypertrophy. II. Pathologic hypertrophy. J Am Coll Cardiol 5:869–874
43. Spann JF, Buccino RA, Sonnenblick EH, Braunwald E (1967) Contractile state of cardiac muscle obtained from cats with experimentally produced ventricular hypertrophy and heart failure. Circ Res 21:341
44. Strauer BE (1979) Myocardial oxygen consumption in chronic heart disease: role of wall stress, hypertrophy and coronary reserve. Am J Cardiol 44:730–740
45. Strauer BE (1979) Ventricular function and coronary hemodynamics in hypertensive heart disease. Am J Cardiol 44:999–1006
46. Strauer BE, Mahmoud MA, Bayer F, Bohn J, Motz U (1984) Reversal of left ventricular hypertrophy and improvement of cardiac function in man by nifedipine. Eur Heart J 5 [Suppl F]:53–60
47. Strauer BE, Bayer F, Brecht HM, Motz W (1985) The influence of sympathetic nervous activity on regression of cardiac hypertrophy. J Hypertension 3 [Suppl 41]:S39–S44
48. Strauer BE (1990) The significance of coronary reserve in clinical heart disease. J Am Coll Cardiol 15:775–783
49. Tomanek RJ, Palmer PY, Pfeiffer GL, Schreiter KL, Eastham CL, Marcus M (1986) Morphometry of canine coronary arteries, arterioles and capillaries during hypertension and left ventricular hypertrophy. Circ Res 58:38–46
50. Vogt M, Motz W, Strauer BE (1989) Coronary flow reserve in arterial hypertension. Scand J Clin Lab Invest 49 [Suppl 196]: 7–15
51. Vogt M, Motz W, Pölitz B, Scheler S, Strauer BE (1991) Improvement of coronary reserve by chronic treatment with ACE-inhibitors. Circulation 84:III–136
52. Weber KT, Janicki JS, Shroff SG, Pick R, Chen RM, Backey RI (1988) Collagen remodeling of the pressure-overloaded, hypertrophied nonhuman primate myocardium. Circ Res 62:757–765
53. Weber KT (1989) Cardiac interstitium in health and disease: remodelling of the fibrillar collagen matrix. J Am Coll Cardiol 13:1637–1652
54. Weber KT, Pick R, Silver MA et al. (1990) Fibrillar collagen and remodeling of dilated canine left ventricle. Circulation 82:1387–1401
55. Weber KT, Brilla CG (1991) Pathological hypertrophy and the cardiac interstitium: fibrosis and the renin-angiotensin-aldosterone system. Circulation 83:1849–1865
56. Weber KT, Anversa P, Armstrong PW et al. (1992) Remodeling and reparation of the cardiovascular system: an international perspective. J Am Coll Cardiol (submitted)
57. Wollam GL, Hall DW, Porter VD et al. (1983) Time course of regression of left ventricular hypertrophy in treated hypertensive patients. Am J Med 75 [Suppl]:100A–110A
58. Yamada H, Fabris B, Allen AM, Jackson B, Johnston CI, Mendelsohn FAO (1991) Localization of angiotensin converting enzyme in rat heart. Circ Res 68:141–149

VII. Ionic Channels and Signal Gating

β-Adrenergic Receptor Dysfunction in the Pathogenesis of Idiopathic Dilated Cardiomyopathy

M.R. Bristow[1], M. Movsesian[2], and A.M. Feldman[3]

Introduction

A major mechanism for regulation of contractility in human heart is the β receptor–G protein–adenylyl cyclase (RGC) pathway, shown in Fig. 1. Over the past 10 years five discrete abnormalities of the RGC pathway (Table 1) have been identified in the failing human heart. Four of these abnormalities are present in end-stage heart failure from idiopathic dilated cardiomyopathy (IDC). The purpose of this paper is to summarize the abnormalities in IDC, to speculate on the signals and molecular mechanisms which led to their creation, and to discuss their pathophysiologic significance.

Abnormalities in the RGC Complex in IDC

β_1 Receptor Down-regulation

Down-regulation of β-adrenergic receptors was the first RGC complex abnormality described in the failing human heart [1, 2]. The down-regulation was subsequently shown to be confined to the β_1 subtype [3–5]. In IDC down-regulation of β_1-adrenergic receptors is related to the severity of heart failure, and is present even in mild heart failure [6]. β_1-Receptor down-regulation is present in all examined types of human heart muscle disease characterized by systolic dysfunction and ventricular dilatation [3–5, 7–9].

Down-regulation of β_1-adrenergic receptors is produced by local mechanisms and is not the result of systemic/hormonal processes such as elevated circulating catecholamines [9]. This can be seen in Fig. 2, where down-regulation of β_1 receptors in subjects with advanced right heart failure from primary pulmonary hypertension is confined to the failing right ventricle and is not present in membranes prepared from the corresponding nonfailing left ventricles.

[1] University of Colorado Health Sciences Center, 4200 East Ninth Avenue, Denver, CO 80262, USA
[2] University of Utah Medical Center, 50 Medical Drive, Salt Lake City, UT 84132, USA
[3] The Johns Hopkins University, 720 Rutland, Baltimore, MD 21205, USA

Figulla et al. (Eds.)
Idiopathic Dilated Cardiomyopathy

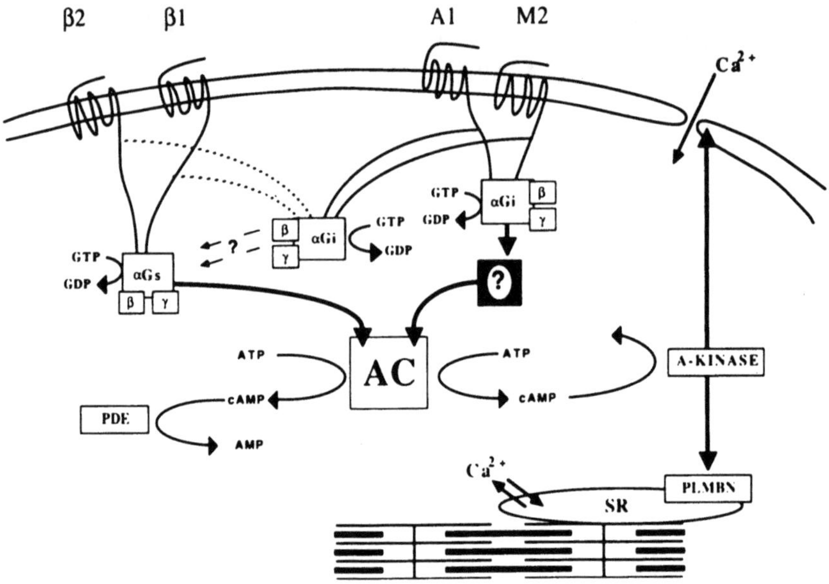

Fig. 1. Schema of the receptor–G protein–adenylyl cyclase (RGC) system in a human ventricular myocardial cell. *AC*, Adenylyl cyclase; *PDE*, phosphodiesterase; *PLMBN*, phospholamban; *SR*, sarcoplasmic reticulum

Table 1. Abnormalities of the β receptor–G protein–adenylyl cyclase system in the failing human heart

Abnormality	Degree of change (0–3+)	Type of heart muscle disease
β_1 Receptor down-regulation	++	IDC
	+	ISCDC
	+++	PPH (RV only)
β_2 Receptor uncoupling	+	IDC
	++	ISCDC
	+	PPH (RV only)
β_1 Receptor uncoupling	+	ISCDC
Increased functional activity	+	IDC
of G_i	+	ISCDC
Decreased activity of	+	IDC (RV only)
AC catalytic unit	+	PPH (RV only)

AC, Adenylyl cyclase; IDC, idiopathic dilated cardiomyopathy; ISCDC, PPH, primary pulmonary hypertension; RV, right ventricle.

The down-regulated β_1-adrenergic receptors are lost from all cellular compartments in the failing heart and are not simply internalized [9]. Therefore, the general mechanism is a loss in receptor protein. This in turn could occur as a result of a decrease in steady-state levels of β_1 receptor mRNA [10, 11], a decrease in mRNA translation frequency, or an increase

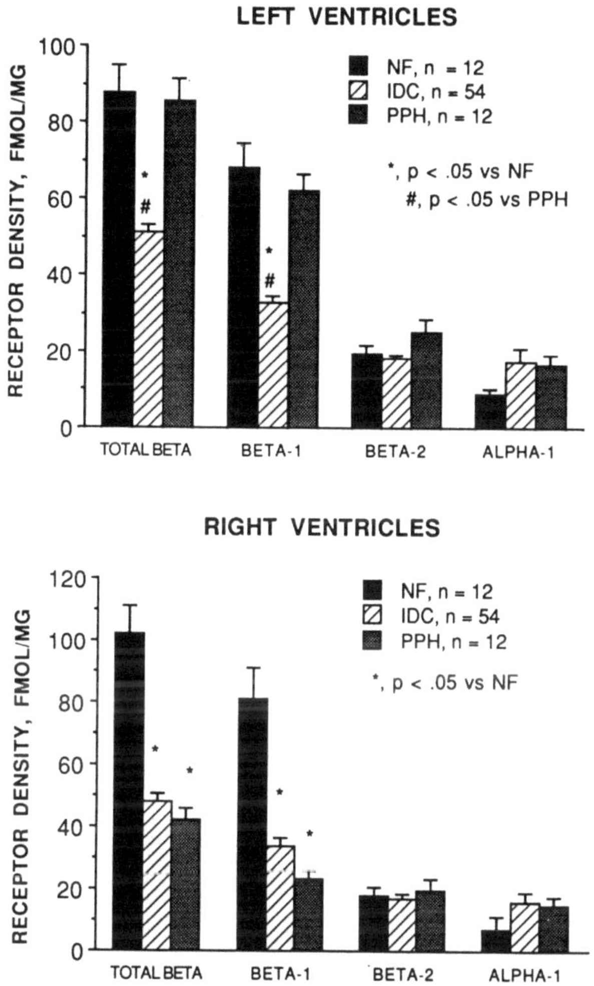

Fig. 2. Adrenergic receptor neasurements from left and right ventricles taken from nonfailing (*NF*) hearts, hearts failing with idiopathic dilated cardiomyopathy (*IDC*), and subjects with isolated right ventricular failure from primary pulmonary hypertension (*PPH*)

in receptor protein turnover. The latter could potentially occur as a result of formation of autoantibodies to the β_1 receptor [12, 13], as antibodies are known to facilitate breakdown of other types of receptors [14].

The initiating signal for β_1 receptor down-regulation is probably exposure to neurotransmitter norepinephrine accumulating in locally high amounts in the failing ventricle [15]. As increased adrenergic drive occurs in failing human ventricular myocardium tissue, norepinephrine stores are depleted [16], probably as a result of decreased neuronal reuptake [17]. Thus, tissue norepinephrine depletion is a marker of increased adrenergic drive. As can be seen in Fig. 3 tissue norepinephrine depletion is confined to

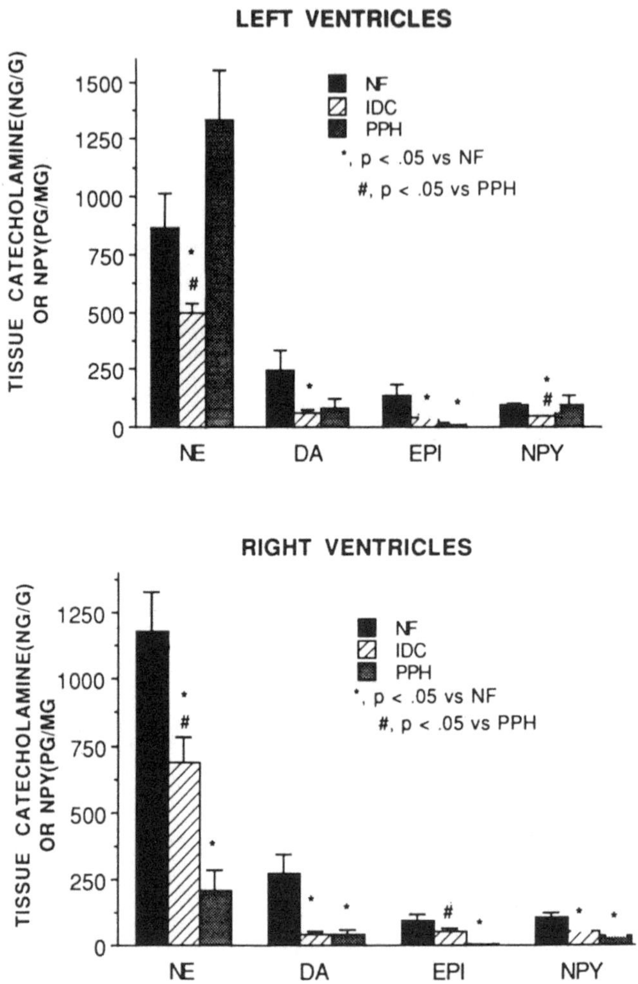

Fig. 3. Tissue catecholamine and NPY levels from same subjects as in Fig. 2

chambers which are failing, and norepinephrine depletion is not found in a nonfailing chamber when the clinical syndrome of heart failure is present. This means that the signal for β_1 receptor down-regulation is likely to be exposure to locally released adrenergic neurotransmitter.

Uncoupling of β_2-Adrenergic Receptors

The second abnormality to be identified in IDC ventricles was uncoupling of β_2-adrenergic receptors [18]. In human ventricular myocardium β_2 receptors are coupled to adenylyl cyclase [18] and a positive inotropic response [3, 19, 20], and in failing hearts β_2 receptors comprise 30%–40% of the total [3–5, 8, 9, 18]. Despite not decreasing in density in heart failure, β_2 receptors exhibit an approximate 30% decline in their ability to mediate adrenergic signals for stimulation of either adenylyl cyclase [18] or muscle contraction [3].

The molecular basis for β_2 receptor uncoupling is not known. The possibilities include receptor phosphorylation [21–23], receptor sequestration [24], and increased functional activity of the inhibitory G protein (G_i) (see below). Since uncoupling of β_2 receptors can be produced by catecholamine exposure [25], increased adrenergic catecholamine levels in failing heart are a likely explanation for the initiating signal of β_2 receptor uncoupling in the failing human heart. This hypothesis is further strengthened by data in isolated tissues taken from subjects receiving β-blocking agents where selective β_1 antagonists have been shown to potentiate β_2 agonist responses [26, 27], apparently through some form of cross-regulatory [28] phenomenon.

Increased Activity of αG_i

As assessed by pertussis toxin catalyzed ADP ribosylation [8, 29–31] and functional assays [8, 29], the activity of G_i is increased in the human heart failing as a result of IDC. Recent data indicate that the αG_i subunit may be directly coupled to β_2 receptors [32], which provides an attractive explanation for both β_2 receptor uncoupling and the functional significance of increased G_i. The latter is important inasmuch as responses mediated by both A_1-adenosine [33, 34] and M_2-muscarinic [35, 36] receptors are not altered in failing human hearts, despite the fact that both receptors are coupled to G_i. It may be that the increased G_i detected by ADP ribosylation assays is a separate species or separate compartment of G_i directly linked to β-adrenergic receptors.

Decreased Catalytic Activity of Adenylyl Cyclase

As assessed by Mn^{2+} stimulation, this abnormality is confined to pressure-overloaded myocardium and is present to a mild degree in IDC in right but not left ventricles [8, 9]. This indicates that the abnormality is not the result of the underlying myopathic process of IDC but rather is produced by superimposed pressure overload from secondary pulmonary hypertension. This abnormality is probably only a "biochemical" defect which can be demonstrated in isolated membrane preparations, as in isolated human heart it does not appear to account for a decline in receptor-mediated muscle contraction beyond that explained by β_1 receptor down-regulation and β_2 receptor uncoupling [9].

Constituents of the RGC Pathway that Appear To Be Unaltered

The constituents of the β receptor pathways which appear to be normal or near normal in the IDC left ventricle include the stimulatory G protein (G_s) [8, 29], the catalytic unit of adenylyl cyclase (C) [8, 9], basal [37] and phospholamban [38] mediated Ca^{2+} uptake in SR, SR-associated phos-

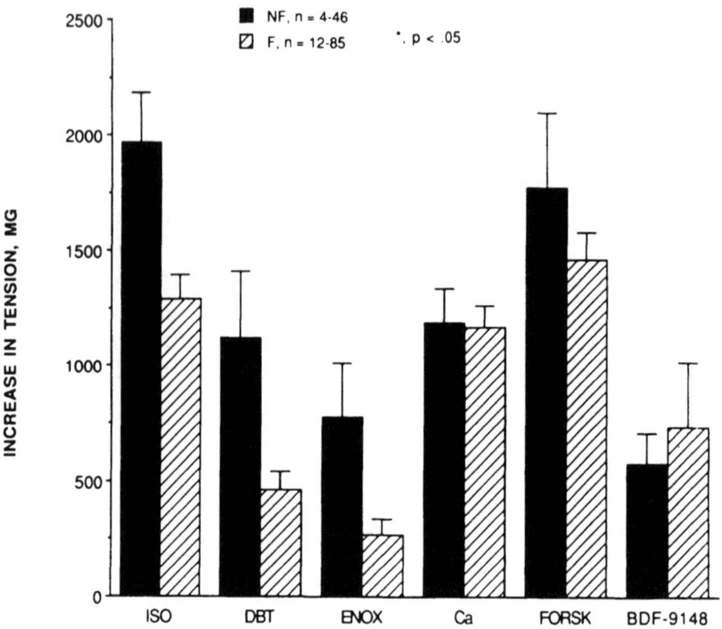

Fig. 4. Response to various agonists of isolated right ventricular trabeculae removed from nonfailing and failing human hearts. *ISO*, Isoproterenol; *DBT*, dobutamine; *ENOX*, enoximone; *Ca*, calcium; *FORSK*, forskolin

pholamban context [39], the α subunit of the sarcolemma-associated Ca^{2+} channel [40], and the myofilaments themselves as assessed by maximal systolic tension response to Ca^{2+} [41, 42] or measurement of contractile proteins [2]. As shown in Fig. 4, the unchanged basal activity and phospholamban mediated Ca^{2+} uptake in IDC is in disagreement with respective mRNA measurements of SR-associated Ca^{2+} ATPase [43] and phospholamban [44]. This suggests that the translation frequency of these SR-associated proteins is increased in heart failure or that a post-translational modification has increased the specific activity of the two enzymes.

Creatinine kinase per gram wet weight, which is used routinely as a marker of viable myocardium and is an important enzyme for energy generation, is decreased by 10%–15% in failing human ventricular myocardium as compared to nonfailing organ donor controls [2, 8, 9, 29]. However, when compared to age-matched control left ventricles not taken from brain-dead organ donors, creatinine kinase activity appears to decrease to a lesser, statistically nonsignificant level [9].

Functional Consequences of RGC Abnormalities in IDC

The receptor down-regulation and uncoupling and increased functional activity of G_i noted above leads to a selective loss in the ability of β agonists

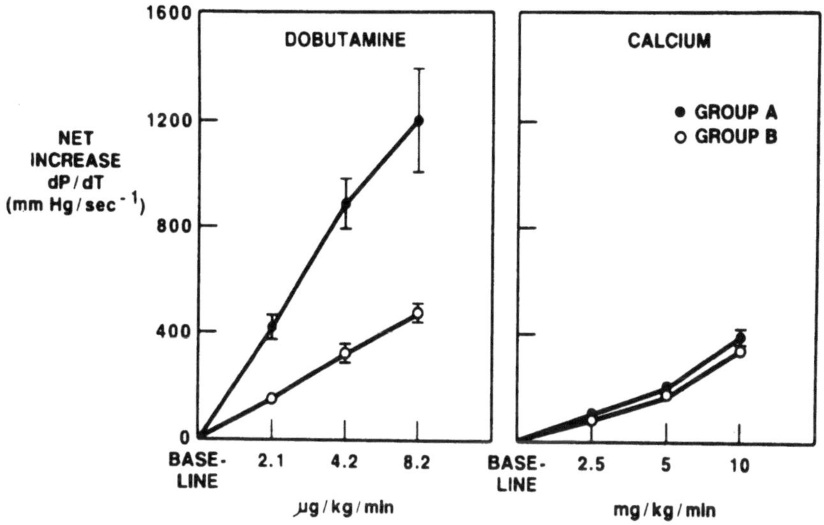

Fig. 5. Effect of dobutamine followed by calcium infusions on increase in d*P*/d*T* in subjects with no or minimal myocardial dysfunction (group A) versus those with advanced heart failure and ejection fractions <20% (group B)

and phosphodiesterase inhibitors to increase maximal systolic force, as noted in Figs. 4 and 5. Figure 4 shows that isolated right ventricular trabeculae have a decreased response to the full, nonselective β agonist isoproterenol, the partial [45], mildly selective β_1 agonist [46] dobutamine, and the phosphodiesterase inhibitor enoximone [47]. On the other hand, the maximal contractile response to calcium [41, 42], forskolin [48], and the sodium channel agonist BDF-9148 [49] is entirely preserved in isolated preparations of failing human heart. Figure 5 demonstrates that *intact* failing human heart also exhibits a selective loss in RGC pathway-mediated responses, as the d*P*/d*T* response of dobutamine is markedly decreased in subjects with IDC and ejection fractions of less than 0.25, compared to normal/near-normal subjects with ejection fractions over 0.50. In contrast, the response to calcium in these same subjects was not different [6].

In contrast to the abnormalities in RGC pathway-mediated systolic function, agonist-mediated increases in diastolic function appears unaltered in heart failure [50]. Colucci et al. have shown that the dobutamine-mediated increase in diastolic relaxation is not decreased in the failing human heart [50]. This is consistent with the distal components of the β receptor–SR pathway being normal in the failing heart, but it is not consistent with the receptor abnormalities noted above. It may be that a discrete pool of receptors couple normally to phospholamban phosphorylation in the failing heart, that there are spare β receptors coupled to phospholamban, or that an increase in specific activity of distal enzymatic components of the pathway in heart failure cancels the receptor desensitization present in proximal pathway constituents. Similarly, β-adrenergic receptors coupled to vascular smooth muscle dilation appear to be unaltered in heart failure [51].

Clinical Importance of β Receptor Pathways in the Failing Heart

Based on observations of prospectively created heart muscle damage [52], it can be shown that subjects with the clinical syndrome of heart failure from diffuse heart muscle disease pass through three stages. Stage 1 is the event that initially damages the contractile apparatus, which is by definition unknown in IDC. Stage 2, which may last for weeks to years, consists of activation of powerful compensatory mechanisms that stabilize central blood pressure and cardiac output distributed to organs with predominantly autoregulated blood flow, namely brain and heart. During this phase the heart muscle disease is subclinical, and subjects are relatively asymptomatic. However, the natural history of heart failure is progressive in at least 80% of subjects. Stage 3 of the syndrome begins when symptoms become progressive; this ends in either death from progressive pump dysfunction or sudden death, usually in the context of progressive pump dysfunction.

Why the natural history of heart failure is progressive is a very important question. One possibility is the underlying pathogenic insult continues, such as when additional myocardial infarctions occur, hypertension goes untreated, valvular heart disease is uncorrected, or a toxic substance such as an anthracycline antibiotic or alcohol continues to be administered. The second possibility is that the compensatory processes becomes less effective over time, due to attenuation of one or more of them. This certainly occurs for β-adrenergic mechanisms and their support of exercise responses. A third possibility, which forms a crucial and testable hypothesis, is that the compensatory processes in and of themselves are harmful. That

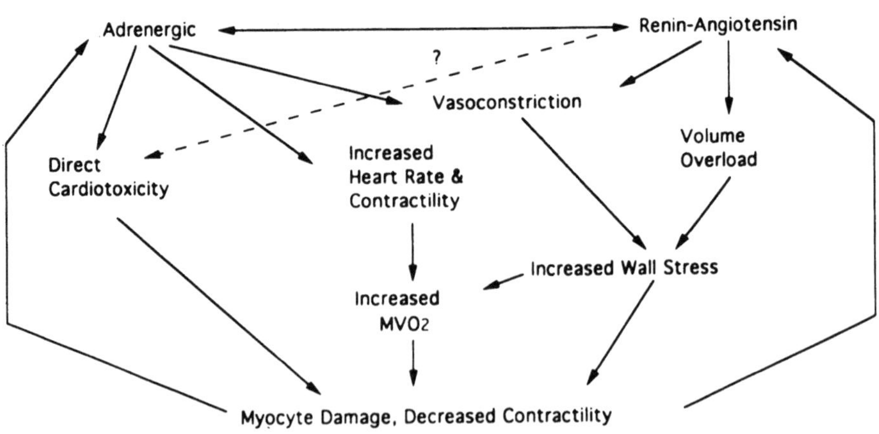

Fig. 6. Adverse myocardial effects of the major heart failure compensatory mechanisms

is, activation of the compensatory processes produces a "secondary" form of myocardial injury that accounts for the progressive nature of the natural history of pump dysfunction and heart failure.

Figure 6 outlines the relationship and functional consequences of the two most powerfully activated compensatory mechanisms in heart failure. When pump failure occurs the renin-angiotensin and adrenergic nervous systems are instantaneously activated, in order to stabilize cardiac output and central blood pressure. Aside from their salutary initial effects, activation of these mechanisms produces potentially deleterious effects of increased diastolic and systolic wall stress, increased heart rate, and increased contractility. Since these three processes are the major determinants of myocardial oxygen consumption, in the face of an already compromised blood supply [53] the increased oxygen demand may not be met. The result may be a deficit of myocardial energetics, resulting in myocardial injury. In addition to this "indirect" means of producing myocardial damage, adrenergic [54] and possibly angiotensin-II [55] stimulation may produce cardiac injury through direct cardiotoxic mechanisms.

As shown in Fig. 7, the positive aspects of the adrenergic compensatory mechanisms are partially attenuated by receptor desensitization. This means that the support of cardiac function provided by the adrenergic nervous system is ultimately diminished. How attenuation of β-adrenergic function may, on the one hand, limit exercise responses and yet not prevent the development of secondary myocardial injury is best illustrated by examining the "natural" role of adrenergic mechanisms in supporting cardiac function, and the results of a clinical experiment where a drug converted normal diurnal adrenergic drive into continuous low-level β_1-adrenergic stimulation. In the normal situation adrenergic drive is not required for resting cardiac function [56, 57], and when increased demand for cardiac output is encountered β-adrenergic mechanisms are capable of increasing cardiac output by two- to fivefold [57, 58]. In heart failure cardiac resting adrenergic drive is increased at rest [59, 60], while the ability to increase cardiac output is blunted [61]. β-Adrenergic desensitization accounts for at least part of the attenuated exercise response in heart failure [62], and since heart failure may be defined as the inability to perform exercise [63], β-adrenergic desensitization is a contributor to the natural history of the heart failure clinical syndrome.

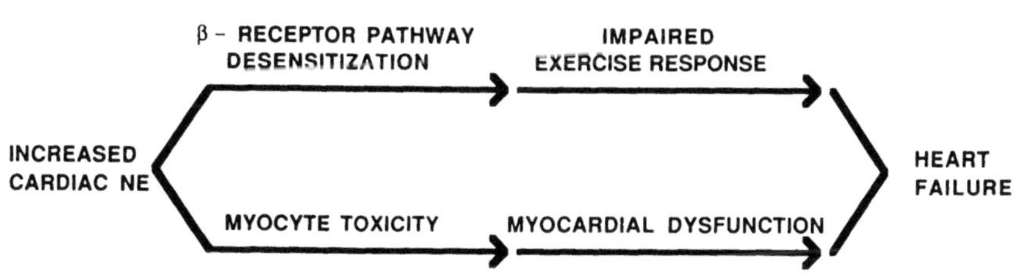

Fig. 7. The parallel effects of increased cardiac adrenergic drive in producing β-receptor pathway desensitization and myocardial damage through cardiotoxicity

On the other hand, β-adrenergic pathway desensitization and increased adrenergic drive in heart failure produce what may be the ultimate maladaptive response in the failing heart. The heart is subjected to a continuous, low-level, constant β_1-adrenergic stimulus that has been shown to deleterious to long-term outcome. That is, chronic low-level β_1-adrenergic stimulation by the partial agonist xamoterol [64] increases mortality in heart failure subjects by threefold [65]. This chronic, continuous low-level stimulation of adrenergic drive in moderately severe heart failure simulates conditions in advanced heart failure and converts the outcome to that more typical of late class IV failure [65].

Finally, the ubiquitous nature of β_1 receptor down-regulation in heart failure produces a phenotypic marker of failing heart muscle. β_1 receptor down-regulation therefore joins expression of the ventricular ANP gene [66] as biochemical markers of failing human ventricular myocardium.

References

1. Bristow MR, Ginsburg R, Sageman WS, Billingham ME, Stinson EB (1981) Analysis of the β-adrenergic receptor pathway in myocardial disease. Circulation 64/2:IV–286
2. Bristow MR, Ginsburg R, Minobe WA, Cubicciotti RS, Sageman WS, Lurie K, Billingham ME, Harrison DC, Stinson EB (1982) Decreased catecholamine sensitivity and β-adrenergic receptor density in failing human hearts. N Engl J Med 307:205–211
3. Bristow MR, Laser JA, Minobe W, Ginsburg R, Fowler MB, Rasmussen R (1984) Selective down-regulation of β_1 adrenergic receptors in the failing human heart. Circulation 70 [Suppl II]:65
4. Bristow MR, Ginsburg R, Fowler M, Minobe W, Rasmussen R, Zera P, Menlove R, Shah P, Stinson E (1986) β_1 And β_2-adrenergic receptor subpopulations in normal and failing human ventricular myocardium: coupling of both receptor subtypes to muscle contraction and selective β_1 receptor down-regulation in heart failure. Circ Res 59:297–309
5. Brodde O-E, Schüler S, Kretsch R, Brinkmann M, Borst HG, Hetzer R, Reidemeister JC Warnecke H, Zerkowski H-R (1986) Regional distribution of β-adrenoceptors in the human heart: coexistence of functional β_1 and β_2-adrenoceptors in both atria and ventricles in severe congestive cardiomyopathy. J Cardiovasc Pharmacol 8:1235–1242
6. Fowler MB, Laser JA, Hopkins GL, Minobe W, Bristow MR (1986) Assessment of the β-adrenergic receptor pathway in the intact failing human heart: progressive β-receptor down-regulation and specific pharmacologic subsensitivity to agonist response. Circulation 74/6:1290–1302
7. Brodde O-E, Zerkowski H-R, Doetsch N, Motomura S, Khamssi M, Michel MC (1989) Myocardial beta-adrenoceptor changes in heart failure: concomitant reduction in Beta$_1$- and beta$_2$-adrenoceptor function related to the degree of heart failure in patients with mitral valve disease. J Am Coll Cardiol 14:323–331
8. Bristow MR, Anderson FL, Port JD, Skerl L, Hershberger RE, Larrabee P, O'Connell JB, Renlund DG, Volkman K, Murray J, Feldman AM (1991) Differences in β-adrenergic neuroeffector mechanisms in ischemic vs. idiopathic dilated cardiomyopathy. Circulation 84:1024–1039
9. Bristow MR, Minobe W, Rasmussen R, Larrabee P, Skerl L, Klein JW, Anderson FL, Murray J, Mestroni L, Karwande SV, Fowler M, Ginsburg R (1992) β-Adrenergic neuroeffector abnormalities in the failing human heart are produced by local, rather than systemic mechanisms. J Clin Invest 89:803–815
10. Hadcock JR, Malbon CC (1988) Down-regulation of β-adrenergic receptors: Agonist-induced reduction in receptor mRNA levels. Proc Natl Acad Sci USA 85:5021–5025

11. Bristow MR, Durham C, Kline J, Rasmussen R, Port JD, Gesteland RR, Barry WH, Feldman AM (1991) Down-regulation of β-adrenergic receptors and receptor mRNA in heart cells chronically exposed to norepinephrine. Clin Res 39:256A
12. Limas CJ, Goldenberg IF (1989) Autoantibodies against cardiac β-adrenoreceptors in human dilated cardiomyopathy. Circ Res 64:97–103
13. Magnusson Y, Marullo S, Hoyer S, Waagstein F, Andersson B, Vahine A, Guillet J-G, Strosberg AD, Hjalmarson Å, Hoebeke J (1990) Mapping of a functional autoimmune epitope on the β_1-adrenergic receptor in patients with idiopathic dilated cardiomyopathy. J Clin Invest 86:1658–1663
14. Shimoyama R, Shelmet JJ, Savage CR, Boden G (1986) Effects of anti-insulin receptor antibodies (AIRA) on downregulation and turnover of insulin receptors on cultured hepatocytes. Diabetes 35:28–32
15. Bristow MR, Sandoval AB, Gilbert EM, Deisher T, Minobe W, Rasmussen R (1988) Myocardial alpha and beta-adrenergic receptors in heart failure: is cardiac derived norepinephrine the regulatory signal? Eur Heart J 9:35–40
16. Anderson FL, Port JD, Reid BB, Larrabee P, Hanson G, Bristow MR (1992) Myocardial catecholamines and neuropeptide Y depletion in the failing ventricles of patients with idiopathic dilated cardiomyopathy: correlation with beta adrenergic down-regulation. Circulation 85:46–53
17. Sandoval AB, Gilbert EM, Rose CP, Bristow MR (1989) Cardiac norepinephrine uptake and release is decreased in dilated cardiomyopathy. Circulation 80/4 [Suppl II]:393
18. Bristow MR, Hershberger RE, Port JD, Rasmussen R (1989) β_1 and β_2 adrenergic receptor mediated adenylate cyclase stimulation in nonfailing and failing human ventricular myocardium. Mol Pharmacol 35:295–303
19. Ginsburg R, Bristow MR, Zera P (1984) β_2 receptors are coupled to muscle contraction in human ventricular myocardium. Circulation 70/2 [Suppl II]:267
20. Kaumann AJ, Lemoine H (1987) β_2-Adrenoceptor-mediated positive inotropic effect of adrenaline in human ventricular myocardium: quantitative discrepancies with binding and adenylate cyclase stimulation. Naunyn Schmiedebergs Arch Pharmacol 335:403–411
21. Stadel JM, Nambi P, Shorr RG, Sawyer DF, Caron MG, Lefkowitz RJ (1983) Catecholamine-induced desensitization of turkey erythrocyte adenylate cyclase is associated with phosphorylation of the β-adrenergic receptor. Proc Natl Acad Sci USA 80:3173–3177
22. Sibley DR, Benovic JL, Caron MG, Lefkowitz RJ (1987) Regulation of transmembrane signaling by receptor phosphorylation. Cell 48:913–922
23. Lefkowitz RJ, Caron MG (1986) Regulation of adrenergic receptor function by phosphorylation. J Mol Cell Cardiol 18:885–895
24. Strader CD, Sibley DR, Lefkowitz RJ (1984) Association of sequestered beta-adrenergic receptors with the plasma membrane: a novel mechanism for receptor down regulation. Life Sci 35:1601–1610
25. Scarpace PJ, Abrass IB (1982) Desensitization of adenylate cyclase and down regulation of beta adrenergic receptors after in vivo administration of beta agonist. J Pharmacol Exp Ther 223:327–331
26. Hall JA, Kaumann AJ, Brown MJ (1990) Selective β_1-adrenoceptor blockade enhances positive inotropic responses to endogenous catecholamines mediated through β_2-adrenoceptors in human atrial myocardium. Circ Res 66:1610–1623
27. Hall JA, Petch MC, Brown MJ (1991) In vivo demonstration of cardiac β_2-adrenoreceptor sensitization by β_1-antagonist treatment. Circ Res 69:959–964
28. Hadcock JR, Port JD, Malbon CC (1991) Cross-regulation between G-protein-mediated pathways: activation of the inhibitory pathway of adenylylcyclase increases the expression of β_2-adrenergic receptors. J Biol Chem 266/18:11915–11922
29. Feldman AM, Cates AE, Veazey WB, Hershberger RE, Bristow MR, Baughman KL, Baumgartner WA, Van Dop C (1988) Increase of the M_r 40000 pertussis toxin substrate in the failing human heart. J Clin Invest 82:189–197
30. Neumann J, Scholz H, Döring V, Schmitz W, von Meyerinck L, Kalmár P (1988) Increase in myocardial G_i-proteins in heart failure. Lancet 2:936–937
31. Böhm M, Gierschik P, Jakobs K-H, Pieske B, Schnabel P, Ungerer M, Erdmann E (1990)

Increase of $G_{i\alpha}$ in human hearts with dilated but not ischemic cardiomyopathy. Circulation 82:1249–1265
32. Okamoto T, Murayama Y, Hayashi Y, Inagaki M, Ogata E, Nishimoto I (1991) Identification of a G_s activator region of the β_2-adrenergic receptor that is autoregulated via protein kinase A-dependent phosphorylation. Cell 67:723–730
33. Böhm M, Pieske B, Ungerer M, Erdmann E (1989) Characterization of A_1 adenosine receptors in atrial and ventricular myocardium from diseased human hearts. Circ Res 65:1201–1211
34. Hershberger RE, Feldman AM, Bristow MR (1991) A_1 adenosine receptor inhibition of adenylate cyclase in failing and nonfailing human ventricular myocardium. Circulation 83:1343–1351
35. Deighton NM, Motomura S, Borquez D, Zerkowski H-R, Doetsch N, Brodde O-E (1990) Muscarinic cholinoceptors in the human heart: demonstration, subclassification, and distribution. Naunyn Schmiedebergs Arch Pharmacol 341:14–21
36. Hershberger RE, Kimball JA, Minobe W, Rasmussen R, Bristow MR (1990) Muscarinic receptor pathways in nonfailing and failing human heart. J Am Cell Cardiol 15:171A
37. Movsesian MA, Bristow MR, Krall J (1989) Ca^{2+} uptake by cardiac sarcoplasmic reticulum from patients with idiopathic dilated cardiomyopathy. Circ Res 65:1141–1144
38. Movsesian MA, Colyer J, Wang JH, Krall J (1990) Phospholamban-mediated stimulation of Ca^{2+} uptake in sarcoplasmic reticulum from normal and failing hearts. J Clin Invest 85:1698–1702
39. Movsesian MA, Leveille C, Krall J, Colyer J, Wang JH, Campbell KP (1990) Identification and characterization of proteins in sarcoplasmic reticulum from normal and failing human left ventricles. J Mol Cell Cardiol 22:1477–1485
40. Rasmussen RP, Minobe W, Bristow MR (1990) Calcium antagonist binding sites in failing and nonfailing human ventricular myocardium. Biochem Pharmacol 39/4:691–696
41. Ginsburg R, Bristow MR, Billingham ME, Stinson EB, Schroeder JS, Harrison DC (1983) Study of the normal and failing isolated human heart: decreased response of failing heart to isoproterenol. Am Heart J 106:535–540
42. Gwathmey JK, Copelas L, MacKinnon R, Schoen FJ, Feldman MD, Grossman W, Morgan D (1987) Abnormal intracellular calcium handling in myocardium from patients with end-stage heart failure. Circ Res 61:70–76
43. Mercadier J-J, Lompré A-M, Duc P, Boheler KR, Fraysse J-B, Wisnewsky C, Allen PD, Komajda M, Schwartz K (1990) Altered sarcoplasmic reticulum Ca^{2+}-ATPase gene expression in the human ventricle during end-stage heart failure. J Clin Invest 85:305–309
44. Feldman AM, Ray PE, Silan CM, Mercer JA, Minobe W, Bristow MR (1991) Selective gene expression in failing human heart. Quantification of steady-state levels of messenger RNA in endomyocardial biopsies using the polymerase chain reaction. Circulation 83:1866–1872
45. Wollmering MM, Wiechmann RJ, Port JD, Hershberger RE, Focaccio A, Bristow MR (1991) Dobutamine is a partial agonist with an intrinsic activity of 0.5 in human myocardium. J Am Cell Cardiol 17:283A
46. Maccarrone C, Malta E, Raper C (1984) β-Adrenoceptor selectivity of dobutamine: in vivo and in vitro studies. J Cardiovasc Pharmacol 6:132–141
47. Gilbert EM, Hershberger RE, Mealey PL, Volkman K, Wiechmann RJ, Menlove RL, Movsesian MA, Bristow MR. Pharmacologic and hemodynamic effect of combined β agonist stimulation and phosphodiesterase inhibition in the failing human heart Anesthesiology (in press)
48. Böhm M, Gierschik P, Jakobs KH, Schnabel P, Kemkes B, Erdmann E (1989) Localization of a "postreceptor" defect in human dilated cardiomyopathy. Am J Cardiol 64:812–814
49. Schwinger RH, Böhm M, Mittmann C, LaRossee K, Erdmann E (1991) Evidence for a sustained effectiveness of sodium-channel activators in failing human myocardium. J Mol Cell Cardiol 23/4:461–471
50. Parker JD, Landzberg JS, Bittl JA, Mirsky I, Colucci WS (1991) Effects of β-adrenergic stimulation with dobutamine on isovolumic relaxation in the normal and failing human left ventricle. Circulation 84:1040–1048

51. Creager MA, Quigg RJ, Ren CJ, Roddy M-A, Colucci WS (1991) Limb vascular responsiveness to β-adrenergic receptor stimulation in patients with congestive heart failure. Circulation 83:1873–1879
52. Bristow MR, Mason JW, Billingham ME, Daniels JR (1981) Dose-effect and structure function relationship in doxorubicin cardiomyopathy. Am Heart J 102:709–718
53. Braunwald E (1969) 13th Bowditch lecture: the determinants of myocardial oxygen consumption. Physiologist 12:65–93
54. Mann DL, Kent RL, Parsons B, Cooper G (1992) Adrenergic effects on the biology of the adult mammalian cardiocyte. Circulation 85:790–804
55. Gavras H, Kremer D, Brown JJ, Gray B, Lever AF, MacAdam RF, Medina A, Morton JJ, Robertson JIS (1975) Angiotensin- and norepinephrine-induced myocardial lesions: experimental and clinical studies in rabbits and man. Am Heart J 89:321–332
56. Hallman GL, Leatherman LL, Leachman RD, Cooley DA (1970) Performance of the transplanted heart. J Cardiovasc Surg 11/1:1–6
57. Harrison DC (1978) Effects of beta-blockade on circulatory dynamics. In: Braunwald E (ed) Beta-adrenergic blockade: a new era in cardiovascular medicine: proceedings of an international symposium. Excerpta Medica, Amsterdam, pp 59–73
58. Ross J, Gault JH, Mason DT, Linhart JW, Braunwald E (1966) Left ventricular performance during muscular exercise in patients with and without cardiac dysfunction. Circulation 34:597–608
59. Swedberg K, Viquerat C, Rouleau J-L, Roizen M, Atherton B, Parmley WW, Chatterjee K (1984) Comparison of myocardial catecholamine balance in chronic congestive heart failure and in angina pectoris without failure. Am J Cardiol 54:783–786
60. Rose CP, Burgess JH, Cousineau D (1983) Reduced aortocoronary sinus extraction of epinephrine in patients with left ventricular failure secondary to long-term pressure or volume overload. Circulation 68/2:241–244
61. Weber KT, Kinasewitz GT, Janicki JS, Fishman AP (1982) Oxygen utilization and ventilation during exercise in patients with chronic cardiac failure. Circulation 65/6:1213–1223
62. White M, Gilbert EM, Yanowitz F, Mealey P, Olsen S, Larrabee P, Bristow MR (1991) Relation of β-adrenergic receptor status to peak exercise in idiopathic dilated cardiomyopathy. Circulation 84:II–53
63. Criteria Committee of the New York Heart Association (1964) Diseases of the heart and blood vessels; nomenclature and criteria for diagnosis, 6th edn. Little and Brown, Boston
64. Focaccio A, Larrabee P, Wollmering MM, Bristow MR (1991) Xamoterol has agonist activity in end-stage failing human heart. J Am Cell Cardiol 17:250A
65. Xamoterol in Severe Heart Failure Study Group (1990) Xamoterol in severe heart failure. Lancet 336:1–6
66. Saito Y, Nakao K, Arai H, Sugawara A, Morii N, Yamada T, Itoh H, Shiono S, Mukoyama M, Obata K, Yasue H, Ohkubo H, Nakanishi S, Imura H (1987) Atrial natriuretic polypeptide (ANP) in human ventricle: increased gene expression of ANP in dilated cardiomyopathy. Biochem Biophys Res Commun 148/1:211–217

Na^+ Channels as Molecular Substrate of Cardiac Excitability: Elementary Properties and Responsiveness to Drugs

M. Kohlhardt

Introduction

The heart beat is one of the most fascinating biological events. The ionic basis of cardiac excitation, its hormonal and metabolic regulation, and its susceptibility to drugs have been evaluated in great detail over the past four decades, but basic research reached the molecular level only in recent years. The breakthrough resulted from a new, revolutionary methodology. The patch clamp technique developed by Neher and Sakmann in the late 1970s allows single-channel analysis and determination of the elementary biophysical properties of ionic channels. The sequence analysis of channel proteins introduced by Numa's group defined the primary structure of a number of ionic channels and provides data for channel modeling. Major aspects of the cardiac excitation process can now be interpreted in molecular terms.

Rhythmic impulse generation in primary sinus node cells triggers an excitation wave which spreads over the atria and ventricles to initiate heart muscle contraction. Physiological pacemaker cells in the sinus node and in the AV node share the peculiarity of generating a particular type of action potential, slow responses, with a low rate of rise, which depend crucially on transmembrane Ca^{2+} currents. The excitation wave spreads in a very complex fashion over the myocardium to activate synchronously the overwhelming majority of ventricular cells. The excitation process in ventricular myocardium, however, is Na^+-dependent and depends basically on the occurrence of a fast transmembrane Na^+ current. Any influence capable of affecting the size of I_{Na} secondarily alters conduction velocity. In contrast to ventricular myocardium, cardiac Purkinje fibers conduct impulses with far greater speed. This well-established difference is not related to structural peculiarities of Purkinje fibers but most likely relies on their ability to generate transmembrane Na^+ currents of higher density. At least in papillary muscles, which have served in the past three decades as standard preparation in conventional cardiac electrophysiology, the maximal rate of rise of the action potential may vary significantly from one cell to another. It remains to be established whether this heterogeneity in Na^+ current size is

Physiological Institute of the University, Hermann-Herder-Str. 7, W-7800 Freiburg, Fed. Rep. of Germany

Figulla et al. (Eds.)
Idiopathic Dilated Cardiomyopathy

restricted to papillary muscles, or whether similar regional differences exist in other regions of the ventricles as well. Major regional differences in Na^+ current size predict a nonuniform conduction velocity within the ventricular wall which might well be of functional importance for the synchronous activation of the cardiac syncytium.

Molecular Structure of Na^+ Channels

Tetrodotoxin (TTX) sensitive Na^+ channels represent the prototype of a voltage-gated ionic channel and are a very characteristic constituent of excitable membranes of nerves, skeletal muscle, heart muscle, and several endocrine glands. They are a glyoprotein with a molecular weight of about 260 kDa and consist of 1800–1900 amino acids [31] organized in three subunits, a large α-subunit (230 kDa) and two (β_1 and β_2) small subunits. Their role for the functional properties, i.e., Na^+ permeation, activation, and inactivation, has been elaborated in rat brain and skeletal muscle Na^+ channels (for review see [8]). The membrane-spanning α-subunit provides the pathway for Na^+ permeation and contains most if not all functional channel properties. From the primary structure, Noda et al. [31] proposed a model of the transmembrane arrangement of the α-subunit consisting of four membrane-spanning domains, each consisting of six segments. Three cytoplasmic loops link the four domains with each other. They have been recognized as important regions, being involved in controlling the gating process and in modulating channel activity. Each domain has six segments which are long enough to span the membrane. Noda et al. [31] suggested that the S1 and S2 segments (hydrophobic with some hydrophilic residues) and the S5 and S6 segments (uniformly hydrophobic) form the transmembrane structure of the channel protein. Segment 4 is outstanding in that it is both hydrophobic and positively charged. Its particular structure resides on a characteristic amino acid arrangement, namely a positively charged residue at every third position provided by arginine and lysine. Catterall [7] first proposed the S4 segment to act as the voltage sensor and, by responding to a voltage gradient across the membrane as on membrane depolarization, to slide by a few Angstroms along its helical axis thereby forming a central pore for Na^+ permeation. Site-directed mutagenesis experiments [46] revealed evidence that the S4 segment is involved in the voltage-sensing mechanism of channel activation: a reduction of the number of positive charges in the S4 segment shifts the voltage dependence of Na^+ activation (m_∞) – precisely the predicted result when the S4 segment is acting as the voltage sensor.

Since the α-subunit has been highly conserved during evolution, the structure of the cardiac Na^+ channel is basically homologous with that of Na^+ channels of other origin. An outstanding discriminating property of the cardiac Na^+ channel is its low affinity for marine toxins such as TTX and saxitoxin (STX). Noda et al. [32] demonstrated that in single point mutation of the rat Na^+ channel II the replacement of glutamic acid 387 by glutamine converts this Na^+ channel from naturally high TTX sensitivity to low TTX sensitivity. Glutamic acid 387 is located between the S5 and S6 segments of domain I, a region assigned to the extracellular side of the membrane, whose occupancy with TTX or STX may be supposed to cause, finally, channel blockade. This very restricted region seems to be differently organized in cardiac Na^+ channels. In fact, Rogart et al. [36] identified a cardiac-specific Na^+ channel exon which encodes a distinct TTX-resistant isoform.

Elementary Properties of Cardiac Na^+ Channels

The biophysical properties of single cardiac Na^+ channels have been analyzed in several preparations, including neonatal and adult cardiocytes, isolated cells from disaggregated cardiac Purkinje's fibers, and frog myocardium. Suprathreshold membrane depolarizations to about −70 mV or more positive cause Na^+ channels to undergo a conformational change. As a result, they leave their rested state and attain the conducting state, which allows Na^+ permeation along the transmembrane electrochemical Na^+ gradient. Channel kinetics can be described by a Markovian reaction scheme according to

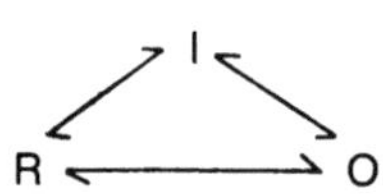

and comprise several closed states [27, 38] (combined as R) but only one open state (O), at least in normal cardiac Na^+ channels with operating inactivation. The absorbing [1] inactivated state (I) can be reached from R or O. Since inactivation develops rapidly during membrane depolarization, channel openings cluster in the very first few milliseconds after a voltage step. Consequently, open probability declines rapidly with voltage-dependent kinetics (time constants in the range of 0.5–5 ms, at 19°C) which mirrors in the fast I_{Na} decay. Late openings can occur but with a low likelihood even in the near-threshold range and are underlying a second I_{Na} decay component with severalfold larger time constants [5, 21, 27, 33]. The channel kinetics are voltage-dependent. Activation and inactivation become accelerated at positive membrane potentials but channel open time is almost voltage-insensitive.

Cardiac Na^+ channels can open more than once during membrane depolarization. Reopening appears as burst-like activity and is noteworthy in that the channel will not leave the O state to reach inevitably the I-state. Reopening becomes facilitated in a near-threshold potential range where inactivation develops rather slowly. It appears as a kinetic property, being quite differently pronounced. In neonatal cardiocytes, for example, only 15%–20% of the cell-attached patches were reported to contain Na^+ channels with a high reopening [21]. Low reopening with a mean number of opennings per burst of 2.5 at −50 mV dominated by far in this preparation. It is not yet clear whether poorly bursting Na^+ channels represent the type specifically present in neonatal cardiocytes. If this were the case, cardiocytes would express a channel subtype in an early developmental stage not existent in the adult stage.

Ultralong reopening [21] and spontaneous failure of inactivation [19] are kinetic episodes occurring with a very low likelihood, which can be explained by heterogeneous gating properties or by a modal gating behavior [29]. Spontaneous failure of inactivation becomes manifest as long-lasting burst activity persisting for hundreds of milliseconds and showing a several-fold prolonged and strongly voltage-dependent life time which is indistinguishable from modified Na^+ channel activity, as it occurs after the treatment with glutaraldehyde [24] and other chemical or pharmacological tools capable of removing Na^+ inactivation. These episodes can be observed in cell-free patches when Na^+ channels are isolated from their natural cytosolic environment but in situ as well. It is tempting to see spontaneous failure of inactivation as due to a transitory structural alteration in the linker region between domains III and IV identified as important for Na^+ inactivation [46, 49]. Such an intrinsic event is, however, too rare to gain functional significance for the shape and duration of the cardiac action potential.

Important permeation properties are the deviation from the independence principle and the saturation of cationic fluxes through Na^+ channels. There is circumstantial evidence that Na^+ channels represent a one-ion pore, i.e., containing only one ion at a time when in the open configuration. Although Na^+ channels can be passed by more than ten kinds of cations, Na^+ ions penetrate the pore preferentially. Sheets et al. [42] found a permeability ratio in single Na^+ channels from cardiac Purkinje fibers for P_{Ca}/P_{Na} of 0.02 and for P_K/P_{Na} of 0.094, thus confirming a very dominating Na^+ permeability as is firmly established in other excitable membranes (for review see [12]). Hille [10] proposed a narrow part of a few Angstroms in diameter located near the outer entrance of the pore to act as a selectivity filter which simultaneously hinders cations from free diffusion. To overcome this diffusion barrier, the hydrated Na^+ ion must interact with negative groups, most probably carboxyl groups. This part of the permeation pathway is extremely short and can be modeled to consist of four [10] or three energy [3] barriers of different height.

Low-affinity binding is important for cationic permeation. The K_D for Na^+ is in the range of 300–400 mmol/l. Cations with a high affinity for free negative groups within the pore cannot permeate but block the channel. This mechanism provides an explanation for the interference of bivalent cations with Na^+ currents. External Ca^{2+} and Mg^{2+} exert, at physiologically relevant concentrations, a depressant effect on cardiac excitability by reducing the elementary Na^+ current size in the near-threshold potential range [29, 42]. Internal Mg^{2+} has the same effect when Na^+ currents are outwardly directed, at membrane potentials positive to E_{Na} [34]. The reduced elementary current size arises from a fast flicker blockade: by competing with Na^+ for negative sites, Ca^{2+} and Mg^{2+} block repetitively the permeation pathway.

Na^+ Channel Distribution in the Sarcolemma

Since high-affinity markers of cardiac Na^+ channels are lacking so far, their distribution (homogeneous or clustered) in the sarcolemma, the control of their density (estimated to be two to ten Na^+ channels per square micrometer of cell surface), and their turnover are unknown. In cultured rat skeletal muscle cells, a pharmacological blockade of spontaneous electrical activity by local anesthetics, TTX or KCl depolarization caused the number of STX binding sites to increase [44] as if depressed excitability were followed by an increase of Na^+ channel density. Although Sherman and Catterall [44] proposed internal Ca^{2+} to have a controlling influence, the cellular signal transducing system involved in regulating synthesis of Na^+ channel protein and its incorporation in the sarcolemma remains to be elucidated. The average life time of the voltage-dependent Na^+ channels was found to be 3.1 days in cultured Schwann cells [35] but only 26 h in neuroblastoma cells [50]. In heart muscle, a detailed analysis of Na^+ channel turnover and its cellular control is particularly challenging. It would be highly interesting to see whether certain pathophysiological conditions or a pharmacologically modulated cardiac excitability influences Na^+ channel density.

Receptor Properties of Cardiac Na^+ Channels

Na^+ channels can be affected by a great variety of organic compounds with largely diverging chemical structures. Binding sites located at the external surface are freely accessible from the extracellular space for both hydrophilic and hydrophobic agents. The latter can reach putative binding sites at the cytoplasmic channel surface via a passage through the lipid matrix of the membrane surrounding the Na^+ channel. Receptor occupation may directly or allosterically evoke the biophysical response.

TTX, sea anemona toxin (ATX II), and α-scorpion toxin (ScTX) each find a specific binding site at the external channel surface. The TTX and ScTX binding sites have been localized in domain I between S5 and S6 [8, 32]. At least the TTX binding site is structurally not identical in the Na^+ channel family. Cardiocytes, embryonic [53], and denervated skeletal muscle cells express Na^+ channels having a severalfold reduced TTX affinity with a K_D in the micromolar range, in contrast to the nanomolar K_D for TTX in the neuronal relatives. A lipid-soluble group of compounds comprise batrachotoxin, aconitine, veratrum alkaloids, insecticides such as pyrethroids, and a novel drug family represented by the piperazinylindole derivative DPI 201-106 and the related BDF 9148. Structure and localization of their binding sites remain to be elucidated, but there is circumstantial evidence that the targets for these compounds are not in the external channel surface.

Local anesthetics and the related class 1 antiarrhythmics are of particular interest. Although their general use in cardiac arrhythmias became a matter of debate and seems questionable, they represent valuable tools in evaluating Na^+ channel properties. At least in neuronal tissue, they must pass the surrounding lipid matrix before they can interact with the Na^+ channel. Again, the nature and localization of the drug target are unknown. This is also true for a channel-associated binding site which interacts with phenylalkylamines and dihydropyridines. These Ca^{2+} blockers have recently also been shown to possess a considerable affinity for Na^+ channels.

Pharmacological interventions can provoke three distinct responses: (a) block as caused by TTX and local anesthetics, (b) activation which can be induced by Ca^{2+}-agonistic dihydropyridines; and (c) modification as evoked by ATX II, ScTX, plant alkaloids, insecticides, DPI 201-106, and BDF 9148. In the latter case, Na^+ channels loose some elementary properties.

Blockade of Na^+ Channels

Tonic and phasic I_{Na} depression represent the two manifestations of a pharmacologically induced Na^+ channel blockade. Lidocaine, mexiletine, quinidine, propafenone and derivatives, and all other antiarrhythmic drugs classified as class 1 agents are less effective, if at all, in rested cardiac preparations but gain effectiveness on repetitive stimulation. Use dependence of drug action is determined by both, the drug-specific dissociation kinetics and the duration of the interstimulus interval where Na^+ channels dwell in a less drug-sensitive, rested state and therefore loose drug occupancy. With rapidly dissociating drugs such as lidocaine, much higher driving rates are needed for an accumulation of blocked channels and, thus, to produce I_{Na} depression than with slowly dissociating drugs such as prajmalium [17]. Phasic I_{Na} depression differs from tonic I_{Na} depression by a less pronounced voltage-dependence and by a lower sensitivity to external Na^+ variations (for review see [16]). This puzzling block phenomenon can be explained by the modulated receptor hypothesis [11, 14], with a state-dependent binding site whose drug affinity varies with the channel configuration and increases when the Na^+ channels leaves its rested state and attains the activated or inactivated configuration. Most of the experimental results also can be fitted to an alternative model, that of the guarded drug receptor [45]. In the latter hypothesis, the channel configuration is proposed to control access to and egress from the binding site of the drug molecules. Both models are based on a single local anesthetic receptor. Since the permanently charged lidocaine derivative QX 314, in contrast to neuronal tissues, depresses I_{Na} in cardiac Purkinje's fibers not only when administered from inside but also from outside, a second local anesthetic binding site located externally seems to exist in cardiac Na^+ channels. This additional local anesthetic binding site would be another example for tissue-specific peculiarities which discriminate cardiac Na^+ channels from other members of this channel family. But what might be the purpose of this diversity? Since endogeneous ligands of the local anesthetic binding site are unknown, the differential drug susceptibility seems meaningless in controlling cardiac excitability.

Local anesthetics and related drugs do not necessarily act stereoselectively in depressing I_{Na}. The (+)- and (−)-enantiomers of disopyramide [28] and diprafenone [22] or the geometrical isomers of quinidine [28] possess equal blocking potency. Tocainide acts stereoselectively in that the (+)-enantiomer was found in Na^+ channels of adult skeletal muscle to be about four times more potent than the (−)-enantiomer [48] while in cardiac Na^+ channels the (−)-enantiomer has been reported to be the more effective form [43]. The tertiary amine local anesthetic RAC 109 may likewise be classified as acting partially stereoselectively [9].

Ca^{2+}-antagonistic drugs, including phenlyalkylamines [2] and 2.4 dihydropyridines [55] or the (+)-enantiomer of DPI 201-106, can also interfere in micromolar concentrations with Na^+ channels, thereby causing a time- and voltage-dependent I_{Na} depression as do local anesthetics. This is note-worthy since Ca^{2+}-antagonistic drugs, DPI 201-106, and local anesthetics are structurally unrelated to each other. The susceptibility to Ca^+-antagonistic drugs may be related to the substantial amino acid sequence similarity between the Ca^{2+} channel α-subunit and the Na^+ channel α-subunit [47]. Possibly, both ionic channels have a homologuous region capable of interacting with these agents. Since many local anesthetics block preferentially Na^+ channels and have less affinity, if at all, for L-type Ca^{2+} channels, it is tempting to speculate that Na^+ channels have two distinct binding sites, one for local anesthetics and the other for Ca^{2+}-antagonistic agents whose drug occupancy has the same functional consequence and causes blockade.

Lidocaine has been shown in gating current measurements in axons and other excitable membranes to induce a charge immobilization (for review see [15]). This mechanism might be the common mode of action of blocking drug molecules. Consequently, they would hinder the Na^+ channels in undergoing a conformational change, perhaps by preventing the S4 segment from moving along its helical axis. Plugging of the open pore is an alternative explanation, but of course this is inconsistent with gating current changes in the presence of lidocaine. Either model postulates an all-or-none blockade of the Na^+ channel. All-or-none blockade was verified in single Na^+ channel experiments in neonatal cardiocytes with amiodarone, propafenone, and lidocaine [23]. Block becomes manifest as a time- and voltage-dependent decline in open probability. Unblocked Na^+ channels which lost drug occupation show normal elementary properties and dwell as long in the open and bursting states as under control conditions, before drug administration. Even quinidine, supposed to interact to a certain degree with activated Na^+ channels, failed to influence the open-state kinetics [4], at least under normal conditions when Na^+ channels attain a single open state.

Many ionic channels respond to the presence of blocking drug molecules with a flicker blockade. Most probably by competing with the permeant ion for its permeation pathway, the drug molecule blocks repetitively the open pore with concentration-dependent rate constants. Na^+ channels behave exceptionally since they are resistant to flicker blockade unless their inactivation process is removed. Only in this particular case does their open state become drug sensitive [4] to several antiarrhythmics, including propafenone, diprafenone, and prajmalium [25]. Although flicker blockade is meaningless for I_{Na} depression in drug-treated cardiac tissue in situ, it attracts interest for theoretical reasons. Na^+ inactivation somehow protects the open state against these drugs. Association rate constants as calculated from the flicker block kinetics are in the order of magnitude of $10^7\,mol^{-1}\,s^{-1}$ for structurally quite heterogeneous agents such as 9-aminoacridine [54] propafenone, and prajmalium [25]. Recent single-channel experiments with lidocaine have revealed, surprisingly, that this classical local anesthetic, in concentrations effective to reduce open probability significantly, fails to flicker-block DPI-modified cardiac Na^+ channels (I. Benz and M. Kohlhardt, in preparation). This and the differential voltage dependence of flicker blockade and conventional I_{Na} depression cast serious doubt on the view that a common molecular mechanism underlies both forms of Na^+ channel blockade.

Activation of Na^+ Channels

Na channel activation can be caused by several 2,4-dihydropyridines, including (−)-Bay K 8644 and CGP 28392 [26, 56], established to act as Ca^{2+} agonists. In Na^+ channels these drugs can exert their promoting effect on open probability without a concomitant increase in burst activity and slightly prolong the open state. Drug-treated Na^+ channels do not necessarily attain two conducting states. This clearly contrasts with the response of L-type Ca^{2+} channels in heart muscle, which can no longer be considered a specific target for Ca^{2+} agonists. Apart from a much more pronounced increase in open probability, (−)-Bay K 8644 provokes a severalfold increased, long-lasting second open state [13].

Modification of Na^+ Channels

As first identified with DPI 201-106, diphenylpiperazinylindole derivatives represent a novel class of organic agents acting at Na^+ channels as modifiers. The highly lipophilic DPI 201-106 strongly retards I_{Na} decay [6] and consequently prolongs the cardiac action potential significantly [39] which can be reversed by TTX. DPI-bound single cardiac Na^+ channels were identified as being prevented from inactivation [18]. They show sustained burst activity for hundreds of milliseconds and, as the other kinetic peculiarity, attain a severalfold prolonged and strongly voltage-dependent open state. Although substate openings occasionally occur [19, 30, 41], openings with a normal conductance of about 15 pS dominate by far. DPI-modified Na^+ channels thus largely preserve their normal permeation properties, in striking contrast to veratridine-modified Na^+ channels [51]. BDF 9148, likewise in micromolar concentrations, exerts the same modifying effect [51]. DPI- and BDF-modified Na^+ channels resemble in many aspects the properties of BTX-modified Na^+ channels and can be fitted to a Markovian reaction scheme by assuming that the inactivated state cannot be reached. In removing inactivation, DPI finds a channel-associated, stereoselective binding site which interacts with the (−)-enantiomer [37]. Wang et al. [52] concluded from I_{Na} measurements in whole-cell clamped cardiocytes that the (−)-DPI binding site is allosterically coupled with an (+)-DPI binding site the latter being involved in the blocking effect of (+)-DPI. Evidence was provided from action potential analysis in papillary muscles that the (−)-DPI binding site is distinct from the binding site for the modifying ATX II and for veratridine [40].

The nature and localization of the (−)-DPI binding site are unknown. It is tempting to speculate that (−)-DPI and related drugs bind to the loop linking domain III with domain IV since this structure was shown in mutagenesis experiments [46] and with site-directed antibodies [49] to be involved in channel inactivation. However, protein reagents such as the oxidizing iodate which preferentially reacts with S-S bonds between cysteine and cysteine and, after cytoplasmic administration, acts as channel modifier, also provoke differential channel kinetics. Iodate-modified cardiac Na^+ channels attain two open states and leave the bursting state more rapidly when compared with (−)-DPI-modified Na^+ channels [24].

(−)-DPI prolongs the cardiac action potential in a voltage-dependent fashion [20]. In partially depolarized myocardium this drug effect is only slightly pronounced, but it becomes accentuated on membrane hyperpolarization. This is because the number of (−)-DPI-bound Na^+ channels is a function of voltage and increases with membrane hyperpolarization. A plausible explanation is that (−)-DPI molecules have a greater chance of finding a binding site with removal of resting inactivation when more Na^+ channels are available. The blocking (+)-enantiomer has the opposite voltage dependence and acts less effectively in cardiocytes with normal, i.e., high resting potential but gains strongly in potency with membrane depolarization.

Although action potential prolongation represents, because of the concomitant increase in refractoriness, an antiarrhythmic principle, the therapeutic importance of Na^+ channel modifiers as class 3 antiarrhythmic agents remains to be defined. Arrhythmogeneity may emerge from a regionally different responsiveness of the cardiac action potential to (−)-DPI and related drugs so that a preexisting spatial heterogeneity in absolute refractoriness could be potentiated.

References

1. Aldrich RW, Corey DP, Sevens CF (1983) A reinterpretation of mammalian sodium channel gating based on single channel recording. Nature 306:436–441

2. Bayer R, Kalusche D, Kaufmann R, Mannhold R (1975) Inotropic and electrophysiological actions of verapamil and D 600 in mammalian myocardium. III. Effects of optical isomers on transmembrane action potentials. Naunyn Schmiederbergs Arch Pharmacol 290:81–97
3. Begenisish TB, Cahalan MD (1980) Sodium channel permeation in squid axons. II. Non-independence and current-voltage relations. J Physiol (Lond) 307:243–257
4. Benz I, Kohlhardt M (1991) Responsiveness of cardiac Na^+ channels to antiarrhythmic drugs: the role of inactivation. J Membr Biol 122:267–278
5. Brown AM, Lee KS, Powell T (1981) Sodium currents in single rat heart cells. J Physiol (Lond) 318:479–500
6. Buggisch D, Isenberg G, Ravens R, Scholtysik G (1985) Effects of the novel cardiotonic compound DPI 201-106 on contractility, membrane potentials and membrane currents in isolated mammalian heart preparations. Naunyn Schmiedebergs Arch Pharmacol 329:R53
7. Catterall WA (1986) Voltage-dependent gating of sodium channels: correlating structure and function. Trends Neurosci 9:7–10
8. Catterall WA (1988) Structure and function of voltage-sensitive ion channels. Science 242:50–61
9. Clarkson CW (1989) Stereoselective block of cardiac sodium channels by RAC109 in single guinea pig ventricle myocytes. Circ Res 65:1306–1323
10. Hille B (1975) Ionic selectivity, saturation, and block of sodium channels. A four barrier model. J Gen Physiol 66:535–560
11. Hille B (1977) Local anesthetics: hydrophilic and hydrophobic pathways for the drug-receptor reaction. J Gen Physiol 69:497–515
12. Hille B (1984) Ionic channels of excitable membranes. Sinauer, Sunderland
13. Hess P, Lansman JB, Tsien RW (1984) Different modes of Ca channel gating behaviour favoured by dihydropyridine Ca agonists and antagonists. Nature 311:538–544
14. Hondeghem LM, Katzung BG (1977) Time- and voltage-dependent interactions of antiarrhythmic drugs with cardiac sodium channels. Biochim Biophys Acta 472:373–398
15. Khodorov BI (1981) Sodium inactivation and drug-induced immobilization of the gating charge in nerve membrane. Prog Biophys Mol Biol 37:49–89
16. Kohlhardt M (1984) Block of sodium current by antiarrhythmic agents: analysis of the electrophysiological effect of propafenone in heart muscle. Am J Cardiol 54:13D–19D
17. Kohlhardt M, Seifert C (1985) Properties of $\dot{V}_{max}$ block of I_{Na}-mediated action potentials during combined application of antiarrhythmic drugs in cardiac muscle. Naunyn Schmiedebergs Arch Pharmacol 330:235–244
18. Kohlhardt M, Fröbe U, Herzig JW (1986) Modification of single cardiac Na^+ channels by DPI 201-106. J Membr Biol 89:163–172
19. Kohlhardt M, Fröbe U, Herzig JW (1987) Properties of normal and non-inactivating single cardiac Na^+ channels. Proc R Soc Lond [B]232:71–93
20. Kohlhardt M, Fröbe U, Herzig JW (1987) Removal of inactivation and blockade of cardiac Na^+ channels by DPI 201-106: different voltage-dependencies of the drug actions. Naunyn Schmiedebergs Arch Pharmacol 335:183–188
21. Kohlhardt M, Fichtner H, Fröbe U (1988) Predominance of poorly reopening single Na^+ channels and lack of slow Na^+ inactivation in neonatal cardiocytes. J Membr Biol 103:283–291
22. Kohlhardt M, Fichtner H (1988) Inhibitory effects of diprafenone stereoenantiomers on cardiac Na^+ channels. Eur J Pharmacol 156:55–62
23. Kohlhardt M, Fichtner H (1988) Block of single cardiac Na^+ channels by antiarrhythmic drugs: the effect of amiodarone, propafenone and diprafenone. J Membr Biol 102:105–119
24. Kohlhardt M, Fichtner H, Fröbe U (1989) Gating in iodate-modified single cardiac Na^+ channels. J Membr Biol 112:67–78
25. Kohlhardt M, Fichtner H, Fröbe U, Herzig JW (1989) On the mechanism of drug-induced blockade of Na^+ currents: interaction of antiarrhythmic compounds with DPI-modified single cardiac Na^+ channels. Circ Res 64:867–881
26. Kohlhardt M, Fichtner H, Herzig JW (1989) The response of single cardiac sodium channels in neonatal rats to the dihydropyridines CGP 28392 and (−)-Bay K 8644. Naunyn Schmiedebergs Arch Pharmacol 340:210–218

27. Kunze DL, Lacerda AE, Wilson DL, Brown AM (1985) Cardiac Na^+ currents and the inactivating, reopening, and waiting properties of single cardiac Na^+ channels. J Gen Physiol 86:691–719
28. Mirro MJ, Watanabe AM, Bailey JC (1981) Electrophysiological effects of the optical isomers of disopyramide and quinidine in the dog. Dependence on stereochemistry. Circ Res 48:867–874
29. Nilius B (1988) Calcium block of guinea pig heart sodium channels with and without modification by the piperazinylindole DPI 201-106. J Physiol (Lond) 399:537–558
30. Nilius B, Vereecke J, Carmeliet E (1989) Properties of the bursting Na^+ channel in the presence of DPI 201-106 in guinea pig ventricular myocytes. Pflugers Arch 413:234–241
31. Noda M, Shimizu S, Tanabe T, Takai T, Kavano T, Ikeda T, Takahashi H, Nakayama H, Kanaoka Y, Minamino N, Kangawa K, Matsuo H, Raftery MA, Hirose T, Inayama S, Hayashida H, Miyata T, Numa S (1984) Primary structure of *Electrophorus electricus* sodium channel deduced from cDNA sequence. Nature 312:121–127
32. Noda M, Suzuki H, Numa S, Stühmer W (1989) A single point mutation confers tetrodotoxin and saxitoxin insensitivity on the sodium channel II. FEBS Lett 259:213–216
33. Patlak JB, Ortiz M (1985) Slow currents through single sodium channels of the adult rat heart. J Gen Physiol 86:89–104
34. Pusch M, Conti F, Stühmer W (1989) Intracellular magnesium blocks sodium outward currents in a voltage- and dose-dependent manner. Biophys J 55:1267–1271
35. Ritchie JM (1988) Sodium channel turnover in rabbit cultured Schwann cells. Proc R Soc Lond [B]233:423–430
36. Rogart RB, Cribbs LL, Muglia LK, Kephart DD, Kaiser MW (1989) Molecular cloning of a putative tetrodotoxin-resistant rat heart Na^+ channel isoform. Proc Natl Acad Sci USA 86:8170–8174
37. Romey G, Quast U, Pauron D, Frelin C, Renaud CF, Lazdunski M (1987) Na^+ channels as sites of action of the cardiotonic agents DPI 201-106 with agonist and antagonist enantiomers. Proc Natl Acad Sci USA 84:896–900
38. Scanley BE, Hanck DA, Chay T, Fozzard HA (1990) Kinetic analysis of single sodium channels from canine cardiac Purkinje cells. J Gen Physiol 95:411–437
39. Scholtysik G, Salzmann R, Berthold R, Herzig JW, Quast U, Markstein R (1985) DPI 201-106, a novel cardioactive agent: combination of cAMP-independent positive inotropic, negative chronotropic, action potential prolonging and coronary dilatory properties. Naunyn Schmiedebergs Arch Pharmacol 329:316–325
40. Scholtysik G, Quast U, Schaad A (1986) Evidence for different receptor sites for the novel cardiotonic S-DPI 201-106, ATX II and veratridine on the cardiac sodium channel. Eur J Pharmacol 125:111–118
41. Schreibmayer W, Tritthart HA, Schindler H (1989) The cardiac sodium channel shows a regular substate pattern indicating synchronized activity of several ion pathways instead of one. Biochim Biophys Acta 986:172–186
42. Sheets MF, Scanley BE, Hanck DA, Makielski JC, Fozzard HA (1987) Open sodium channel properties of single canine cardiac Purkinje fibres. Biophys J 52:13–22
43. Sheldon RS, Cannon NJ, Nies AS, Duff HJ (1988) Stereospecific interaction of tocainide with the cardiac sodium channel. Mol Pharmacol 33:327–331
44. Sherman SJ, Catterall WA (1984) Electrical activity and cytosolic calcium regulate levels of tetrodotoxin-sensitive sodium channels in cultured rat muscle cells. Proc Natl Acad Sci USA 81:262–266
45. Starmer CF, Grant AO, Strauss HC (1984) Mechanism of use-dependent block of sodium channels in excitable membranes by local anesthetics. Biophys J 46:15–27
46. Stühmer W, Conti F, Suzuki H, Wang X, Noda M, Yahagi N, Kubo H, Numa S (1989) Structural parts involved in activation and inactivation of the sodium channel. Nature 339:597–603
47. Tanabe T, Takeshima H, Mikami A, Flockerzi V, Takahashi H, Kangawa K, Kojima M, Matsuo H, Hirose T, Numa S (1987) Primary structure of the receptor for calcium channel blockers from skeletal muscle. Nature 328:313–318
48. Tricarico D, Fakler B, Spittelmeister W, Ruppersberg JP, Stützel R, Franchini C, Tortorella V, Conte-Camerino D, Rüdel R (1991) Stereoselective interaction of tocainide

and its chiral analogs with the sodium channels in human myoballs. Pflugers Arch 418:234–237
49. Vassilev P, Scheuer T, Catterall WA (1989) Inhibition of inactivation of single sodium channels by a site-directed antibody. Proc Natl Acad Sci USA 86:8147–8151
50. Waechter CJ, Schmidt JW, Catterall WA (1983) Glycosylation is required for maintenance of functional sodium channels in neuroblastoma cells. J Biol Chem 258:5117–5123
51. Wang G, Dugas M, Armah BI, Honerjäger P (1990) Sodium channel comodification with full activator reveals veratridine reaction dynamics. Mol Pharmacol 37:144–148
52. Wang G, Dugas M, Armah BI, Honerjäger P (1990) Interaction between DPI 201-106 enantiomers at the vardiac sodium channel. Mol Pharmacol 37:17–24
53. Weiss RE, Horn R (1986) Functional differences between two classes of sodium channels in developing rat skeletal muscle. Science 233:361–364
54. Yamamoto D, Yeh JZ (1984) Kinetics of 9-aminoacridine block of single Na^+ channels. J Gen Physiol 84:361–377
55. Yatani A, Brown AM (1985) The calcium channel blocker nitrendipine blocks sodium channels in neonatal rat cardiac myocytes. Circ Res 56:868–875

The Involvement of G_s in Regulation of the L-Type Voltage-Gated Ca^{2+} Channel

L. Birnbaumer[1,2], J. Codina[2], R. Mattera[2], A. Yatani[1], Y. Imoto[1], A.M. Brown[1], M.-J. Hawkes[1], S.L. Hamilton[1], L. Toro[1], and E. Stefani[1]

Introduction

Our interest in calcium channels was prompted by the results of studies published by Hescheler and collaborators at the beginning of 1987, in which these authors showed in whole cell recordings that the inhibitory regulation of the voltage gate calcium channels in NG108-15 cells in response to opioids is reconstituted after pertussis toxin (PTX) treatment with G_o as well as G_i [1]. We attempted to obtain similar results by adding activated PTX-sensitive G proteins (G_o or G_i) to inside-out membrane patches of cardiac ventricle cells, but we were faced with the fact that single calcium channel currents in inside-out membrane patches are extremely unstable, inactivating within 15–30 s from the time of patch excision. By studying ways to stabilize calcium channel function in inside-out membrane patches we noted that these channels could be stabilized by activating what appeared to be G_s. The following is an account of some of these experiments and the research that followed.

Stimulation of Ca^{2+} Channel Activity by G Protein

Two lines of evidence, one obtained with guinea pig ventricle membrane patches in the inside-out configuration [2], the other with cardiac sarcolemmal and skeletal muscle T tubule membranes incorporated into planar lipid bilayers [2–5], indicate that dihydropyridine (DHP) sensitive voltage-gated Ca^{2+} channels of these tissues are positively modulated by a G protein, and that this G protein is G_s, the stimulator of adenylyl cyclases.

Experiments with Cardiac Membrane Patches

In contrast to K^+ channels which survive patch excision very well, allowing regulatory studies to be done by direct application of the test compounds to the cytoplasmic face of the membrane/channel, voltage-gated Ca^{2+} channels of the kind putatively regulated in intact cells by G protein-coupled re-

[1] Department of Molecular Physiology and Biophysics, and
[2] Department of Cell Biology, Baylor College of Medicine, Houston, TX 77030, USA

Figulla et al. (Eds.)
Idiopathic Dilated Cardiomyopathy

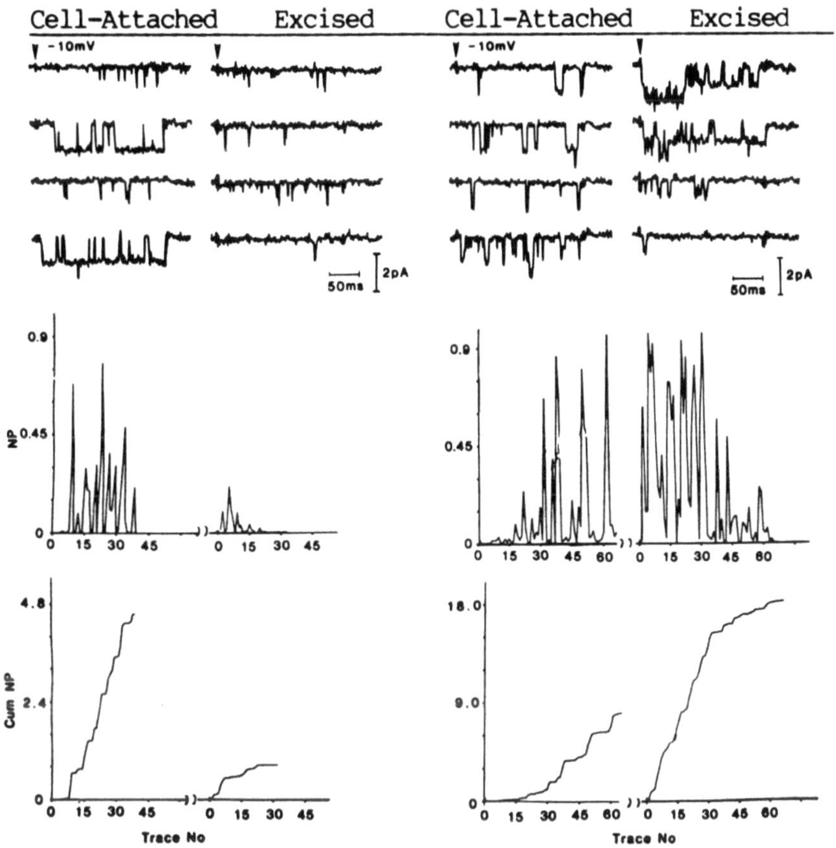

Fig. 1. Transient survival of heart ventricle voltage-gated Ca^{2+} channels after membrane patch excision and stabilization through isoproterenol-dependent activation of a G protein with GTPγS. Guinea pig cardiac ventricle cells were patch-clamped in the cell-attached mode with isoproterenol in both bath and pipette. Single Ca^{2+} channel current was recorded during 200-ms test depolarizations from holding potentials of −90 to −20 mV applied at a frequency of 2 Hz. *Left panels*, bath solution without GTPγS; *top*, representative single-channel current activities recorded during test potentials before and after excision of a membrane patch into bathing solution without GTPγS; *bottom*, NP diaries of the Ca^{2+} channel activities of the same membrane patch during 45 successive test periods immediately before and after excision. *Right panels*, bath solution with 100 μM GTPγS; *top*, representative single-channel Ca^{2+} currents as obtained in test cycles before and after excision of a representative membrane patch; *bottom*, NP diaries of Ca^{2+} channel activates in the same membrane patch before and after its excision into a solution with GTPγS. On the average, the presence of GTPγS had two effects. (a) it increased the number of test cycles during which single Ca^{2+} channel opening was observable, and (b) it increased the cumulative NP valuees as quantified during the first 30 test cycles. (Adapted from [2])

ceptors are extremely unstable after patch excision [6]. The left panels of Fig. 1 illustrate this point for the cardiac guinea pig ventricle Ca^{2+} channels. The channels are readily seen in the cell-attached patch throughout repeated cycles of activation by the depolarizing test potential. However, after patch

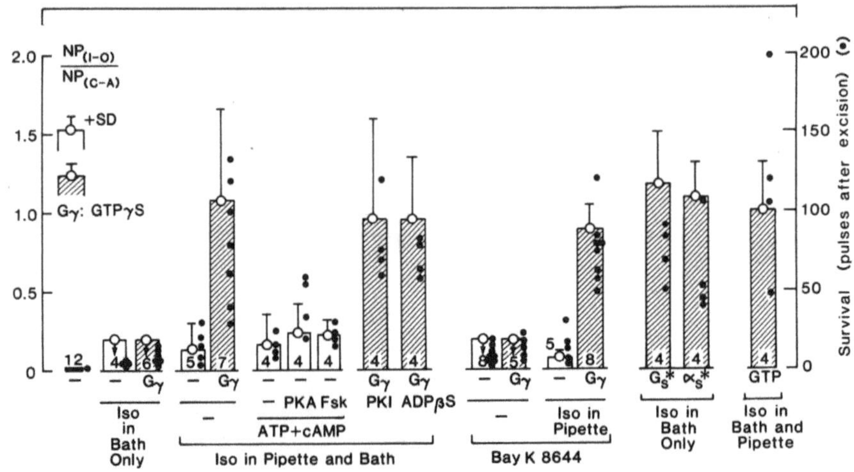

Fig. 2. Summary of experiments with guinea pig cardiac ventricle voltage-gated, DHP-sensitive Ca^{2+} channels, such as shown in Fig. 1, in which additions to pipette and bath solutions were varied. H.P. = −60 mV; 200-ms pulses to −10 mV at 0.5 Hz. *Bars*, reatio between cumulative NP values during the first 30 test cycles after patch excision and those prior to patch excision (*left ordinate*); *closeed circles*, last test cycle during which after patch excision at least one single-channel opening was observed (*right ordinate*). *Iso*, Isoproterenol; *PKA*, catalytic unit of cAMP-dependent protein kinase; *PKI*, inhibitor of PKA; *lpt*, leupeptin. When added, GTPγS was 100 μ*M*; GTP, 1 m*M*; ATP, 1 m*M*; cAMP, 500 μ*M*; Bay K8644, 3 μ*M*; iso, 10^{-5} *M*; GTPγS-activated G_s (G_s*) or GTPγS-activated α_s (α_s*) 100–200 p*M*. (Adapted from [2])

excision, activity was observed only during the initial tests and rapidly extinguishes thereafter. The particular experiment shown was obtained by including the β-adrenergic receptor agonist in the bath and pipette. In absence of the agonist, cell-attached Ca^{2+} channel activity induced by depolarization was reduced by about 50%–70%, a reflection of the loss of stimulatory effect of isoproterenol. However, survival after patch excision was essentially nil (not shown). The right panels of Fig. 1 show that when GTPγS was added to the bath into which the membrane patch was excised, this resulted in a significant increase in both average activity during the depolarizing test periods and the number of tests during which depolarization was able to lead to channel activation. As summarized in Fig. 2, this stabilizing effect could not be mimicked by protocols that increased adenylyl cyclase activity, increased cAMP levels, or provided for phosphorylation by the catalytic unit of cAMP-dependent protein kinase (protein kinase A). Also, inhibition of protein kinase A by addition of excess protein kinase inhibitor (PKI) did not inhibit the effect of isoproterenol plus GTPγS to stabilize these Ca^{2+} channels [2].

The requirement for isoproterenol in the pipette to obtain the effect of GTPγS suggested that the responsible G protein might be G_s, with β-adrenergic receptors in the membrane patch acting locally to accelerate the activation of G_s before irreversible inactivation of the channel molecules would occur. Indeed addition of GTPγS-activated G_s, but not GTPγS-activated G_k, substituted for the combination isoproterenol in the pipette and GTPγS or GTP in the bath. Pretreatment of ventricle cells with isoproterenol or, in its defect, pretreatment of the cells with the DHP agonist Bay K8644 was still necessary.

It has been well established that cardiac ventricle dihydropyridine-sensitive Ca^{2+} currents (the so-called slow inward Ca^{2+} current of cardiac cells) is stimulated about two- to threefold by adrenergic stimuli via a protein kinase A-mediated phosphorylation reaction [7]. It is mimicked quantitatively by injection of pure active catalytic subunit of protein kinase A, and it is inhibited by injection of PKI. However, the experiments presented above indicated that the same cardiac DHP-sensitive Ca^{2+} channels can also be modulated positive by G_s, provided the channels are

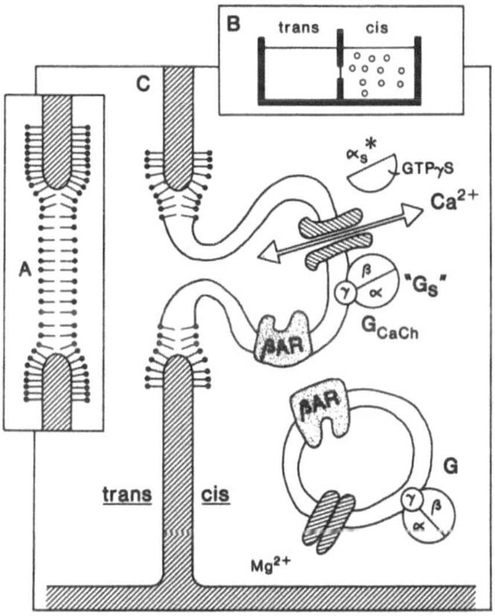

Fig. 3. Strategy of incorporation of DHP-sensitive Ca^{2+} channels into planar phospholipid bilayers to study their direct regulation by a G protein. Phosphatidylcholine and phosphatidylethanolamine in decane were painted onto a hole 0.1–0.3 mm in diameter in a partition separating two chambers of 0.15–0.5 ml referred to as *cis* and *trans*, with respect to the side from which membranes were forced to be fused with the lipid bilayer. Prior to membrane addition the *trans* chamber was filled with buffer A (50 m*M* NaCl, 2 m*M* $MgCl_2$, and 5 m*M* HEPES, pH 7.0) and the *cis* chamber with buffer A containing 100 m*M* $BaCl_2$. Incorporation of ionic channels into the lipid bilayer was obtained after addition of membrane vesicles in 10- to 20-μl aliquots to the *cis* chamber. After incorporation had occurred and to prevent further incorporations from occurring during the course of the experiments, one of two protocols was followed. In one, the liquid in the *cis* chamber was extensively exchanged with the same solution, buffer A plus 100 m*M* $BaCl_2$, free of membranes. In the other, the liquids of both chambers were exchanged, the *trans* chamber receiving the original *cis* solution (buffer A plus 100 m*M* $BaCl_2$) and the *cis* chamber receiving buffer A. In both cases Ba^{2+} was used as the charge carrier, the difference being that single-channel currents were in the *cis* to *trans* direction in the first and in the *trans* so *cis* direction in the second of these protocols. The figure depicts a membrane vesicle that is inside-out with respect to the orientation of its cytoplasmic side, as is the case with skeletal muscle T, tubules, before and after its fusion to the lipid bilayer, and illustrates the sidedness that results

prestimulated either by physiological means as obtained with isoproterenol or by pharmacological means as obtained with Bay K8644. Although on rare occasions it was possible to restimulate single-channel Ca^{2+} currents with GTPγS-activated G_s after they had ceased to respond to the test potential, this was not a consistent findings. We therefore asked whether the G protein actually stimulated or simply stabilized channel molecules. To obtain an answer to this question it was necessary to obtain a cell-free system in which DHP-sensitive Ca^{2+} channels were more stable. Incorporation of Ca^{2+} channels into planar phospholipid bilayers [8, 9] (schematized in Fig. 3) provided such a system.

Experiments with Native Membranes Incorporated into Lipid Bilayers

The results obtained indicated that Ca^{2+} channels are stimulated under steady-state conditions by G_s or a molecule very similar to it [3]. Initial experiments were with Ca^{2+} channels of bovine cardiac sarcolemmal vesicles. Conclusive data were obtained with Ca^{2+} channels from skeletal muscle T tubules.

Skeletal muscle T tubule membranes are inside-out vesicles with respect to the orientation of their cytoplasmic surface and incorporate into bilayers exposing their extracellular surface to the *trans* chamber [9]. After incorporation in the presence of Bay K8644 and using Ba^{2+} as the charge carrier, the DHP-sensitive Ca^{2+} channels exhibit an activity at 0 mV that, while being cyclical and occurring in bursts and clusters of bursts, occurs nevertheless in a stochastical manner and is on the average stable over long periods of up to 30 m in or until bilayer breakdown (middle left panel of Fig. 4). Addition of GTPγS-activated G_s or GTPγS, which activates endogenous coincorporated skeletal muscle T tubule G protein, results in clear increases in NPo, i.e., in the product of the opening probability of incorporated channels times the number of channels incorporated (lower two panels of Fig. 4). This can be mimicked by cholera toxin treated G_s in the presence of GTP (right middle panel of Fig. 4) and by the resolved GTPγS-activated α subunit of G_s (not shown). Stimulations by the variously activated G_s preparations averaged two- to threefold as seen in over 30 independent experiments. Importantly, stimulation of incorporated Ca^{2+} channels is not dependent on presence of Bay K8644 (top two panels of Fig. 4). Indeed, in one series of 12 experiments without Bay K8644, the relative increases in NPo values averaged over 2–5 min were 18-fold.

The above results were all obtained in the absence of ATP and hence not mediated by channel phosphorylation. This indicated that these Ca^{2+} channels, like the muscarinic K^+ channels and adenylyl cyclase, are under direct control of a G protein, which, as is shown below, is undoubtedly G_s. In agreement with whole-cell recordings [7] from ventricle cells, as well as single channel current records from inside-out membrane patches of GH_3 cells [10], addition to the *cis* chamber of catalytic unit of protein kinase A and ATP or ATPγS also stimulated single Ca^{2+} channel currents by increasing NPo values, which in a series of four experiments in the presence of Bay K8644 averaged twofold. Figure 5 presents a diagram that summarizes

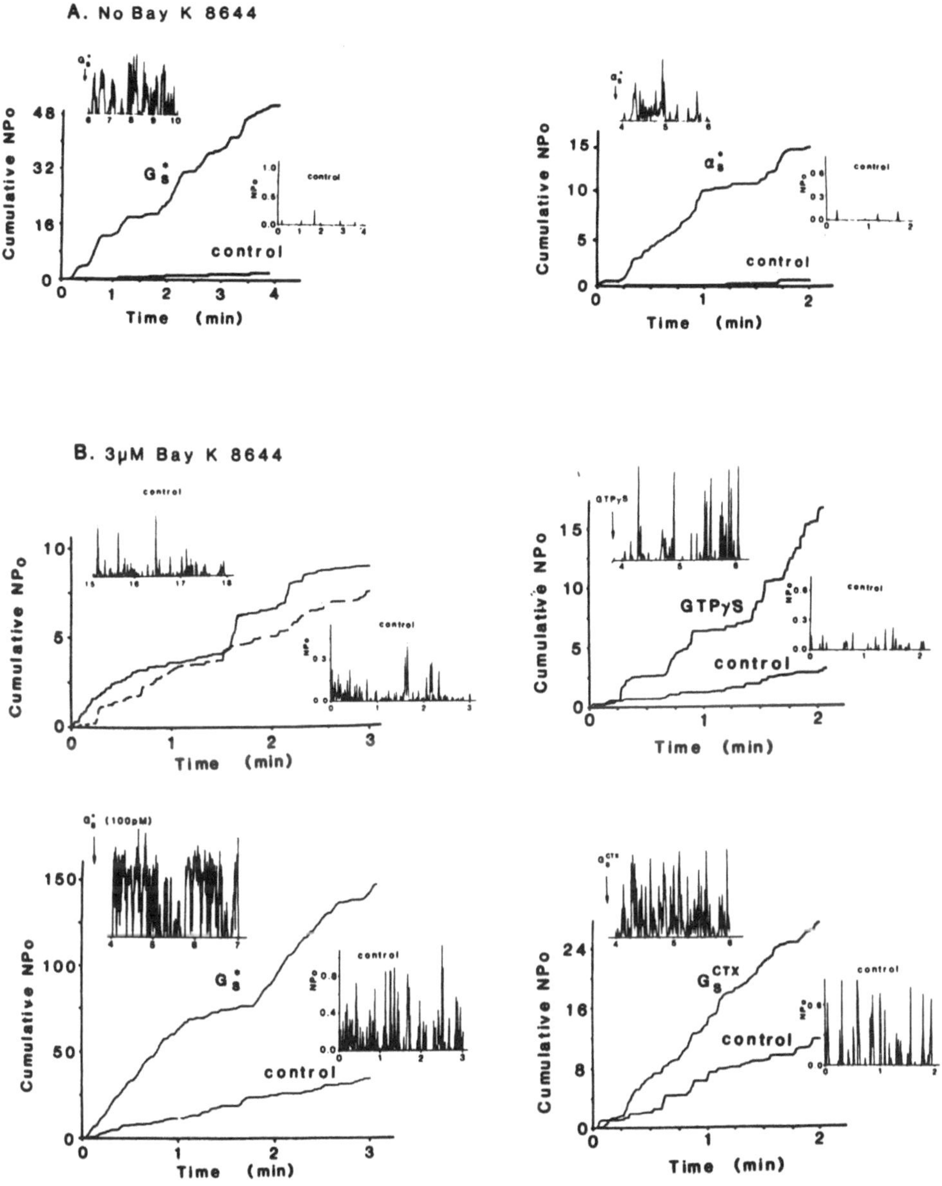

Fig. 4A,B. Representative experiments showing direct regulation of the skeletal muscle T tubule DHP-sensitive Ca^{2+} channel by purified human erythrocyte G_s. Experiments were carried out both in the absence (**A**) and the presence of the DHP agonist Bay K8644 in the *cis* chamber (**B**). Shown are the NPo diaries prior and after addition of GTPγS-activated G_s, GTPγS-activated α_s, CTX-treated G_s (in the presence of GTP) or GTPγS and the corresponding cumulative NPo curves. The holding potentials were either 0 or +20 mV, and all experiments were performed with $BaCl_2$ in the *cis* chamber. Average stimulations of activity were between 10- and 20-fold in the absence and 2- to 3-fold in the presence of $3\,\mu M$ Bay K8644. Note that the Ca^{2+} channels incorporated into the bilayers could be stimulated either by exogenously activated G_s or by activation of an endogenous T tubule G protein, presumably also G_s, coincorporated with the Ca^{2+} channel. In other experiments it was established that GTPγS, to be effective, had to be added to the *cis* chamber, confirming the sidedness of the T tubule membranes. (Adapted from [3])

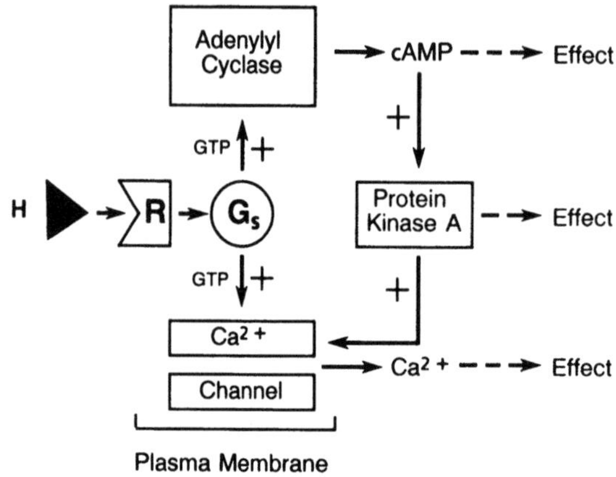

Fig. 5. Diagram of dual regulation of the DHP-sensitive Ca^{2+} channel by G_s and phosphorylation by cAMP-dependent protein kinase

the concept that emerged as to how G_s is involved in a dual manner in positively modulating the activity of voltage-gated Ca^{2+} channels.

Identity of the Stimulatory G Protein: Bacterially Expressed α Subunit of G_s Stimulates the Voltage-Gated Ca^{2+} Channel

It has been well established that the cardiac voltage-gated DHP-sensitive Ca^{2+} channel is positively regulated by cAMP-dependent phosphorylation [7]. In agreement with this it was shown that incubation of inside-out membrane patches from GH_3 cells can lead to recovery of Ca^{2+} channel activity that had been lost on patch excision [10]. The finding described above that G_s preparations appear to stimulate the Ca^{2+} channel directly was therefore surprising, for it means that these channels are apparently under dual and redundant regulation by G_s. We therefore asked ourselves whether indeed it was G_s that caused the stimulation of Ca^{2+} channel activity, or whether the effect had been due to the action of a closely related but nevertheless distinct G protein. The ability to express recombinant α_s in bacteria that is active as stimulator of adenylyl cyclase activity allowed us to test whether this protein also stimulates the activity of Ca^{2+} channels incorporated into lipid bilayers. Figure 6 illustrates that indeed recombinant α_s, but not recombinant α_{i3}, stimulates the activity of the skeletal muscle T tubule Ca^{2+} channel.

Association of G_s with the Skeletal Muscle DHP-Sensitive Ca^{2+} Channel

The above experiments indicated that a G protein, identified as G_s is able to stimulate the activity the DHP-sensitive Ca^{2+} channel present in intact

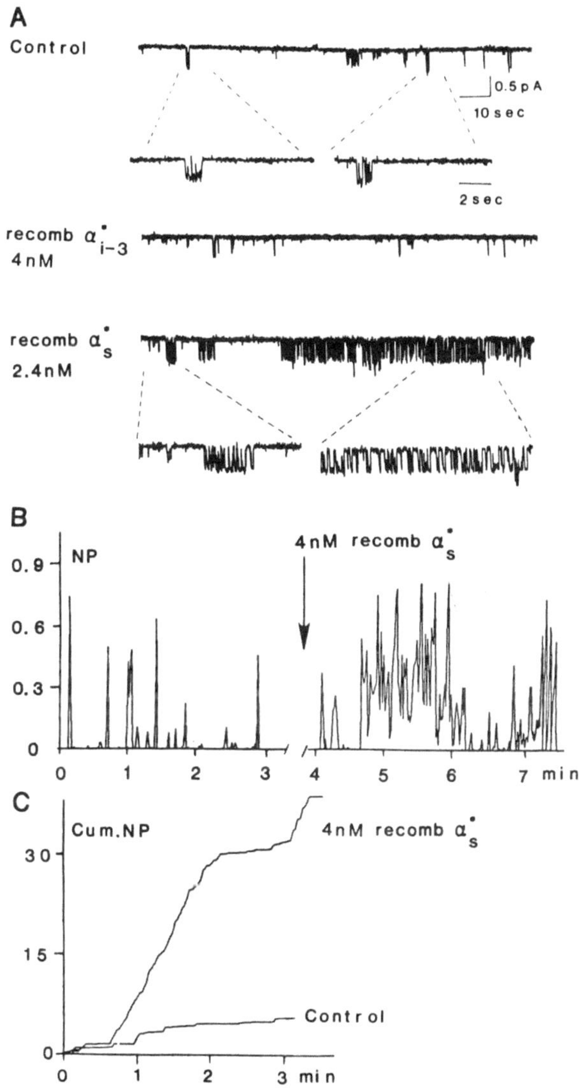

Fig. 6A–C. Stimulation of skeletal muscle T tubule DHP-sensitive Ca^{2+} channel incorporated into lipid bilayers by recombinant α_s. Recombinant α_s (Ser^-, Asp^{71}-Glu^{72} version) and recombinant α_{i3} were prepared in their GTPγS-activated form and partially purified by DEAE-Sephacel chromatography using the two-plasmid approach (for details see [3, 18]). **A** Single-channel Ba^{2+} currents (*trans* to *cis*) before and after addition first of GTPγS-activated recombinant α_{i3} and then of GTPγS-activated recombinant α_s. **B**, **C**, NPo diaries of Ca^{2+} channel activities before and after addition of GTPγS-activated recombinant α_s to the *cis* chamber (**B**) and cumulative NPo values as a function of time of the same experiment (**C**)

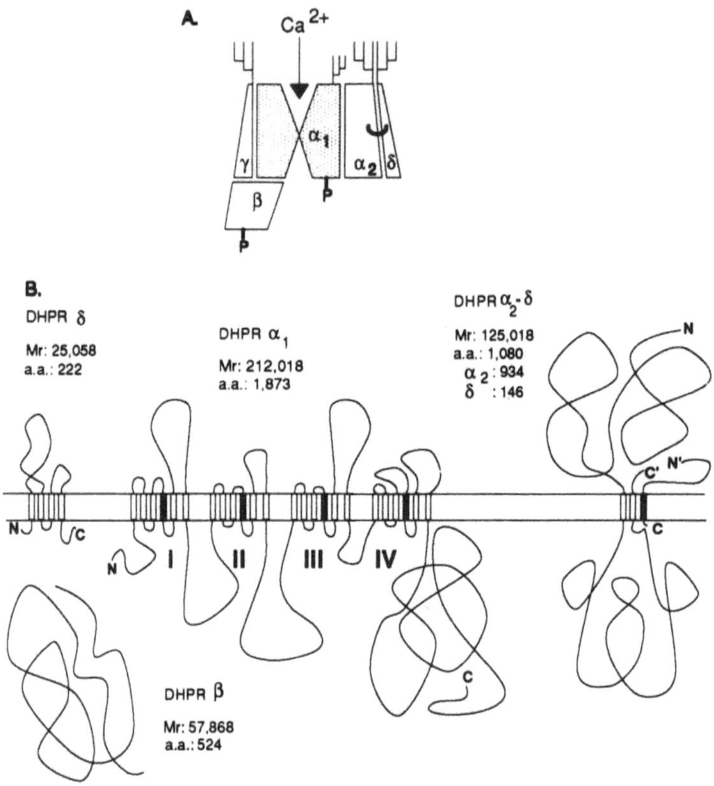

Fig. 7A,B. Primary structure of the subunits of the skeletal muscle DHP-sensitive L-type Ca^{2+} channel. **A** As derived from peptide and size analysis (adapted from [19]). **B** As deduced from their cDNAs [25]

skeletal muscle membrane vesicles. The data were consistent with the idea that this effect of G_s is due to direct G protein–Ca^{2+} channel protein interaction rather than being mediated by a membrane or submembrane delimited yet indirect pathway. We next sought more direct, i.e., biochemical, evidence for the direct G–protein Ca^{2+} channel interaction.

The skeletal muscle and quite likely also the cardiac DHP-sensitive Ca^{2+} channel is formed of five polypeptides, α_1, α_2, β, γ, and δ, having apparent molecular weights of 210000, 140000, 52000, 32000, and 25000, as schematized in Fig. 7. Of these, α_2 and δ are disulfide bonded and the proteolytic cleavage product of a single precursor. The α_1 carries the Ca^{2+} channel function and DHP-binding activity [11], which is regulated by the other subunits. Upon sucrose density gradient centrifugation the DHP-sensitive Ca^{2+} channel complex–also referred to as DHP receptor–migrates, with an apparent M_r of ca. 450000. In contrast, the G_s G protein is formed of three subunits, α_s (M_r 45000), β (M_r 35000), and γ (M_r 7000). In sucrose density gradients the complex migrates as a protein of ca. 100000

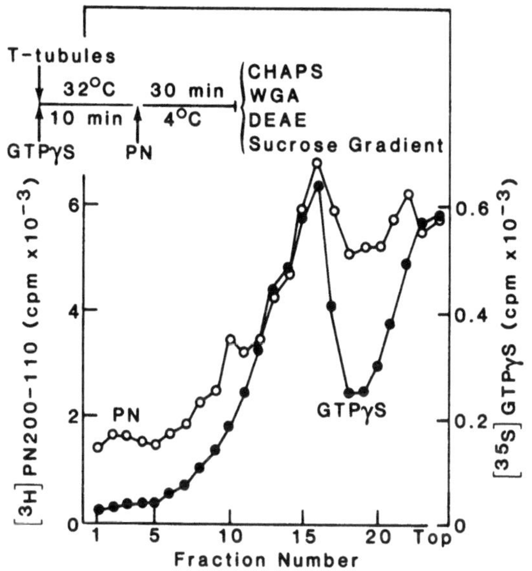

Fig. 8. Activation of T tubule G proteins in the presence of β-adrenergic receptor agonist (Iso) leads to association of a GTPγS-liganded molecule with a molecular structure of the size and chromatographic as well as sedimentation behavior of the DHP-receptor/Ca^{2+} channel complex. (Adapted from [12])

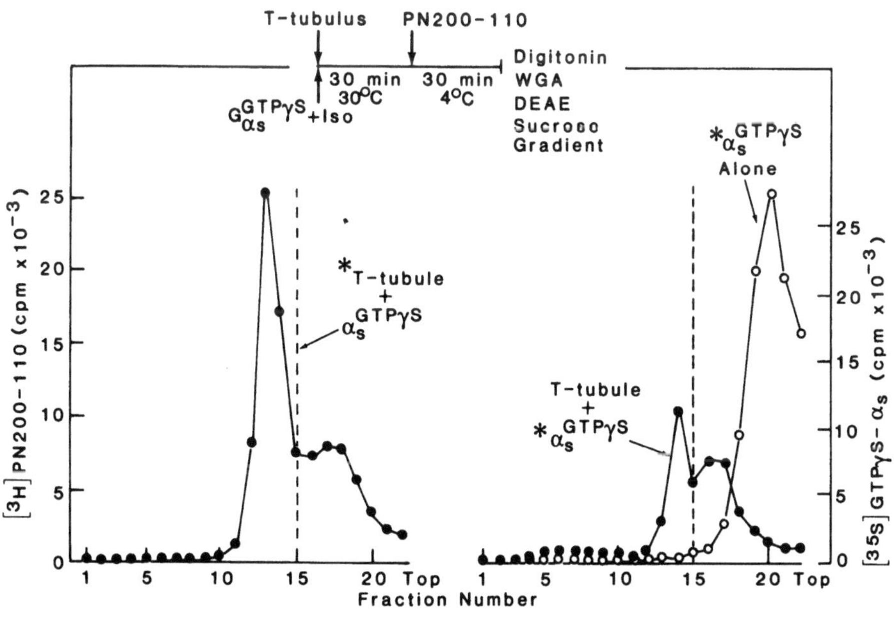

Fig. 9. Incubation of T tubule membranes with pure GTPγS-activated α subunit of G_s leads to its association with a molecular structure of the size and chromatographic as well as sedimentation behavior of the DHP-receptor/Ca^{2+} channel complex. (Adapted from [12])

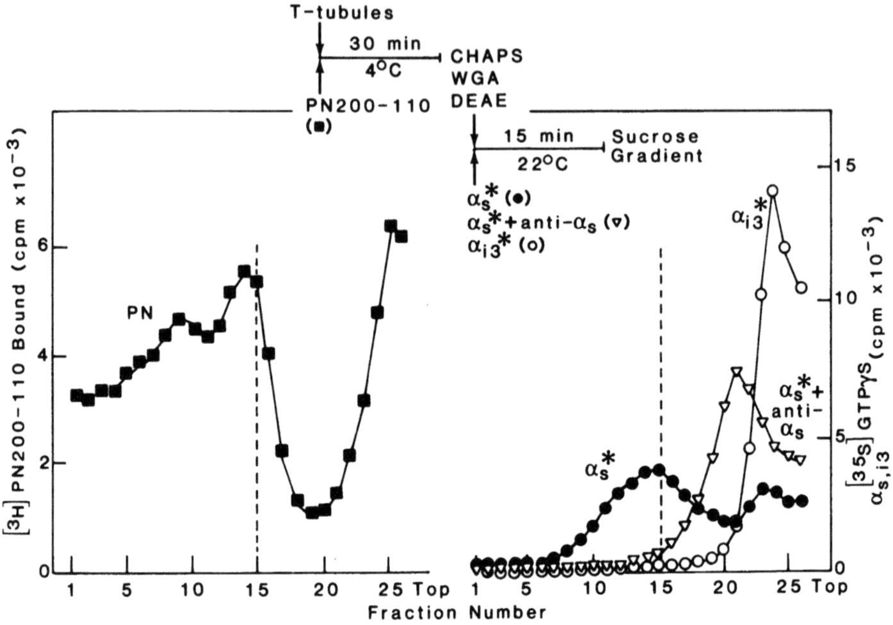

Fig. 10. Incubation of solubilized and partially purified DHP-receptor/Ca^{2+} channel complex from skeletal muscle T tubules with GTPγS-activated α_s, but not with GTPγS-activated α_{i3}, leads to its selective association with a molecular structure of the size and chromatographic as well as sedimentation behavior of the DHP-receptor/Ca^{2+} channel complex. (Adapted from [12])

and the GTPγS-activated form, α_s*, behaves as a correspondingly smaller molecule of ca. 45 000, easily distinguishable from that of the DHP-sensitive Ca^{2+} channel complex.

Figures 8–10 present the results from three types of experiments, in each of which association of the GTPγS-activated α subunit of G_s with the DHP-sensitive Ca^{2+} channel complex was tested after partial purification [12]. Partial purification consisted in extraction of the DHPP binding complex from T tubule membranes with either CHAPS or digitonin, two sequential chromatographies, first over wheatgerm agglutinin (WGA; elution with *N*-acetyl-glucosamine) and then over DEAE-Tris acryl M (elution with NaCl), followed by sucrose density gradient centrifugation. Interaction of G_s with the Ca^{2+} channel was scored as comigration of ^{35}S-labeled (and activated) α_s with the DHP ([^{3}H]PN200-110)-liganded Ca^{2+} channel complex (for details see [12]).

In the first of the three experimental approaches it was found that a protocol for activation of G proteins in T tubules that favors activation of G_s (due to addition of isoproterenol) prior to detergent extraction results in comigration of [^{35}S]GTGPγS and liganded [^{3}H]PN200-110 (Fig. 8). In the second of the experimental approaches it was found that incubation of T tubule enriched membranes prior to detergent extraction with purified

α_s activated with [^{35}S]GTPγS under conditions in which the G_s system is turning over, i.e., in the presence of isoproterenol, now leads to α_s* to comigration with the Ca^{2+} channel complex (Fig. 9). Finally, in the third experimental approach incubation of the [^{3}H]PN200-110-labeled Ca^{2+} complex with [^{35}S]GTPγS-activated α_s and [^{35}S]GTPγS-activated α_{i3} after the channel had been extracted and chromatographed over WGA and DEAE resulted a selective association of the PN200-110-labeled protein with α_s but not α_{i3} (Fig. 10).

While these experiments do not prove direct interaction between the G_s protein and one of the components of the Ca^{2+} channel complex, they are certainly totally consistent with this possibility. Other experiments for example showed that gradient fractions with both the DHP-labeled complex and the γ^{32}P-labeled α_s are devoid of adenylyl cyclase activity, indicating that the macromolecular behavior of α_s is to association to complexes other than adenylyl cyclase. Experiments are currently ongoing in which the pure α_s is mixed with purified skeletal muscle Ca^{2+} channel complex–also DHP receptor–to determine the subunit with which α_s is likely to interact.

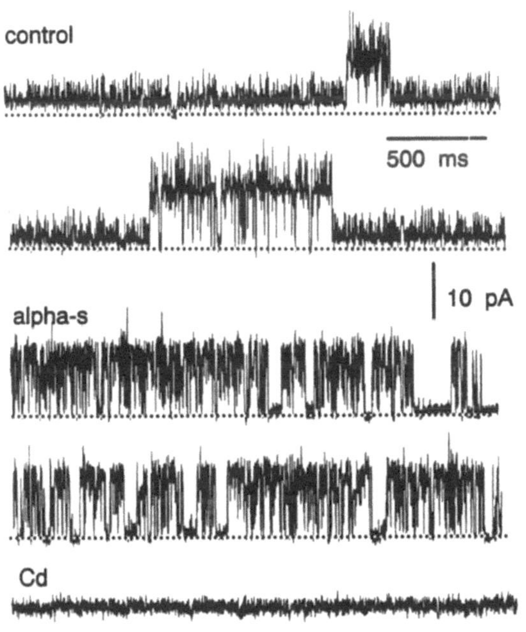

Fig. 11. Regulation of gating behavior of L-type Ca^{2+} channel currents in lipid bilayers by GTPγS-activated α_s (*alpha-s*). Rabbit skeletal muscle T tubule DHP-receptor/Ca^{2+} channel complexes pretreated with Bay K8644 were solubilized and partially purified by chromatography over wheat germ agglutinin and DEAE and centrifugation through a sucrose density gradient. Proteins of aliquots of sucrose density gradient fractions were incorporated into lipid bilayers and tested for L-type Ca^{2+} channel activity in the presence of 1 μM Bay K8644 using Na^+ as the charge carrier (for details see [12]). *Control*, representative records of single-channel currents before addition of GTPγS-activated α_s; *Alpha-s*, representative records after addition of 100 pM human erythrocyte GTPγS-α_s complex; *Bottom panel*, after addition of 1 mM Cd^{2+}. (Adapted from [12])

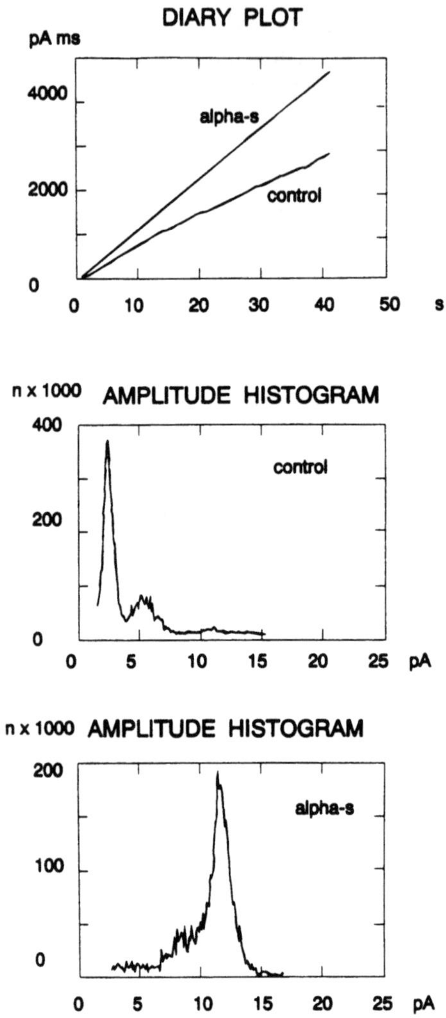

Fig. 12. Cumulative NPo and amplitude frequency histograms of single-channel currents of partially purified Bay K8644-stimulated rabbit skeletal muscle L-type Ca^{2+} channels in lipid bilayers. (Adapted from [12])

Regulation by α_s* of the Activity of the Partially Purified Ca^{2+} Channel Complex Incorporated into Lipid Bilayers

The experiments shown above predict that a Ca^{2+} channel complex that has been purified through the WGA, DEAE, and sucrose density gradient steps should show regulation by activated α_s, provided the Ca^{2+} channel aspect of the protein complex is still active. To test this hypothesis T tubules were treated with the agonist Bay K8644, and the Ca^{2+} channel completely extracted with CHAPS and subjected to WGA and DEAE chromatography followed by sucrose density gradient centrifugation. Fractions were then tested for Ca^{2+} channel activity by incorporation into lipid bilayers. Single Ca^{2+} channel currents were indeed obtained, using either Ba^{2+} or Na^+ as the charge carrier. As shown in Figs. 11 and 12, the gating behavior of the Ca^{2+} channel currents obtained in lipid bilayers with the partially purified Ca^{2+} channel complex are dramatically affected upon the addition of GTPγS-activated α_s (alpha-s in the figure). Records of single channel currents, which in the absence of α_s are predominantly of a low amplitude and long lasting (upper panel of Fig. 11, quantified in the middle panel of Fig. 12), are dominated after α_s addition by frequent openings and closing of a large amplitude (middle panel of Fig. 11, quantified in the bottom panel of Fig. 12). As a consequence, the cumulative NPo of these Bay K8644-simulated channels increased by a factor of about 2 (top panel of Fig. 12). These currents were fully blocked by Cd^{2+} (lower panel of Fig. 11) and had unitary current amplitudes and voltage dependence that were very similar to those obtained under the same conditions by incorporation of T tubule vesicles instead of the soluble protein into the lipid bilayers [12]. They differed however, in that the low-amplitude openings that predominated in the absence of α_s with the partially purified protein were very infrequent when T tubule membranes were used. Studies are underway to investigate further the basis for the existence of apparent subconductance states and factors that influence the frequency of their occurrence.

Conclusions

The electrophysiological data presented here were obtained over a period of 4 years and involved two laboratories. They strongly suggest that cardiac

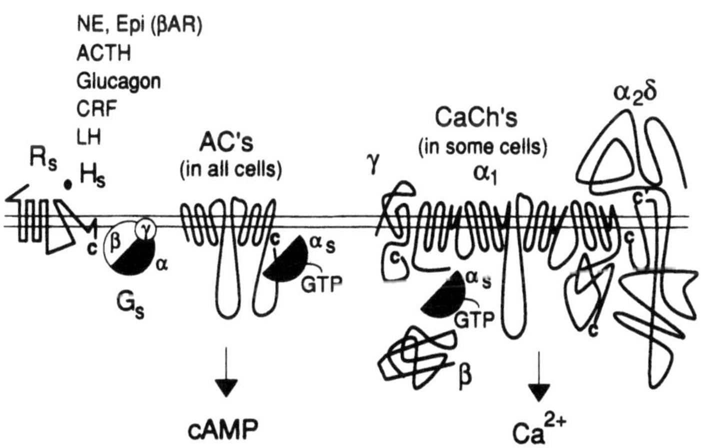

Fig. 13. Diagram of dual role of G_s to regulate adenylyl cyclase and one of the components of the DHP-receptor/Ca^{2+} complex

and skeletal muscle Ca^{2+} channels are under the regulation of G_s by a membrane-delimited phosphorylation independent mechanism. The biochemical studies summarized above are clearly consistent with the idea that this mechanism is likely to involve direct protein–protein interaction with G_s. Thus, data are consistent with our view that the G_s protein through its α subunit regulates at least two membrane functions, adenyl cyclase and the L-type Ca^{2+} channel (Fig. 13). This of course does not detract from the regulation by cAMP-dependent protein kinase mediated phosphorylation such as originally shown by Kameyama et al. [7] and since then confirmed by many others including Armstrong and Eckert [10] and several of us [3, 13, 14] in heart and Garcia et al. [15] in skeletal muscle. The study of the skeletal muscle Ca^{2+} channels is of particular interest, for it had shown under whole-cell recording conditions the existence of both a cAMP-dependent protein kinase A mediated and a protein kinase A independent but GTPPγS-mediated regulation of the Ca^{2+} current, in that GTPγS was able to stimulate gating currents which PKA could protein kinase A not. Both agents stimulated permeation, as seen by increased Ca^{2+} channel inward currents [15]. As a consequence it would appear that even though the regulatory effect of G_s is not always apparent (see [16, 17]) the L-type Ca^{2+} channel system is indeed the subject of dual regulation by phosphorylation and G_s (Fig. 5).

References

1. Hescheler J, Rosenthal W, Trautwein W, Schultz G (1987) The GTP-binding protein N_o regulates neuronal calcium channels. Nature 325:445–477
2. Yatani A, Codina J, Imoto Y, Reeves JP, Birnbaumer L, Brown AM (1987) A G protein directly regulates mammalian cardiac calcium channels. Science 238:1288–1292
3. Yatani A, Imoto Y, Codina J, Hamilton SL, Brown AM, Birnbaumer L (1988) The stimulatory G protein of adenylyl cyclase G_s directly stimulates dihydropyridine-sensitive skeletal muscle Ca^{2+} channels. Evidence for direct regulation independent of phosphorylation by cAMP-dependent protein kinase. J Biol Chem 263:9887–9895
4. Imoto Y, Yatani A, Reeves JP, Codina J, Birnbaumer L, Brown AM (1988) α Subunit of G_s directly activates cardiac calcium channels in lipid bilayers. Am J Physiol 255:H722–H728
5. Mattera R, Graziano MP, Yatani A, Zhou Z, Graft R, Codina J, Birnbaumer L, Gilman AG, Brown AM (1989a) Individual splice variants of the α subunit of the G protein G_s activate both adenylyl cyclase and Ca^{2+} channels. Science 243:804–807
6. Cavalié A, Ochi R, Pelzer D, Trautwein W (1983) Elementary currents through Ca^{2+} channels in guinea pig myocytes. Pflugers Arch 398:284–297
7. Kameyama M, Hescheler, J, Hofmann F, Trautwein W (1986) Modulation of Ca current during the phosphorylation cycle in the guinea pig heart. Pflugers Arch 407:121–128
8. Mueller P, Rudin D, Tien HT, Westcott WC (1962) Reconstitution of excitable cell membrane structure in vitro. Circulation 26:1167–1171
9. Coronado R, Affolter H (1986) Insulation of the conduction pathway of muscle transverse tubule calcium channels from the surface charge of bilayer phospholipid. J Gen Physiol 87:933–953
10. Armstrong D, Eckert R (1987) Voltage-activated calcium channels that must be phosphorylated to respond to membrane depolarization. Proc Natl Acad Sci USA 84:2518–2522

11. Perez-Reyes E, Kim HS, Lacerda AE, Horne W, Wei X, Rampe D, Campbell KP, Brown AM, Birnbaumer L (1989) Induction of calcium currents by the expression of the α_1-subunit of the dihydropyridine receptor from skeletal muscle. Nature 340:233–236
12. Hamilton SL, Codina J, Hawkes MJ, Yatani A, Sawada T, Strickland FM, Froehner SC, Spiegel AM, Toro L, Stefani E, Birnbaumer L, Brown AM (1991) Evidence for direct interaction of $G_s\alpha$ with the Ca^{2+} channel of skeletal muscle. J Biol Chem 266:19528–19535
13. Yatani A, Brown AM (1989) Rapid β-adrenergic modulation of cardiac calcium channel currents by a fast G protein pathway. Science 245:71–74
14. Pelzer S, Shuba YM, Asai T, Codina J, Birnbaumer L, McDonald T, Pelzer D (1990) Direct stimulation of heart cell calcium current by β-adrenergic signal transducing G_s protein. Am J Physiol 259:H264–H267
15. Garcia J, Gamboa-Aldeco R, Stefani E (1989) GTPγS increases charge movement and calcium currents in skeletal muscle fibers. Pflügers Arch 417:114–116
16. Hartzel CH, Méry P-F, Fischmeister R, Szabo G (1991) Sympathetic regulation of cardiac calcium current is due exclusively to cAMP-dependent phosphorylation. Nature 351:573–576
17. Cavalié A, Allen TJA, Trautwein W (1991) Role of the GTP-binding protein G_s in the β-adrenergic modulation of cardiac Ca^{2+} channels. Pflugers Arch 419:433–443
18. Mattera R, Yatani A, Kirsch GE, Graf R, Olate J, Codina J, Brown AM, Birnbaumer L (1989b) Recombinant α_i-3 subunit of G Protein activates G_k-gated K^+ channels. J Biol Chem 264:465–471
19. Catterall WA (1988) Structure and function of voltage-sensitive ion channels. Science 242:50–61
20. Tanabe T, Takeshima H, Mikami A, Flockerzi V, Takahashi H, Kangawa K, Kojima M, Masuo H, Hirose T, Numa S (1987) Primary structure of the receptor for calcium channel blockers from skeletal muscle. Nature 328:313–318
21. Ellis SB, Williams ME, Ways NR, Brenner R, Sharp AH, Leung AT, Campbell KP, McKenna E, Koch WJ, Hui A, Schwartz A, Harpold MM (1980) Sequence and expression of mRNAs encoding the α_1 and α_2 subunits of a DHP-sensitive calcium channel. Science 241:1661–1664
22. Ruth P, Röhrkasten A, Biel M, Bosse E, Regulla S, Meyer HE, Flockerzi V, Hofmann F (1989) Primary structure of the β subunit of the DHP-sensitive calcium channel from skeletal muscle. Science 245:1115–1118
23. Bosse E, Regulla S, Biel M, Ruth P, Meyer HE, Flockerzi V, Hofmann F (1990) The cDNA and deduced amino acid sequence of the γ subunit of the L-type calcium channel from rabbit skeletal muscle. FEBS Lett 267:153–156
24. Jay SD, Ellis SB, McCue AF, Williams ME, Vedvick TS, Harpold MM, Campbell KP (1990) Primary structure of the gamma subunit of the DHP-sensitive calcium channel from skeletal muscle. Science 248:490–492
25. DeJongh KS, Warner C, Catterall WA (1990) Subunits of purified calcium channels. J Biol Chem 265:14738–14741

Inositol Phosphates in Cardiac Regulation

G.W. Mayr

Introduction

In recent years it has been clearly established (see reviews [24, 66]) that the myocardium contains all the elements of the inositide signaling system. It has also been established that a number of agonists activate the key enzyme of inositide signaling, phospholipase C (PLC) at physiological concentrations leading to a release of the Ca^{2+}-mobilizing second messenger inositol 1,4,5-trisphosphate [Ins(1,4,5)P_3] and diacyl glycerol (DAG), the latter activating protein kinase C. The functions of protein kinase C in the heart are not discussed here. In a large number of electrically excitable and nonexcitable animal cells, the former messenger is able to mobilize Ca^{2+} from intracellular stores by activating second messenger operated intracellular Ca^{2+} channels. Although all these elements of an inositide-based Ca^{2+}-mobilizing system are definitely present in cardiomyocytes, it is still difficult to assign specific roles to this signaling system since the predominant Ca^{2+}-mediated processes of the heart, excitation-contraction (EC) coupling, and inotropic events appear to be possible without its activation. However, since the myocardial cell has a larger repertoire of mechanisms of cellular Ca^{2+} handling than most other cell types (see reviews [8, 46] and below), it is difficult functionally to "dissect" such a second messenger operated Ca^{2+}-release system from the other systems with which it works in concert.

This chapter presents a brief overview of some recent findings concerning myocardial inositide signaling elements. From the data discussed here it is evident that this system has a "fine tuning" rather than an EC coupling role in cardiac performance. Potential specific roles of Ins(1,4,5)P_3 in myocardial Ca^{2+} handling, a likely cross-talk between glycolysis and this messenger system, and a potential involvement of inositol phosphates in the induction of physiological or pathological cardiac hypertrophy are discussed.

Institut für Physiologische Chemie, Ruhr-Universität Bochum, W-4630 Bochum, Fed. Rep. of Germany

Figulla et al. (Eds.)
Idiopathic Dilated Cardiomyopathy

Phosphoinositide Signaling in Cardiac Cells

Cardiac Phosphoinositides: Abundance and Turnover in the Myocardium

Myocardial cells contain high levels of the inositol phospholipids of the inositide cycle [6]. Our direct mass measurements from rat hearts show that they contain about 150 μM phosphatidylinositol (PtdIns), 18 μM PtdIns4P, and 10 μM PtdIns(4,5)P_2 (unpublished data). A comparison with the masses in skeletal muscle [35] demonstrates similar concentrations in both tissues. Since the latter two inositol phospholipids are almost exclusively located in the sarcolemma [90], the membrane concentrations are in the millimolar range. Regional differences of masses and of agonist- or Ca^{2+}-stimulated turnover rates are not known at present. Detailed corresponding studies are necessary to clarify the importance of this signaling system in different heart regions and to correlate the findings with other regional functions such as atrial natriuretic factor (ANF) secretion, electrical conduction, and inotropic response.

Phospholipase C: At Least Two Cardiac Isoforms

Phosphoinositide-specific phospholipase C (PLC, also termed phosphoinositidase C) is the key enzyme driving the turnover of phosphoinositides and the release of Ins(1,4,5)P_3 and DAG. Recent cloning and northern analysis data have shown that in the rat myocardium at least two isoforms of PLC to be present, γ_2 ($M_r \approx 145$ kDa) and δ ($M_r \approx 85$ kDa) [25], the former dominating in its mRNA level. The former exhibits a high degree of homology with the almost ubiquitous γ_1-isoform. It is not activated by G protein coupling but by a specific tyrosine phosphorylation-dependent translocation to the plasma membrane, the compartment where the substrate PtdIns(4,5)P_2 is located. Translocation is brought about by an interaction of its two *src* homology (SH-2) domains, unique tyrosine-phosphopeptide binding domains, with tyrosine-phosphorylated, membrane-anchored, nonreceptor tyrosine kinases (TKs) or growth factor-activated receptor tyrosine kinases (RTKs), respectively [40]. It is not known whether γ_2 is activated by the same mechanism although the very high degree of homology with γ_1 and the presence of a SH-2 domain makes it likely. It is also unknown which RTKs or membrane-anchored TKs might activate this PLC isoform in the heart. The relatively small δ isoform not containing SH-2 domains might be the enzyme which is coupled to the group of non-TK agonist receptors activating PLC in the heart (see below). However, up to now confirmation of this view is also lacking. In a recent study [80] a further 86-kDa isoform (PLCε) has been postulated as the major G protein regulated enzyme which might be identical with the 85-kDa enzyme islated from heart [43]. In contrast to skeletal muscle, where PLC activation apparently does not involve a G protein and is most likely due to direct Ca^{2+} activation and/or indirect activation by increased substrate availability [e.g., by the removal of "protecting proteins" from PtdIns(4,5)P_2] [40–43],

Table 1. Agonists and stimuli activating cardiac phospholipase C

Agonist	Atrial effect	Ventricular effect
α_1-Adrenergic [9, 18, 23, 45, 47, 58, 64, 82]	Positive inotropic Hypertrophic ANF secretion	Positive inotropic[a] Hypertrophic ANF secretion
Endothelin-1 [33, 72, 83, 84]	Positive inotropic Hypertrophic ANF secretion	Positive inotropic Hypertrophic ANF secretion
Angiotensin II [1, 3, 4, 38]	Hypertrophic ANF secretion	Hypertrophic ANF secretion
A_2-purinergic [37, 59]		Positive inotropic
Histamine (H_1) [63]	Positive inotropic	
Muscarinic [5, 9, 29, 41, 58]	Negative inotropic Negative chronotropic	Positive/negative/not inotropic
A_1-puringergic[b] [29, 37]	Negative inotropic	Not inotropic
K-opioidergic [82]		Negative inotropic
Calcium overload [34, 44, 47, 48, 53, 54]	Positive/negative/inotropic	Positive/negative inotropic
Electrical stimulation [58]	Contraction	Contraction
Myocyte stretch [31]	Hypertrophic ANF secretion	Hypertrophic

[a] Species dependent (see [18]).
[b] Conflicting results (see discussion in [29]).

at least one of the cardiac enzymes appears to be activated via a pertussis toxin insensitive large heterotrimeric G protein [40, 41, 65]. Ca^{2+} activation [43, 67] (see next paragraph) and an indirect activation by the substrate availability [42] may be also important.

Agonists and Other Stimuli Activating Myocardial PLC

A number of agonists are able to activate cardiac PLC. Among them are α_1-adrenergic agonists, muscarinic cholinergic agonists, endothelin, histamine, A_2-purinergic (in two reports also A_1-purinergic) agonists, opioid agonists, and angiotensin II (AT II). In Table 1 their atrial and ventricular effects are compiled. All these agonists except the A_1-purinergic have been demonstrated in other cell types to activate PLC and to mobilize intracellular Ca^{2+}. In the heart, most studies have demonstrated a release of inositol phosphates or an increase in inositide turnover by these agonists. A Ca^{2+}-mobilizing effect in isolated cardiomyocytes has up to now been shown only for stimulation by α_1-adrenergic agonist [9], endothelin [83], ATP [59], and AT II [1]. However, detailed data about accompanying systolic-diastolic Ca^{2+} changes are still lacking. α_1-Adrenergic agonists act in a positive inotropic way in atria and ventricles of most species investigated (see Table 1). Likewise, endothelin has marked positive inotropic actions both in atria and ventricles. These and the other agonists of the first group listed in Table 1 appear to act in cardiac muscle as expected, and as they do in smooth muscle, i.e., they mobilize Ca^{2+} and thus increase contractility [10, 74].

In contrast, muscarinic or A_1-purinergic stimulation of atria has a markedly negative inotropic and negative chronotropic effect and, depending on species and experimental conditions, only a slightly positive inotropic effect in ventricles. The reason for this inconsistency of actions of cardiac PLC-activating agonists appears to be the existence of two or more cholinoceptor and A_1-purinoceptor signal transduction paths coupling the agonist actions to partially opposing cellular events. Muscarinic cholinergic agonists also decrease cardiac cAMP by the inhibitory receptor pathway, i.e., inhibit adenylyl cyclase via an inhibitory receptor R_i and G_i (see chapter by Böhm et al., this volume). The negative inotropic and chronotropic effect of muscarinic agonists in atria appears to be due mainly to direct G protein coupled potassium channel activation [56] in which a further atrial type of cholinoceptor may be involved. Therefore, the PLC-activating and Ca^{2+}-mobilizing potency of cholinergic agonists may be overridden by the latter effects. A_1-purinergic agonists are not positive inotropic in papillary muscles and negative inotropic in atria. However, in two studies a PLC activation has been reported [29, 37]. A predominance of the activation of a potassium conductance over the Ca^{2+}-mobilizing effects of $InsP_3$ release has also been discussed for these agonists [29]. For the remaining $InsP_3$-mobilizing agonists little is known about their physiological role in adult hearts. In fetal or newborn hearts or myocardial cells AT II is known to stimulate cell growth [3]. A_2-purinergic agonists [37, 59] and histamine [63] were found to act in a positive inotropic way in ventricles.

As mentioned above, there are a number of experimental reports indicating that cardiac PLC may also be directly activated by Ca^{2+}. Different regimens leading to cellular Ca^{2+} overload, such as doxorubicin intoxication [34], ouabaine treatment [53], activation of sodium channels [44], and postischemic conditions [47, 48, 54], lead to an increase in phosphoinositide turnover and the myocardial concentration of inositol phosphates. Since an increase in Ca^{2+} is the triggering event and not a consequence of PLC activation in these cases, inositol phosphate release might serve other functions than releasing Ca^{2+} and/or facilitate a more pronounced Ca^{2+} release.

A physical stimulus inducing release of $InsP_3$ is mechanical stretching of cardiomyocytes [31]. As demonstrated for skeletal muscle fibers [20], mechanically operated sarcolemmal Ca^{2+} channel activation or a direct "stretch receptor" induced activation of PLC [31] could be the triggering event. In the former case PLC appears, again, to be activated subsequent to an initial increase of Ca^{2+}.

An interesting feature is that myocardial ischemia leads to a decrease in the concentration of $InsP_3$ and myocardial phosphoinositides. One possible explanation is that the strong reduction in the energy charge in ischemic myocardium reduces the activities of one or several of the phosphate-transferring enzymes involved in the inositide cycle, such as DAG kinase or the phosphoinositide kinases. Upon postischemic reperfusion an increase in inositol phosphates and the turnover of phosphoinositides was observed [47, 48, 54]. The results suggest that the reperfusion-induced Ca^{2+} influx [47] also activates PLC as a secondary event.

Evidence for Phosphoinositide 3-Kinase: Is it Linked to Growth Factor Action?

In many cell types the role of 3-phosphorylated inositides [PtdIns3P, PtdIns(3,4)P_2, PtdIns(3,4,5)P_3] in cell growth and cytoskeletal rearrangements has recently been demonstrated (reviewed in [12]). Whether these inositides are present in myocardial cells has not yet been investigated. However, in a recent study the noncatalytic 85-kDa subunit of phosphoinositide 3-kinase, the key enzyme responsible for the formation of these compounds, has been found in cardiac muscle [73]. Like PLCγ, this enzyme contains two SH-2 domains. Activation appears to be brought about by tyrosine phosphorylation-dependent translocation of the cytosolic enzyme to the plasma membrane employing the same mechanism as is effective in PLCγ_1 (see above). It is unknown how the enzyme is activated in cardiomyocytes. However, insulin receptors and other growth factor receptors of the RTK type (e.g., basic fibroblast growth factor receptors; see chapter by. Cummins, this volume, and [36]) being expressed in cardiomyocytes could be involved.

Postreceptor Events in Cardiac Inositide Signaling

Ins(1,4,5)P_3 Release and Metabolism: Dephosphorylation Is the Major Metabolic Fate in Adult Heart

Cardiac muscle contains resting levels of Ins(1,4,5)P_3 between 0.4 and 3.5 μM (unpublished data). Under control conditions the total concentration of this $InsP_3$ isomer is normally lower than 3 μM, whereas upon stimulation of PLC by one of the above agonists or by Ca^{2+} overload concentration of up to 8 μM can be attained. A very low concentration of a second isomer of $InsP_3$, Ins(1,3,4)P_3 is present. The latter is due to dephosphorylation of Ins(1,3,4,5)P_4, which is formed only to a low degree from Ins(1,4,5)P_3 in adult cardiac muscle after PLC stimulation (see below for its importance). Therefore, Ins(1,4,5)P_3 is eliminated preferentially via dephosphorylation by 5-phosphatase. The cardiac enzyme has not yet been characterized in detail or isolated. Ins(1,4)P_2, the product of this dephosphorylation reaction has correspondingly high basal levels (1.7–4 μM) which increase further upon PLC activation.

The Cardiac Ins(1,4,5)P_3 Receptor: A Second Ca^{2+} Release Channel in Addition to the Ryanodine Receptor?

Unlike skeletal muscle, cardiac muscle contains a higher amount of the Ins(1,4,5)P_3 receptor (IP_3R) Ca^{2+} channel protein ($M_r \approx 260$ kDa; reviewed in [70]), likely the isoforms S_I and S_{IIABC^-}, [49]) in addition to the ryanodine receptor protein ($M_r \approx 565$ kDa [39]), the major triadic/dyadic Ca^{2+} release channel-forming protein of cardiomyocytes. Both types of Ca^{2+} release channel proteins form large squarelike homotetrameric structures about 25 nm in side length in the SR/ER membrane. The C-terminal membrane-spanning domains (four to ten transmembrane helices) form in a ligand-dependent manner a large central nonspecific cation channel. Each protomer contains a large N-terminal cytoplasmic domain, which in the former case contains one high-affinity $InsP_3$ binding, as well as Ca^{2+}-calmodulin binding, nucleotide binding, and phosphorylation domains and in the latter contains a ryanodine binding site instead of the high-affinity $InsP_3$ binding site (reviewed in [2, 49]). The footlike appearance of these domains gives the ryandine receptor the name "foot protein." The myocardial and subcellular distribution of the IP_3R channels is not yet known. One might speculate that the corbular SR [75], a specific SR structure observed only in cardiomyocytes containing footlike cytosolic projections but forming no triad/dyad junctions, is the structure containing this $InsP_3$-induced Ca^{2+} release (IP_3ICR) system. The coexistence of two Ca^{2+} release systems in one cell type has been shown in cerebellar Purkinje's neurons [87]. Indirect evidence for a similar coexistence in cardiomyocytes comes from a recent study in atrial myocytes. Endothelin exerts an inotropic effect [by Ins(1,4,5)P_3 sensitive Ca^{2+} mobilization from intracellular stores], and this inotropic response is not sensitive to ryanodine or caffeine [83]. Even in skeletal muscle, where the level of IP_3R mRNA is very low, an apparent coexistence of these two types of Ca^{2+} channels has been observed in a single-channel study with SR vesicles [78]. IP_3ICR may operate independently of the major cardiomyocyte Ca^{2+} release system which appears to be a Ca^{2+}-induced Ca^{2+} release (CICR) system [19], and unlike in skeletal muscle not a depolarization-induced Ca^{2+} release (DICR) system [2]. However, since intimate contacts are also observed in cardiomyocytes between dihydropyridine (DHP) receptor proteins (forming the L-type Ca^{2+} channels) and ryanodine receptors by electron microscopy of triad/dyad structures [26], a role of DICR in addition to CICR is possible. Consequently, the three systems increasing sarcoplasmic Ca^{2+} by influx (the L-type and T-type Ca^{2+} channels and the Na/Ca exchanger) are joined by two or even three systems releasing Ca^{2+} from SR stores. Combined with three systems extruding Ca^{2+} from this compartment under physiological conditions, the Ca^{2+} ATPase of the SR, the Na/Ca exchanger, and the sarcolemmal Ca^{2+}

ATPase, a complex multimodulated Ca^{2+}-handling system, is operative in cardiomyocytes. This system is undoubtedly dominated by L-type Ca^{2+} channels, ryanodine receptor channels (these two channels contributing most to CICR), the Na/Ca exchanger, and the Ca^{2+}-ATPase of the SR. The other components including the IP_3ICR may, however, contribute to and modulate this system to a significant extent. The complexity of the system and its multimodulation make it difficult experimentally to insulate the action of IP_3ICR. For both the cerebellar and the skeletal muscle IP_3ICR systems a bell-shaped dependence curve on the cytoplasmic Ca^{2+} concentration of the opening probability of the channel has been found [7, 78]. $InsP_3$ opens the channel only when the free Ca^{2+} at the cytoplasmic face is between $0.03\,\mu M$ and $10-100\,\mu M$. The affinity of $InsP_3$ for the receptor depends strongly on Ca^{2+}. At a Ca^{2+} concentration of about $0.1\,\mu M$ the $InsP_3$ binding and activation of the receptor from skeletal muscle SR [78] is at its maximum.

Higher Phosphorylated Inositols in the Myocardium: What Is Their Role?

Adult myocardium (rat, pig) contains only a low concentration of Ins(1,3,4,5)P_4 ($<0.25\,\mu M$) (unpublished data). The enzyme responsible for its formation is a Ca^{2+} calmodulin dependent Ins(1,4,5)P_3 3-kinase. Its activities in adult cardiac muscle are comparable [5, 22, 61, 77, 82] to those in skeletal muscle [42] and are much lower than in most nonmuscle cell types. Fetal or newborn rat hearts and cultured cardiac cells have more $InsP_3$ 3-kinase than adult hearts [77] and correspondingly higher concentrations of Ins(1,3,4,5)P_4. Stimulation by α_1-adrenergic agonists [22, 77, 82], cholinergic agonists [5, 61], endothelin [57], and opioid agonists [82] induces a transient increase in Ins(1,3,4,5)P_4 and its dephosphorylation product Ins(1,3,4)P_3. In all kinetic investigations the increase in Ins(1,3,4,5)P_4 followed that of Ins(1,4,5)P_3 and lasted longer than the increase in $InsP_3$. A role for Ca^{2+} homoeostasis of Ins(1,3,4,5)P_4 as postulated for some other cell types (see [60]) is unlikely. However, it is key metabolite for the biogenesis of all other higher phosphorylated inositols in animal cells [16, 71]. Rat Myocardium contains significant concentrations of several other higher phosphorylated inositols (unpublished data), two further $InsP_4$ isomers [Ins(1,3,4,6)P_4 and Ins(1,4,5,6)P_4 and/or Ins(3,4,5,6)P_4, each about $0.4\,\mu M$], three $InsP_5$ isomers [about $0.4\,\mu M$ Ins(1,2,3,4,5)P_5, up to $1\,\mu M$ Ins(1,2,4,5,6)P_5, $5-7\,\mu M$ Ins(1,3,4,5,6)P_5] and $InsP_6$ ($8-13\,\mu M$). Ins(1,3,4,6)P_4 is believed to be the slowly formed metabolic intermediate in the synthesis of Ins(1,3,4,5,6)P_5 from Ins(1,3,4)P_3 [71]. The metabolic interrelationship of the other isomers is much less understood. The roles of these inositol phosphates in myocardial cells as well as in many other cell types are not yet clarified. Due to a significantly slower metabolism compared with Ins(1,4,5)P_3 or Ins(1,3,4,5)P_4 (see [71] for a detailed discussion) only long-term regulatory functions might be exerted if any of these compounds has signal function.

Potential Roles of Inositol Phosphates in Myocardial Function

Fine Tuning of the Myocardial Ca^{2+} Handling by Ins(1,4,5)P_3

As in skeletal muscle, an essential role of Ins(1,4,5)P_3 in a triadic/dyadic EC-coupling is unlikely although not totally ruled out. In smooth muscle, however, contraction can be brought about only by InsP_3-mediated Ca^{2+} release from the ER without the necessity of a Ca^{2+} influx [74]. Earlier data leading to the conclusion of such effect of InsP_3 (e.g., [85]) in striated muscles were not confirmed by subsequent, more sophisticated analyses [55, 86]. The most important finding against such a mechanism was that, under all conditions investigated hitherto, the velocity of an Ins(1,4,5)P_3-induced Ca^{2+} release, if observed, is too slow to account for the delay times involved in striated muscle EC coupling [55]. On the other hand, the triadic/dyadic ryanodine receptor as the Ca^{2+} release channel responsible for EC coupling does not bind Ins(1,4,5)P_3 at all [13]. Nonetheless, the marked effects of a prolonged lithium exposure on Ca^{2+} transients in frog muscle fibers and its reversal by *myo*-inositol [91] strongly imply that the turnover of phosphoinositides is somehow involved in modulating (rather than mediating) EC coupling in skeletal muscle.

Potentially important for a physiological function of InsP_3 in the myocardial Ca^{2+} handling is the bell-shaped cytoplasmic Ca^{2+} dependence curve of the IP_3R Ca^{2+} channel (see above). When InsP_3 is sufficiently high to saturate the receptor binding sites, the channel might contribute to CICR, i.e., to the systolic Ca^{2+} increase, with a similar Ca^{2+} dependence as the ryanodine receptor channel [78]. Although its kinetics of Ca^{2+} release may be slower than that of the ryanodine receptor, it might contribute significantly to the peak systolic Ca^{2+} and thereby to the peak force of contraction. A Ca^{2+}-induced secondary activation of PLC would be sufficient for such a potentiating mechanism.

At diastolic Ca^{2+} concentrations (0.12–0.20μM [14, 15, 88]) the InsP_3-saturated channel is almost maximally open, so that InsP_3 can induce a diastolic release of intracellular Ca^{2+}. A prerequisite for this to occur is that diastolic InsP_3 levels are high enough. Although not yet determined in a phase-dependent manner in the systole and the diastole, most studies mentioned above have shown that agonist-induced InsP_3 increases are maintained over many beats in the heart or in cardiomyocytes.

Such agonist-induced increase of diastolic sarcoplasmic Ca^{2+} might serve several purposes. Myocardial fiber stiffness and thus the myocardial mechanics could be influenced. The Na/Ca exchanger [8], sarcolemmal Ca^{2+} channels [81], sarcolemmal calcium pump [11], and ryanodine receptor [89] respond to submaximal sarcoplasmic Ca^{2+} and thus could be modulated in their activation-inactivation characteristics. Further functions likely to depend on diastolic Ca^{2+}, such as ANF secretion and possibly cardiac protein synthesis, could be thereby initiated or maintained. An indication of diastolic triggering of ANF secretion comes from its diastolic stretch-induced induction [31]. Apparently, a stretch-induced Ca^{2+} channel or PLC activity (see above) leads to a diastolic Ca^{2+} influx and/or Ca^{2+} release insufficient for contraction but sufficient for ANF synthesis and secretion.

Sophisticated measurements of diastolic Ca^{2+} in myocardial cells or carefully isolated ventricular trabeculae in conjunction with detailed inositol phosphate measurements appear essential to confirm these possible implications. Due to the predominating role of β-adrenergic mechanisms in positive inotropism and chronotropism it may be anticipated that systolic effects of PLC-activating agonists are less important than diastolic ones. The most important role of the inositide signaling system in the heart thus may be a "trophotropic" rather than an "ergotropic" one.

Cross-talk Between Glycolysis Activation and Ins(1,4,5)P_3 Release

In skeletal muscle we have recently pointed toward a type of cross-talk between glycolysis regulation and the inositol phosphate signaling system [42]. Aldolase A acts as a high capacity $InsP_3$ "buffering system" since it binds Ins(1,4,5)P_3 and other inositol phosphates [32] competitively with Fru(1,6)P_2 at its catalytic site. Thus, most of the total $InsP_3$ present in skeletal muscle is buffered away under conditions where Fru(1,6)P_2 is significantly below the concentration of aldolase binding sites, i.e., under resting conditions. When a fast-twitch muscle contracts tetanically, Fru(1,6)P_2 increases within a few seconds (from below $100\,\mu M$ to above $1\,mM$) and equally rapidly displaces bound $InsP_3$ from aldolase. Thus, even without the necessity of a rise in total $InsP_3$, its free concentration can be amplified within seconds by up to two orders of magnitude (from less than $50\,nM$ to above $1\,\mu M$). The myocardium also contains high concentrations of aldolase A and exhibits marked changes in Fru(1,6)P_2 [28, 50, 52] due to changes in cardiac performance. Marked changes in the free sarcoplasmic Ins(1,4,5)P_3 may thus be brought about. If effective in the heart, such mechanism could also explain the observed reduction in the basal and stimulated levels of total $InsP_3$ and inositides in the ischemic heart (see above). Due to anaerobic glycolysis, strongly increased Fru(1,6)P_2 might displace most of the $InsP_3$ from aldolase sites into the sarcoplasm. The resulting enhanced IP_3ICR might be a beneficial compensatory inotropic mechanism in brief periods of ischemia. Since Ins(1,4,5)P_3 5-phosphatase works below its K_m value [71], i.e., is activated by increasing free substrate, $InsP_3$ would be degraded more rapidly, leading to a decrease in its total concentration. After a period with elevated free $InsP_3$ and increased phosphoinositide turnover, the cycling rate may be reduced due to an energetic insufficiency to resynthesize phosphoinositides (see below).

A Role of Phosphoinositide-Derived Signals in Myocardial Hypertrophy

It is remarkable that the long-term actions of three of the ligands activating PLC and IP_3ICR, namely α_1-adrenergic agonists, endothelin, and AT II, are strictly correlated with an increase in the synthesis and secretion of ANF, a process known to involve Ca^{2+} (see above and [62, 76]). Such increases in

ANF mRNA and ANF secretion [68] indicate necessarily cardiomyocyte hypertrophy, i.e., they are always associated with increased biosynthesis of mRNA for contractile proteins (see chapter by Chien, this volume). The finding that long-term treatment with an α_1-adrenoceptor blocker leads to significant regression in cardiac hypertrophy in spontaneously hypertensive rats [79] further argues for an essential role of α-adrenoceptor activation in cardiomyocyte hypertrophy.

On the basis of our present knowledge, one or several of the following cellular messages could be the intracellular signal(s) for the initiation and maintenance of cardiomyocyte hypertrophy. (a) A long-term increase in diastolic sarcoplasmic Ca^{2+}. Such increase could be obtained by IP_3ICR (discussed above), by activation of hypothetic stretch-induced Ca^{2+} channels or by an inhibition of Ca^{2+}-sequestering or -extrusion systems (see above). It must be the diastolic increase in Ca^{2+} since many regimens increasing the systolic Ca^{2+} do not induce hypertrophy. (b) An increased level of one or a combination of the intracellular messengers formed by activation of PLC. This could be $Ins(1,4,5)P_3$ itself, a certain isomer or isomers of the longer lived $InsP_3$-derived higher phosphorylated inositols, or DAG. The importance of an increased level of DAG and the resulting activation of protein kinase C in cardiomyocyte hypertrophy has recently been shown by several groups [27, 31, 57, 68]. (c) Changes in one or several of the sarcolemmal inositol phospholipids due to a stimulated inositide cycling rate or to activation of phosphoinositide 3-kinase. Insulin, which has long been known to be involved in an increase in myocardial protein synthesis [69] and potentially other cardiac growth factors activating receptors of the RTK type (see above) and being able to induce cardiomyocyte hypertrophy might exert their actions via an activation of phosphoinositide 3-kinase (see above). (d) An increase in intracellular pH due to activation of Na^+-H^+ exchanger secondary to the stimulation of PLC and PKC is also discussed to be an essential signal in initiating cardiac protein synthesis [21, 33].

An interesting model assigning the inositide signaling system a potential importance in the development of dilative cardiomyopathy (DC) is based on the above discussed cross-talk between glycolysis activation and inositol phosphate release. Since many if not all cases of DC are accompanied by a deficiency of cardiac energetics (see chapter by Katz, this volume), the glycolytic flux and thus $Fru(1,6)P_2$ is increased chronically and, in parallel, the level of ATP decreased. The elevated level of $Fru(1,6)P_2$ may mobilize bound $Ins(1,4,5)P_3$ and thus release Ca^{2+} in the diastole via IP_3ICR. An effective removal of the released sarcoplasmic Ca^{2+} might be prevented by the changed kinetics and reduced overall amount of SR Ca^{2+} ATPase (see Katz, this volume). In concert with further of the above cellular signals this might, on the one hand, lead to the beneficial consequence of "compensatory" cardiomyocyte hypertrophy. On the other hand, the chronic deficiency of ATP combined with a permanent hyperstimulation of the energy-dependent inositide cycle bears the risk of an exhaustion of its metabolic and signaling elements (phosphoinositides, DAG, and inositol phosphates). In fact, a decreased level of DAG was reported in myopathic heart failure in hamsters [51], and a decreased rate of inositide cycling and a lower level of Ins(1,4,5)P_3 was observed in ischemic hearts (see above). As a fatal consequence of such exhaustion of cellular signaling elements, the response of cardiomyocytes to hypertrophic and "trophotropic" stimuli might be severely impaired or even lost. In particular in a state of permanently increased mechanical and energetic "stress," a situation characteristic for heart diseases leading to DC, this inability to adequately respond to such adaptive stimuli might be lethal for the cells.

Concluding Remarks

Inotropic responses and potentially the diastolic calcium appear to be modulated via phosphoinositide-derived signals. Recent experimental data have highlighted a correlation between the action of certain positive inotropic agonists stimulating the phosphoinositide cycle, ANF secretion, and cardiomyocyte hypertrophy. Future studies must define precisely the involved intracellular signaling events linking receptor activation to a highly specific expression of certain cardiomyocyte genes. Only a sophisticated molecular cardiology employing microanalytical, cell biochemical, cell physiological, and recombinant DNA techniques in isolated and cultured cardiomyocytes and in transgenic animals will be able to clarify the underlying molecular mechanisms. This knowledge will help us to understand what occurs in physiological and pathological cardiac hypertrophy and in heart failure and may lead to new regimens of DC treatment.

Acknowledgement. This work was supported by the Deutsche Forschungsgemeinschaft by grants Ma 989/1 and Ma 989/2.

References

1. Allen IS, Cohen NM, Dhallan RS, Gaa ST, Lederer WJ, Rogers TB (1988) Angiotensin II increases spontaneous contractille frequency and stimulates calcium current in cultured neonatal rat heart myocytes. Insights into underlying biochemical machanisms. Circ Res 62:524–534
2. Ashley CC, Mulligan IP, Lea TJ (1991) Ca^{2+} and activation mechanisms in skeletal muscle. Q Rev Biophys 24:1–73
3. Baker KM, Aceto JF (1990) Angiotensin II stimulation of protein synthesis and cell growth in chick heart cells. Am J Physiol 259:H610–H618
4. Baker KM, Singer HA (1988) Identification and characterization of guinea pig angiotensin II ventricular and atrial receptors coupling to inositol phosphate production. Circ Res 62:896–904
5. Berg I, Guse AH, Gercken G (1989) Carbamoylcholine-induced accumulation of inositol mono-, bis-, tris-, and tetrakisphosphates in isolated cardiac myocytes from adult rats. Biochim Biophys Acta 1010:100–107
6. Berridge MJ, Irvine RF (1989) Inositol phosphates and cell signalling. Nature 341:197–205
7. Bezprozvanny I, Watras J, Ehrlich BE (1991) Bell-shaped calcium-response curves of Ins(1,4,5)P_3- and calcium-gated channels from endoplasmic reticulum of cerebellum. Nature 351:751–754

8. Blaustein MP (1988) Sodium/calcium exchange and the control of contractility in cardiac muscle and vascular smooth muscle. J Cardiovasc Pharmacol 12 [Suppl 5]:56–68
9. Brown JH, Buxton ILO, Brunton LL (1985) α_1-Adrenergic and muscarinic cholinergic stimulation of phosphatidylinositol hydrolysis in adult rat cardiomyocytes. Circ Res 57:532–537
10. Brown RD, Berger KD, Taylor P (1984) α_1-Adrenergic receptor activation mobilizes cellular Ca^{2+} in a muscle cell line. J Biol Chem 259:7554–7562
11. Carafoli E (1991) Calcium pump of the plasma membrane. Physiol Rev 71:129–153
12. Carpenter CL, Cantley LC (1990) Phosphoinositide kinases. Biochemistry 29:11147–11156
13. Chadwick CC, Saito A, Fleischer S (1990) Isolation and characterization of the inositol trisphosphate receptor from smooth muscle. Proc Natl Acad Sci USA 87:2132–2136
14. Cobbold PH, Bourne PK (1984) Aequorin measurements of free calcium in single heart cells. Nature 312:444–446
15. Doeller JE, Wittenberg BA (1990) Intracellular calcium and high-energy phosphate in isolated cardiac myocytes. Am J Physiol 259:H1851–H1859
16. Downes CP, MacPhee CH (1990) *myo*-Inositol metabolites as cellular signals. Eur J Biochem 193:1–18
17. Eckel J, Gerlach-Eskuchen E, Reinauer H (1991) Alpha-adrenoceptor-mediated increase in cytosolic free calcium in isolated cardiac myocytes. J Mol Cell Cardiol 23:617–625
18. Endoh M, Hiramoto T, Ishihata A, Takanashi M, Inui J (1991) Myocardial α_1-adrenoceptors mediate positive inotropic effect and changes in phosphatidylinositol metabolism. Species differences in receptor distribution and the intracellular coupling process in mammalian ventricular myocardium. Circ Res 68:1179–1190
19. Fabiato A (1989) Appraisal of the physiological relevance of two hypotheses for the mechanism of calcium release from the mammalian cardiac sarcoplasmic reticulum. Calcium-induced release versus charge-coupled release. Mol Cell Biochem 89:135–140
20. Franco A Jr, Lansman JB (1990) Calcium entry through stretch-inactivated ion channels in mdx myotubes. Nature 344:670–673
21. Fuller SJ, Gaitanaki CJ, Hatchett RJ, Sugden PH (1991) Acute α_1-adrenergic stimulation of cardiac protein synthesis may involve increased intracellular pH and protein kinase activity. Biochem J 273:347–353
22. Guse AH, Berg I, Gercken G (1989) Metabolism of inositol phosphates in α_1-adrenoceptor-stimulated and homogenized myocytes of adult rats. Biochem J 261:89–92
23. Heathers GP, Corr PB, Rubin LJ (1988) Transient accumulation of inositol 1,3,4,5-tetrakisphosphate in response to α_1-adrenergic stimulation in adult cardiac myocytes. Biochem Biophys Res Commun 156:485–492
24. Heller Brown J, Jones LG (1986) Phosphoinositide metabolism in the heart. In: Phosphoinositides and receptor mechanisms. Liss, New York, pp 245–270
25. Homma Y, Takenawa T, Emori Y, Sorimachi H, Suzuki K (1989) Tisue and cell type-specific expression of mRNAs for four types of inositol phospholipid-specific phospholipase C. Biochem Biophys Res Commun 164:406–412
26. Inui M, Saito A, Fleischer S (1987) Isolation of the ryanodine receptor from cardiac sarcoplasmic reticulum and identity with the feet structures. J Biol Chem 262:15637–15642
27. Kariya K, Karns LR, Simpson PC (1991) Expression of a constitutively activated mutant of the β-isozyme of protein kinase C in cardiac myocytes stimulates the promoter of the β-myosin heavy chain isogene. J Biol Chem 266:10023–10026
28. Katz LA, Swain JA, Portman MA, Balaban RS (1989) Relation between phosphate metabolites and oxygen consumption of heart in vivo. Am J Physiol 256:H265–H274
29. Kohl C, Linck B, Schmitz W, Scholz H, Scholz J, Toth M (1990) Effects of carbachol and (−)-N^6-phenylisopropyladenosine on myocardial inositol phosphate content and force of contraction. Br J Pharmacol 101:829–834
30. Kohl C, Schmitz W, Scholz H (1990) Positive inotropic effect of carbachol and inositol phoshate levels in mammalian atria after pretreatment with pertussis toxin. Circ Res 66:580–583
31. Komuro I, Katoh Y, Kaida T, Shibazaki Y, Kurabayashi M, Hoh E, Takaku F, Yazaki Y (1991) Mechanical loading stimulates cell hypertrophy and specific gene expression in cultured rat cardiac myocytes. J Biol Chem 266:1265–1268

32. Koppitz B, Vogel F, Mayr GW (1986) Mammalian aldolases are isomer-selective high-affinity inositol poyphosphate binders. Eur J Biochem 161:421–433
33. Krämer BK, Smith TW, Kelly RA (1991) Endothelin and increased contractility in adult rat ventricular myocytes. Role of intracellular alkalosis induced by activation of the protein kinase C-dependent Na^+-H^+ Exchanger. Circ Res 68:269–279
34. Kusuoka H, Futaki S, Koretsune Y, Kitabatake Y, Suga H, Kamada T, Inoue M (1991) Alterations of intracellular calcium homeostasis and myocardial energetics in acute Adriamycin-induced heart failure. J Cardiovasc Pharmacol 18:437–444
35. Lagos N, Vergara J (1990) Phosphoinositides in frog skeletal muscle. A quantitative analysis. Biochim Biophys Acta 1043:235–244
36. Lee PL, Johnson DE, Cousens LS, Fried VA, Williams LT (1989) Purification and complementary DNA cloning of a receptor for basic fibroblast growth factor. Science 245:57–60
37. Legssyer A, Poggioli J, Renard D, Vassort G (1988) ATP and other adenine compounds increase mechanical activity and inositol trisphosphate production in rat heart. J Physiol Lond 401:185–199
38. Leung E, Johnston CI, Woodcock EA (1986) Stimulation of phosphatidylinositol metabolism in atrial and ventricular myocytes. Life Sci 39:2215–2220
39. MacLennan DH (1990) Molecular tools to elucidate problems in excitation-contraction coupling. Biophys J 58:1355–1365
40. Martin TF (1991) Receptor regulation of phosphoinositidase C. Pharmacol Ther 49:329–345
41. Masters SB, Martin MW, Harden TK, Brown JH (1985) Pertussis toxin does not inhibit muscarinic receptor-mediated phosphoinositide hydrolysis or calcium mobilization. Biochem J 227:933–937
42. Mayr GW, Thieleczek R (1991) Masses of inositol phosphates in resting and tetanically stimulated vertebrate skeletal muscles. Biochem J 280:631–640
43. McDonald LJ, Mamrack, MD (1989) Purification and characterization of bovine heart phosphoinositide-specific phospholipase C. Kinetic analysis of the Ca^{2+} requirement and La^{3+} inhibition. Biochemistry 28:9926–9932
44. McDonough PM, Goldstein D, Heller Brown J (1988) Elevation of cytoplasmic calcium concentration stimulates hydrolysis of phosphatidylinositol bisphosphate in chick heart cells. Effect of sodium channel activators. Mol Pharmacol 33:310–315
45. Meji JTA, Lamers JMJ (1989) Alpha-1-adrenergic stimulation of phosphoinositide breakdown in cultured neonatal rat ventricular myocytes. Mol Cell Biochem 88:73–75
46. Morgan JP (1991) Abnormal intracellular modulation of calcium as a major cause of cardiac contractile dysfunction. N Engl J Med 325:625–632
47. Mouton R, Huisamen B, Lochner A (1991) Increased myocardial inositol trisphosphate levels during α_1-adrenergic stimulation and reperfusion of ischemic rat heart. J Mol Cell Cardiol 23:841–850
48. Mouton R, Huisamen B, Lochner A (1991) The effect of ischemia and reperfusion on sarcolemmal inositol phospholipid and cytosolic inositol phosphate metabolism in the isolated perfused rat heart. Mol Cell Biochem 105:127–135
49. Nakagawa T, Okano H, Furuichi T, Aruga J, Mikoshiba K (1991) The subtypes of the mouse inositol 1,4,5-trisphosphate receptor are expressed in a tissue-specific and developmentally specific manner. Proc Natl Acad Sci USA 88:6244–6248
50. Neely JR, Denton RM, England PJ, Randle PJ (1972) The effects of increased heart work on the tricarboxylate cycle and its interaction with glycolysis in perfused rat heart. Biochem J 128:147–159
51. Okumura K, Yamada Y, Kondo J, Hashimoto H, Ito T, Kitoh J (1991) Decreased 1,2-diacylglycerol levels in myopathic hamster hearts during the development of heart failure. J Mol Cell Cardiol 23:409–416
52. Opie LH (1984) Substrate and energy metabolism in the heart. In: Sperelakis N (ed) Physiology and pathophysiology of the heart. Nijhoff, Boston, pp 301–326
53. Otani H, Otani H, Morita M, Das DK (1989) Effect of calcium overload on the phosphoinositide breakdown in the left ventricular papillary muscle. Mol Cell Biochem 90:111–120

54. Otani H, Prasad MR, Engelman RM, Otani H, Cordis GA, Das DK (1988) Enhanced phosphodiesteratic breakdown and turnover of phosphoinositides during reperfusion of ischemic rat heart. Circ Res 63:930–936
55. Pape PC, Konishi M, Baylor SM, Somlyo AP (1988) Excitation-contraction coupling in skeletal muscle fibres injected with the $InsP_3$ blocker, heparin. FEBS Lett 235:57–62
56. Pfaffinger PJ, Martin JM, Hunter DD, Nathanson NM, Hille B (1985) GTP-binding proteins couple cardiac muscarinic receptors to a K channel. Nature 317:536–538
57. Pitkänen M, Mäntymaa P, Ruskoaho H (1991) Staurosporine, a protein kinase C inhibitor, inhibits atrial natriuretic peptide secretion induced by sarafotoxin, endothelin and phorbol ester. Eur J Pharmacol 195:307–315
58. Poggioli J, Sulpice JC, Vassort G (1986) Inositol phosphate production following α_1-adrenergic, muscarinic or electrical stimulation in isolated rat heart. FEBS Lett 206:292–298
59. Puceat M, Clement O, Scamps F, Vassort G (1991) Extracellular ATP-induced acidification leads to cytosolic calcium transient rise in single rat cardiac myocytes. Biochem J 274: 55–62
60. Putney JW (1991) Receptor-regulated calcium entry. Pharmacol Ther 48:427–434
61. Renard D, Poggioli J (1987) Does the inositol tris/tetrakisphosphate pathway exist in rat heart? FEBS Lett 217:117–123
62. Ruskoaho H, Toth M, Ganten D, Unger T, Lang RE (1986) The phorbol ester-induced atrial natriuretic peptide secretion is stimulated by forskolin and BAY K8644 and inhibited by 8-bromo-cyclicGMP. Biochem Biophys Res Commun 139:266–274
63. Sakuma J, Gross SS, Levi R (1988) Positive inotropic effect of histamine on guinea pig left atrium. H_1-receptor-mediated stimulation of phosphoinositide turnover. J Pharmacol Exp Ther 247:466–472
64. Schmitz W, Scholz H, Scholz J, Steinfath M (1987) Increase in IP_3 precedes α-adrenoceptor-induced increase in force of contraction in cardiac muscle. Eur J Pharmacol 140:109–111
65. Schmitz W, Scholz H, Scholz J, Steinfath M, Lohse M, Puurunen J, Schwabe U (1987) Pertussis toxin does not inhibit the α_1-adrenoceptor-mediated effect on inositol phosphate production in the heart. Eur J Pharmacol 134:377–378
66. Scholz J (1989) Inositol trisphosphate, a novel second messenger for positive inotropic effects in the heart? Klin Wochenschr 67:271–279
67. Schwertz DW, Halverson J (1989) Characterization of phospholipase C-mediated polyphosphoinositide hydrolysis in rat heart ventricles. Arch Biochem Biophys 269:137–147
68. Sei CA, Irons CE, Sprenkle AB, McDonough PM, Heller Brown J, Glembotski CC (1991) The α-adrenergic stimulation of atrial natriuretic factor expression in cardiac myocytes requires calcium influx, protein kinase C, and calmodulin-regulated pathways. J Biol Chem 266:15910–15916
69. Sender PM, Garlick PJ (1973) Synthesis rates of protein in the Langendorff perfused rat heart in the presence and absence of insulin, and in the working heart. Biochem J 132:603–608
70. Shears SB (1991) Structural and functional interrelationship of the intracellular receptor for inositol 1,4,5-trisphosphate. Cancer Cells 3:97–99
71. Shears SB (1991) Regulation of the metabolism of 1,2-diacylglycerols and inositol phosphates that respond to receptor activation. Pharmacol Ther 49:79–104
72. Shubeita HE, McDonough PM, Harris AN, Knowlton KU, Glembotski CC, Heller Brown J, Chien KR (1990) Endothelin induction of inositol phospholipid hydrolysis, sarcomere assembly, and cardiac gene expression in ventricular myocytes. J Biol Chem 265:20555–20562
73. Skolnik EY, Margolis B, Mohammadi M, Lowenstein E, Fischer R, Drepps A, Ullrich A, Schlessinger J (1991) Cloning of PI_3 kinase-associated p85 utilizing a novel method for expression/cloning of target proteins for receptor tyrosine kinases. Cell 65:83–90
74. Somlyo AP, Himpens B (1989) Cell calcium and its regulation in smooth muscle. FASEB J 3:2266–2276
75. Sommer JR, Johnson EA (1979) Ultrastructure of cardiac muscle. In: Berne RM (ed)

Handbook of physiology: 2. The cardiovascular system. American Physiological Society, Bethesda, pp 113–186
76. Sonnenberg H, Veress AT (1984) Cellular mechanism of release of atrial natriuretic factor. Biochem Biophys Res Commun 124:443–449
77. Steinberg SF, Kaplan LM, Inouye T, Zhang JF, Robinson RB (1989) Alpha-1 adrenergic stimulation of 1,4,5 inositol trisphosphate formation in ventricular myocytes. J Pharmacol Exp Ther 250:1141–1148
78. Suarez-Isla BA, Acayage C, Marengo JJ, Bull R (1991) Activation of inositol trisphosphate-sensitive Ca^{2+} channels of sarcoplasmic reticulum from frog skeletal muscle. J Physiol (Lond) 441:575–591
79. Takeda N, Nakamura I, Ohkubo T, Iwai T, Tanamura A, Nagano M (1991) Effects of long term treatment with an α_1 adrenoceptor blocker on cardiac hypertrophy, contractility, and myosin isoenzymes in spontaneously hypertensive rats. Cardiovasc Res 25:565–567
80. Thomas GMH, Geny B, Cockroft S (1991) Identification of a novel cytosolic polyphosphoinositide-specific phospholipase C (PLC-86) as the major G-protein regulated enzyme. EMBO J 10:2507–2512
81. Tseng GN, Boyden PA (1991) Different effects of intracellular Ca and protein kinase C on cardiac T and L Ca currents. Am J Physiol 261:H364–H379
82. Ventura C, Guarnieri C, Stefanelli C, Cirielli C, Lakatta EG, Capogrossi MC (1991) Comparison between alpha-adrenergic- and K-opioidergic-mediated inositol 1,4,5-P_3/ inositol 1,3,4,5-P_4 formation in adult cultured rat ventricular cardiomyocytes. Biochem Biophys Res Commun 179:972–978
83. Vigne P, Breittmayer JP, Marsault R, Frelin C (1990) Endothelin mobilizes Ca^{2+} from a caffeine and ryanodine-insensitive intracellular pool in rat atrial cells. J Biol Chem 265:6782–6787
84. Vigne P, Lazdunski M, Frelin C (1989) The inotropic effect of endothelin-1 on rat atria involves hydrolysis of phosphatidylinositol. FEBS Lett 249:143–146
85. Volpe P, Salviati G, DiVirgilio F, Pozzan T (1985) Inositol 1,4,5-trisphosphate induces calcium release from sarcoplasmic reticulum of skeletal muscle. Nature 316:347–349
86. Walker JW, Somlyo AV, Goldmann YE, Somlyo AP, Trentham DR (1987) Kinetics of smooth and skeletal muscle activation by laser photolysis of caged inositol 1,4,5-trisphosphate. Nature 327:249–252
87. Walton PD, Sutko JL, Beck CF, Mignery GA, Südhof TC, Deerink TJ (1991) Ryanodine and inositol trisphosphate receptors coexist in avian cerebellar pukinje neurons. J Cell Biol 113:1145–1157
88. Wendt-Gallitelli MF, Isenberg G (1991) Total and free myoplasmic calcium during a contraction cycle. X-ray micronanalysis in guinea-pig ventricular myocytes. J Physiol (Lond) 435:349–372
89. Witcher DR, Kovacs RJ, Schulman H, Cefali DC, Jones LR (1991) Unique phosphorylation site on the cardiac ryanodine receptor regulates calcium channel activity. J Biol Chem 266:11144–11152
90. Wolf RA (1990) Synthesis, transfer, and phosphorylation of phosphoinositides in cardiac membranes. Am J Physiol 259:C987–C994
91. Zhu PH, Fu DX (1990) Effect of prolonged in vitro lithium treatment on calcium transients in frog twitch muscle fibres and its reversal by exogenous *myo*-inositol. Neuroscience 39:271–278

VIII. Virus and Cardiomyopathy

Enterovirus-Induced Cardiomyopathy: Molecular Analysis of Acute and Persistent Myocardial Infections

R. Kandolf, K. Klingel, A. Canu, R. Zell, H.-C. Selinka, F. McPhee, W.M. Franz, J. Gulizia, A. Heim, U. Fortmüller, C. Hohenadl, U. Raab, G. Mall, and B. McManus

Introduction

Enteroviruses of the Picornaviridae, such as coxsackieviruses (CV) of group B (CVB), are the most common cause of viral myocarditis [1, 12, 18, 23, 34]. Other members of the human enterovirus group [24] comprising at present about 70 serotypes (e.g., various coxsackie A viruses and echoviruses) have also been associated with human viral heart disease.

The enterovirus genome is contained in a single-stranded RNA molecule of positive polarity which is about 7500 nucleotides in length and covalently linked to a small virus-encoded protein, VPg (3B), at the 5′ end. The infectious viral RNA is polyadenylylated at the 3′ end and serves as messenger for the synthesis of virus-directed proteins and as template for replication [29]. Four structural proteins, VP1–VP4 (1A–1D), are generated from the amino-terminal part of the precursor polyprotein, whereas the nonstructural proteins such as the RNA-dependent RNA polymerase ($3D^{pol}$), VPg (3B), two proteinases ($2A^{pro}$ and $3C^{pro}$), and several polypetides of as yet unknown function are released from the carboxy-terminal part. During replication the genomic viral plus-strand RNA serves as a template for transcription of a minus-strand RNA intermediate by $3D^{pol}$. Minus-strand RNA, which is synthesized in acutely infected cells in a low copy number, is transcribed again by $3D^{pol}$ to generate large amounts of viral plus-strand RNA genomes which are packaged into new virions. Cell lysis and virus release usually complete the infectious cycle of acute forms of the infection.

Diagnosing viral heart disease by clinical features is an exercise of uncertain validity [2]. In practice, the measurement of enterovirus serotype-specific serum antibody titers seldom proves useful diagnostically, except for cases that occur as part of an epidemic. In addition, the histological features of biopsy specimens of patients with a clinical suspicion of viral myocarditis are not virus specific and are indistinguishable from those seen in inflammatory heart disease of nonviral origin. However, confirmation of the clinical suspicion of enterovirus-induced cardiomyopathy demands the demonstration of replicating virus in myocardial cells, which is exceedingly difficult by conventional methods, such as isolation of infectious virus

Institute of Pathology, University of Tübingen, W-7400 Tübingen and Max Planck Institute for Biochemistry, W-8033 Martinsried, Fed. Rep. of Germany

Figulla et al. (Eds.)
Idiopathic Dilated Cardiomyopathy

or immunoflourescent staining of serotype-specific viral antigens. These classical virological techniques are usually successful only in fulminant and letal cases of neonatal enterovirus myocarditis.

Molecular genetic techniques have provided investigators with powerful tools for defining the role of enteroviruses in the induction and maintenance of the disease [9, 13, 31] as well as to study the molecular basis of pathogenicity. This review focuses on enterovirus group-specific in situ hybridization [12, 16, 20], which is currently the method of choice for diagnosing myocardial enterovirus infections. Using in situ hybridization, myocardial enterovirus infection is detectable in all stages of acute and chronic enterovirus-induced myocarditis and in end-stage dilated cardiomyopathy, indicating that chronic myocardial injury is associated with persistent enterovirus infection [12, 14]. The discovery of possible enterovirus persistence in the human heart is supported by our recent finding of enterovirus persistence in various murine models of chronic myocarditis, demonstrating that CVB3, typically a cytolytic virus, is capable of evading the immunological surveillance in a host-dependent fashion [20]. Furthermore, antisera raised against bacterially synthesized distinct CVB3 proteins are described, which allow the study of enteroviral protein expression in acute and persistent myocardial infections. In addition, in vitro experiments are reviewed, demonstrating the potent antiviral activity of type I and type II interferons (IFN) in cultured human heart cells. Finally, we summarize studies addressing the characterization of protective enteroviral B-cell epitopes by construction of antigen chimeras using CVB3 as an RNA virus vector.

Cloned Coxsackievirus B3 cDNA as a Diagnostic Reagent for the Detection of Enteroviruses

One major prerequisite for the introduction of in situ hybridization for the diagnosis of enteroviral heart disease was the molecular cloning and characterization of the single-stranded genomic RNA of a cardiotropic CVB3 [13] which has been propagated in cultured human heart cells [15]. Full-length, reverse-transcribed CVB3 cDNA generated replication-competent CVB3 upon transfection of recombinant viral cDNA into mammalian cells, demonstrating the molecular cloning of a faithful transcript of the viral RNA. As expected, in vitro synthesized RNA transcripts were also found to be infectious. Furthermore, the complete nucleotide sequence of the infectious CVB3 cDNA was determined [21], offering unique opportunities for the genetic analysis of this virus.

From the diagnostic point of view, cloned CVB3 cDNA may be used to obtain a virus group-specific diagnosis of enterovirus infections. Because of the high degree of genetic identity shared among nonstructural gene regions of the human enterovirus group, detection of the agents most commonly associated with viral heart disease, including the groups A and B CVs and the echoviruses, is possible in a single hybridization assay [16]. The broad spectrum of enteroviruses detected by cloned CVB3 cDNA greatly facilitates diagnosis of myocardial enterovirus infection since, from the clinical point of view, serotypic identification of the implicated enterovirus is of secondary importance. In theory this could be carried out by bybridization with serotype-specific probes although such probes are available only for a limited number of molecularly cloned enterovirus serotypes.

Highly specific hybridization conditions have been established for the detection of enteroviruses. When radioactively labeled cloned CVB3 cDNA corresponding to 95.4% of the viral genome (nucleotides 66–7128) was hybridized to electrophoretically resolved CVB3 cDNA and to total RNA from cultured human heart cells, specific hybridization was found for the viral RNA and not for human myocardial RNA [13].

The feasibility of using the in situ hybridization technique [4, 33] to detect enterovirus RNA was first established in cell culture systems and then applied to myocardial tissue of athymic mice persistently infected with CVB3 [16]. In situ hybridization proved to be a powerful tool not only for establishing an unequivocal diagnosis of myocardial enterovirus infection but also for understanding its pathogenesis. Myocardial enterovirus infection was found to be multifocal and randomly distributed. Infected myocytes were often found in clusters within areas of severe myocardial lesions and fibrosis. Furthermore, progression of the infection could be observed from areas with myocardial injury and fibrosis to as yet uninfected myocytes, demonstrating the possible cell to cell spread of the virus. Infected myocytes, however, were also found in apparently normal myocardial tissue, which was observed primarily in the early stage of the infection. Labeled myocardial cells were not found when infected myocardial tissue was hybridized with the radiolabeled plasmid p2732B control DNA probe or with the cloned *Eco*RI J fragment of the genetically unrelated cytomegalovirus, demonstrating the specificity of the in situ hybridization technique for the presence of enteroviral RNA [16].

A high sensitivity of detection was achieved by several improvements in methodology, including optimized hybridization conditions to prevent nonspecific binding and the use of radiolabeled cloned cDNA fragments about 100 nucleotides in length [16]. Quantification of the CVB3 copy number in infected Vero cells by RNA blot analysis and comparison with the results of in situ hybridization to cells from the same culture indicated that as few as 20 viral copies are easily detectable within 2 weeks of autoradiographic exposure. Clearly positive hybridization signals with acutely infected mouse myocardial tissue were observed after only 2 days of exposure to the ^{35}S-labeled cDNA probe, indicating a high copy number of about 5×10^4 viral genomes in acutely infected myocardial cells.

Moreover, the application of radioactively labeled, strand-specific RNA probes for in situ detection of both plus-strand and minus-strand enteroviral RNA in myocardial tissue has been established [8]. This in situ hybridization technique provides a valuable tool not only for unequivocal diagnosis of myocardial enterovirus infections but also for the analysis of acute and chronic forms of the disease at the level of single-stranded RNA.

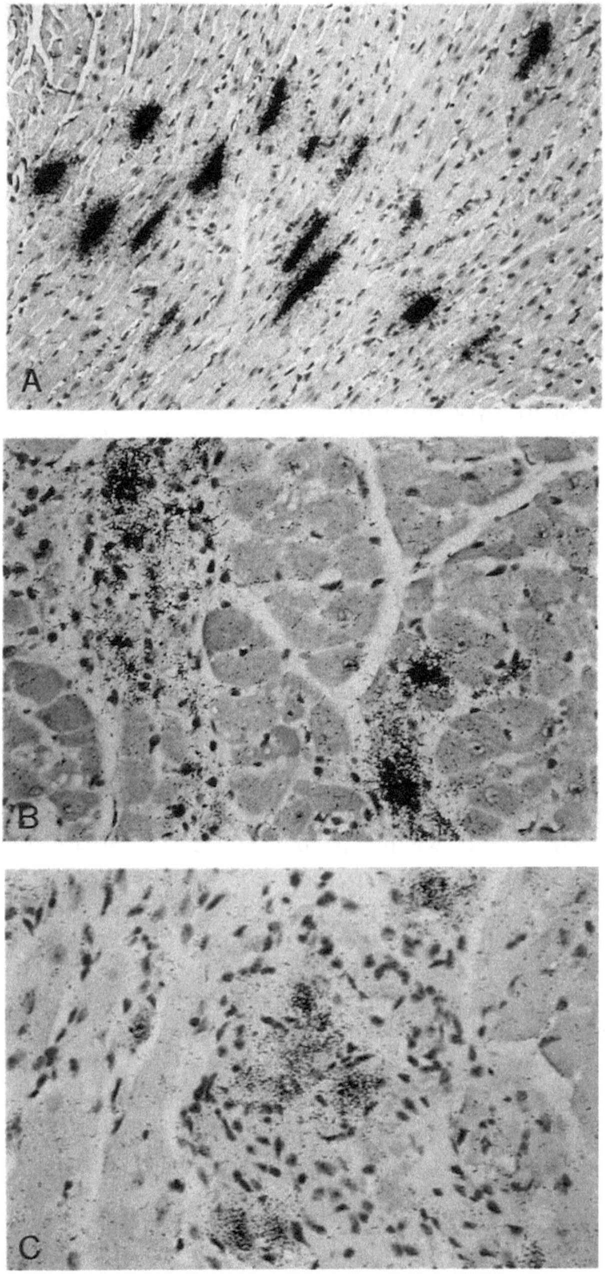

Fig. 3A–C. In situ patterns of acute and persistent myocardial infection in CVB3-induced acute and chronic myocarditis of immunocompetent A.CA/SnJ mice. Days after infection were 3 (**A**), 12 (**B**) and 30 (**C**). Despite the prominent cellular immune response (**B**) the virus was found to be capable of evading immunological surveillance, thus inducing a persistent infection. No inflammatory lesions were found to develop in the absence of virus replication during the course of acute and chronic disease, indicating that immune-mediated processes are maintained by persistent infection. (From [20])

In Situ Detection of Myocardial Enterovirus Infection in Patients with Myocarditis and Dilated Cardiomyopathy

In situ hybridization has proved to be a valuable tool for assessing the presence of enterovirus RNA in endomyocardial biopsy samples obtained from patients with suspected myocarditis and dilated cardiomyopathy [12, 14]. Replicating enterovirus RNA was found to be present in 23 of 95 patients (24%) with a clinical suspicion of acute myocarditis, including 10 of 33 patients (30%) with congestive heart failure of recent onset. All 53 patients of a pathological control group with other specific heart muscle diseases not consistent with a primary viral etiology (e.g., ischemic, hypertrophic, or metabolic cardiomyopathies) were negative when myocardial tissue was examined using in situ hybridization.

Figure 1A illustrates the typical pattern of acute myocardial infection as observed in the endomyocardial biopsy of an adult patient with severe congestive heart failure of recent onset. The autoradiographic silver grains, indicating hybridization between viral RNA and a radiolabeled plus-strand specific RNA probe, are clearly localized to infected myocytes, thereby providing an etiologic diagnosis of myocardial enterovirus infection. Infected myocytes are easily identified because of their characteristic size and morphology. Infected myocytes are also seen adjacent to mononuclear inflammatory cells, which provide the basis for the histopathological diagnosis of myocarditis.

Moreover, enterovirus RNA was detected not only in myocardial biopsy samples of patients with acute and chronic myocarditis but also in patients with chronic dilated cardiomyopathy, suggesting persistence of the virus in the human heart. Of 47 patients with chronic idiopathic dilated cardiomyopathy, 8 patients (17%), including 4 of 19 patients undergoing heart transplantation because of end-stage dilated cardiomyopathy, were found to have myocardial enterovirus infection. Figure 1B shows, as an example, in situ detection of myocardial enterovirus infection in the endomyocardial biopsy of a patient with chronic dilated cardiomyopathy. This patient had developed a dilated type of cardiomyopathy over a period of 3 years following an initial episode of clinically suspected acute myocarditis. Typically, a reduced number of infected myocardial cells consistent with a persistent type of infection is observed in chronic enterovirus-induced cardiomyopathy, indicating that restricted replication plays a role in the development of ongoing disease. The concept of possible enterovirus persistence in chronic dilated cardiomyopathy, evolving from acute and possibly subacute infections, was further substantiated by our finding of enterovirus persistence in serial biopsies obtained from patients with ongoing disease as well as in explanted hearts and in autopsy hearts of patients with end-stage congestive heart failure. Figure 1C illustrates a typical pattern of obviously persistent enterovirus infection in the autopsy heart of a patient who died of end-stage dilated cardiomyopathy.

In contrast to the acute infection (Fig. 1A), which is characterized by virus-induced myocytolysis in the presence of inflammatory cells, persistent enterovirus infection in dilated cardiomyopathy (Fig. 1B,C) was found to be associated with degeneration of myofibers, hypertrophy, and interstitial fibrosis, usually in the absence of a light microscopic apparent cellular immune response. Importantly, direct cell-to-cell spread of the virus, or possibly viral RNA, was also observed in patients with patterns of persistent enterovirus infection, suggesting that active replication is involved in the pathogenesis of the ongoing disease. Moreover, enterovirus RNA was found to be present not only in myocytes but also in small interstitial cells, possibly myocardial fibroblasts. These observations agree with our previous in vitro findings in cultured human heart cells and persistently infected human myocardial fibroblasts [15, 17].

Examination of autopsy hearts of patients who died of acute enterovirus myocarditis associated with severe congestive heart failure revealed a multifocal and disseminated infection of the human heart with involvement of both atrium and ventricle walls and the interventricular septum. Figure 2A illustrates a typical in situ pattern of acute myocardial enterovirus infection in the heart of an adult patient who died of acute enterovirus myocarditis. Numerous myocardial cells replicate high copy numbers of the viral genome as indicated by the high density of autoradiographic labeling, and histologic acute myocarditis is evident from myocardial injury and inflammation. Thus far, the most extensive patterns of acute myocardial enterovirus infection were observed in fatal cases of enterovirus-induced neonatal myocarditis. Figure 2B shows the dramatic pattern of acute enterovirus infection in the heart of an infant who died of suspected fulminant enterovirus myocarditis. Infected myocytes are seen adjacent to inflammatory cells and progression of the infection can be observed from areas of inflammation to as yet uninflamed myocardial tissue, demonstrating the cell-to-cell spread of the enterovirus. Moreover, infected myocardial cells are also observed in areas without inflammation, indicating that the extent of heart muscle infection increases during the acute stage of the disease.

Figure 2C illustrates the negative controls which are carried out as a matter of routine to control the specificity of in situ hybridization results. No labeled cells are observed when serial myocardial tissue sections of the same patient are probed with the radiolabeled plasmid vector control DNA or RNA probe, demonstrating the specificity of the in situ hybridization technique for the presence of enteroviral RNA.

Quantitative Characterization of Acute and Persistent Myocardial Infection in the Murine Model of Ongoing Enteroviral Heart Disease

Experimentally, acute and chronic forms of enterovirus-induced myocarditis have been described in various murine models [7], which exhibit histopathologic features resembling heart muscle lesions observed in humans.

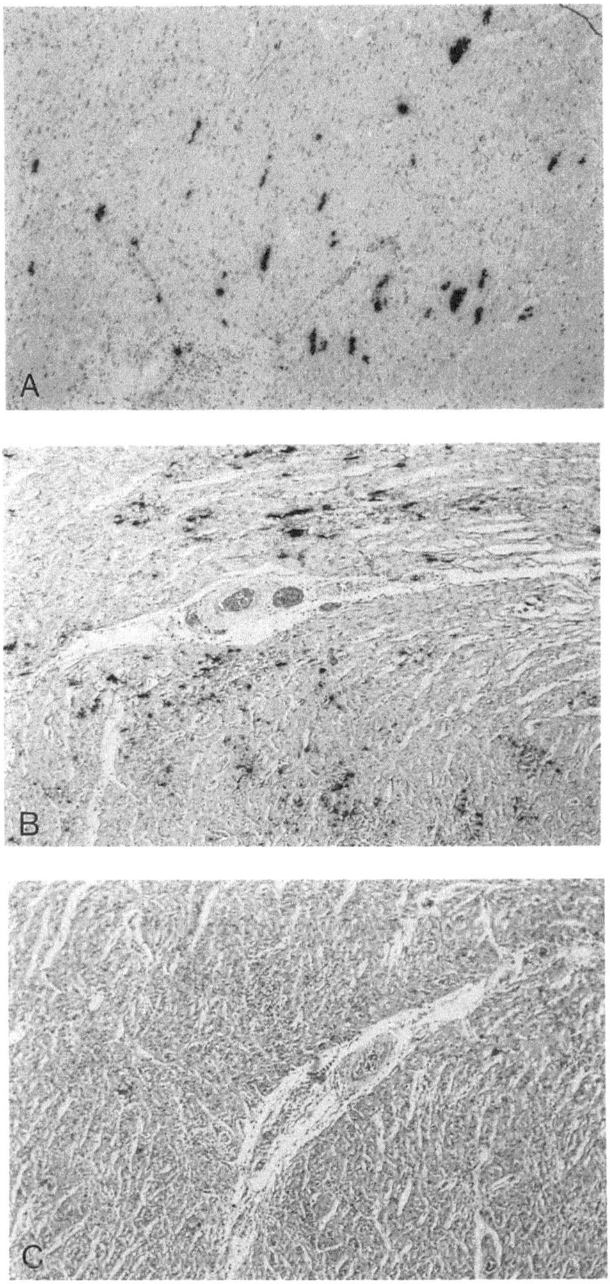

Fig. 2A–C. Patterns of acute myocardial enterovirus infection in fatal cases of adult (**A**) and neonatal (**B**,**C**) myocarditis. The autoradiographic silver grains, which indicate hybridization between viral RNA and the ^{35}S-labeled enterovirus-specific cDNA, are clearly localized to distinct infected myocardial cells, allowing an unequivocal diagnosis of myocardial enterovirus infection. No labeling of myocardial cells is observed when myocardial tissue sections are hybridized with a ^{35}S-labeled unrelated control DNA probe (**C**), demonstrating the specificity of in situ hybridization

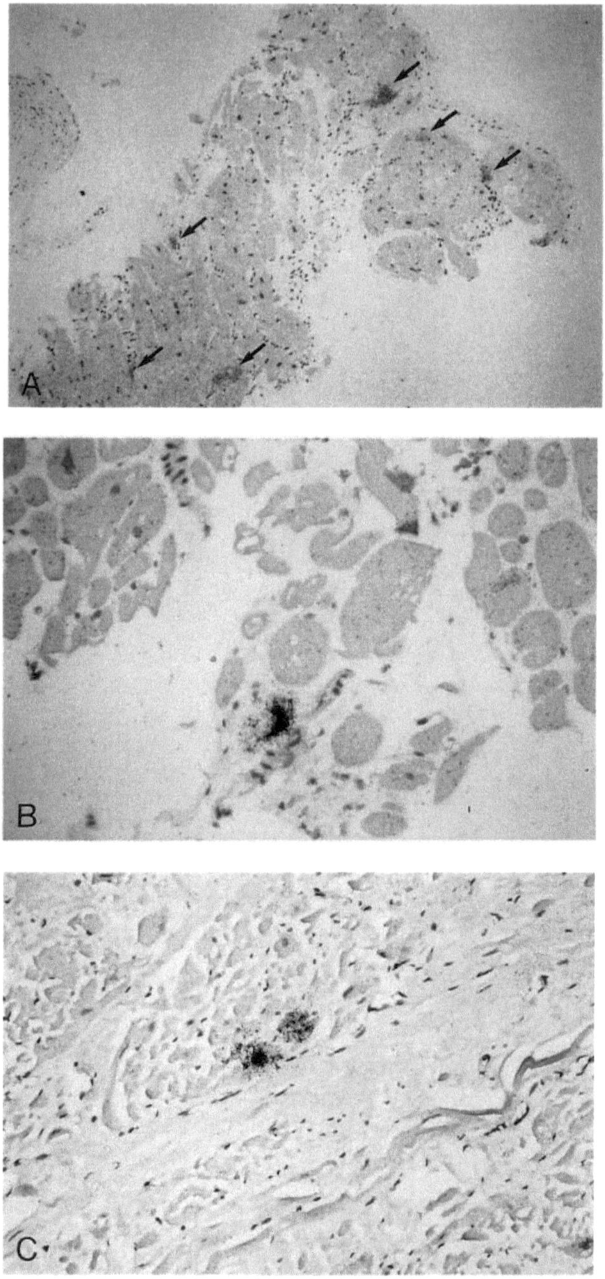

Fig. 1A–C. In situ hybridization of the ^{35}S-labeled enterovirus-specific plus-strand RNA probe (**A**) and the ^{35}S-labeled enterovirus-specific cDNA probe (**B**,**C**) to endomyocardial biopsies (**A**,**B**) in acute myocarditis (**A**) and dilated cardiomyopathy (**B**), and to myocardial tissue of the autopsy heart of a patient with end-stage disease (**C**). In contrast to the acute myocardial enterovirus infection (**A**), characterized by numerous infected myocytes (*arrows*) in the presence of inflammation, a decreased number of enterovirus-positive cells is observed in patients with idiopathic dilated cardiomyopathy

Based on our findings in humans, we followed the natural course of CVB3-induced myocarditis in different immunocompetent mouse strains using in situ hybridization for the detection of viral RNA. Importantly, we found that CVB3 can induce a persistent infection in A.CA/SnJ(H-2^f), A.BY/SnJ(H-2^b) as well as in SWR/J(H-2^q) mice but not in DBA/1J(H-2^q) mice, indicating that host genetics are a major determinant for chronic enteroviral heart disease [20]. Figure 3 illustrates the development of a persistent type of CVB3 infection for the representative immunocompetent mouse strain A.CA/SnJ. Two to three days postinfection (p.i.) individual in situ-positive myocytes were found randomly distributed throughout the right and left ventricle, indicating hematogenous infection of myocytes during viremia (Fig. 3A). The high local autoradiographic grain densities in hybridization-positive cells demonstrate high copy numbers of replicating viral genomes in acutely infected myocardial cells. Between days 12 and 15 p.i., when the maximum of cellular infiltration was observed, the number of infected myocardial cells decreased (Fig. 3B), indicating benefical effects of the cellular immune response. Nonetheless, CVB3 was found to be capable of inducing a persistent infection in a host-dependent fashion. Figure 3B represents the typical in situ pattern of persistent myocardial CVB3 infection as observed between days 30 and 42 p.i. This pattern of infection is characterized by ongoing inflammation in the presence of infected myocardial cells, which exhibit a decreased copy number of viral RNA (as reflected by the decreased autoradiographic grain densities) when compared with acute infection. Persistently infected myocardial cells were found to be located primarily within and adjacent to foci of chronic myocardial lesions containing replacement fibrosis, degenerated myocytes, and mononuclear cell infiltrates.

The morphometrical quantification of myocardial infection following digital image processing revealed increasing area fractions of infected cardiac tissue until day 9 after CVB3 infection, when up to 13% of the myocardial cells were found to be infected [20]. Thereafter, decreasing area fractions of infected myocardial tissue were measured, indicating a reduced number of infected myocardial cells at later stages of the disease. During persistent infection, e.g., 30 days p.i., approximately 0.01% of myocardial cells were found to be infected, which is obviously sufficient to sustain myocardial inflammation.

Concurrent with increasing area fractions of infected myocardial tissue, an increase in numerical densities of macrophages and T lymphocytes was observed during acute infection. In early infection (up to day 6 p.i.), the number of cytotoxic/suppressor T lymphocytes exceeded the number of T helper cells. The maximum densities of mononuclear infiltrates were reached 12 days p.i. and coincide with the beginning decrease in the area fractions of in situ positive cells, indicating that macrophages and T lymphocytes are crucial for limiting virus dissemination in the heart. Despite the prominent cell-mediated immune response, the virus was found to be capable of evading the immunological surveillance in different immunocompetent mouse strains, thus inducing a persistent infection.

A strong correlation, both spatial and temporal, between viral replication and development of myocardial tissue lesions was observed not only during acute myocardial infection but also during chronic disease. No inflammatory lesions were found to evolve in the absence of virus replication in the course of the disease. Our finding in DBA/1J mice that inflammatory myocardial processes can be terminated by eliminating the virus of the heart implies that the development of chronic disease is dependent on viral persistence. In addition, our previous observations of ongoing severe myocardial injury in persistently CVB3-infected T-cell deficient NMRI (nu/nu) mice [16] support the concept that myocardial injury is the consequence of virus-induced cytopathology. Thus, it can be concluded that immune-mediated processes described for chronic enterovirus-induced myocarditis [22] are triggered by persistent infection.

Generation of Enterovirus Group-Specific Antisera

Although a number of cross-relationships exist between different enteroviruses, the use of group-specific antigen detection has met only modest success in the identification of cardiotropic enteroviruses because of the antigenic heterogeneity among and within the serotypes. Compared with hyperimmune antisera obtained by immunization of rabbits with purified enteroviral virions [25], antisera raised against bacterially synthesized CVB3 proteins revealed a much broader spectrum of cross-reactivity within the enteroviruses [32]. Several subgenomic fragments from infectious recombinant CVB3 cDNA were inserted into the expression plasmid pPLc24 [27] within the coding sequence of the replicase gene of the MS2-phage, thereby achieving a fusion of both reading frames. Using this approach, plasmids were constructed which expressed either the structural proteins VP4, VP2, and VP3 (p1A–1C) or VP1 (p1D) or the RNA-dependent RNA-polymerase ($p3D^{pol}$) of CVB3 [12, 32]. Polyclonal antisera raised in rabbits against the purified expression products of structural proteins allow enterovirus-specific identification of various commonly implicated agents of viral heart disease, including groups A and B CV and echoviruses [32]. Using enterovirus-specific antisera, an immunohistochemical technique was developed capable of detecting the presence of viral capsid proteins in formalin-fixed paraffin-embedded murine hearts [20]. This technique was also tested on autopsy heart and pancreas from 21 infants who were thought to have died of acute CV myocarditis [3]. Cardiac myocytes were positive for VP1 capsid protein in 12 of 20 cases in which the heart was available for study. Insulitis was present in the pancreas in seven of these cases, and in all seven islet endocrine cells containing VP1 were found. In heart and pancreas, cells shown to contain viral protein usually showed signs of necrosis. With use of a double-labeling assay, the simultaneous in situ detection of enterovirus RNA and the demonstration of distinct viral proteins by immunohistochemistry is currently being persued.

Antiviral Activity of Interferon in Cultured Human Heart Cells

Effective antiviral therapy has not yet been established in the treatment of viral heart disease. The ultimate goal of antiviral treatment, to restore function to the infected cell, usually appears unattainable. A realistic goal would be the inhibition of viral replication, thus preventing virus spread to uninfected cells. This is achieved, in part, in many natural infections by the endogenous IFN system [11]. Regarding the potential clinical application of exogenous IFN in enterovirus-induced cardiomyopathy, one major prerequisite is the demonstration that the virus to be treated is susceptible in vitro to the action of a given type of IFN in the specific host cell.

The protective role of natural human fibroblast IFN-β in CVB3-infected, enriched human myocytes has been established [15]. Myocytes protected by IFN-β continued to beat rhythmically as long as IFN-β was given with exchanges of medium every 2 days. In contrast, virus replication in unprotected cultures was accompanied by loss of spontaneous contractility within 9h, followed by complete lysis of myocytes within 20h of infection.

The discovery of persistently infected CVB3 carrier cultures of human myocardial fibroblasts [15] provided another useful test system for studying the activity of antiviral agents. In this type of infection, only a small proportion of the cell population is productively infected. Virus titers of 10^7 plaque-forming units (PFU) of infectious CVB3 per milliliter of culture medium are obtained. Treatment of these cultures with IFN-β at a dose of 300IU/ml culture medium every 24h eliminated infectious progeny virus within 11–13 days. An important finding is the observation that this potent in vitro activity of IFN-β is completely blocked in the presence of 10μM prednisolone [17]. In contrast, azathioprine and cyclosporine do not interfere with the antiviral activity of IFN-β.

Recently, we have evaluated the synergistic interaction between IFN-β and IFN-γ in persistently CVB3-infected carrier cultures of human myocardial fibroblasts with regard to low-dose combination treatment [6]. In general, a synergistic combination schedule of two antiviral agents allows the use of lower doses of both agents, thereby reducing potential side effects of either drug. To evaluate synergistic activity, dose-effect data were obtained on IFN-β and IFN-γ, and the mode of interaction was quantitatively assessed with the median-effect principle and the isobologram equation. Synergism was observed with concentrations as low as 30IU IFN-β plus 10IU IFN-γ/ml reducing mean viral titers within 21 days of administration from 6×10^7 to 1.3×10^4 PFU per milliliter and number of infected cells from 14% to 0.1% as determined by quantitative in situ hybridization. Although the 30IU IFN-β plus 10IU IFN-γ combination was not able to suppress infectious virus progeny completely within 21 days, this combination schedule resulted in a more than 4000-fold reduction of virus titers. This degree of antiviral activity might be sufficient to limit virus replication in vivo and thereby limit the disease.

Higher concentrations of IFN (both ⩾30IU/ml) were associated with an initial transient antagonism followed by antiviral synergism later on. With 100 IU IFN-β plus 30 IU IFN-γ/ml, elimination of infectious virus progeny was consistently achieved, whereas at least threefold higher concentrations were required with single drugs [6]. Suppression of infectious virus progeny could be sustained for a period of 6 weeks after cessation of IFN administration. However, using the sensitive method of in situ dybridization, CVB3 RNA could be still detected in a few cells (<0.05%) of these cultures. This finding indicates the possibility of virus persistence in the absence of biologically assayable infectious virus and explains the phenomenon of virus reactivation observed in this study 6 weeks after cessation of IFN administration. Therefore, sustained IFN treatment, even after successful elimination of infectious virus, should be considered. In summary, these in vitro findings on the antiviral activity of interferons support a concept of low-dose IFN combination schedules which might become useful in the treatment of life-threatening enterovirus infections.

Construction of Enterovirus Antigen Chimeras Using Coxsackievirus B3 as an RNA Virus Vector

Although the humoral immune response is essential for protection against enterovirus infections, very little is known about the viral epitopes mediating B cell response. The high homology in tertiary structure among picornaviruses [28] and the degree of sequence identity of CVs and polioviruses [21] suggest that the virus capsid structure of CVs is similar to the poliovirus capsid structure and that homologous loops contribute to antigenic sites. Based on this assumption, we recently identified neutralizing antigenic site I (N-Ag I) of CVB4 by construction of antigen chimeras between CVB3 and CVB4 [26]. One chimera, designated CVB3/4, was constructed by inserting five amino acids (ESNNL) of the putative BC loop of the structural protein VP1 of CVB4 into the corresponding loop of CVB3 (GAKR) by site-directed mutagenesis of infectious recombinant CVB3 cDNA. The resulting chimeric viurs CVB3/4 was neutralized and precipitated by CVB4 and CVB3 serotype-specific polyclonal antisera, demonstrating that it unifies antigenic properties of both CV serotypes. In addition, the chimera elicitated antibodies in rabbits capable of neutralizing the two CV serotypes CVB3 and CVB4. These findings gave evidence for the presence of a neutralizing antigenic site in the BC loop of VP1 of CVB4 (amino acids 81–89).

In addition, CVB3/4 antigen chimeras were designed by molecular modeling studies in which disruption of predicted hydrogen bond interactions, suggested between the BC and neigboring HI loop in a nonviable construct, would result (McPhee et al., in preparation). Significant neutralization responses to CVB4 serotype-specific antibodies were observed in viable chimeras even when amino acids Gly 83 and/or Ala 84, at the carboxyl-terminus of the VP1 BC loop, were deleted. However, exchange of amino acid Ser 86, in the middle of the BC loop, led to complete loss of antigenicity, demonstrating the importance of Ser 86 for binding of CVB4-neutralizing antibodies. These studies suggest that the construction of enteroviral antigen chimeras provides a powerful tool for identifying antigenic sites of cardiotropic enteroviruses, which is also potentially useful in the development of new polyvalent enterovirus vaccines. Molecular modeling is currently being employed as an aid to locate neutralizing antigenic site II of various CVs by construction of antigen chimeras.

Conclusions and Prospects

In situ hybridization studies in endomyocardial biopsy specimens of patients have proved that enterovirus infection is detectable in all stages of acute and chronic enterovirus myocarditis, and also in a significant proportion of patients with dilated cardiomyopathy, indicating the possibility of enterovirus persistence in the human heart. In patients with acute onset of congestive heart failure, the most dramatic manifestation of myocarditis beyond sudden death, the incidence of myocardial enterovirus infections is approximately 30 per cent. Progress is currently being made in unraveling the molecular mechanisms of enterovirus persistence, the diversity of host and virus genetics and their impact on the nature and severity of the disease.

The number of infected myocardial cells appears to be related to the severity of clinical symptoms at least during the acute phase of enterovirus infection. Consequently, the most extensive patterns of myocardial enterovirus infections with numerous infected myocardial cells were observed in fulminant cases of the disease resulting in early death. Typical patterns of persistent enterovirus infections with a reduced number of infected cells were assessed in patients with chronic dilated cardiomyopathy, indicating that restricted replication plays a major role in the mechanism of virus persistence. In situ hybridization with strand-specific RNA probes in persistently CVB3-infected immunocompetent A.CA/SnJ mice clearly demonstrated that replication in persistent enterovirus infection is restricted at the level of genomic plus-strand RNA synthesis [8, 20]. In contrast to the acute infection of myocardial cells, which is characterized by the synthesis of viral plus-strand RNA in great excess, persistently infected cells were found to contain plus- and minus-strands of viral RNA in approximately similar amounts. Restricted replication may explain the common failure to isolate infectious virus from patients after the acute phase of virus replication. In addition, reactivation of infectious virus replication might explain the clinical phenomenon of recurrence and the progression of enteroviral infection to long-term cardiac disease.

Regarding the pathogenesis of enteroviral heart disease, a controversy exists as to whether myocardial injury is due to virus-induced pathobiological events or results from autoimmune processes initially triggered by viral infection of the heart [22]. If viral products are required for the development and sustainment of myocardial lesions, a strong correlation between viral replication and tissue injury should exist. On the other hand, if autoimmune mechanisms are crucial for the formation of lesions, heart tissue damage should develope independently from virus replication. As discussed in this review, we have addressed this question in different immunocompetent mouse strains. Patterns of acute and persistent myocardial infection were quantitatively assessed in the representative mouse strain A.CA/SnJ,H-2^f by application of computer-assisted digital image processing and related to the extent of myocardial tissue damage and inflammation [20]. Acute and chronic myocardial lesions were found to be consistently associated with infected myocardial cells, indicating the importance of viral replication in the pathogenesis of ongoing heart muscle disease. CVB3 was found to be capable of inducing a persistent infection not only in A.CA/SnJ mice but also in A.BY/SnJ and SWR/J mice. In contrast, DBA/1J mice were found to be able to eliminate CVB3 after the acute infection. Consequently no chronic disease was observed in this strain. The different course of the disease in SWR/J and DBA/1J mice, which share the same H-2 haplotype (H-2^q), suggests that the resistance to the development of persistent heart muscle infection is not linked to the H-2 locus of the host. In addition to persistent heart muscle infection, spleen and lymph nodes were also found to exhibit persistent infection, which might play a crucial role in dissemination of the virus or maintenance of a noncardiac viral reservoir [20]. The identification of persistently infected lymphoid or dentritic cells of spleen and lymph nodes is currently being persued. Regarding host-specific determinants of enterovirus-induced cardiomyopathy, work is in progress to molecularly clone and characterize the CVB receptor as well as to assess the role of endogenous IFNs and other regulatory cytokines in chronic inflammatory heart muscle disease.

The in situ hybridization technique is now also available at the electron microscopic level (Klingel et al., in preparation). The predominant ultrastructural pathologic changes so far observed in acute myocardial enterovirus infection were disintegration of myofibrils together with the formation of virus-induced vesiculated regions which are both the sites of viral replication and the ultrastructural reflection of the virus-induced cytopathic effect.

To investigate further the molecular mechanisms of enterovirus replication and persistence, the infectious CVB3 cDNA is used as a source for the generation of distinct viral proteins in different expression systems. Particularly intriguing are studies addressing the interference of viral structural and nonstructural proteins with myocyte function in transgenic mice. So far, a 2.1-kb fragment of the 5′ upstream regulatory sequence of the myosin light chain 2 (MLC-2) was used to target the firefly luciferase as a reporter gene to the heart (Franz et al., in preparation). Heart muscle of transgenic mouse lines was found to express exclusively the transgene in a developmentally upregulated fashion. This in vivo model is expected to provide a powerful means for selective high-level expression of viral gene products in the heart. Early prenatal MLC-2 promoter activity, which coincides with the education period of B- and T-cells, should result in a "self-acceptance" of viral products by the murine immune system [5]. This would imply that the immune system does not interfere with pathogenic effects of viral gene products. By this approach, distinct viral gene regions of CVB3 are being expressed in a heart-specific manner to unravel interference with myocyte function.

Regarding the expression of viral proteins in the infected host, polyclonal antisera raised against bacterially synthesized CVB3 proteins offer the possibility of enterovirus group-specific identification of the various commonly implicated agents of viral heart disease by antigen detection [32]. These antisera proved useful in demonstrating the presence of distinct viral capsid proteins in autopsy hearts of neonates and infants who were thought to have died of acute CVB myocarditis [3]. Furthermore, viral capsid protein expression was studied by these reagents in the murine model of chronic CVB3-induced myocarditis [20]. The results indicate that the amount of enteroviral capsid protein synthesized is decreased in persistently infected myocardial cells, thus refelcting restricted viral RNA replication. Antisera raised against distinct viral gene regions also provided the basis for the development of an enterovirus-specific enzyme-linked immunosorbent assay which was found to be useful in the serodiagnosis of patients (to be published elsewhere). Such an assay provides a first step in the evaluation of patients suspected of having enteroviral heart disease.

In addition to in situ hybridization, the recently developed polymerase chain reaction (PCR) [30] makes it possible to amplify distinct enteroviral gene regions in vitro and in situ following cDNA synthesis. So far, we chose primers corresponding to highly conserved sequences of the 5′ nontranslated region of various enteroviruses. The enterovirus PCR was found to be sensitive and specificity of amplified sequences was shown by hybridization to an internal, virus-specific oligonucleotide (unpublished results). Direct sequencing of enzymatically amplified gene regions [10] may allow typing of an etiologically implicated enterovirus strain and the identification of specific mutations which might be implicated in enterovirus persistence.

Apart from providing an unequivocal diagnosis of myocardial infection, there are therapeutic implications from the demonstration of replicating enterovirus RNA in myocardial cells. Clearly, corticosteroids are contraindicated in the presence of myocardial infection since increased viral replication [19] and inhibition of the endogenous IFN system may follow their use [17]. It is hoped that antiviral agents, for example, IFNs, might provide protection against the effects of myocardial enterovirus infection. The protective role of natural fibroblast IFN-β in CVB3-infected cultured human heart cells has been described [15, 17]. In addition, we have assessed the synergistic interaction between IFN-β and IFN-γ in persistently infected cultured human myocardial fibroblasts with regard to a low-dose combination treatment [6]. Although the relationship between the in vitro effects of IFNs and their potential in vivo activities needs to be confirmed in the established murine models, these in vitro observations are an important prerequisite for correct therapeutic use of exogenous IFNs in enterovirus infections.

Regarding the prevention of enterovirus-induced cardiomyopathies, it is hoped that the characterization of protective enteroviral B-cell epitopes by construction of antigen chimeras may also facilitate the development of enterovirus vaccines.

Acknowledgements. The authors wish to thank Prof. Dr. P.H. Hofschneider and Prof. Dr. G. Riecker and all clinical collegues for providing the impetus to persue this topic over the years. The work discussed in this review was supported in part by Grant Ka 593/2-2 from the Deutsche Forschungsgemeinschaft, by grant 321-7291-BCT-0370 "Grundlagen und Anwendungen der Gentechnologie" from the German Ministry for Research and Technology, by the Hermann and Lilly Schilling Foundation for Medical Research (R.K.), and by a Max Planck Research Award (R.K. and B.M.M.).

References

1. Abelmann WH (1973) Viral myocarditis and its sequelae. Annu Rev Med 24:145–152
2. Dec GW, Palacios IF, Fallon JT, Aretz HT, Mills J, Lee DCS, Johnson RA (1985) Active myocarditis in the spectrum of acute dilated cardiomyopathy. N Engl J Med 312:885–890
3. Foulis AK, Farquharson MA, Cameron SO, McGill M, Schönke H, Kandolf R (1990) A search for the presence of the enteroviral capsid protein VPI in pancreases of patients with type 1 (insulin-dependent) diabetes and pancreases and hearts of infants who died of coxsackieviral myocarditis. Diabetologia 33:290–298
4. Haase A, Brahic M, Stowring L, Blum H (1984) Detection of viral nucleic acids by in situ hybridization. Methods Virol 7:189–226
5. Hannahan D (1989) Transgenic mice as probes into complex systems. Science 246: 1265–1275
6. Heim A, Canu A, Kirschner P, Simon T, Mall G, Hofschneider PH, Kandolf R (1992) Synergistic interaction of interferon-beta and interferon-gamma in coxsackievirus B3 infected carrier cultures of human myocardial fibroblasts. J Infect Dis 166:958–965
7. Herskowitz A, Wolfgram LJ, Rose NR, Beisel KW (1987) Coxsackivirus B3 murine myocarditis – histopathologic spectrum in genetically defined inbred strains. J Am Coll Cardiol 9:1311–1319
8. Hohenadl C, Klingel K, Mertsching J, Hofschneider PH, Kandolf R (1991) Strand-specific detection of enteroviral RNA in myocardial tissue by in situ hybridization. Mol Cell Probes 5:11–20
9. Hyypiä T, Stålhandske P, Vainionpää R, Pettersson U (1984) Detection of enteroviruses by spot hybridization. J Clin Microbiol 19:436–438
10. Innis MA, Myambo KB, Gelfand DH, Brow MAD (1988) DNA sequencing with Thermus aquaticus DNA polymerase and direct sequencing of polymerase chain reaction-amplified DNA. Proc Natl Acad Sci USA 85:9436–9440
11. Joklik WK (1986) Interferons. In: Fields BN, Knipe DM (eds) Fundamental virology. Raven, New York, pp 281–307
12. Kandolf R (1988) The impact of recombinant DNA technology on the study of enterovirus heart disease. In: Bendinelli M, Friedman H (eds) Coxsackieviruses – a general update. Plenum, New York, pp 293–318
13. Kandolf R, Hofschneider PH (1985) Molecular cloning of the genome of a cardiotropic coxsackie B3 virus: full-length reverse-transcribed recombinant cDNA generates infectious virus in mammalian cells. Proc Natl Acad Sci USA 82:4818–4822
14. Kandolf R, Hofschneider PH (1989) Enteroviral heart disease. Springer Semin Immunopathol 11:1–13
15. Kandolf R, Canu A, Hofschneider PH (1985) Coxsackie B3 virus can replicate in cultured human foetal heart cells and is inhibited by interferon. J Mol Cell Cardiol 17:167–181

16. Kandolf R, Ameis D, Kirschner P, Canu A, Hofschneider PH (1987) In situ detection of enteroviral genomes in myocardial cells by nucleic acid hybridization: an approach to the diagnosis of viral heart disease. Proc Natl Acad Sci USA 84:6272–6276
17. Kandolf R, Kirschner P, Ameis D, Canu A, Hofschneider PH (1987) Cultured human heart cells: a model system for the study of the antiviral activity of interferons. Eur Heart J 8 [Suppl J]:453–456
18. Kawai D, Matsumori A, Fugiwara H (1987) Myocarditis and dilated cardiomyopathy. Annu Rev Med 38:221–239
19. Kilbourne E, Horsfall F (1951) Lethal infections with coxsackievirus of adult mice given cortisone. Proc Soc Exp Biol Med 77:135–138
20. Klingel K, Hohenadl C, Canu A, Albrecht M, Seemann M, Mall G, Kandolf R (1992) Ongoing enterovirus-induced myocarditis is associated with persistent heart muscle infection: quantitative analysis of virus replication, tissue damage, and inflammation. Proc Natl Acad Sci USA 89:314–318
21. Klump WM, Bergmann I, Müller BC, Ameis D, Kandolf R (1990) Complete nucleotide seqeunce of infectious coxsackievirus B3 cDNA: two initial 5′uridine residues are regained during plus-strand RNA synthesis. J Virol 64:1573–1583
22. Leslie K, Blay R, Haisch C, Lodge A, Weller A, Huber S (1989) Clinical and experimental aspects of viral myocarditis. Clin Microbiol Rev 2:191–203
23. McManus BM, Kandolf R (1991) Evolving concepts of cause, consequence, and control in myocarditis. Curr Opin Cardiol 6:418–427
24. Melnick JL (1989) Enteroviruses. In: Evans AF (ed) Viral infections of humans. Plenum, New York, pp 191–263
25. Mertens T, Pika U, Eggers HJ (1983) Cross antigenicity among enteroviruses as revealed by immunoblot technique. Virology 129:431–442
26. Reimann BY, Zell R, Kandolf R (1991) Mapping of a neutralizing antigenic site of coxsackievirus B4 by construction of an antigen chimera. J Virol 65:3475–3480
27. Remaut E, Tsao H, Fiers W (1983) Improved plasmid vectors with a thermoinducible expression and temperature-regulated runanway replication. Gene 22:103–133
28. Rossmann MG, Johnson JE (1989) Icosahedral RNA virus structure. Annu Rev Biochem 58:533–573
29. Rueckert RR (1990) Picornaviridae and their replication. In: Fields BN, Knipe DM (eds) Virology, vol 1. Raven, New York, pp 507–548
30. Saiki RK, Gelfand DH, Stoffel S, Scharf SJ, Higuchi R, Horn GT, Mullis KB, Erlich HA (1988) Primer-directed enzymatic amplification of DNA with a thermostable DNA polymerase. Science 239:487–491
31. Tracy S (1984) A comparison of genomic homologies among the coxsackievirus B group: use of fragments of the cloned coxsackievirus B3 genome as probes. J Gen Virol 65:2167–2172
32. Werner S, Klump WM, Schönke H, Hofschneider PH, Kandolf R (1988) Expression of coxsackievirus B3 proteins in *Escherichia coli* and generation of virus-specific antisera. DNA 7:307–316
33. Wolf H, zur Hausen H, Becker V (1973) EB viral genomes in epithelial nasopharyngeal carcinoma cells. Nature New Biol 244:245–247
34. Woodruff JF (1980) Viral myocarditis – a review. Am J Pathol 101:427–479

Direct Damage of Myocardium by Enterovirus

B. McManus[1], L.H. Chow[2], J.E. Wilson[1], D.R. Anderson[1], and R. Kandolf[3]

Introduction

Human idiopathic myocarditis has long troubled those interested in establishing cause(s) and identifying efficacious therapies. To date, neither the cause nor specific therapies are known. Perhaps these difficulties relate to the ambiguities of clinical presentation, the poorly characterized natural history, and the lack of distinctive therapeutic targets. Nonetheless, continued efforts worldwide to understand inflammatory heart disease have reestablished a view that viruses, and particularly enteroviruses, play a pathogenetic role in at least some human myocarditides. Summary experience from several investigators suggests that approximately 25% of inflamed failing and/or arrhythmic hearts are attributable to enterovirus infections [1, 2]. The frequency of other community-acquired viral infections involving the heart is much less clear. Even more uncertain is the manner in which viruses may effect injury in the human heart. Hypotheses abound, but little in the way of solid data is available. The role of virus in direct injury of the myocardium has appeared more convincing in fulminant infantile myocarditis where enterovirus is often isolated and/or demonstrated by molecular techniques (McManus, unpublished data). Autoimmune phenomena are often concomitants of human myocarditis and dilated cardiomyopathy [3], but their pathogenetic significance has not been established.

The murine model of enterovirus-induced myocarditis has been the model of choice for most investigators daunted by the stubborn obscurity of the human disease. As detailed recently [4], several lines of evidence indicate that damage of the myocardium in various murine hosts infected with different coxsackievirus variants is largely if not solely due to enterovirus replication and cell cytotoxicity. This perspective has been overshadowed by the belief that cellular and humoral phenomena triggered by virus lead to myocardial damage during early and late periods following infection. In review of previous studies on tissue injury in the setting of enteroviral infections, light and electron microscopic evidence has been

[1] Cardiovascular Registry, Department of Pathology and Microbiology, University of Nebraska Medical Center, Omaha, NE, USA
[2] Robarts Research Institute, University of Western Ontario, London, ON, Canada, and the
[3] Max Planck Institute for Biochemistry, W-8033 Martinsried, Fed. Rep. of Germany

Figulla et al. (Eds.)
Idiopathic Dilated Cardiomyopathy

steadily accruing which points to the now reemergent concept that the murine model of enterovirus myocarditis is largely one of viral injury. The secondarily elicited host immune response may participate in lysis of infected cells, its role in damage of otherwise normal tissue, and in cardiac dysfunction being indeterminant at this time.

The present paper details further morphological observations which strongly imply that most, if not all, direct injury in the murine model of myocarditis is due to viral infection of cardiocytes. The relationship of these findings to previous classical studies and to current molecular findings is discussed briefly.

Methods and Materials

Human fetal cardiac myocytes were cultured as described [5]. Aseptic, disaggregated, and filtered cardiocytes were depleted of endothelial and fibroblastic cells by adsorption and seeded in complete Dulbecco's modified Eagle's minimum medium at 37°C in 5% CO_2. For certain experiments, plaque-purified coxsackievirus B3 was added at high multiplicity of infection (50 pfu/cell) to the myocyte cultures 8 days following culture initiation. Phase contrast and electron microscopy were utilized to assess the cytopathological effects of the virus. Cells fixed in 2% glutaraldehyde and postfixed in 1% osmium tetroxide with cacodylate buffer were dehydrated, embedded in Epon, sectioned en face, thin sectioned, and stained with uranyl acetate and lead citrate. Sections were examined with an electron microscope.

In separate experiments [6], adolescent male mice of different H-2 haplotypes were infected with 2×10^5 $TCID_{50}$ intraperitoneally with myocarditic variants of coxsackievirus B3. Heart tissues taken at representative time points postinfection were evaluated in blinded fashion for histopathological injury lesions and inflammation as seen in hematoxylin and eosin stained paraffin sections. Selected animals had ultrastructural studies carried out on the heart muscle in a manner similar in the approach described above. In situ hybridization was carried out on paraffin sections mounted on silanated slides with well-characterized, enterovirus-specific, ^{35}S-labeled cDNA probes according to our previously described technique [7, 8] following prehybridization. After washes, the slides were immersed in NTB2 nuclear track emulsion (Eastman Kodak), and then exposed until development and counterstaining.

Results

Fetal cardiac myocytes (Fig. 1) were readily infected with coxsackievirus, and the apparent proliferation of membranes in the cytoplasm of infected cells was accompanied by viral lattice or arrays, signal to the infection (Fig. 2). Clearing of the cytoplasm occurred in proximity to the lattice and

Fig. 1. Electron micrograph of cultured, coxsackievirus B3-infected human fetal cardiac myocytes with myofilaments, Z bands, and cytoplasmic particles consistent with ribosomes or enterovirus. This in vitro system was utilized to examine the evolution of viral-induced cell injury. ×21 000

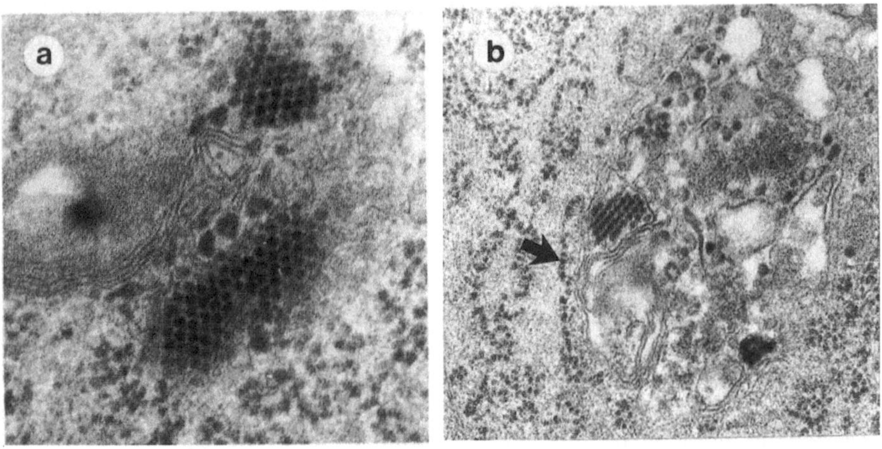

Fig. 2a,b. Electron micrographs of coxsackievirus crystalline lattice associated with excess smooth membranes in in vitro infected human fetal cardiac myocytes. **a** The viral arrays are contrasted to the neighboring particles which could be ribosomes, viroplasm, or virions. **b** Suggestions of early cytopathic effect are noted in the region of the lattice, as well as lineal ribosomes (*arrow*) along endoplasmic reticulum. **a** ×35 000, **b** ×24 000

probable viral particles. These in vitro features have been noted previously to progress until the point of myofibrillar dissolution and accumulation of agglutinated ribosomes and clustered viroplasm. Similar features can be documented in vivo. The juxtaposition of necrotic, infected myocytes and viable myocytes is common in the hearts of infected animals (Fig. 3). The necrotic myocytes have many dense deposits in mitochondria (Figs. 3, 4). The particulate material in the myocytes of infected hearts most likely

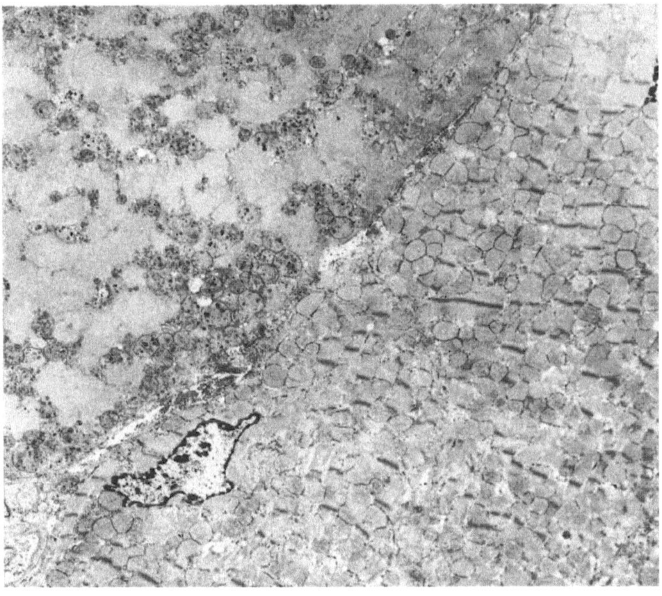

Fig. 3. Electron micrograph of ventricular cardiac myocytes from coxsackievirus B3-infected adolescent male A/J mouse 4 days postinfection illustrating juxtaposition of a viable myocyte (*lower right*) and a necrotic myocyte (*upper left*). The severely damaged myocyte has light-stained, coagulated myofilament bundles with mitochondria containing many dense mineral deposits. ×9000

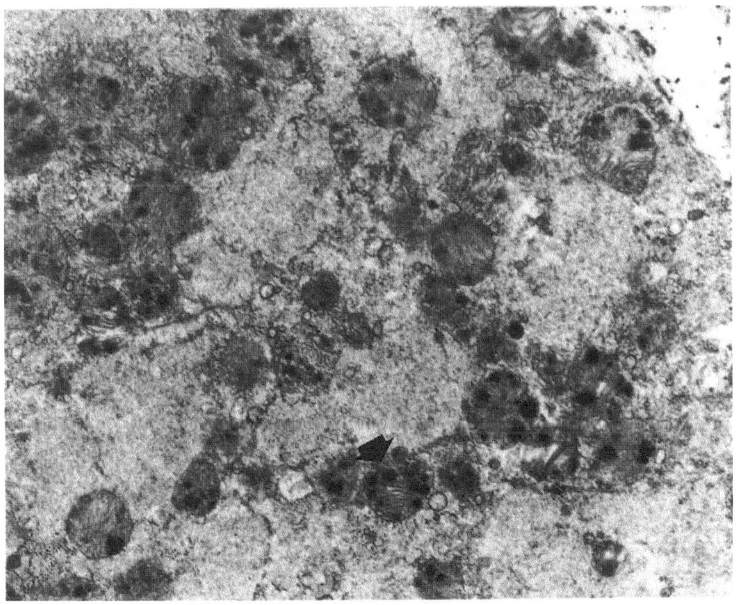

Fig. 4. Electron micrograph of necrotic ventricular myocyte from coxsackievirus B3-infected adolescent male A/J mouse 5 days postinfection illustrating the coagulated myofilament bundles and the extent of dense deposits in degenerate mitochondria (*arrow*). ×30000

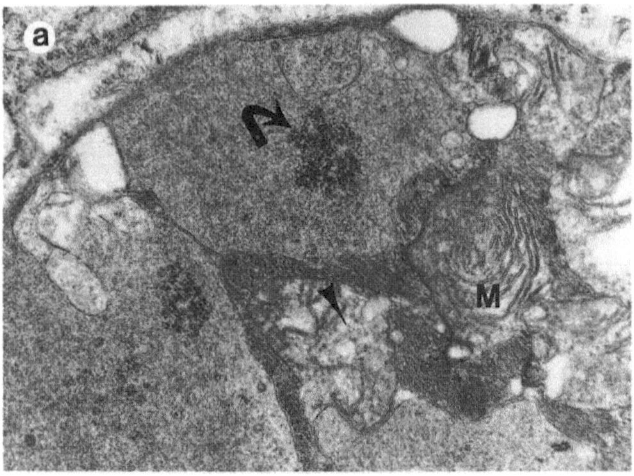

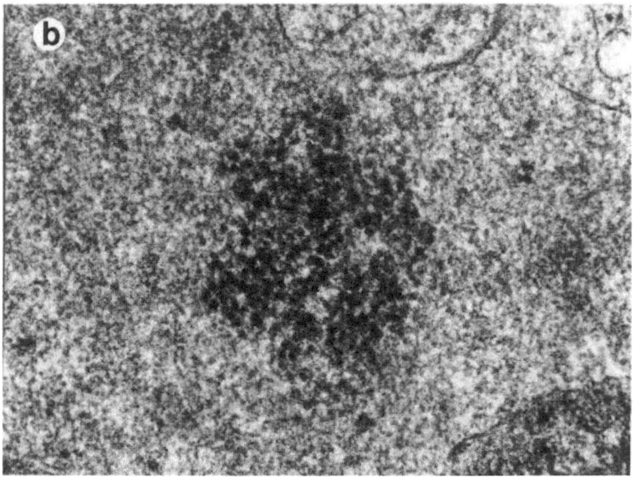

Fig. 5a,b. Electron micrographs of dying cardiac myocyte from ventricular myocardium of coxsackievirus B3-infected adolescent male A/J mouse 4 days postinfection illustrating coagulated myofilament bundles with associated particulate material (*curved arrow*, **a**) consistent with agglomerated ribosomes, viroplasm, and/or virions. The comparability of size between ribosomes and enterovirus particles has long prevented definitive identification of virus within infected cells in the absence of viral lattice. More intact mitochondria (*M*) and those with cytopathic effects (*arrowhead*) are present. **a** ×20 000, **b** ×45 000

includes both ribosomes and viral particles (Fig. 5), the distinction of which can be achieved by immunoelectron microscopy [9]. Cytopathic effects seen at the light microscopic level (Fig. 6) are well seen at various stages ultrastructurally (Fig. 7). The myriad microscopic features of virus-induced cell death are reinforced by the localization of hybridization signals over dying myocytes with active replication of positive strand enteroviral genome (Fig. 8). All of the light and electron microscopic features of active infection from

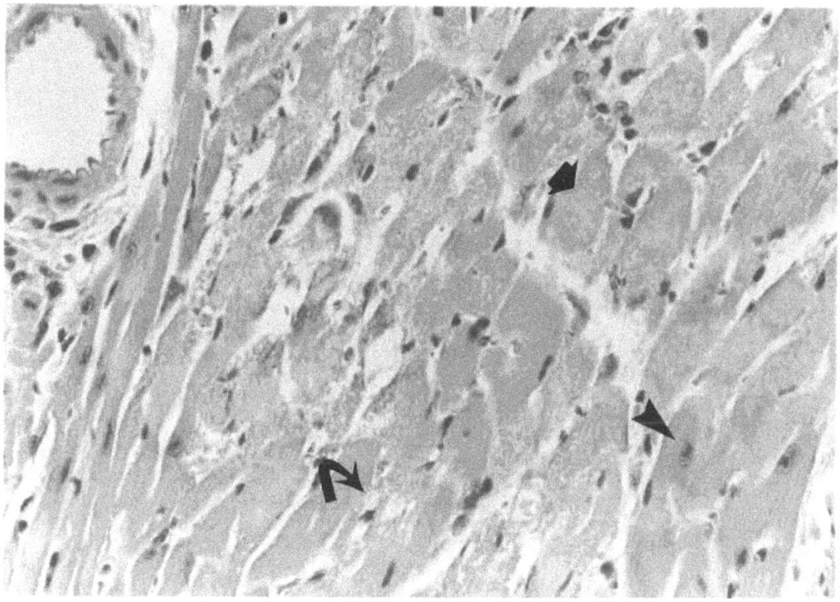

Fig. 6. Photomicrograph of ventricular myocardium from coxsackievirus B3-infected adolescent male mice illustrating the extent of damage which is typically present by days 4–5 postinfection. Many coagulated myocytes (*arrow*) and those with cytopathic vacuoles (*curved arrow*) are present without an inflammatory infiltrate. A normal-appearing myocyte is denoted (*arrowhead*). Hematoxylin & eosin stain, ×330

day 3 to 5 postinfection in susceptible hosts occur in the absence of an inflammatory infiltrate.

Discussion

The present imagery from in vitro tissue culture and in vivo cardiac infections reinforces the cumulative data from several experimental approaches which support a predominant role for coxsackievirus B3 in the injury and necrosis of cardiomyocytes in enteroviral myocarditis models. In early classical work on the cytopathology of enteroviral infections, Godman [10] noted that most histopathological studies of experimental coxsackievirus infection had dealt with "relatively late or regressive cellular changes and host-tissue reactions to these." He thus emphasized the difficulty in drawing comparisons between observations in vitro regarding cell necrosis and cytopathic effects and observations in vivo. Indeed, he indicated the need for studies of the early cellular changes in vivo in enteroviral infection of animals. Such a focus in our work has allowed rediscovery of the pervasive cytopathic effects which occur in myocardial cells in vivo [4]. The determinants of these early effects can now be sought, and the pathogenetic basis and significance of viral tropism [11] in vivo can be better appreciated.

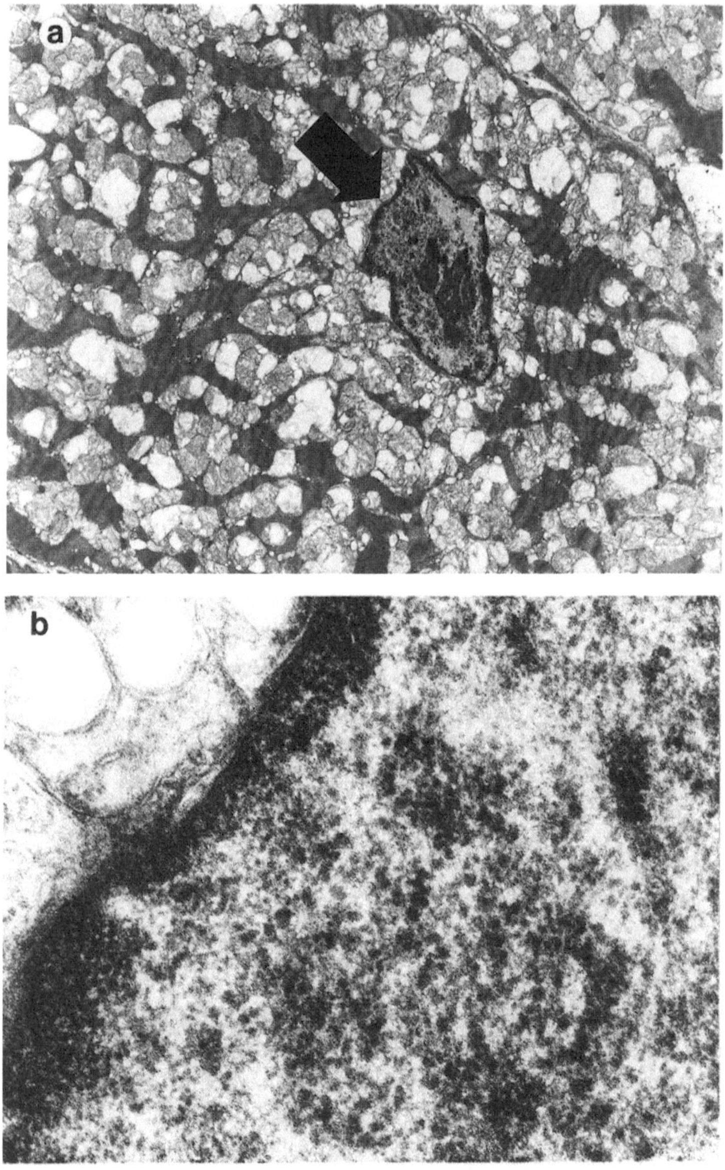

Fig. 7a,b. Electron micrographs of a severely damaged cardiac myocyte from ventricular myocardium of coxsackievirus B3-infected male adolescent mouse 5 days postinfection illustrating widespread vacuolar changes with clustered and degenerative mitochondria, contraction bands and a nucleus (*arrow* in a depicted region shown at higher power in **b**) with early dissolution of nucleoplasm (*light areas*) and peripheral coarse chromatin. **a** ×9 000, **b** ×45 000

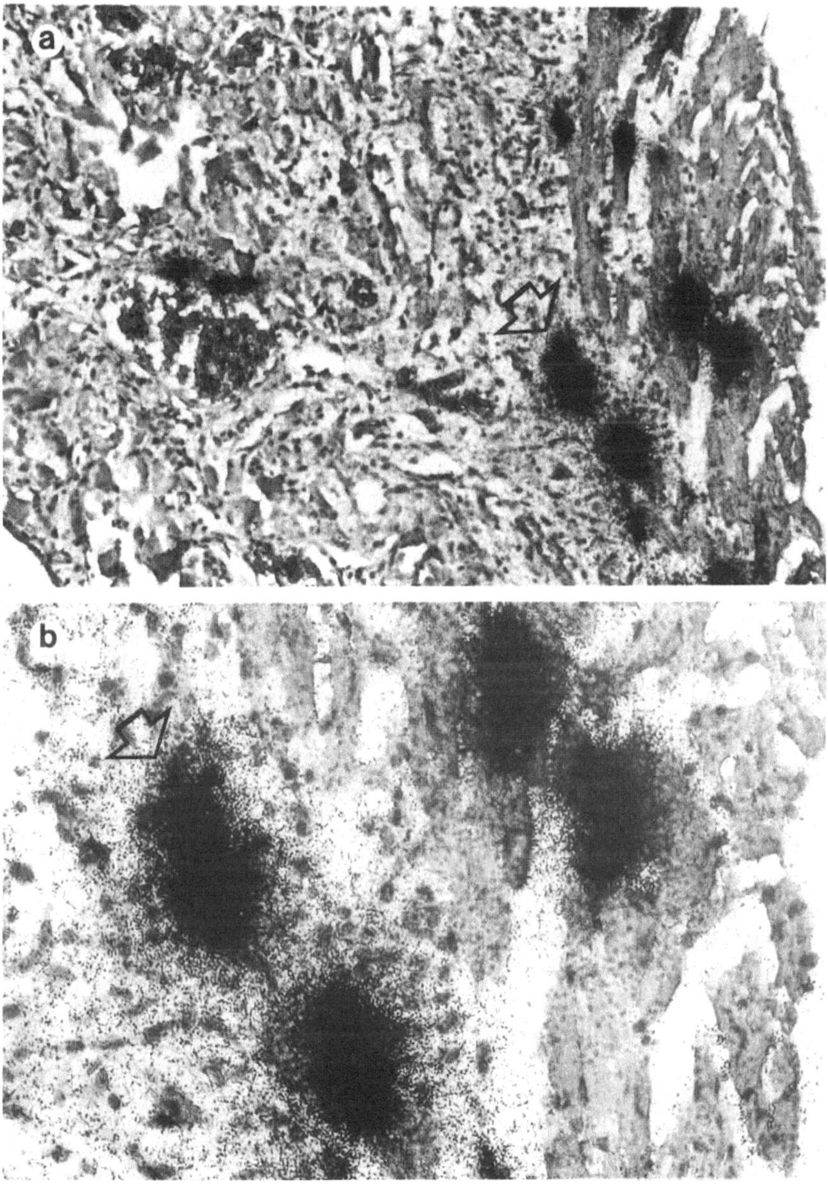

Fig. 8a,b. Photomicrographs of ventricular myocardium from coxsackievirus B3 (SH) -infected adolescent male BALB/c mouse which died spontaneously 3 days postinfection. The markedly necrotic and already calcified tissue (seen transmurally in **a**) has been in situ hybridized with a ^{35}S-labeled cDNA probe to conserved regions of the coxsackievirus B3 (RK) genome. Numerous autoradiographic signals overlie infected cardiocytes (*arrow* of same focus a low and high power in **a** and **b**, respectively). These replicative foci of virus largely conform to the contour of cardiac myocytes and are seen to be such on plasmid controls. Loci of viral genome are often at the margin of evolving necrotic lesions. **a** ×125, **b** ×500

Light microscopy performed by Gifford and Dalldorf in 1951 [12] on mice under 10 days of age infected with a number of coxsackieviruses in group B indicated that in myocardial lesions with diffuse and irregular borders myofibers appeared faded and hyaline. Fibers were fragmented and lost over large areas, and hydropic degeneration was noted at the margin of lesions. Many spindled interstitial cells were noted and believed to be connective tissue elements. Inflammatory cells were absent or inconspicuous despite prominent necrotic lesions. Of related interest, these workers also observed that group A coxsackieviruses caused loss of skeletal muscle detail, hyaline degeneration, fragmentation, and clumping with few inflammatory cells present during the reparative stage. Similarly, Rabin and colleagues [13] described light microscopic evidence of extensive myocardial injury in coxsackievirus-infected 2- to 3-week-old mice in which damage could be seen by the 2nd day postinfection, and progressive swelling, effacement and fragmentation of myocytes with accompanying nuclear changes were seen when "inflammatory cells were sparse or had not yet appeared." Thus, disruption of cardiac myocytes was considered an early manifestation of injury, and myocardial damage was attributed to viral replication within myocytes.

Ultrastructural studies in mice infected with coxsackievirus B3 (Nancy strain) [13, 14] and coxsackievirus B4 [15], as well as in hamsters infected with coxsackievirus B3 [16], have cumulatively provided a portrait of myofilament contraction band formation and destruction, apparent membrane proliferation, cytoplasmic vacuolation, and mitochondrial swelling, disruption and dense deposits. Mitochondrial deposits were demonstrated to be calcium by X-ray spectrometric analysis. Nuclei often had accumulated coarse and dense chromatin and enlarged nucleoli with prominent granules. Sarcoplasmic reticula were dilated and perinuclear space was increased. Occasionally, crystalline lattice or arrays were observed [14], however particles suspicious for being virus were not clearly distinguishable from ribosomes. Further, ribosomes appear increased in the infected cells. Macrophages were frequently seen to have consumed cellular debris [14, 16]. Nearly all authors emphasized that injury was maximum by 5–6 days postinfection, a time point before significant cellular infiltration by immunocytes.

The concept supported by the work presented in this paper and recently elsewhere [4, 8], and bolstered by earlier classical studies, that enteroviruses inflict most direct damage to the myocardium in murine models of acute enteroviral heart disease, appears to be soundly based. If correct, and if the same is true for any time period following infection, then the investigation of immune responses, cellular or humoral, may be cast in the context of host protection and not host injury. The autoimmunity of postmyocarditic mice has not yet shown itself as important in the actual pathogenesis of the murine disease. The implications for humans remain speculative.

Acknowledgements. This work was conducted under grants-in-aid from the American Heart Association–Nebraska Affiliate (B.M.M., L.H.C.), a Senior Fogarty International Fellowship–

National Institutes of Health (B.M.M.), and a Max Planck Research Prize (B.M.M., R.K.). The authors appreciate the technical help of Patricia Bestmann, B.A.

References

1. McManus BM (1990) Progress in resolving diagnostic, etiogenetic and therapeutic issues in myocarditis. In: Fenoglio-Prieser C (ed) Advances in pathology. Year Book Medical Publishers, Chicago, pp 139–171
2. McManus BM, Kandolf R (1991) Evolving concepts of cause, consequence, and control in myocarditis. Curr Opin Cardiol 6:418–427
3. Caforio ALP, Grazzini M, Mann JM et al. (1992) Identification of alpha- and beta-cardiac myosin heavy chain isoforms as major autoantigens in dilated cardiomyopathy. Circulation 85:1734–1742
4. McManus BM, Chow LH, Wilson JE et al. (1992) Direct myocardial injury by enterovirus: a central role in the evolution of murine myocarditis. In: Rose N (ed) Immune-mediated heart disease. Springer, Berlin Heidelberg New York, (in press)
5. Kandolf R, Canu A, Hofschneider PH (1985) Coxsackie B3 virus can replicate in cultured human fetal heart cells and is inhibited by interferon. J Mol Cell Cardiol 17:167–181
6. Chow LH, Beisel K, McManus BM (1991) Differentiation of viral and immune factors in pathogenesis of myocarditis: immune deficient and normal mice. Lab Invest 64:17A
7. Kandolf R, Ameis D, Krischner P, Canu A, Hofschneider PH (1987) "In situ" detection of enteroviral genomes in myocardial cells by nucleic acid hybridization: an approach to the diagnosis of viral heart disease. Proc Natl Acad Sci USA 84:6272–6276
8. Chow LH, Beisel KW, McManus BM (1992) Enteroviral infection of mice with severe combined immunodeficiency: evidence for direct viral pathogenisis of myocardial injury. Lab Invest 66:24–31
9. Hofschneider PH, Klingel K, Kandolf R (1990) Toward understanding the pathogenesis of enterovirus-induced cardiomyopathy: molecular and ultrastructural approaches. J Struct Biol 104:1–6
10. Godman GC (1966) The cytopathology of enteroviral infection. Int Rev Exp Pathol 5:67–110
11. Mims CA (1989) The pathogenetic basis of viral tropism. Am J Pathol 135:447–455
12. Gifford R, Dalldorf G (1951) The morbid anatomy of experimental coxsackie virus infection. Am J Pathol 27:1047–1063
13. Rabin ER, Hassan SA, Jenson AB, Melnick JL (1964) Coxsackie virus B3 myocarditis in mice: an electron microscopic, immunofluorescent and virus-assay study. Am J Pathol 44:775–797
14. Deguchi H (1981) Ultrastructural alterations of the myocardium in coxsackie B-3 virus myocarditis in mice: 18 month follow-up study by transmission and analyticl electron microscopy. Jpn Circ J 45:695–712
15. RS Sohal, Burch GE (1969) Ultrastructural lesions of the myocardial cell in coxsackie B4 virus infected mice. Virchows Arch [A] 346:361–373
16. Morita H (1981) Experimental coxsackie B3 virus myocarditis in golden hamsters. Jpn Circ J 45:713–729

Natural History of Enterovirus Persistence in the Murine Model of Coxsackievirus-Induced Myocarditis: A Quantitative In Situ Hybridization and Immunohistochemistry Study

G. Mall, K. Klingel, M. Albrecht, M. Seemann, T. Mandel, and R. Kandolf

Introduction

It is generally accepted that enteroviruses, especially coxsackieviruses of group B (types 1–5), play an important role in acute myocarditis of humans [1]. These RNA viruses, which contain a single-stranded RNA genome of positive polarity can induce dilated cardiomyopathy of acute onset or lead to life-threatening arrhythmias and sudden death.

A controversy exists on the pathogenesis of this heart disease as to whether myocardial injury results from virus-induced pathobiological events of autoimmune processes initially triggered by a viral infection of the heart [3, 7]. If viral products essentially sustain myocardial lesions, a correlation between viral replication, tissue injury, and possibly inflammation should exist [2, 6, 7]. In contrast, if autoimmune mechanisms are a major determinant for lesion formation, damage and inflammation may be observed in the absence of virus replication. Recently, an in situ hybridization technique was developed using recombinant coxsackievirus B3 (CVB3) cDNA as an enterovirus-specific probe [4]. This new technique allows enterovirus RNA detection in myocardial biopsy samples of patients with acute and chronic myocarditis and also in patients with end-stage dilated cardiomyopathy. This indicates that chronic myocardial injury may be associated with persistent enterovirus infection [5].

We report here on persistent myocardial enterovirus infection in various immunocompetent mouse strains. By means of quantitative in situ hybridization, quantitative immunochemistry and histomorphometry of myocardial damage, the possible correlation between virus replication, inflammation, and progression of myocardial damage was investigated.

Material and Methods

Several inbred mouse strains (ACA/SnJ, ABY/SnJ, PL/J, DBA/2J, SWR/SnJ, DBA/IJ, SJL/J, ASW/SnJ) were studied. Four-week-old male animals were inoculated intraperitoneally with 6 × 10^4 plaque-forming units of

Max-Planck-Institut für Biochemie, W-8033 Martinsried, Fed. Rep. of Germany
Pathologisches Institut, Städtische Kliniken, W-6100 Darmstadt, Fed. Rep. of Germany

Figulla et al. (Eds.)
Idiopathic Dilated Cardiomyopathy

CVB3. The hearts were aseptically removed after 6–40 days postinfection (p.i.). Tissue samples were either quick frozen in liquid nitrogen or fixed for 2h at 4°C by immersion in 1.5% formaldehyde, 1.5% glutardialdehyde, 0.1 *M* sodium phosphate buffer, pH 7.2, and embedded in paraffin. For in situ hybridization, serial sections (4 μm) were mounted on microscopic slides that had been cleaned in 10% Extran MA 01 (Merck) and coated with 3-aminopropyl-3 ethoxysilane. Dewaxed paraffin heart tissue sections were hybridized basically as described elsewhere [6]. Hybridization mixture contained the ^{35}S-labeled CVB3 cDNA probe. Myocardial slide preparations were autoradiographed as described [7] and stained with hematoxylin-eosin. For characterization of the inflammatory cells, cryostate heart tissue sections were incubated with a panel of rat anti-mouse monoclonal antibodies recognizing Macl (macrophages) Lyt2 cytotoxic/suppressor T-lymphocytes L3T4 (helper T cells). Positive cells were visualized by a biotin-streptavidin-immunoperidase technique [8].

Myocardial in situ autoradiographs were processed by means of the automatic image analysing system IBAS II (Kontron, Zurich). Myocardial slide preparations were examined with a Leitz microscope and a black-and-white videocamera at a primary magnification of ×40; the video signals were transformed to a digital format under interactive visual control. Area fractions of infected myocardial tissue were computed on transverse sections of both heart chambers. The extent of myocardial lesions was quantified on hematoxylin-eosin stained transverse sections of the heart chambers. Damage was defined as myocardial cell necrosis, inflammation, or scarring; the area fraction of damaged myocardium was calculated with point counting according to standard morphometric procedures. The number of immunochemically stained inflammatory cells was counted and referred to 1 mm^2 of sectional area.

Some results were extrapolated in the three-dimensional space, using basic stereological principles [8]. The number of infected cells was derived from the following principles and assumptions. (a) The area fraction on a random histological section is an unbiased estimate of the volume fraction. (b) A rough estimate of the mean volume of myocytes can be obtained easily if the cylindrical shape of myocytes is assumed. From the diameters of single myocytes (transverse sections) and the lengths of myocytes (longitudinal sections) the average volume of myocytes can be calculated. (c) The volume fraction of myocytes in the myocardium was determined according to the well-known point counting procedure.

From the volume fraction of infection, the volume fraction of myocytes and the average volume of myocytes the number of infected myocytes per unit reference volume can be calculated. During persistent infection the in situ hybridization signals no longer resemble the contours of normal myocytes. This is caused by persistent infection of degenerated myocytes and even by infection of interstitial cells. The number of infected cells was estimated from the numerical density of signals per unit sectional area, the volume fraction of infection, and the assumption of ellipsoid shapes with a ratio of axes of 1.2:1.

The number of inflammatory cells in the three-dimensional space was estimated from the number of nuclei on area which were enveloped by a positively immunostained cytoplasm, the mean diameter of nuclei, the loss of nuclear caps by tangent sectioning, and the section thickness of frozen sections [8].

One-way analysis of variance and Scheffe's test were used to detect significant differences at indicated time points. Values of $p < 0.05$ were considered statistically significant.

Results and Discussion

In ACA/SnJ mice, morphometrical quantification of myocardial infection by means of digital image processing [6] revealed an increased area fraction of infected cardiac tissue until day 9 after CVB 3 infection, when up to 13% of the myocardial cells were infected. Thereafter, decreased area fractions

Table 1. CVB3 myocarditis in ACA/SnJ mice

Number of cells/ mm^3 myocardium	Acute infection (9 days p.i.)	Persistent infection (18 days p.i.)
Infected cells	3 000	30
Macrophages (Mac1)	14 000	5000
T lymphocytes (Lyt2)	9 000	3500

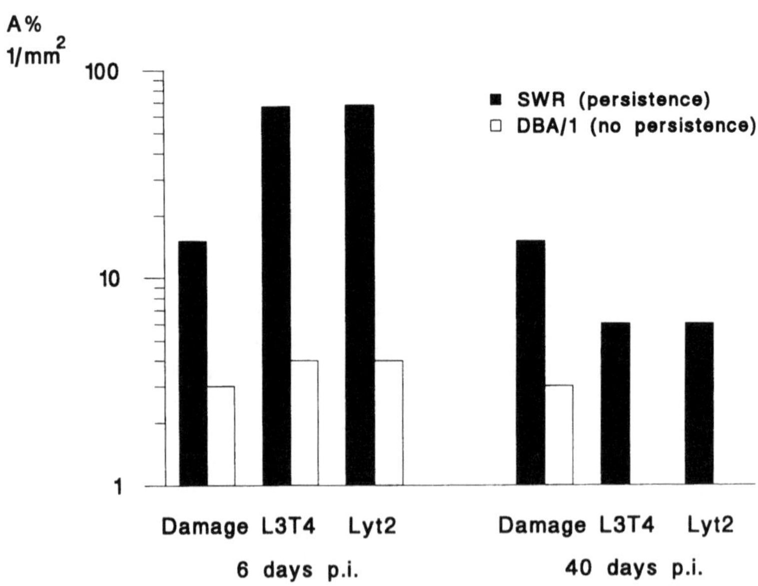

Fig. 1. Myocardial damage and lymphocytic infiltrates in CVB3 myocarditis

of infected myocardial tissue occurred, reflecting a significantly reduced number of infected myocardial cells at later stages of the disease. During persistent infection, for example, 30 days p.i., ca. 0.01% of myocardial cells were found to be infected. The progression of acute infection was characterized by rapid exacerbation of myocardial lesions (up to day 12

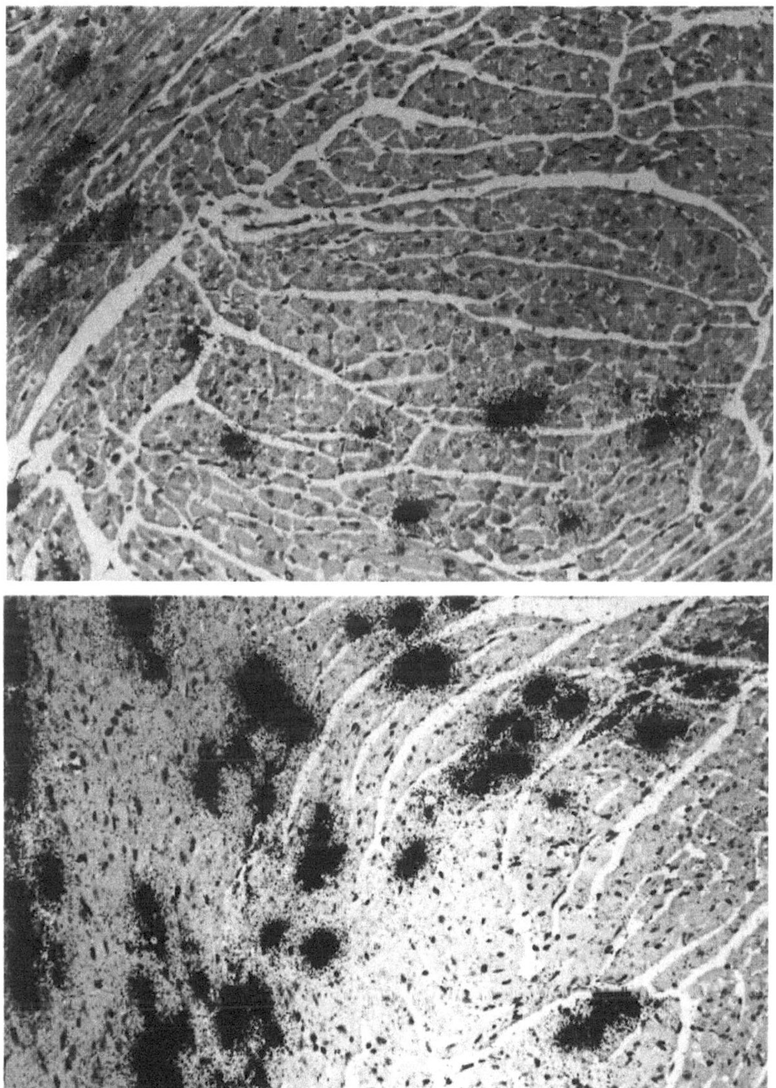

a

b

Fig. 2a–d. Histological sections of CVB3-infected mice after in situ hybridization to detect enteroviral RNA. In DBA1 mice number and areal signals in acute infection (**a**) were less prominent than in SWR/SnJ mice (**b**). This corresponds to the absence of persistent infection in DBA1 (**c**) and the presence of persistent infection in SWR/SnJ (**d**) 40 days after infection

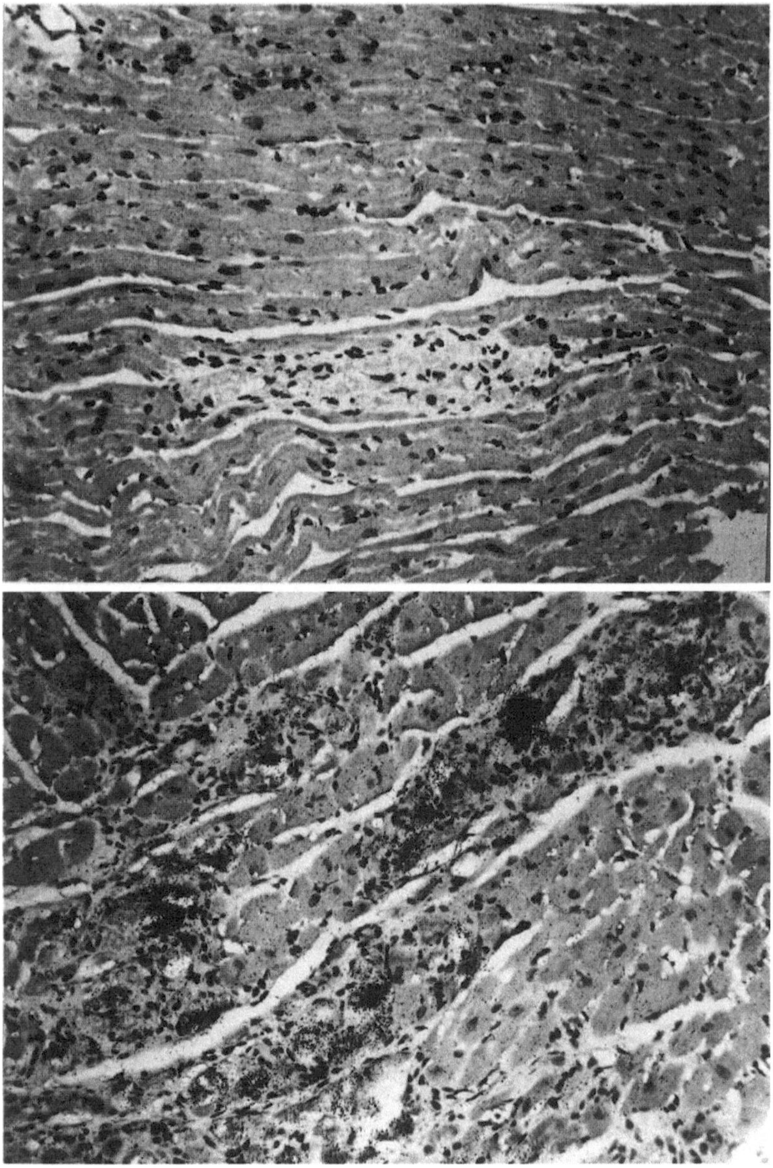

c

d

Fig. 2. *Continued*

p.i.) and by increasing inflammatory cells densities (up to day 12 p.i.). From the mean volume of myocytes ($12\,000/\mu m^3$) and the 85% volume fraction of myocytes in the normal heart, and the area fraction of in situ hybridization-positive myocardial cells it could be estimated that $1\,mm^3$ of myocardial tissue contained an average of 3000 infected myocytes on day

9 p.i. However, morphometrical evaluation of myocardial damage suggested myocytolysis and necrosis of ca. 11 000 myocytes/mm^3 of myocardium at this time point. This observation is related to the relatively short mean survival time of the acutely infected myocytes. Provided that the expansion of myocardial injury between day 3 and 9 p.i. can be described as an exponential function, infected myocytes survived about 24 h during acute replication. This agrees with earlier observations in CVB3-infected cultured myocytes. During chronic disease, up to 30 cardiac cells were infected in 1 mm^3 of heart tissue, which was obviously sufficient to sustain myocardial inflammation. The ratio of T lymphocytes (T helper and T cytotoxic/suppressor) to infected cells was 30 times higher during persistent infection compared with the acute state (Table 1). Here it should be emphasized that in other experiments persistent inflammation (ongoing myocarditis) was strongly correlated with the presence of virus replication (Fig. 1).

For example, DBA mice in which persistent infection was not found by in situ hybridization (Fig. 2a,b), the number of T lymphocytes decreased to normal values 40 days p.i. In contrast, the persistently infected hearts of SWR/SnJ mice (Fig. 2c,d) showed ongoing myocarditis with infiltrates of T lymphocytes (Fig. 1). An expansion of myocardial damage could not be established independently of viral persistence after more than 12 days p.i. This was most probably due to the slow progression of lesions in chronic disease, which is likely to be smaller than the interanimal variation.

The question arises of what factors determine the occurrence of viral persistence. For this purpose, the extent of myocardial damage in acute myocarditis of several strains was correlated with the occurrence and extent of persistent infection 42 days p.i. (Table 2). Spearman's rank coefficient of correlation indicated positive association of the variables ($p < 0.05$).

The present experimental investigations provide evidence that ongoing enterovirus-induced myocarditis is based on persistent infection. Parallel studies with strand-specific probes showed restriction of virus replication in persistent infection [6]. Besides the fact of restricted replication, viruses are known to establish persistence by infection of cellular constituents of the

Table 2. Histopathological damage score in acute CVB3 myocarditis versus persistent CVB3 infection score

Strain (mice)	Damage (rank)	Persistent infection (rank)
DBA/1 J	1	No (2)
PL/J	2	No (2)
SJL/J	3	No (2)
SWR/SnJ	4	Yes (8)
ACA/SnJ	5	Yes (4.5)
DBA/2J	6.5	Yes (6)
ASW/SnJ	6.5	Yes (4.5)
ABY/SnJ	8	Yes (7)

12 days p.i. versus 42 days p.i.; $r = 0.63$; $p < 0.05$.

immune system. Preliminary observations in CVB3-infected mice show that during the chronic phase of the disease, in addition to heart muscle cells, spleen and lymph nodes exhibited persistent infection of lymphoid cells, which might play a role in dissemination of virus or maintenance of unknown cardiac viral reversal.

The availability of the described murine models of persistent CVB3 heart muscle infection could prove useful for further studies on viral and immunological mechanisms of enterovirus-induced cardiomyopathy, for example, with regard to identification of host genes involved in the development of chronic disease.

References

1. Abelmann WH (1971) Virus and the heart. Circulation 44:950–956
2. Chow LH, Beisel KW, McManus BM (1992) Enteroviral infection of mice with severe combined immunodeficiency. Evidence for direct viral pathogenesis of myocardial injury. Lab Invest 66:24–31
3. Huber SA, Lodge PA (1986) Coxsackievirus B3 myocarditis. Identification of different pathogenic mechanisms of DBA/2 and BALB/c mice. Am J Pathol 122:284–291
4. Kandolf R, Ameis D, Kirschner P, Canu A, Hofschneider PH (1987) In situ detection of enteroviral genomes in myocardial cells. An approach to the diagnosis of viral heart disease. Proc Natl Acad Sci USA 84:6272–6276
5. Kandolf R, Hofschneider PH (1985) Molecular cloning of the genome of a cardiotropic Coxsackie B3 virus. Full-length reverse-transcribed recombinant cDNA generates infectious virus in mammalian cells. Proc Natl Acad Sci USA 82:4818–4822
6. Klingel K, Hohenadl C, Canu A, Albrecht M, Seemann M, Mall G, Kandolf R (1992) Ongoing enterovirus-induced myocarditis is associated with persistent heart muscle infection. Quantitative analysis of virus replication, tissue damage, and inflammation. Proc Natl Acad Sci USA 89:313–318
7. Mall G, Klingel K, Albrecht M, Seemann M, Rieger P, Kandolf R (1991) Natural history of coxsackievirus B3-induced myocarditis in ACA/Sn mice: viral persistence demonstrated by quantitative in situ hybridization histochemistry. Eur Heart J 12:121–123
8. Weibel ER (1979) Stereological methods I. Academic, London

Detection and Typing of Enteroviruses

T. Hyypiä

Introduction

Enteroviruses are a large group of human pathogens which cause a great variety of clinical illnesses. They consist of more than 60 serotypes subgrouped in polioviruses, coxsackie A and B viruses, echoviruses, and enteroviruses 68–71 (Table 1). Central nervous system involvement is a relatively frequent complication in infections caused by virtually any membr of the enterovirus group. Other diseases caused by enteroviruses include rash illnesses, respiratory infections, and epidemic myalgia. Especially coxsackie B viruses are among pathogens often associated with acute myocarditis, and there is also evidence of their etiological role in cardiomyopathy.

Diagnosis of enterovirus infections is usually carried out by isolation of the virus in tissue culture followed by typing or by demonstrating an increase in viral antibody levels between serum samples taken in the early and convalescent phase of infection. In both cases the currently used methods are laborious and time consuming, and the diagnosis cannot be obtained at the acute phase of the disease. Recent developments in molecular biology have offered new, highly sensitive methods for diagnostic purposes and increased our understanding of the properties of enteroviruses. It has been shown that some parts of the genome are highly similar between different serotypes and can be used for group-common detection of enteroviruses. On the other hand, detailed information on the viral proteins and their relationships may also make it possible to develop rapid, standardized assays for serological diagnosis of enterovirus infections.

Classical Methods in Enterovirus Diagnosis

Virus Isolation

In virus isolation, samples containing clinical material (cerebrospinal fluid, stool, tissue) are inoculated in cell cultures which are periodically examined for viral cytopathic effect. Because changes in cell morphology are not usually specific for any virus group, a confirmatory assay is needed for

Department of Virology, University of Turku, SF-20520 Turku, Finland

Figulla et al. (Eds.)
Idiopathic Dilated Cardiomyopathy

Table 1. Enterovirus subgroups

Subgroup	No of serotypes	Examples of clinical diseases
Polioviruses	3	Paralysis
Coxsackie A viruses	23	Hand, foot and mouth disease
Coxsackie B viruses	6	Epidemic myalgia, myocarditis
Echoviruses	31	Encephalitis, exanthema
Enteroviruses 68–71	4	Epidemic conjunctivitis, paralysis

further typing of an isolated virus strain. Most of the enterovirus serotypes grow well in standard cell cultures, with the exception that some coxsackie A viruses can be isolated only in suckling mice. The growth properties of enteroviruses are variable, and cytopathic effect can sometimes be observed in 1 day, but often much longer time periods are required. In spite of the relatively high sensitivity of virus isolation as a diagnostic method, the need of cell cultures and repeated examination of the cytopathic effect together with the long duration of the procedure makes the technique rather cumbersome.

Isolation of an enterovirus strain is followed by neutralization typing. This is carried out by incubating the isolate with a panel of antiserum pools and by subsequent evaluation of the decrease in virus growth. By this means it is usually possible to confirm the isolation result and to identify the virus serotype. The assay system gives detailed information for epidemiological purposes, but because another cycle in cell culture is needed it increases the time needed for diagnosis.

In myocarditis patients, isolation of enteroviruses directly from the heart muscle, in rare cases where samples have been available, has usually been negative. On the other hand, isolation of the pathogen from stool does not give firm evidence of disease association since enteroviruses are circulating continuously in the population, and subclinical infections are common. It is even more unlikely to expect positive isolation findings with etiological significance in cases of persistent infections, and therefore other diagnostic methods are needed in these disease conditions.

Serological Diagnosis

The most specific method for serological diagnosis of enterovirus infections is based on differences in neutralization titers between serum samples from acute and convalescent phases of the disease. If it is necessary to study the serological response against all the enteroviruses, this requires usage of more than 60 virus serotypes and testing the reduction of infectivity in cell cultures and in suckling mice. Therefore this assay system is restricted to cases in which the serotype during an epidemic period is known with high probability, or to detailed epidemiological analysis of selected serum materials. Cross-reactivity between enterovirus antigens can be obtained by partial denaturation, and this phenomenon has been used in complement fixation assays in which evidence of enterovirus infection can be obtained without more specific information of the serotype responsible for infection. Again, a serum pair, taken at 2- to 3-week intervals, is required, and the assay system is rather difficult to standardize. For these reasons, there is an obvious need for improvements in serological assays for enterovirus infections to increase the accuracy and the speed of laboratory diagnosis.

New Trends in Identification of Enteroviruses

Detection of Viral Proteins

Demonstration of viral proteins directly in clinical samples using immunoassays is a widely used diagnostic method, for example, for hepatitis B virus infections and rotavirus gastroenteritis. The assay is rapid, specific, and relatively sensitive, and therefore possibilities to develop similar tests for enteroviruses have been explored. Yolken and Torsch [20] described an enzyme immunoassay for coxsackie A viruses which could identify virtually all the serotypes. It was also possible to demonstrate viral antigens in 8 of 11 isolation positive stool samples analyzed. Ukkonen et al. [16] have used the same test principle for polioviruses in which the sensitivity was about 5 ng purified virus. When clinical specimens were analyzed, only 10% of poliovirus 3 isolation positive stool samples were detected. This seems to be a general problem in many enterovirus infections, and usually only a small proportion of isolation positive samples can be found positive by immunoassays, limiting the use of immunoassays mostly to cell culture grown material. Viral proteins can also be detected in tissue sections by immunohistochemistry using antiserum, as has been reported by Foulis et al. [4] for heart muscle and pancreas of infants who died of coxsackievirus myocarditis. When used together with RNA detection methods, this approach makes it possible to study viral macromolecule synthesis in detail in infected cells, which is very important when mechanisms behind persistent infections in humans are explored.

Another possibility for demonstration of viral protein structures is to use virus receptors as detecting reagents. Receptors are cell surface molecules which have important functions in the normal life course of the cells. Viruses have adapted themselves to utilize these structures as vehicles in their entry into the cell. Because this step is always needed to initiate the infection cycle, the presence of a receptor is an important, but not the only, determinant of tissue tropism. It is known that the division of enteroviruses into subgroups follows quite well the division in receptor groups [3], and the usage of receptors as diagnostic tools might correlate well with the ability of viruses to infect different cell types. The receptor for the major group of rhinoviruses, another clinically important picornavirus group, is intercellular adhesion molecule-1, and the soluble form of this protein has already been used to detect rhinoviruses in an assay system comparable with immunoassays [9]. Although the sensitivity of this assay may not be sufficient for

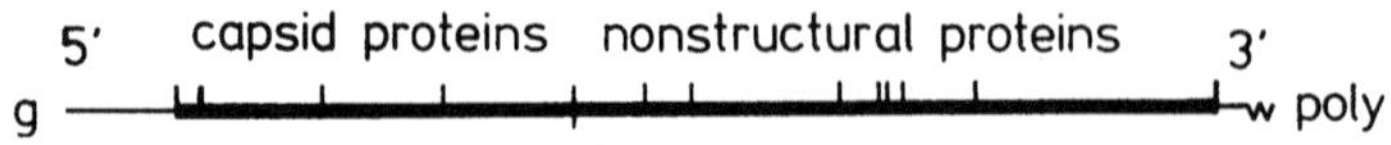

Fig. 1. Organization of the enterovirus genome. The 5′ untranslated region, representing one tenth of the approximately 7500-nucleotide-long genome, precedes the long open reading frame coding for a large polyprotein. The capsid proteins, found in the virus particle, are in the amino-terminal half of the polyprotein followed by nonstructural proteins which are responsible for functions needed in replication

direct detection of enteroviruses in clinical samples, it offers an interesting alternative for the presently used typing system after cell culture isolation.

Detection of Viral RNA

Progress in molecular virology and introduction of new assay methods for the detection nucleic acids have provided diagnostic laboratories with new alternatives for enterovirus identification. The enterovirus genome is an approximately 7500-nucleotide-long, single-stranded RNA molecule which directly acts as a messenger RNA in the cells, giving rise to a large polyprotein which is subsequently autocatalytically processed [15] (Fig. 1). At the 5′ end of the genome there is an about 750 nucleotide long, untranslated region which contains sequences highly similar between virtually all the enteroviruses. This is followed by an open reading frame coding for both capsid proteins, found in the virus particle, and nonstructural proteins which are enzymes and other components needed in the replication cycle. In general, the capsid protein region is the most different part among the serotypes, which can be easily understood because antigenic properties of the virus particle are the determinants used in typing. On the other hand, many regions in the nonstructural protein region are relatively well conserved due to their important functions in virus replication. Mainly conserved sequences, found in the 5′ untranslated region of the genome, have been used in diagnostic approaches since they give an opportunity to detect the whole group using only one set of reagents. Other parts of the genome also have diagnostic potential, and together with increasing sequence data they may allow clinically useful subgrouping of enteroviruses for molecular diagnostics. A sequence comparison based on the presently known enterovirus sequences is shown in Fig. 2 to illustrate molecular relationships.

Spot Hybridization. Among the first attempts to use molecular biology in laboratory diagnosis of enterovirus infections was the spot hybridization assay, in which virus RNA is first immobilized onto membrane filters and then detected using radioactively labeled cDNA probes [6, 12]. It was shown that sequnces from both coxsackievirus B3 nonstructural protein gene region and the 5′ untranslated region of poliovirus 1 can be used in group-common detection of enteroviruses. Bowles et al. [1] used ratios between the signal obtained with a coxsackievirus B2 probe and a control probe as the interpretation criteria for positivity in myocardiac samples. They found evidence

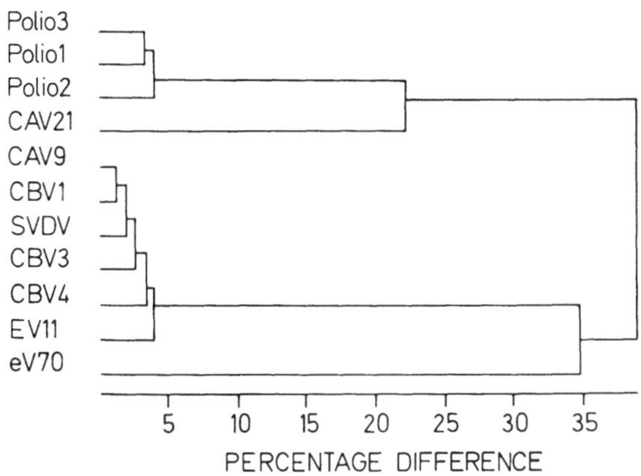

Fig. 2. Molecular relationships between enteroviruses. The dendrogram is based on amino acid differences observed in the nonstructural polypeptide 2C. Two major clusters of related serotypes can be observed. *Polio*, Poliovirus; *CAV*, coxsackie A virus; *CBV*, coxsackie B virus; *SVDV*, swine vesicular disease virus; *EV*, echovirus, *eV70*, enterovirus 70

for the presence of viral RNA in 9 of 17 patients with myocarditis or with dilated cardiomyopathy with inflammatory changes. However, the relatively low sensitivity of this method when compared to in situ hybridization and polymerase chain reaction (PCR) generally restricts the use of spot hybridization in cell culture amplified material.

In Situ Hybridization. Assays for in situ hybridization combine specific detection of enterovirus RNA by labeled probes with visualization of the tissue. This, of course, has the great advantage that the presence and localization of viral genomes can be directly correlated with histopathological changes. The method is highly sensitive, especially in cases in which only a small number cells in tissue are infected but contain sufficient copy number of viral nucleic acid to produce a signal which exceeds the background level. A disadvantage is that the assay procedure is laborious, time consuming, and requires specialized laboratories. Largely due to the long exposure times, the time required to complete the test may be several weeks if high sensitivity is required. In the near future, it may become possible to combine in situ hybridization with amplification of the target nucleic acid and thus shorten the time needed without reducing sensitivity. In situ hybridization technique has been succesfully applied in analysis of the presence of enterovirus RNA in heart tissue. These studies, in which direct evidence of the involvement of viruses in cardiomyopathy has been obtained, have shed new light on the etiology and pathogenesis of the disease and are discussed in detail elsewhere in this volume.

Polymerase Chain Reaction. The sensitivity problem in identification of nucleic acids was largely solved when the enzymatic in vitro amplification method, PCR, was developed [14]. Using this method it is possible to amplify any nucleic acid region, where sufficient sequence data is available, to a detectable level in a test tube reaction within a few hours. When applied to enterovirus detection, cDNA is first generated using a specific primer and viral RNA as a template in a reaction catalyzed by reverse trascriptase. This product is further amplified using repeated cycles of DNA synthesis initiated by specific primers and heat stable polymerase. In the assays described so far, primers have been mainly selected from conserved regions of the 5′ untranslated region of the viral genome [2, 7, 11, 13, 21] (Fig. 3). Using these reagents it is possible to detect virtually all the enterovirus serotypes by using one pair of primer reagents. The specificity of the assay can be further increased using labeled oligonucleotide probes with limited reactivity inside the amplified region. Furthermore, it has been possible to design reagents which recognize specifically poliovirus strains [19], which is important because this enterovirus subgroup needs special control. In addition to rapid detction of enteroviruses in clinical material, PCR makes sequencing of the amplified product possible, thus enabling very detailed identification and comparison of virus strains present in the samples.

The method has been applied to the myocardial biopsy specimens by Jin et al. [8], who analyzed samples from 48 patients with clinically suspected myocarditis or cardiomyopathy. Five specimens were positive, and two of these represented myocarditis while three were from cases of cardiomyopathy. This study used broadly reacting primers and probes from the 5′ untranslated region of the genome. Weiss et al. [17] used more specific primers from the 2C nonstructural polypeptide region, and these reagents were shown to detect only coxsackievirus B3 of the serotypes analyzed. One of five myocarditis specimens was positive with this set of reagents whereas none of 11 specimens from dilated cardiomyopthy cases gave a signal. The low

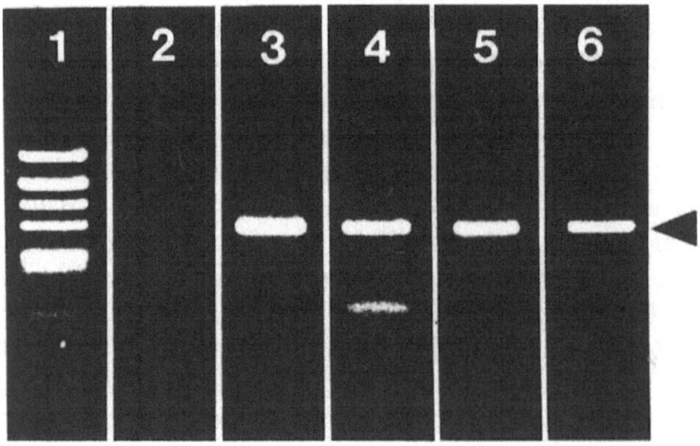

Fig. 3. Detection of enterovirus RNA by PCR using primers from the conserved regions in the 5′ untranslated part of the genome. The amplification products were analyzed using agarose gel electrophoresis; *arrow* shows the size of the specific fragment. *1*, Molecular weight markers; *2*, uninfected control cells; *3*, samples containing enteroviruses

number of enterovirus-positive samples in this study can be explained largely by the limited reactivity of the assay compared with other studies. These studies show that myocardiac biopsy specimens can be used as PCR samples, and it may also be possible to obtain sensitive detection of viral RNA from paraffin blocks. Another potential possibility is to use blood as a sample, especially in acute infections in which viruses may still be present at the time of disease onset.

Detection of Enterovirus Antibodies

Virus infection can often be diagnosed by demonstrating IgM class antibodies against specific antigens in a single serum sample taken even in the acute phase of the illness. Several assay systems for enteroviruses, and especially for coxsackie B viruses, have been described, and they have been used in a number of studies to analyze the etiological role of enteroviruses in myocarditis and cardiomyopathy. The proportion of positive cases varies depending on the test system used and on the epidemiological situation during the collection of samples. For instance, Mokhtar et al. [10] used coxsackievirus B1–5 antigens in an enzyme immunoassay to detect specific IgM class antibodies and observed that approximately 37% of patients with acute myocarditis or pericarditis were positive, as were also two of 28 cases with congestive cardiomyopathy. Frisk et al. [5] found that in their myopericarditis patient group 49% of patients had coxsackie B virus specific IgM.

Increased knowledge on the molecular properties of enteroviruses gives also an opportunity to use standardized antigens, synthetic peptides, or proteins expressed using recombinant DNA technology, in serological assays. It has already been shown that bacterially expressed VP1 capsid polypeptide of coxsackievirus B3 [18] is broadly reactive and has a potential as a diagnostic reagent. We have recently analyzed synthetic peptides representing different regions of coxsackievirus proteins and observed that some of them are widely reactive among enteroviruses while others are serotype specific (Auvinen et al., unpublished). Further studies using different approaches are still needed to find optimal reagents for serological diagnosis of enterovirus infections. Development of standardized, generally available reagents would be most important also for improved diagnosis of viral heart disease.

Conclusions and Future Prospects

Presently used methods for the diagnosis of enterovirus infections (virus isolation and serological assays) are usually not satisfactory in viral heart disease. Recent developments in molecular biology have made it possible to use genotypic, rather than antigenic, properties for identification of enteroviruses in heart muscle. Highly similar regions between virtually all the enteroviruses are found in certain regions of the genome allowing group-common detection by using hybridization methods and PCR. Increased information on the identities and differences in enterovirus proteins between serotypes gives also detailed data for development of new serological assays.

When enterovirus etiology of viral heart disease is studied, the following principles should be concerned. In acute myocarditis, detection of specific IgM in serum samples is the optimal approach since the specimens are easily available, and usually no invasive studies are carried out. In persistent infections, for example, cardiomyopathies, the interpretation of serological assay results in individual cases is difficult, and in these patients myocardial biopsy samples are presently often available. They can be tested for the presence of enterovirus RNA either by in situ hybridization or by PCR. The latter method allows analysis of large numbers of samples in a sensitive manner as soon as some of the technical problems have been solved, and it may become a useful screening test. When more detailed information on the localization of viral RNA in tissue is needed, in situ hybridization method should be used. Both of these methods are also suitable for identification of other microbial agents in the same sample material.

References

1. Bowles NE, Richardson PJ, Olsen EGJ, Archard LC (1986) Detection of coxsackie-B-virus-specific RNA sequences in myocardial biopsy samples from patients with myocarditis and dilated cardiomyopathy. Lancet i:1120–1123
2. Chapman NM, Tracy S, Gauntt CJ, Fortmueller U (1990) Molecular detection and identification of enteroviruses using enzymatic amplification and nucleic acid hybridization. J Clin Microbiol 28:843–850
3. Colonno RJ (1986) Cell surface receptors for picornaviruses. BioEssays 5:270–274
4. Foulis AK, Farquharson MA, Cameron SO, McGill M, Schönke H, Kandolf R (1990) A search for the presence of the enteroviral capsid protein VP1 in pancreases of patients with type 1 (insulin-dependent) diabetes and pancreases and hearts of infants who died of coxsackieviral myocarditis. Diabetologia 33:290–298
5. Frisk G, Torfason EG, Diderholm H (1984) Reverse radioimmunoassays of IgM and IgG antibodies to coxsackie B viruses in patients with acute myopericarditis. J Med Virol 14:191–200
6. Hyypiä T, Ståhlhandske P, Vainionpää R, Pettersson U (1984) Detection of enteroviruses by spot hybridization. J Clin Microbiol 19:436–438
7. Hyypiä T, Auvinen P, Maaronen M (1989) Polymerase chain reaction for human picornaviruses. J Gen Virol 70:3261–3268
8. Jin O, Sole MJ, Butany JW, Chia W-K, McLaughlin PR, Liu P, Liew C-C (1990) Detection of enterovirus RNA in myocardial biopsies from patients with myocarditis and cardiomyopathy using gene amplification by polymerase chain reaction. Circulation 82:8–16
9. Last-Barney K, Marlin SD, McNally EJ, Cahill C, Jeanfavre D, Faaness RB, Merluzzi VJ (1991) Detection of major group rhinoviruses by soluble intercellular adhesion molecule-1 (sICAM-1). J Virol Methods 35:255–264
10. Mokhtar M, El-Hagrassy O, Banatvala JE (1980) Coxsackie-B-virus-specific IgM responses in patients with cardiac and other diseases. Lancet ii:1160–1162
11. Olive DM, Al-Mufti S, Al-Mulla W, Khan MA, Pasca A, Stanway G, Al-Nakib W (1990) Detection and differentiation of picornaviruses in clinical samples following genomic amplification. J Gen Virol 71:2141–2147
12. Rotbart HA, Levin MJ, Villarreal LP (1984) Use of subgenomic poliovirus DNA hybridization probes to detect the major subgroups of enteroviruses. J Clin Microbiol 20:1105–1108

13. Rotbart HA (1990) Enzymatic RNA amplification of the enteroviruses. J Clin Microbiol 28:438–442
14. Saiki RK, Gelfand DH, Stoffel S, Scharf SJ, Higuchi R, Horn GT, Mullis KB, Erlich HA (1988) Primer-directed enzymatic amplification of DNA with a thermostable DNA polymerase. Science 239:487–491
15. Stanway G (1990) Structure, function and evolution of picornaviruses. J Gen Virol 71:2483–2501
16. Ukkonen P, Huovilainen A, Hovi T (1986) Detection of poliovirus antigen by enzyme immunoassay. J Clin Microbiol 24:954–958
17. Weiss LM, Movahed LA, Billingham ME, Cleary ML (1991) Detection of coxsackievirus B3 RNA in myocardial tissues by the polymerase chain reaction. Am J Pathol 138:497–503
18. Werner S, Klump WM, Schönke H, Hofschneider PH, Kandolf R (1988) Expression of coxsackievirus B3 capsid proteins in *Escherichia coli* and generation of virus-specific antisera. DNA 7:307–316
19. Yang C-F, De L, Holloway BP, Pallansch MA, Kew OM (1991) Detection and identification of vaccine-related polioviruses by the polymerase chain reaction. Virus Res 20:159–179
20. Yolken RH, Torsch VM (1981) Enzyme-linked immunosorbent assay for detection and identification of coxsacakieviruses A. Infect Immun 31:742–750
21. Zoll GJ, Melchers WJG, Kopecka H, Jambroes G, van der Poel HJA, Galama JMD (1992) General primer-mediated polymerase chain reaction for detection of enteroviruses: application for diagnostic routine and persistent infections. J Clin Microbiol 30:160–165

Specific and Nonspecific Heart Defenses in Enteroviral Infections

C.J. Gauntt, A.L. Higdon, H.M. Arizpe, E.A. Maull, C.W. Lutton, M.A. Beck, N.M. Chapman, B. McManus, J.C. Mullican, and S.M. Tracy

Introduction

Enterovirus infections of humans are common [31], with the majority being asymptomatic [6]. These viruses, particularly the coxsackieviruses group B (CVB) are associated with diseases of the heart [26, 27], with acute myocarditis occurring in up to 5% of infected individuals [26, 40]. Neonatal myocarditis is a severe disease and contributes to mortality in infants and young children [6, 26, 27]. Among the CVB, coxsackievirus B3 (CVB3) is most frequently associated with acute myocarditis [34]. Acute disease can progress to chronic myocarditis, and clinical laboratory data suggest that some cases of chronic disease develop into idiopathic dilated cardiomyopathy [26, 27, 33, 35].

The ubiquity of the CVB among humans is well documented [21, 30]; neutralizing antibodies to several of five CVB serotypes are found in most individuals [21, 31]. Homotypic neutralizing IgM or IgG antibodies are present in sera of humans infected with a single enterovirus; heterotypic IgM antibodies to several enteroviruses are detected in many of these sera by ELISA [9, 32]. Heterotypic anti-CVB antibodies are not neutralizing and recognize a CVB group-reactive epitope on capsid polypeptide VP1 [25, 32]. Not all antiviral antibodies are of benefit to the host. A subpopulation of viral antibodies may have pathologic consequences for the host, as recent studies of CVB-mouse models of infection/disease suggest that CVB infections induce antibodies which cross-react with normal cells [14, 15, 26, 34, 40]. Infections of humans [3] or mice [2] by a single enterovirus also results in sensitization of T lymphocytes which proliferate in vitro in response to homologous/heterologous enteroviral antigens, suggesting T cell recognition of a common enteroviral antigen [3, 4].

The induction, development, and molecular mechanisms of coxsackievirus-induced acute and chronic myocarditis have been studied in a number of CVB3-mouse models of disease [7, 24, 26, 27]. These excellent

[1] Department of Microbiology, The University of Texas Health Science Center at San Antonio, 7703 Floyd Curl Drive, San Antonio, TX 78284-7758, USA
[2] Department of Neurology, University of Utah Medical Center, Salt Lake City, UT 84132, USA
[3] Department of Pathology and Microbiology, University of Nebraska, Medical Center, 600 S 42nd Street, Omaha, NE 68198-6495, USA

Figulla et al. (Eds.)
Idiopathic Dilated Cardiomyopathy

models exhibit many parallel events which occur in humans. Age of a mouse experiencing a CVB3 infection is a major determinant of disease: younger mice die from infections by temperature-sensitive mutants, amyocarditic variants, or myocarditic variants of CVB3 [10, 37]. In young mice, immunogenetic background determines rate of production of protective antibodies, a response which significantly affects the extent of cardiopathology induced [23, 39]. Genetic background of a mouse can also predict death from a CVB3 infection; several inbred strains die within a week of challenge [7, 26]. Adolescent/adult mice of inbred strains develop acute myocarditis due to a myocarditic CVB3 variant, but extent of cardiopathology depends upon determinants located outside of the 5′ terminus nontranslated region [37] and genetic background of the murine strain [7, 10, 24, 26]. Murine strains which succumb to development of CVB3-induced chronic myocarditis possess certain major histocompatibility complex (MHC) and non-MHC determinants which are currently being defined [5, 27]. Thus age and genetic background are major host factors predisposing mice to chronic myocarditis due to infection by a myocarditic CVB3 variant.

During initial infection of mice with a myocarditic CVB3 variant, several nonspecific host defense mechanisms respond within 2–4 days postinoculation (p.i.), including synthesis of arachidonic acid cascade products [13, 16], lymphokines and cytokines [6], interferon-β and -γ [12, 18], activation of natural killer (NK) cells [18], and adsorption of virus by leukocytes [12]. In acute myocarditis, arachidonic acid cascade products [13, 16] and interferon-γ [12] continue to be produced, and a small number of NK cells are activated. These nonspecific responses have not been described in tissues during chronic disease.

The cell-mediated immune system responds to a CVB3 infection by production of T lymphocytes specific for antigens on cultures of virus-infected and uninfected cells [24, 26, 38]. In vitro, cytotoxic T lymphocytes lyse virus-infected and uninfected targets and proliferate in response to a group-specific enterovirus antigen(s) [1–4] found on several different enterovirus particles [1]. T lymphocytes are found in myocarditic heart tissues of infected mice [19]. It is not known which subset(s) of T lymphocytes [24, 38] is present in heart tissues of mice with CVB3-induced chronic myocarditis [5, 23, 34, 40].

The humoral immune system responds rapidly to a CVB3 infection in mice: protective antiviral antibodies can be detected by day 3 p.i. [5, 26, 39], and titers remain high during acute and chronic diseases [5, 40]. Some murine strains with CVB3-induced chronic myocarditis produce antibodies which bind antigens on normal heart tissues [34, 39, 40]. A portion of these antibodies recognize epitopes on cardiac myosin, an immunogen which can induce an acute myocarditis in certain inbred strains [30, 34], data suggesting potential cross-reacting epitopes on heart tissues. Hyperimmune anti-CVB3 antisera contains antibodies capable of exacerbating CVB3-induced acute myocarditis [17]. Studies of anti-CVB3 neutralizing monoclonal antibodies (mAbs) show that some mAbs can participate in potential proinflammatory reactions in vitro, and other mAbs can induce cardiopathologic alterations in vivo [11, 14]. Thus, published data show that autoimmune responses, with cardiopathologic consequences, can be attributed to both cell-mediated and humoral immune systems. These data imply that immune systems recognition of viral epitopes shared with heart tissues can lead to autoimmune (chronic) myocarditis in mice, and perhaps in humans, of specific genetic backgrounds.

Materials and Methods

$CVB3_m$, a highly myocarditic variant of CVB3, was propagated and assayed in HeLa cells as previously described [18], as well as other CVB serotypes whose origins have been described [20]. ^{35}S-labeled coxsackievirus B2 (CVB2) was prepared as follows. HeLa cells were incubated with 50–100 plaque-forming units (PFU) of virus/cell for 1 h at 37°C. MEM containing 1% fetal bovine sera was added and the cells incubated for an additional 2 h at 37°C. The culture fluids were replaced by methionine-free MEM with 1% fetal bovine serum for 45 min. Trans-^{35}S-label (ICN Biomedicals, Irvine, CA; Sp. Act. 1085 Ci/mmol) was added at 100 μCi/ml and the cells incubated for 15–18 h at 37°C in a humidified CO_2 incubator. Culture fluids and cell debris were processed for virus particles, including banding of particles in CsCl gradients, as previously described [18]. Purified ^{35}S-labeled CVB2 particles were dialyzed versus Dulbecco's phosphate-buffered saline (DPBS), pH7.4, and stored at −20°C. Specific infectivity of several preparations varied from 500 to 2253 PFU/cpm. Experiments on rate zonal centrifugation of ^{35}S-labeled CVB2 particles ± an anti-$CVB3_m$ neutralizing mAb were performed as follows. mAb-virus particle mixtures were layered over preformed 10%–40% sucrose gradients (sucrose was dissolved in DPBS) and centrifuged for 2.5 h at 32K rpm in a Beckman SW50.1 rotor at 5°C. Fractions were collected from the bottom and assayed for infectivity by a plaque method, cpm by a β spectrometer, and percentage sucrose by an Abbe 3L refractometer (Bausch and Lomb).

Hybridomas producing neutralizing anti-$CVB3_m$ mAbs were previously described [11] and were propagated in Dulbecco's MEM containing 10% fetal bovine serum and antibiotics. All mAb preparations were assayed for anti-$CVB3_m$ neutralizing antibody content by a cytopathic effects (CPE) inhibition assay [11], and those with titers of at least 8 were used. Isotyping was performed with commercial kits. mAbs were purified by affinity chromatography using affinity-purified goat anti-mouse IgG bound to agarose beads (Sigma Chemical, St. Louis, MO). IgG concentrations were determined by a dye-binding method (BioRad Laboratories, Richmond, CA). Protein A-purified mAb HB79 (anti-DNP) was a gift from Judy Teale, University of Texas Health Science Center.

Adolescent (4–6 weeks old) CD-1 or C3H/HeJ mice were purchased from Charles River Breeding Laboratories (Boston, MA) or Jackson Laboratories (Bar Harbour, ME), respectively. Mice were given water and mouse chow ad libitum. Mice were inoculated with virus by intraperitoneal route. Sacrifice of deeply anesthesized (ether) mice was by cervical dislocation. Hearts were quickly removed and fixed in 10% buffered formalin, and coronal sections were stained with hematoxylin/eosin. Sections were examined for focal lesions containing necrotic myocytes and inflammatory leukocytes at 40 and 100 [18].

Results

Virus-neutralizing antibodies in serum are a major factor in terminating primary infections/diseases and in providing protection against subsequent illnesses/diseases induced by the same virus serotype through reducing spread of virus within the host. Our findings with two anti-CVB3 neutralizing mAbs show that such mAbs can participate in other reactions which are not in the best interests of the host.

Properties of Anti-CVB3$_m$ Neutralizing mAbs 14 and 24

Both mAbs are of the IgG2a subclass. mAb 24 but not mAb 14 participates in complement-mediated lysis of normal mouse fibroblasts [11, 14, 15]. Both mAbs induce production of a soluble chemoattractant for unelicited macrophages from normal mouse fibroblasts [11]. Intraperitoneal inoculations (nine inoculations every other day with $\approx 1\,\mu g$ mAb) of either mAb into normal adolescent male C3H/HeJ mice can result in pathologic alterations in heart tissues [11, 14]. mAb 24 but not mAb 14 binds to rabbit skeletal muscle myosin [11].

Binding of mAbs 14 and 24 to CVB2 Particles

Virus-neutralization assays performed with these mAbs and all six CVB serotypes confirmed neutralization of only CVB3 [11]. However, in these CPE inhibition assays it was observed that HeLa cells incubated with CVB2-mAb 14 or CVB2-mAb 24 complexes exhibited CPE many hours in advance of cultures challenged with only CVB2. These reproducible observations led to the following experiments.

Incubating CVB2 virions with these mAbs prior to challenge of HeLa cells enhanced yields in comparison to cells challenged with CVB2 alone (Table 1). In the initial experiment, either mAb enhanced yields 18- to 65-fold. In experiment 2, different concentrations of purified mAbs were adsorbed to CVB2 prior to challenge of cells. At $125\,\mu g/ml$ mAb, yields were at least 100-fold above that of the virus control and 12.5 to $3.1\,\mu g/ml$ mAb generally enhanced yields by approximately ten-fold above the virus control. HeLa cells incubated with either mAb alone prior to challenge with CVB2 gave virus yields similar to the virus control. These data suggest that mAb-induced enhancement of virus yields was due to increased adsorption of CVB2 virions.

Mechanisms by which these mAbs might enhance yields of CVB2 from

Table 1. Effect of anti-CVB3$_m$ neutralizing mAbs on CVB2 yields in HeLa cells: virus yield as fold-increase above virus control at mAb concentrations (μg/ml)

Experiment 1			(1–5)			
mAb 14			65			
mAb 24			18			
Experiment 2	(25)	(12.5)	(6.2)	(3.1)	(1.5)	(0.75)
mAb 14	125	9	14	20	2	0.3
mAb 24	240	19	14	7	4	3

mAb- or DMEM-CVB2 (0.4–1 × 10^5 PFU) mixtures were incubated at 37°C for 45 min and added to HeLa cell monolayers at 1 PFU/25–50 cells. After 1 h at 37°C the monolayers were washed three times and incubated in MEM (1% fetal bovine serum + antibiotics) for 24 h (experiment 1) or 8 h (experiment 2). Cells were frozen, thawed three times, and assayed for virus by the plaque method. Virus titers in control cultures were 5.5 × 10^5 (experiment 1) or 2.4 × 10^2 (experiment 2) PFU/ml.

Table 2. Effect of mAb 24 on binding of CVB2 to HeLa cell cultures

		Percentage cpm in virus bound to HeLa cells after incubation with:		
Experiment no.	Adsorption conditions	mAb24	mAb HB79	Medium
1	4°C, 2 h	1.0 ± 0.1	2.7 ± 0.1	3.0 ± 0.1
2	37°C, 15 min	3.9 ± 0.3	2.0 ± 0.2	2.4 ± 0.1
3	37°C, 1 h	13.8 ± 2.5[a]	5.8 ± 1.8	4.4 ± 0.4

Antibodies, mAb 14 (IgG2a, 1–5 μg/ml) or mAb HB79 (anti-DNP, IgG2a, 2 μg/ml), or hybridoma culture medium were incubated with ≈100000 cpm of CsCl-banded ^{35}S-labeled CVB2 at various times and temperatures, with shaking. Triplicate monolayer cultures of HeLa cells (2.5 × 10^5/well) were challenged with ≈20000 cpm in 0.2 ml under the specified conditions above. The inocula were removed and the cultures washed three times with medium, collected, and counted in a β spectrometer.

[a] Significantly different from either percentage value obtained with mAb HB79 or medium (Student's *t* test, $p > 0.05$).

HeLa cells were examined as follows. Data on mAb 24 increasing adsorption of ^{35}S-labeled CVB2 to monolayer cultures of HeLa cells are presented in Table 2. At 4°C, very little CVB2 was bound, whether mAb was present or not. At 37°C, CVB2 adsorption was enhanced by mAb 24 during 1 h of incubation, providing one explanation for the more rapid CPE observed in HeLa cells challenged with CVB2-mAb complexes compared with cells challenged with CVB2 alone.

Rate zonal ultracentrifugation analyses of mAb-purified ^{35}S-labeled CVB2 particle complexes in dense buffered sucrose gradients suggested a second mechanism to explain mAb-increased virus yields. During ultracentrifugation in sucrose gradients, purified CVB2 particles were detected by ^{35}S cpm and infectivity in two populations; one population sedimented as a band at approximately 35% sucrose and the second population at approximately 26% sucrose. Studies of many picornavirus particles separated

Table 3. Rate zonal sedimentation analysis of ^{35}S-labeled CVB2 particle-anti-CVB3 neutralizing mAb complexes in sucrose density gradients

	CVB2 particles sedimenting at ≈35% sucrose			CVB2 particles sedimenting at ≈26% sucrose		
CVB2 particles incubated with:	Total infectivity recovered ($\times 10^6$)	Total cpm recovered	Specific infectivity (PFU/cpm)	Total infectivity recovered ($\times 10^5$)	Total cpm recovered	Specific infectivity (PFU/cpm)
mAb14	189	28659	6595	176	5038	3493
mAb24	92	25805	3565	95	4406	2156
mAb HB79	41	11873	3453	39	10970	356
(anti-DNP) medium	76	11617	6542	26	10329	252

CsCl-banded ^{35}S-labeled CVB2 particles were incubated with 1–5 μg mAb 14 or mAb 24 or 2 μg mAb HB79 in hybridoma culture medium or hybridoma culture medium alone for 1 h at 37°C with shaking prior to centrifugation into dense sucrose gradients, as described in "Materials and Methods."

Table 4. Radioimmunoprecipitation of ^{35}S-labeled CVB2 by anti-$CVB3_m$ neutralizing mAbs

^{35}S-labeled CVB2 particles reacted with:	cpm ppt.[a] (% total)	cpm ppt. by mAb/cpm ppt. by antisera E
mAb 14	28.5	55.0
mAb 24	12.3	23.8
Antisera E	51.7	100.0
Medium	3.9	7.5

CsCl-banded ^{35}S-labeled CVB2 particles + mAb (1–5 μg/ml) or antisera E hyperimmune murine anti-$CVB3_m$ (CsCl-banded virus) antisera (1:100 dilution), were incubated for 2 h at 25°C; mixtures were then incubated with affinity-purified goat anti-mouse IgG coupled to agarose for 4 h at 25°C. All precipitates were washed five times with Dulbecco's PBS, and cpm were determined in a β spectrometer. Specific infectivity of ^{35}S-labeled CVB2 was 10.4 PFU/cpm.

[a] Total cpm/sample = 30968.

by this methodology have shown that slower sedimenting virus particles contain mostly empty capsids (no RNA) aggregated with infectious particles, whereas faster sedimenting virus particles are mostly infectious with few empty capsids. The data (Table 3) show that particles (± bound mAb) sedimenting at ca. 35% sucrose in all four gradients had similar specific infectivities (3453–6596 PFU/cpm). CVB2 particles bound to a mAb which sedimented at 26% sucrose had specific infectivities about half of those measured for respective virus particle populations sedimenting at ca. 35% sucrose. In contrast, CVB2 particles incubated with mAb HB79 or medium alone which sedimented at 26% sucrose had specific infectivities 5%–10% of those measured for respective virus populations sedimenting at 35% in these gradients. These data suggest that binding of these two mAbs to

CVB2 particles prevents particle aggregation, thereby increasing infectivity because an aggregate of infectious particles assay as one plaque.

Direct binding of mAb 14 or mAb 24 to CVB2 was demonstrated by immunoprecipitation of purified ^{35}S-labeled CVB2 particles using a mAb and goat anti-mouse IgG coupled to agarose beads (Table 4). Hyperimmune mouse antisera to purified CVB3$_m$ particles (antisera E) and hybridoma medium were included as a positive and negative controls. Using cpm in ^{35}S-labeled CVB2 particles precipitated by polyvalent antisera E as maximum amount bound, mAb 14 bound about one-half and mAb 24 about one-fourth the number of cpm as antisera E, levels three- to sevenfold above background (medium control).

These data suggest a second mechanism by which neutralizing antibodies against one CVB serotype can promote infections/diseases: enhancing infectivity of a different CVB serotype. We are not aware of any published data on this mechanism promoting enterovirus infections/diseases, but it may be operative in the following situation.

Challenge of C3H Mice with CVB2 Particles prior to Infection with CVB3 Exacerbates CVB3-Induced Myocarditis

In an elegant series of experiments, Beck and colleagues [1] showed that sequential CVB infections in young mice can have serious consequences

Table 5. Histopathology in heart tissues of C3H/HeJ mice challenged with CVB2 and subsequently CVB3 (from [1])

Challenge inocula (PFU/mouse)		Mean lesion no./section (day 10)[a]	Comments on histopathology[b]
Primary	Secondary		
None	CVB2 (10^3.or 10^5)	≈3	Small lesions, no calcific deposits, few inflammatory cells/lesion
None	CVB3$_m$ (10^5)	≈10	Small to medium lesions, no calcific deposits, 10–50 inflammatory cells/lesion
CVB2 (10^3 or 10^5)	CVB3$_m$ (10^5)	≈10–25	Small to medium lesions, calcific deposits in many lesions; 50–100 inflammatory cells/some lesions
CVB2 (UV-10^5)	CVB3$_m$ (10^5)	>25	Medium lesions, calcific deposits many lesions; ≥50–100 inflammatory cells/lesion

Male 26-day-old mice were challenged with CVB2; 10 days later some groups were challenged with CVB3$_m$. All mice were 64 days of age at sacrifice.

[a] Day 10 after primary challenge alone or day 10 after secondary challenge (20 days total after primary and secondary challenge).

[b] Lesion size: small, 15–75 μm diameter; medium, 76–200 μm diameter.

above those found with a single infection. Their data [1] (Table 5) show that young male C3H/HeJ mice develop minimal or no myocarditis when challenged with CVB2, but challenge with $CVB3_m$ induces moderate myocarditis. However, a 10-day exposure of mice to CVB2, either infectious or UV-irradiated virus, prior to challenge with $CVB3_m$ significantly exacerbated pathologic alterations in heart tissues. Prior sensitization of animals with CVB2 increased both number of $CVB3_m$-induced myocarditic lesions and lesion size. Exposure of mice to CVB2 prior to $CVB3_m$ also resulted in Ca^{2+} deposition in many lesions, a pathologic feature not found in myocarditic lesions of mice challenged with only $CVB3_m$. Interestingly, while mice inoculated with CVB2 developed significant neutralizing antibody titers to CVB2, challenge of these mice with $CVB3_m$ resulted in a rapid but transient (several days) loss in titer of anti-CVB2 neutralizing antibodies [1]. This finding suggests that anti-CVB2 antibodies may have bound to $CVB3_m$ particles in the challenge inoculum upon entering the bloodstream, possibly contributing to exacerbation of myocardial disease. In contrast, the following data show that mice inoculated with CVB4 before challenge with $CVB3_m$ show some protection against myocardial disease.

Table 6. Effect of sequential CVB4/$CVB3_m$ inoculations upon induction of myocarditis in CD-1 mice

Virus challenge	Day of sacrifice (day p.i.)	Myocarditic lesion number (mean ± SEM)
Experiment 1[a]		
CVB4	7	3.4 ± 1.1
CVB4	14	0.4 ± 0.2
$CVB3_m$	7	35.6 ± 14.1(A)
$CVB3_m$	14	30.3 ± 11.9(B)
$CVB3_m$/CVB4	7	8.1 ± 4.4
CVB4/$CVB3_m$	14	3.3 ± 2.2(A,B)
Experiment 2[b]		
CVB4	14	0
$CVB3_m$	7	11.5 ± 3.9(C)
CVB4/$CVB3_m$	14	0(C)
Experiment 3[c]		
CVB4	7	0
CVB4	14	0
$CVB3_m$	7	31.9 ± 8.4(D)
$CVB3_m$	14	6.1 ± 1.5(E)
$CVB3_m$/CVB4	14	17.9 ± 3.6
CVB4/$CVB3_m$	14	0(D,E)

A,B,C,D, Lesion numbers between sets were significantly different at $p < 0.05$ (Student's *t* test).

[a] Four female mice/group were inoculated by intraperitoneal route with 5×10^6 PFU of primary, then secondary virus on days 0 and 7, respectively.

[b] Six to seven male mice/group were inoculated by intraperitoneal route with 10^7 PFU of primary, then secondary virus on days 0 and 7, respectively.

[c] Six to eight male mice/group were inoculated by intraperitoneal route with 10^6 PFU of primary or secondary virus on days 0 and 7 respectively.

Adolescent CD-1 Mice Exposed to CVB4 Prior to $CVB3_m$ Have Decreased Myocardial Disease

A CVB4 variant with minimal cardiopathogenicity was used as the primary infecting virus (Table 6). At 7 days p.i. of CVB4, CD-1 mice were challenged with $CVB3_m$, and sacrificed 7 days later. Hearts from mice challenged with CVB4 prior to challenge with $CVB3_m$ consistently (three experiments) showed a reduction in lesion number compared to hearts from siblings challenged with only $CVB3_m$. Heart tissue lesions in CVB4/$CVB3_m$-inoculated mice were considerably smaller in size than myocardial lesions induced with $CVB3_m$ (data not shown). These results were surprising in view of the CVB2/$CVB3_m$ data and show that it is premature to draw general conclusions about heart disease induced during sequential infections with two CVB serotypes. The genetic background of the host likely influenced these different results. C3H mice can progress through $CVB3_m$-induced acute myocarditis into chronic myocarditis, whereas CD-1 mice with $CVB3_m$-induced acute myocarditis undergo healing fibrosis.

Discussion

Heterotypic antibody responses to a CVB serotype other than the infecting serotype are common in humans [9, 32]. The molecular basis for this heterologous anamnestic antibody response is not understood, but similar types of heterotypic antibody responses frequently occur in humans experiencing influenza viruses serotype A infections [28]. Our finding that anti-CVB3 neutralizing mAbs could bind to CVB2 particles and promote higher virus yields from HeLa cells suggests a potential pathologic role for these antibodies. Adsorption of CVB2 to the common CVB receptor on HeLa cells has never been demonstrated, and altering several culture conditions which enhanced binding of CVB3 to HeLa cells did not increase binding of CVB2 particles [8]. CVB2 yields from HeLa cells are low, likely reflecting minimal adsorption [8]. Recently, HeLa cells have been shown to possess an F_c receptor for IgG ($F_{C\gamma}$RIII, 36) which may account for binding and uptake of CVB2 complexes. Enhanced infectivity of dengue virus-heterotypic IgG complexes in vivo occurs by this mechanism, resulting in dengue hemorrhagic fever/shock syndrome [22]. Antibody-dependent enhancement of virus infectivity occurs in flaviviruses, alphaviruses, reoviruses, rhabdoviruses, arenanaviruses, coronoviruses, and retroviruses but has not been reported in picornaviruses [29]. This report is the first to show that neutralizing antibody against one enterovirus can enhance infectivity of a different enterovirus, possibly through binding of virus-antibody complexes to F_c receptors and preventing aggregation of particles.

Although multiple infections of humans by different CVB serotypes occur, the influence of a prior CVB infection on subsequent illness/disease induced by a second CVB has been examined in only a single published report [1] and experiments reported herein. Exacerbation of disease by heterologous antibody enhancement of infectivity could explain increased myocarditis in CVB2/CVB3-challenged mice. Increased lesion size in these mice could reflect rapid clonal expansion of anti-CVB T cells sensitized to a common enterovirus group antigen during initial infection [2, 4] and their subsequent infiltration during secondary infection, resulting in an increased number of virus-infected foci and increased lesion size. Thus under two sequential CVB serotype infections, sensitization of both immune systems to a common CVB group antigen could lead to more severe disease.

Protection afforded by sequential CVB infections is hypothesized to occur through binding of nonneutralizing heterotypic antibody to the second virus and promoting clearance of the more virulent virus. Although the authors are not aware of such an example for picornaviruses, studies of nonneutralizing antibodies against alphaviruses, flaviviruses, coronaviruses, orthomyxoviruses, and paramyxoviruses show that such antibodies can provide protection in vivo [29]. Additional studies involving avirulent/virulent CVB which cause myocarditis and other types of disease in mice are clearly warranted to provide further information of future relevance to generation of a CVB vaccine.

Acknowledgements. We thank Elsa Garay for her patience and typing skills during many drafts of this manuscript. This work was supported by grants from the ERACE Foundation, Los Angeles, California, and the National Heart, Lung and Blood Institute, Public Health Service grant HL45979.

References

1. Beck MA, Chapman NM, McManus BM, Mullican JC, Tracy S (1990) Secondary enterovirus infection in the murine model of myocarditis. Am J Pathol 136:669–681
2. Beck MA, Tracy SM (1989) Murine cell-mediated immune response recognizes an enterovirus group-specific antigen(s). J Virol 63:4148–4156
3. Beck MA, Tracy SM (1990) Evidence for a group-specific enteroviral antigen(s) recognized by human T cells. J Clin Microbiol 28:1822–1827
4. Beck MA, Tracy S, Coller BA, Chapman NM, Hufnagel G, Johnson JE, Lomonossoff (1992) Comoviruses and enteroviruses share a T cell epitope. Virology 186:238–246
5. Beisel KW, Traystman MD (1988) Viral myocarditis: immunogenetic and autoimmune aspects. In: Schultheiss HP (ed) New concepts in viral heart disease. Springer, Berlin Heidelberg New York, pp 148–159
6. Cherry JD (1987) Enteroviruses: polioviruses (poliomyelitis), coxsackieviruses, echoviruses, and enteroviruses. In: Feigin RD, Cherry JD (eds) Textbook of pediatric diseases. Saunders, Philadelphia, pp 1729–1790
7. Chow LH, Gauntt CJ, McManus BM (1991) Differential effects of myocarditic variants of coxsackievirus B3 in inbred mice: a pathologic characterization of heart tissue damage. Lab Invest 64:55–64
8. Crowell RL, Landau BJ (1983) Receptors in the initiation of picornavirus infections. In: Fraenkel-Conrat H, Wagner RR (eds) Comprehensive virology, vol 18. pp 1–42
9. Frisk G, Nilsson E, Ehrnst A, Diderholm H (1989) Enterovirus IgM detection: specificity of μ-antibody-capture radioimmunoassays using virions and procapsids of coxsackie B virus. J Virol Methods 24:191–202
10. Gauntt CJ (1988) The possible role of viral variants in pathogenesis. In: Friedman H, Bendinelli M (eds) Coxsackieviruses – a general update. Plenum, New York, pp 159–179

11. Gauntt CJ, Arizpe HM, Higdon AL, Rozek MM, Crawley R, Cunningham MW (1991) Anti-coxsackievirus B3 neutralizing antibodies with pathologic potential. Eur Heart J 12 [Suppl D]:124–129
12. Gauntt CJ, Godeny EK, Lutton CW (1988) Host factors regulating viral clearance. Pathol Immunopathol Res 7:251–265
13. Gauntt CJ, Godeny EK, Lutton CW, Arizpe HM, Chapman NM, Tracy SM, Revtyak GE, Valente AJ, Rozek MM (1988) Mechanism(s) of coxsackievirus-induced acute myocarditis in the mouse. In: de la Maza LM, Peterson E (eds) Medical virology VIII. Erlbaum, Hillsdale, pp 161–182
14. Gauntt GJ, Higdon AL, Arizpe HM, Tamayo MR, Crawley R, Henkel RD, Pereira MEA, Tracy SM, Cunningham MW (1992) Epitopes shared between coxsackievirus B3 (CVB3) and normal heart tissue contribute to CVB3-induced murine myocarditis. In: Rose NR, Baughman KL, Herskowitz A, Newmann DA (eds) immune mediated heart disease. Springer, Berlin Heidelberg New York (in press)
15. Gauntt CJ, Lutton CW, Arizpe HM, Higdon AH, Tracy SM (1991) Autoimmune reactions in coxsackievirus B3-induced murine myocaditis. Life Sci Adv 10:23–31
16. Gauntt CJ, Lutton CW, Revtyak GE, Valente AJ, Fernandes AN, Arizpe HM, Rozek MM (1988) Induction of soluble mediators during coxsackievirus B3 replication in vitro and in vivo. In: Schultheiss HP (ed) New concepts in viral heart disease. Springer, Berlin Heidelbergy New York, pp 93–104
17. Godeny EK, Arizpe HM, Gauntt CJ (1988) Characterization of the antibody response in vaccinated mice protected against coxsackievirus B3-induced myocarditis. Viral Immunol 1:305–314
18. Godeny EK, Gauntt CJ (1986) Involvement of natural killer cells in coxsackievirus B3-induced murine myocarditis. J Immunol 137:1695–1702
19. Godeny EK, Gauntt CJ (1987) In situ immune autoradiographic identification of cells in heart tissues of mice with coxsackievirus B3-induced myocarditis. Am J Pathol 129:267–276
20. Godeny EK, Sprague EA, Schwartz CJ, Gauntt CJ (1986) Coxsackievirus group B replication in cultured fetal baboon aortic smooth muscle cells. J Med Virol 20:135–149
21. Grist NR (1988) Epidemiology and pathogenicity of coxsackieviruses. In: Schultheiss HP (ed) New concepts in viral heart disease. Springer, Berlin Heidelberg New York, pp 26–32
22. Halstead SB (1988) Pathogenesis of dengue: challenges to molecular biology. Science 239:476–481
23. Herskowitz A, Wolfgram LJ, Rose NR, Beisel KW (1987) Coxsackievirus B_3 murine myocarditis-marked strain differences in histopathologic features of early and late phase myocarditis. J Am Coll Cardiol 9:1311–1319
24. Huber SA, Weller A, Herzum M, Lodge PA, Estrin M, Simpson K, Guthrie M (1988) Immunopathogenic mechanisms in experimental picornaviros-induced autoimmunity. Pathol Immunopathol Res 7:279–291
25. Katze MG, Crowell RL (1980) Immunological studies of the group B coxsackieviruses by the sandwich enzyme-linked immunosorbent assay (ELISA) and immunoprecipitation. J Gen Virol 50:357–367
26. Leslie K, Blay R, Haisch C, Lodge A, Weller A, Huber S (1989) Clinical and experimental aspects of viral myocarditis. Clin Microbiol Rev 2:191–203
27. McManus BM, Gauntt CJ, Cassling RS (1987) Immunopathologic basis of myocardial injury. Cardiovasc Clin 18:163–184
28. Mitchell DM, McMichael AJ, Lamb RA (1985) The immunology of influenza. Br Med Bull 41:80–86
29. Murphy BR, Chanock RM (1991) Immunization against viruses. In: Fields BN, Knipe DM (eds) Fundamental virology, 2nd edn. Raven, New York, pp 371–404
30. Neu N, Rose NR, Beisel KW, Herskowitz A, Gurri-Glass G, Craig SW (1987) Cardiac myosin induces myocarditis in genetically predisposed mice. J Immunol 139:3630–3636
31. Pallansch MA (1988) Epidemiology of group B coxsackieviruses. In: Bendinelli M, Friedman H (eds) Coxsackieviruses – a general update. Plenum, New York, pp 399–417
32. Pattison JR (1983) Tests for coxsackie B virus-specific IgM. J Hyg 90:327–332
33. Reyes MP, Lerner AM (1988) Myocarditis. Clinical and experimental correlates. In:

Bendinelli M, Friedman H (eds) Coxsackieviruses – a general update. Plenum, New York, pp 253–270
34. Rose NR, Neu N, Neumann DA, Herskowitz A (1988) Myocarditis: a postinfectious autoimmune disease. In: Schultheiss HP (ed) New concepts in viral heart disease. Springer, Berlin Heidelberg New York, pp 139–147
35. See DM, Tilles JG (1991) Viral myocarditis. Rev Infect Dis 13:951–956
36. Su H, Spangrude GJ, Caldwell HD (1991) Expression of Fcγ RIII on HeLa 229 cells: possible effect on in vitro neutralization of *Chlamydia trachomatis*. Infect Immun 59: 3811–3814
37. Tracy S, Chapman NM, Beck MA (1991) Molecular biology and pathogenesis coxsackie B viruses. Rev Med Virol 1:145–154
38. van Houten N, Huber S (1989) Role of cytotoxic T cells in exptl. myocarditis. Springer Semin Immunopathol 11:61–68
39. Wolfgram LJ, Beisel KW, Herskowitz A, Rose NR (1986) Variations in the susceptibility to coxsackievirus B3-induced myocarditis among different strains of mice. J Immunol 136:1846–1852
40. Wolfgram LJ, Rose NR (1989) Coxsackievirus infection as a trigger of cardiac autoimmunity. Immunol Res 8:61–80

IX. Immunological Factors in Cardiomyopathy

Heart Autoantibodies in Human Myocarditis and Cardiomyopathy: Virus and Skeletal Muscle Cross-reactivity

D.A. Neumann[1], G.S. Allen, C.R. Narins, N.R. Rose, and A. Herskowitz

Introduction

Human myocarditis and idiopathic dilated cardiomyopathy (IDC) are clinical syndromes associated with multiple etiological mechanisms including cardiotropic virus infections, autoimmunity, cardiotoxic drugs, and cardiac metabolic disorders [1, 29, 32]. Diagnosis of myocarditis depends upon histological fulfillment of the Dallas criteria [5], i.e., demonstration of myocardial inflammation with accompanying myocyte necrosis in cardiac tissue obtained by endomyocardial biopsy. IDC may be diagnosed on the basis of clinical presentation, echocardiographic findings, and failure of biopsy specimens to meet the Dallas criteria. Although these two syndromes are frequently considered distinct entities, there is mounting evidence that they represent points on a spectrum of disease leading to congestive heart failure [6, 15, 23].

Aside from clinical correlates between myocarditis and IDC, virology and serology studies provide additional evidence for such a linkage. Cardiotropic viruses, principally enteroviruses such as coxsackie, are associated with a significant proportion of myocarditis cases of known etiology [32]. Enteroviral genomic material has been demonstrated in heart tissue from approximately 40% of myocarditis patients and from a similar proportion of IDC patients [4, 7, 13]. Similarly, the incidence of antibodies to enteroviruses is comparable in myocarditis and IDC patients [24]. These data suggest that cardiotropic virus infection is responsible for some cases of acute myocarditis and may be associated with the progressive or delayed presentation of IDC.

In a serological study of a large number of patients with myocarditis or IDC of undetermined origin, similar profiles of heart-reactive antibodies were observed [18]. This study suggested that whether autoimmune recognition of the myocardium is a principle etiological agent or a consequence of some other initiating event, myocarditis and IDC patients could not be readily distinguished serologically. Similarly, antibodies to an adenine nucleotide translocator (ANT) molecule, to a branched chain ketoacid dehydrogenase (BCKD), and to β-adrenergic receptors are equally common in myocarditis and IDC patients [3, 14, 25, 26].

[1] *Present address:* International Life Sciences Institute, Risk Science Institute, 1126 Sixteenth Street, N.W., Washington, DC 20036, USA

Figulla et al. (Eds.)
Idiopathic Dilated Cardiomyopathy

These observations raise the question of the relationship between virus infection and cardiac autoantibodies. Autoantibodies may be elicited as a consequence of virus-associated myocardial injury [2, 31], through induction of antibodies which cross-react with both pathogen and heart (e.g., [8, 9, 11]), or through induction of antibodies which recognize novel self-antigens expressed in response to virus infection [12, 19]. This study was initiated to determine the incidence of antibodies to coxsackievirus B3 (CB3), an enterovirus commonly associated with myocarditis and IDC [1, 32], and in such patients to determine whether antibodies to CB3 recognize normal heart constituents. These sera were also tested for heart tissue reactive IgG antibodies. Serum from seven patients with active myocarditis was tested for reactivity with sequentially prepared extracts of normal heart before and after absorption with skeletal muscle.

Methods

Serum samples were obtained from patients with active (n = 10) or borderline myocarditis (n = 9), IDC (n = 9), or ischemic heart disease (n = 10). Heart-reactive immunoglobulin G (IgG) antibodies were detected by western immunostaining as previously described [18]. Briefly, a 10% w/v extract of normal human myocardial tissue from the explanted heart of a heart-lung transplant recipient was prepared in saline with detergent. Aliquots of the extract, 5 μg total protein, were separated by polyacrylamide gel electrophoresis (PAGE) on a 3%–16% gradient gel; the heart constituents were electrotransferred onto a polyvinyl difluoride membrane (Immobilon, Enprotech, Hyde Park, MA). The membrane was cut into narrow strips, and individual strips were incubated with serum from individual patients diluted 1:150. After suitable incubation and washing the strips were incubated with a biotinylated goat $F(ab')_2$ anti-human IgG Fc region-specific (Jackson Immunoresearch, West Grove, PA) antibody. Reactivity was detected following subsequent incubation with an Extravidin/biotinylated horseradish peroxidase complex (Sigma Chemical, St. Louis, MO). The molecular weight of the cardiac antigens recognized by the sera were calculated from standard curves based on the positions of molecular weight markers on each membrane.

Serum IgG reactivity with CB3 was determined similarly. CB3 was grown in Vero monkey kidney cells and purified by successive sucrose and cesium chloride gradient centrifugations [10]. The virus was solubilized in sodium dodecyl sulfate (SDS) PAGE buffer and loaded onto 12% PAGE gels. The separated proteins were electrotransferred to Immobilon membranes. Each membrane was incubated in blocking buffer for 1 h at room temperature with constant rocking [18]. After washing for 10 min, each membrane was incubated with 6 ml of a 1:150 dilution of serum from individual patients for 2 h as above. Each membrane was washed as before, and a 2- to 3-mm-wide strip was cut from each and used for immunostaining as described above. CB3 proteins were identified to the basis of the estimated molecular weight of reactive antigens which corresponded to published values [22].

The remainder of each membrane ($\approx$6.6 cm^2) was incubated with 5.5 ml 0.2 *M* glycine containing 0.05% Tween 20 (Sigma), pH 2.5, for 20 min at room temperature to elute IgG from the membrane bound viral antigens. Eluates were neutralized by adding 0.9 ml 0.5 *M* Tris base (Sigma). Each neutralized eluate, ca. 6 ml, was incubated with a strip of membrane bearing heart extract constituents prepared as described above. IgG reactivity was detected as previously described.

Serum from each patient was tested for heart tissue reactive IgG by indirect immunofluorescence assay (IFA) as previously described [18]. Each serum sample demonstrating reactivity upon initial screening was serially diluted, and each dilution was tested for reactivity with normal rat heart tissue. The greatest dilution of each at which a distinct fluorescent signal was observed was taken as the titer.

Serum from seven myocarditis patients was tested for IgG reactivity with each of six sequentially prepared extracts of normal human heart. Extracts were prepared as described by Alvarez et al. [2] using low salt (20 m*M* Tris-HCl, 1 m*M* EDTA), high salt (20 m*M* Tris-HCl, 0.62 *M* KCl, 2 m*M* ATP, 2 m*M* $MgCl_2$), high salt with pyrophosphate (20 m*M* Tris-HCl, 50 m*M* $Na_4P_2O_7 \cdot H_2O$, 2 m*M* ATP, 2 m*M* $MgCl_2$, 1 m*M* DTT), the nonionic detergents Triton (20 m*M* Tris-HCl, 0.6 *M* KCl, 2% Triton X-100) and CHAPS (20 m*M* Tris-HCl, 0.6 *M* KCl, 2% CHAPS), and SDS-PAGE buffers. After initially homogenizing the tissue in low salt buffer and centrifuging, the pellet was extracted with the high salt buffer; the same sequence was followed with the other buffers.

Each extract was loaded into wells of 3%–16% PAGE gels, separated electrophoretically, and electrotransferred to nitrocellulose (Schleicher & Schuell, Keene, NH) membranes. Serum samples were diluted 1:100 and incubated with separate membranes for 2 h. After washing, a biotinylated horse anti-human IgG H- and L-chain-specific antibody (Vector Laboratories, Burlingame, CA) was incubated with each membrane for 30 mins. An avidin/biotinylated horseradish peroxidase complex (Vector) was then incubated with each membrane for 1 h. The membranes were washed, and diaminobenzidine substrate was added to facilitate the detection of antibody/antigen interactions.

Acetone extracts of normal human skeletal muscle were prepared by mincing 2 g tissue, adding 4 volumes of acetone, and mixing for 30 min at 4°C. The tissue was collected by centrifugation, washed 6× with phosphate-buffered saline, reextracted 4× in acetone, washed with distilled water, frozen, and lyophilized. The lyophilized tissue (30 mg) was incubated with serum diluted 1:10 (total volume 750 μl) for 1 h at room temperature. Following centrifugation, sufficient buffer was added to effect a final serum dilution of 1:100 which was incubated with membranes containing each of the six heart extracts. Immunodetection was performed as previously described.

Results

CB3 and Heart Extract Cross-reactivity

IgG bound to CB3 proteins immobilized on a membrane were eluted with little loss of functional activity. The elution process effectively removed all IgG bound to the viral protein, and the eluted antibody reacted with CB3 protein on a second membrane (Fig. 1).

Serum from each of the four patient groups exhibited variable reactivity with extracts of normal human heart. Of patients with active myocarditis, IDC, or ischemia, 80%–90% possessed heart-reactive IgG antibodies; 56% of those with borderline myocarditis had similar antibodies (Table 1).

Antibodies to CB3 were less common among all patient groups, ranging from 10% of the ischemic patients to 60% of patients with active myocarditis (Table 1). In all cases CB3-reactive IgG recognized only one of the five viral proteins present on each of the membranes. Serum from active myocarditis patients bound to VP0, VP1, or VP2/3 while serum from borderline myocarditis patients recognized VP0 and VP2/3 (Table 2). Serum from IDC patients recognized VP0 and VP1; the one ischemic patient with anti-CB3 antibody recognized VP4 (Table 2).

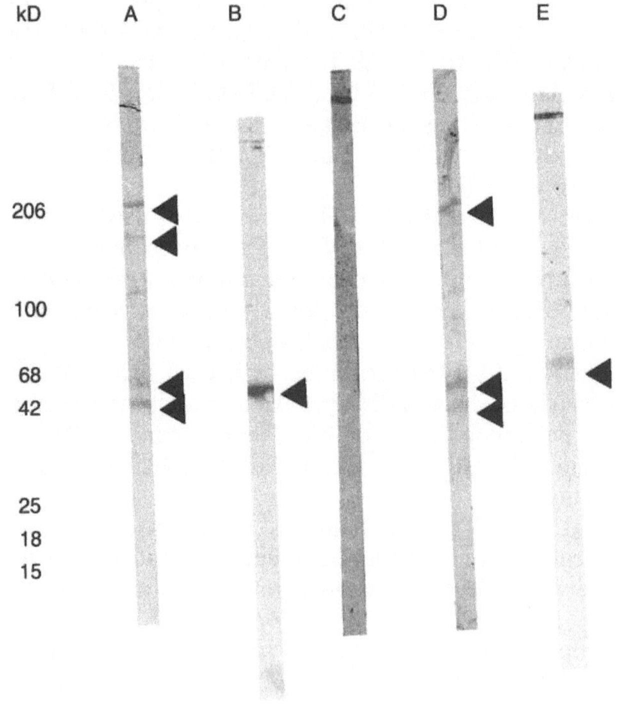

Fig. 1. Representative western blot strips demonstrating human IgG reactivity with normal human heart extract (*A*) and coxsackievirus B3 (*B*). Heart extract (*D*) and CB3 (*E*) were incubated with IgG eluted from strip *B*. Heart extract (*C*) was incubated with serum from the same patient following absorption with CB3 (*B*)

Table 1. Number of patients with IgG antibodies to normal human cardiac antigens, with IgG antibodies to coxsackievirus B3, CB3 antibodies which cross-react with cardiac antigens, and IgG antibodies which recognize heart tissue (values are the number of patients with the respective antibodies over the number tested)

	Active myocarditis	Borderline myocarditis	IDC	Ischemia
Heart extract reactive	8/10	5/9	8/9	8/10
CB3 reactive	6/10	4/9	2/9	1/10
Cross-reactive (IgG eluted from CB3)	6/6	3/4	2/2	1/1
Heart tissue reactive	8/10	5/9	7/9	7/9

Table 2. Number of patients with IgG antibodies to each of four principal coxsackievirus B3 proteins (values are the number of patients with the respective antibodies over the number tested)

Protein	kD[a]	Active myocarditis	Borderline myocarditis	IDC	Ischemia
VPO	43	4/6	2/4	1/2	0/1
VP1	35	1/6	0/4	1/2	0/1
VP2/3	30 and 26	1/6	2/4	0/2	0/1
VP4	7	0/6	0/4	0/2	1/1

Reactivity with viral proteins VP2 and VP3 was indistinguishable and is indicated as VP2/3.

Table 3. Mean number of cardiac antigens recognized by IgG antibodies in serum from patients with active myocarditis, borderline myocarditis, IDC, or ischemia

	Active myocarditis (n = 8)	Borderline myocarditis (n = 5)	IDC (n = 8)	Ischemia (n = 8)
Serum	3.3 ± 1.7	1.4 ± 0.8	2.3 ± 1.2	2.6 ± 1.2
Eluate	2.5 ± 2.1	0.6 ± 0.8	0.8 ± 1.6	0.1 ± 0.3
t	0.778	1.581	2.089	5.601
p	NS	NS	$0.1 > p > 0.05$	0.001

Serum and IgG eluted from CB3 proteins (Eluate) were tested on a normal human heart extract. The mean values were compared by Student's t test.

IgG antibodies eluted from CB3 proteins reacted with normal heart constituents (Table 1). Except for one borderline myocarditis patient with anti-CB3 antibodies that did not cross-react with heart, all CB3 eluates reacted with normal heart constituents (Table 1). For all patient groups, the mean number of cardiac antigens recognized by the eluted IgG was markedly less than the mean number of antigens recognized by unabsorbed serum (Table 3). However, this reduction in the number of recognized antigens was significant ($p < 0.001$) only in the case of the ischemic patients.

The molecular weights of the cardiac antigens recognized by IgG eluted from the CB3 proteins were determined. IgG eluted from VP0 reacted with six cardiac antigens, those eluted from VP1 with five antigens, and those

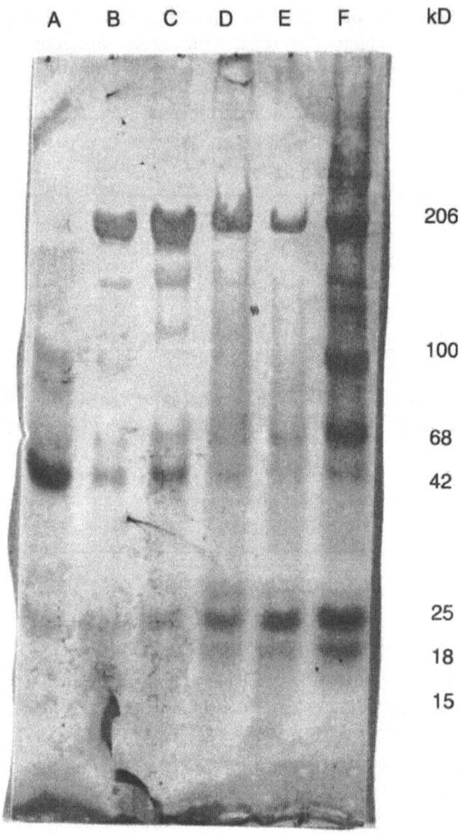

Fig. 2. Representative western blot demonstrating IgG reactivity of serum from a myocarditis patient with each of six sequentially prepared heart extracts

eluted from VP2/3 with four antigens. Cardiac antigens of 166 and 50 kDa, 130 and 112 kDa, and 153 and 80 kDa were uniquely recognized by IgG eluted from VP0, VP1, and VP2/3, respectively. Reactivity with cardiac antigens of 235, 98, 53, and 45 kDa was shared by eluates from two or more of the viral antigens.

Heart Tissue Reactivity

The incidence of heart tissue IgG reactivity, as determined by IFA among each of the patient groups, was similar to that observed for extract reactivity (Table 1). The incidence was lowest (56%) among the borderline myocarditis patients and was virtually identical (77%–80%) among each of the remaining groups. The mean IgG titers were 44, 10, 22, and 26 for patients with active myocarditis, borderline myocarditis, IDC, and ischemia, respectively. The mean titers for each of the patients groups did not differ significantly ($p > 0.05$) by t test. The modal titers were 10, 0, and 40 for the active and borderline myocarditis and the IDC patients, respectively.

Of the ten patients from all four groups who had no tissue-reactive IgG, seven had extract reactivity and four had CB3 reactivity; CB3 eluates from

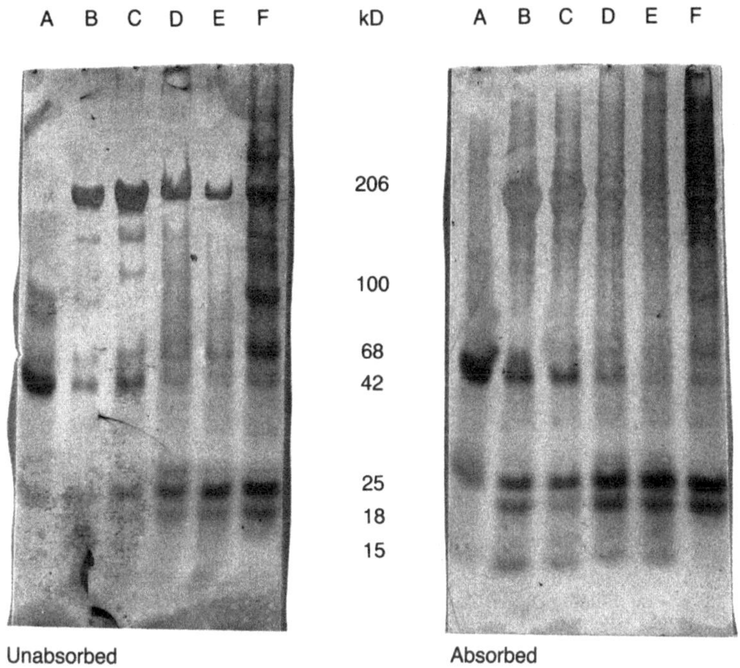

Fig. 3. Representative western blot demonstrating IgG reactivity of serum from a myocarditis patient with each of six sequentially prepared heart extracts before and after absorption with normal human skeletal muscle

each of these four were cross-reactive with heart tissue. Of the nine patients from all four groups who had no extract reactivity, six had heart tissue reactivity, and three had CB3 reactivity.

IgG Reactivity with Sequential Heart Extracts

Serum samples from seven active myocarditis patients were simultaneously tested for IgG reactivity with six sequentially prepared heart extracts. All samples exhibited multiple bands of reactivity with each of the six extracts (Fig. 2). For each extract there appeared to be distinct clusters of cardiac antigen recognition. IgG reacted strongly with cardiac antigens of ca. 25 and 55 kDa in all six extracts. Cardiac antigens of ca. 65 kDa in the high-salt, Triton, CHAPS, and SDS extracts were commonly recognized by IgG from these patients. Antigens of ca. 85 kDa in the low-salt extract, ca. 115 and ca. 195 kDa in the high-salt with pyrophosphate extract, and ca. 75 and ca. 95 kDa in the SDS extract were frequently recognized as well.

Since much of this reactivity is likely attributable to antibodies which cross-react with both cardiac and skeletal muscle antigens, serum from two patients was absorbed with an acetone extract of skeletal muscle prior to testing for IgG reactivity with the sequential heart extracts. For all six extracts, absorption with skeletal muscle eliminated or reduced reactivity with numerous cardiac antigens, particularly those with molecular weights greater than 80 kDa (Fig. 3). Reactivity with lower molecular weight antigens was unchanged or enhanced following absorption with skeletal muscle.

Discussion

These results confirm that heart-reactive IgG antibodies are common in patients with heart disorders of diverse etiologies. Significantly, anti-CB3 antibodies occur in 44% of borderline and 60% of active myocarditis patients but in only 22% of IDC patients and 10% of ischemic patients. This provides additional evidence linking CB3 infection to inflammatory heart disease and supports the concept of a progressive relationship between myocarditis and IDC. The high incidence of heart- but not CB3-reactive antibodies among ischemic patients supports the hypothesis that autoantibodies associated with ischemia likely result from tissue injury without virus involvement.

IgG antibodies which recognize CB3 are largely cross-reactive with a number of cardiac antigens. The relationship between virus infection and the induction of autoimmunity is of considerable interest to immunologists (e.g., [28]). In a murine model of CB3-induced autoimmune myocarditis, Neu et al. [17] reported that absorption of myosin-specific autoantibodies did not abolish CB3 neutralizing antibody activity. Since the ability of an antibody to recognize denatured viral protein may be distinct from its ability to neutralize intact virus, these results do not imply that protective antibodies are responsible for the observed cross-reactivity.

The evidence presented here suggests that antibodies to the major CB3 proteins cross-react with heart antigens, although little of this reactivity appears directed against the myosin heavy chain (ca. 200 kDa). Cross-reactivity with heart constituents of 45 and 50 kDa are suggestive of reactivity with BCKD (46- and 54-kDa subunits) or with calcium channel molecules (52-kDa subunit), two cardiac antigens associated with human myocarditis and IDC [3, 14, 30].

IgG reactivity with these lower molecular weight moieties is retained following the absorption of serum from myocarditis patients with skeletal muscle. The diminution of reactivity with cardiac antigens of greater than 80 kDa suggests that much of the reactivity directed against these larger molecules is attributable to antibodies that cross-react with skeletal muscle and thus may have little diagnostic relevance in humans.

The ability of CB3-reactive IgG antibody to cross-react with multiple heart constituents was unexpected, although Prabhakar et al. [21] indicate that molecular mimicry between infectious agents and self-antigens is common. Examination of nucleic acid and amino acid sequence data has suggested that few homologies exist between CB3 and ANT or myosin [27]. Although the identities of the cross-reactive cardiac antigens are unknown, this activity may reflect recognition of coiled coil epitopes shared by multiple heart constituents [8, 9, 11]. An alternative hypothesis which remains to be tested is that some or all of the cross-reactivity is due to nonspecific IgG binding attributable to Fc or lectin binding mechanisms.

These results indicate that IgG antibodies to CB3 from the majority of myocarditis and IDC patients cross-react with cardiac antigens. Thus heart-reactive antibodies may be elicited by at least two mechanisms: epitopes shared by both virus antigens and self-antigens or by release of self-antigens as a consequence of virus-induced tissue injury. A third possibility is that anti-idiotypic antibodies elicited to antivirus antibodies react with self-antigens [20]. Regardless of the mechanism, a key question which has yet to be addressed is what sustains immune recognition of the myocardium? Virus antigens are generally undetectable in myocardial tissue [16], and in situ hybridization data suggest that only small amounts of viral genomic material are distributed diffusely through the myocardium subsequent to infection [7, 13]. Moreover, myocyte necrosis may be rare and focal in borderline myocarditis and nonexistent in IDC. Thus, while our ability to describe the immunological processes in this spectrum of disease increases and potentially leads to improved noninvasive diagnostic procedures, the mechanisms which elicit immune recognition of the myocardium remain to be elucidated.

References

1. Abelmann, WH (1988) The etiology, pathogenesis and pathophysiology of dilated cardiomyopathies. In: Schultheiss H-P (ed) New concepts in viral heart disease. Springer, Berlin Heidelberg New York, pp 3–21
2. Alvarez FL, Neu N, Rose NR, Craig SW, Beisel KW (1987) Heart-specific autoantibodies induced by coxsackievirus B3: identification of heart autoantigens. Clin Immunol Immunopathol 43:129–139
3. Ansari AA, Herskowitz A, Danner DJ (1988) Identification of mitochondrial proteins that serve as targets for autoimmunity in human dilated cardiomyopathy. Circulation 78 [Suppl]:457
4. Archard L, Freeke C, Richardson P, Meany B, Olsen E, Morgan-Capner P, Rose M, Taylor P, Banner N, Yacoub M, Bowles N (1988) Persistence of enterovirus RNA in dilated cardiomyopathy: a progression from myocarditis. In: Schultheiss H-P (ed) New concepts in viral heart disease. Springer, Berlin Heidelberg New York, pp 349–362
5. Aretz HT, Billingham ME, Edwards WD, Factor SM, Fallon JT, Fenoglio JJ Jr, Olsen EGJ, Schoen FJ (1986) Myocarditis, a histopathologic definition and classification. Am J Cardiovasc Pathol 1:3–14
6. Billingham ME, Tazelaar HD (1986) The morphological progression of viral myocarditis. Postgrad Med J 62:581–584
7. Bowles NE, Richardson PJ, Olsen EGJ, Archard LC (1986) Detection of coxsackie-B-virus-specific RNA sequences in myocardial biopsy samples from patients with myocarditis and cardiomyopathy. Lancet 1:1120–1123
8. Cunningham MW, McCormack JM, Fenderson PG, Ho MK, Beachy EH, Dale JB (1989) Human and murine antibodies cross-reactive with streptococcal M protein and myosin recognize the sequence GLN-LYS-SER-LYS-GLN in M protein. J Immunol 143:2677–2683
9. Fenderson PG, Fischetti VA, Cunningham MW (1989) Tropomyosin shares immunologic epitopes with group A streptococcal M proteins. J Immunol 142:2475–2481
10. Gauntt CJ, Trousdale MD, Lee JC, Paque RE (1983) Preliminary characterization of coxsackievirus B3 temperature-sensitive mutants. J Virol 45:1037–1047
11. Gulizia JM, Cunningham MW, McManus BM (1991) Immunoreactivity of antistreptococcal monoclonal antibodies to human heart valves, evidence for multiple cross-reactive epitopes. Am J Pathol 138:285–301
12. Huber SA, Heintz N, Tracy R (1988) Coxsackievirus B-3-induced myocarditis, virus and

actinomycin D treatment induces novel antigens recognized by cytolytic T lymphocytes. J Immunol 141:3214–3219
13. Kandolf R, Ameis D, Kirschner P, Canu A, Hofschneider PH (1987) In situ detection of enteroviral genomes in myocardial cells by nucleic acid hybridization: an approach to the diagnosis of viral heart disease. Proc Natl Acad Sci USA 84:6272–6276
14. Limas CJ, Goldenberg IF, Limas C (1989) Autoantibodies against beta-adrenoceptors in human idiopathic dilated cardiomyopathy. Circ Res 64:97–103
15. MacArthur CGC, Tarin G, Goodwin JF, Hallidie-Smith KA (1984) The relationship of myocarditis to dilated cardiomyopathy. Eur Heart J 5:1023–1035
16. Morgan-Capner P, Richardson PJ, McSorley C, Daly K, Pattison JR (1984) Virus investigations in heart muscle disease In: Bolte H-D (ed) Viral heart disease. Springer, Berlin Heidelberg New York, pp 99–115
17. Neu N, Craig SW, Rose NR, Alvarez FL, Beisel KW (1987) Coxsackievirus induced myocarditis in mice: cardiac myosin autoantibodies do not cross-react with the virus. Clin Exp Immunol 69:566–574
18. Neumann DA, Burek CL, Baughman KL, Rose NR, Herskowitz A (1990) Circulating heart-reactive antibodies in patients with myocarditis or cardiomyopathy. J Am Coll Cardiol 16:839–846
19. Paque RE, Gauntt CJ, Nealon TJ, Trousdale MD (1978) Assessment of cell-mediated hypersensitivity against coxsackievirus B3 viral-induced myocarditis utilizing hypertonic salt extracts of cardiac tissue. J Immunol 120:1672–1678
20. Paque RE, Miller R (1991) Autoanti-idiotypes exhibit mimicry of myocyte antigens in virus-induced myocarditis. J Virol 65:16–22
21. Prabhakar BS, Srinivasappa J, Beisel KW, Notkins AL (1988) Virus-induced autoimmunity: cross-reactivity of antiviral antibodies with self-components. In: Schultheiss H-P (ed) New concepts in viral heart disease. Springer, Berlin Heidelberg New York, pp 168–177
22. Putnak JR, Phillips BA (1981) Picornaviral structure and assembly. Microbiol Rev 45:287–315
23. Quigley PJ, Richardson PJ, Meany BT, Olsen EGJ, Monahan MJ, Jackson G, Jewitt JE (1987) Long-term follow-up of acute myocarditis, correlation of ventricular function and outcome. Eur Heart J 8 [Suppl J]:39–42
24. Regitz V, Strasser R, Chmielewski G, Blumcke S, Fleck E (1988) Myocarditis in patients with dilated cardiomyopathy – correlation with clinical, hemodynamic, and biochemical findings. In: Schultheiss H-P (ed) New concepts in viral heart disease. Springer, Berlin Heidelberg New York, pp 51–60
25. Schultheiss H-P, Bolte H-D (1985) Immunological analysis of autoantibodies against the adenine nucleotide translocator in dilated cardiomyopathy. J Mol Cell Cardiol 17:603–617
26. Schultheiss H-P (1989) The significance of autoantibodies against the ADP/ATP carrier for the pathogenesis of myocarditis and dilated cardiomyopathy – clinical and experimental data. Springer Semin Immunopathol 11:15–30
27. Schwimmbeck PL, Bland NK, Schultheiss H-P, Strauer BE (1991) The possible value of synthethic peptides in the diagnosis and therapy of myocarditis and dilated cardiomyopathy. Eur Heart J 12 [Suppl D]:76–80
28. Shoenfeld Y, Isenberg D (1989) The mosaic of autoimmunity. Elsevier, Amsterdam
29. Tanigawa G, Jarcho JA, Kass S, Solomon SD, Vosberg H-P, Seidman JG, Seidman CE (1990) A molecular basis for familial hypertrophic cardiomyopathy: an alpha/beta cardiac myosin heavy chain hybrid gene. Cell 62:991–998
30. Ulrich G, Kuhl U, Melzner B, Janda I, Schafer B, Schultheiss H-P (1988) Antibodies against the adenosine di-/triphosphate carrier cross-react with the Ca channel – functional and biochemical data. In: Schultheiss H-P (ed) New concepts in viral heart disease. Springer, Berlin Heidelberg New York, pp 225–235
31. Wolfgram LJ, Beisel KW, Herskowitz A, Rose NR (1986) Variations in the susceptibility to coxsackievirus B3-induced myocarditis among different strains of mice. J Immunol 136:1846–1852
32. Woodruff JF (1980) Viral myocarditis, a review. Am J Pathol 101:427–479

Are Anti-cardiac Antibodies in Dilated Cardiomyopathy "Natural" Antibodies?

B. Maisch, E. Bauer, L. Drude, M. Herzum, G. Hufnagel, U. Schönian, and S. Pankuweit

Introduction

The etiology and pathogenesis of myocarditis, postmyocarditic heart muscle disease, and dilated cardiomyopathy (DC) are still matters of controversy and have remained partly unresolved. Not only enterovirus [1, 11], but also cytomegalovirus [36, 40], herpes virus (unpublished), and rickettsia [14] or borrelia induced [12] infection and persistence as evidenced by in situ hybridization, slot blot, conventional light or electron microscopy have been made responsible for the early and/or for the late alterations in cardiac ultrastructure and hemodynamic performance in acute myocardial inflammation and its sequelae. As derived from animal experiments of murine myocarditis [7–9, 37] and in man cellular [14, 17, 19–22, 27, 32, 35] and humoral autoreactive processes [4–7, 13–35, 39, 41, 42, 45, 46] have been demonstrated to be operative in inflammatory and dilated heart muscle diseases. We were the first to demonstrate cross-reactive, complement-dependent cardiocytolytic anti-sarcolemmal (AsAs) and anti-myolemmal antibodies (AMLAs) in coxsackievirus B myocarditis [23, 35] and recently demonstrated that certain sarcolemmal membrane epitopes cross-react, for example, the 110-, 48-, 35-, 31-, and 28-kDa proteins cross-react with enteroviral proteins [23] in western blot experiments. Apart from the cross-reactive, cardio- and virus-specific antibodies were also described.

Not only in viral but also in bacterial forms of inflammatory heart disease [4, 14, 29, 32, 45], in postpericardiotomy, and postcardiac injury syndromes (see references in [20]) anti-membrane antibodies were found to be either cytolytic to living heart cells at rest [18, 20, 23, 27–29, 31, 32, 34, 35] or during stimulation [6] or spontaneous activity [13] in vitro, thus underlining the potential pathophysiological impact of their presence. Obviously not all membrane antibodies secreted possess a cytolytic function. Positive chronotropic, that is, stimulating, anti-beta-receptor antibodies have been described [46] as well, and protective antibodies have been discovered both in myocarditis and different forms of rejection (unpublished). In the course of a polyclonal immune response inert, protective, or innocent bystander antibodies may be observed, whose functions – if

Department of Internal Medicine – Cardiology, Philipps-University Marburg, W-3550 Marburg, Fed. Rep. of Germany

Figulla et al. (Eds.)
Idiopathic Dilated Cardiomyopathy

present – remain to be defined, as must be done for their cardiospecificity or cardioselectivity.

Until the late 1970s autoantibodies were considered unwanted and a highly pathological phenomenon. Jerne called B-cells and plasma cells producing autoantibodies "forbidden clones" in the framework of his network theory [10]. Even earlier, Ehrlich described them as "horror autotoxicus." Meanwhile both in systemic autoimmune disorders and in normal healthy individuals autoantibodies to different constituents of normal cells and matrix constituents have been demonstrated and referred to as natural antibodies [20, 28, 43, 44]. This may also apply to autoantibodies to the cardiac myocyte [15, 34, 35], conducting tissue [30], cytoskeleton [20, 38], connective tissue [20, 38], extracellular matrix, endothelial cells, and fibroblasts in different heart muscle [22–35] and pericardial diseases. Each part of the cardiac tissue can be a target of cellular and humoral immune reactions [17–20].

We have come to realize that autoantibodies and/or natural antibodies [21, 28, 43, 44] and autoreactive T-cells play a part in the immunological "network" that regulates the interaction of lymphocytes, immune regulation, T-cell traffic, "homing" of cells, and the effector mechanisms. In any one of these conditions not only the cardiac myocyte, endothelial cells, and fibroblasts but also the extracellular matrix are exposed to immunocompetent cells. Myocytes, fibroblasts, and vascular endothelium may be partially altered in inflammatory heart muscle diseases (myocarditis, endocarditis, and rejection after heart transplantation), in ischemic heart disease and in the cardiomyopathies, be it the dilated [19, 20, 27] or the hypertrophic form [22].

Disease-induced changes from toxic viral or bacterial assaults from ischemia or during a reaction episode predispose the heart to be an excellent target for recognition by antigen-presenting or other immunocompetent cells or to the response by humoral immunological effector organs. These may be monitored in the myocardium itself by endomyocardial biopsy [19, 26]. Its fingerprints can be detected even years later in the peripheral blood as circulation autoantibodies. Their incidence in different forms of heart disease has recently been reviewed [20].

From this antagonism the question therefore arises whether the anticardiac antibodies in dilated heart muscle disease are only "natural antibodies," or whether they are truly cardiospecific, functionally relevant immunoglobulins. From recent work in animal models certain characteristics of natural antibodies have been defined. (a) They have specificity for highly conserved structures, for example, xanthinoxidase, myelin basic protein, collagen, DNA, cytoplasmatic filaments, cytoskeleton, thyreoglobulin, myoglobin, sperm proteins, albumin, neural tissue or the Fc fragment of IgG. (b) They increase with age. (c) They may possess cross-reactivity because their respective autoantigens are ubiquitous. Their immune reaction is polyclonal and polyspecific. (d) In experimental animals they are also found in germ-free mice. It is therefore unlikely that they occur primarily as a result of stimulation by bacterial antigens.

It was the purpose of the present investigation to reexamine in primary and secondary dilated heart disease (SDHD) the role of anti-membrane antibodies, ASAs, and AMLAs, in particular with respect to characteristics of natural antibodies.

Materials and Methods

Patients

From our 2100 patients who underwent endomyocardial biopsy over a 10-year-period we reexamined immunohistochemical and immunoserological data from patients with status post-myocarditis ($n = 22$) and status post-perimyocarditis ($n = 15$) without cardiomegaly and 28 patients with post-myocarditic heart muscle disease and 50 patients with idiopathic dilated cardiomyopathy (DC). They were compared with 10 adult patients with coxsackievirus B3 and B4 myocarditis (eight men, two women; mean age 33.8 ± 12.4 years; histologically confirmed as active myocarditis according to the Dallas criteria [42], 10 patients with biopsy-confirmed active myocarditis of unknown origin (seven men, three women; mean age 43 ± 14.2 years), 10 patients with coxsackievirus B3 and B4 myocarditis (eight men, two women; mean age 33.8 ± 12.4 years), and 10 patients with active myocarditis of unknown origin, 50 adult patients with perimyocardits (31 men, 29 women; mean age 44.2 ± 18.7 years, with pericardial effusion and segmental wall motion abnormality or cardiomegaly in levocardiography.

Diagnostic Criteria

DC was defined according to the WHO-ISFC task force as dilated heart muscle disease of unknown cause. Therefore coronary artery disease and extracardiac, valvular, and inflammatory heart muscle disease had to be excluded by coronary angiography, left ventricular cineangiocardiography, and by endomyocardial biopsy. Histological diagnosis of biopsy confirmed myocarditis was based on the Dallas criteria [2]. The essential criteria for definite perimyocarditis diagnosed on clinical grounds included pericardial rubs or effusion with segmental wall motion abnormality by levocardiography and/or two-dimensional echocardiography after exclusion of coronary artery disease by cineangiography [16, 18, 21]. Postmyocarditic dilated SHMD was ascertained when histologically myocarditis had healed (second biopsy) in a dilated heart [2].

Light microscopical investigations, including routine hematoxylin-eosin staining, were carried out at the pathological-anatomical institutes of the respective contributing centers of a multicenter study (Prof. Thomas, Marburg; Prof. E. Olsen, London; Prof. H.K. Müller-Hermelink, Würzburg; Prof. W. Hort, Düsseldorf;), and in our own laboratories in Würzburg and in Marburg.

For immunohistochemistry of endomyocardial biopsies the direct technique as described previously was used [19, 26, 27, 31–33, 35]. In addition, a double-sandwich technique was used for the demonstration of circulating anti-heart antibodies to bind on the same cryostat section at the same antibody binding site using TRITC- and FITC-labeled antibodies [$F(ab)_2$ fragments, dilution 1:100; Medac] [19]. Furthermore binding of the monoclonal antibody directed against the membrane attack complex (C_{5b9}) was analyzed (dilution 1:40) by a peroxidase-coupled second antibody (rabbit anti-mouse IgG) [28].

Circulating anti-heart and non-organ-specific antibodies were determined on cryostat sections of human, rat, and bovine heart and skeletal muscle, thyroid gland, stomach, kidney, and liver [23, 28, 29, 31–35]. Circulating heterologous AMLAs were determined on intact rat cardiocytes isolated by Percoll gradient centrifugation after perfusion with a collagenase (Worthington)-Ringer perfusate [27–29, 35]. Circulating homologous AMLAs were assessed on intact myocytes from human atrial appendages removed during open heart surgery, minced, and incubated in a calcium-free Ringer collagenase solution [composition according to medium 2 [22]] at 37°C, sedimented by centrifugation (80*g*) after addition of EDTA. Fixation was accomplished by acetone added to the Ringer solution in the same volume after centrifugation [28].

The pathogenetic relevance of AMLAs, ASAs, and anti-endothelial antibodies (AEAs) was measured in a microcytotoxicity assay using vital cardiocytes as target cells. Antibody-mediated cytolysis in the presence of complement was assessed by an index comparing the half-lives of cardiocytes in the presence of the patient's serum and with control sera (pool). Indices under 0.75 indicated significant cardiocytolysis [23, 27, 35].

Biochemical and western blot studies were carried out in 28 selected patients with and without AMLAs and ASAs out of the cohort of patients presented above to further characterize the relevant binding sites of membrane antibodies in DC (Maisch et al., submitted) and 10 patients with coxsackievirus B myocarditis [23].

The sarcolemmal vesicles of the heart of Sprague-Dawley rats (150–200 g) immersed in iced TSE medium (sucrose 0.25 *M*, Tris/HCl 20 m*M*. EGTA 1 m*M*, at pH 7.5) and frozen at −78°C, and of the heart of a histopathologically investigated 17-year-old patient (blood group 0) who died after a spinal operation, were isolated according Paris et al. [39]. Electron microscopic evaluation of the final pellet demonstrated highly enriched membrane vesicles with less than 5% contaminants. Endoplasmatic reticulum and mitochondria were ruled out enzymatically [23, 45].

Sodium dodecyl sulfate polyacrylamide gel (12.5%) electrophoresis of cardiac membrane proteins and of coxsackievirus B was performed according to standard procedures. The proteins were transferred quantitatively from the polyacrylamide gel to a nitrocellulose sheet [18, 23, 45]. The unstained electrophoretic blots were soaked in 5% bovine serum albumin in PBS buffer (10.15 *M* NaCl, 10 m*M* Na_2HPO_4, 2.5 m*M* KCl, 1.5 m*M* KH_2PO_4, 0.3% Tween 20, pH7.4) for 16 h at 0°C to saturate nonspecific binding sites and then rinsed twice in PBS. Each sheet was incubated for 2 h with 100 μl patient serum diluted 1:50 in 5 ml PBS containing 1% bovine serum albumin. Alternatively, peroxidase-labeled lectins, HLA and DR antibodies, the previously described monoclonal and monospecific antibodies were tested. The sheets were then washed five times in PBS and incubated with goat anti-human $F(ab)_2$-peroxidase conjugate (Medac), ratio 1:500 in PBS, containing 1% bovine serum albumin. The sheets were rinsed ten times with PBS. For color reaction, the blots were soaked in a freshly prepared, filtered solution of 6 mg 3-amino-9-ethyl-carbazol (Sigma), 7.5 ml DMF (Merck), 0.1 ml HCl 1 *M*, and 0.15 ml 30% H_2O_2 in 150 ml destilled water at pH 5.5–6.0. The reaction was terminated after 5–10 min by washing with water. The blots were dried and stored protected from light.

To account for minor background staining by the human sera, standard control sera were used in each analysis. Positive antibody binding to one of the bands was accepted only when, compared to the control sera, new bands appeared or at least a twofold intensity of immunoglobulin binding was observed.

Results and Discussion

Pro: AMLAs and ASAs Could Be Natural Antibodies in Dilated Cardiomyopathy

It can be demonstrated that the cardiac sarcolemma and myolemma are highly conserved structures. The sarcolemma and myolemma are first-line autoantigens in dilated heart muscle disease, but they are only part of the anatomical constitutents of heart tissue that could give rise to antibody formation. Figure 1 gives a schematic view of all constitutents to which anti-

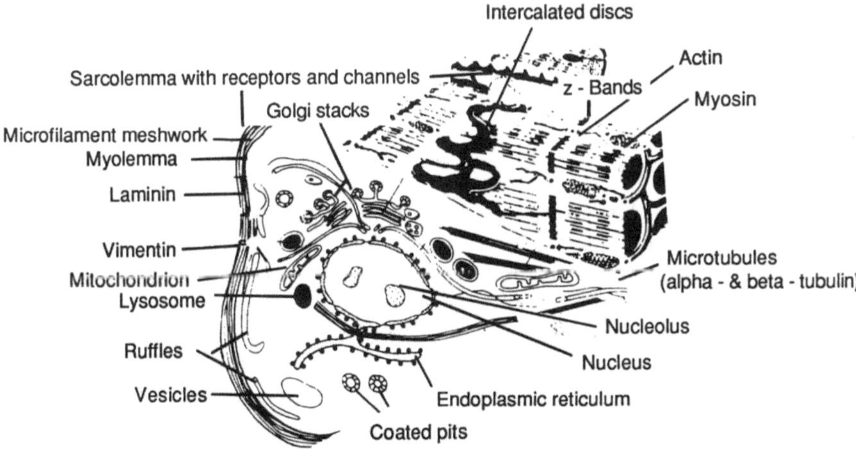

Fig. 1. Cardiac autoantigens. The cardiac sarcolemma and myolemma are highly conserved structures, but only one of the possible cardiac autoantigens

Table 1. Immunohistochemistry in endomyocardial biopsies

Clinical diagnosis	n	Trivalent	IgG	IgM	IgA	C3	C3 or IgM
Myocarditis (active/acute)	20	100[a,b]	90[a,b]	55[a,b]	70[a,b]	70[a,b]	85[a,b]
Perimyocarditis (active/acute)	50	90	90	75	85	80	90
Status postmyocarditis (no cardiomegaly)	22	100	95	32	32	36	45
Status postperimyocarditis (no cardiomegaly)	15	73[a]	60[a]	13	7	33[a]	40[a]
Postmyocarditic SHMD[c] (cardiomegaly)	28	79[a]	75[a]	18	36[a]	61[a,b]	75[a]
Dilated cardiomyopaphy (idiopathic)	50	60[a]	56[a]	48[a,b]	8	12	48[a]
Noncardiac controls		12	12	0	0	0	0
Coronary artery disease	100	43[a]	41[a]	11	20	3	14

[a] $p > 0.05$ by $\hat{\chi}^2$ analysis when compared to noncardiac controls.
[b] $p > 0.05$ by $\hat{\chi}^2$ analysis when compared to coronary artery disease.
[c] Secondary heart muscle disease.

cardiac antibodies can be directed. Table 1 gives the incidence of anti-heart antibodies in the endomyocardial biopsies of patients with DC and SHMD in comparison to healthy controls and patients with active myocarditis. Table 2 summarizes the incidence of circulating antibodies found in patients with DC, myocarditis, and perimyocarditis and in healthy controls from our own patient cohorts and from the literature.

AMLAs are characterized by a linear membrane fluorescence with isolated human myocytes (Fig. 2). The sarcolemmal structure is less easily identified on biopsies, since antibodies against collagen, interstitial tissue, and the extracellular matrix may show similar immunofluorescence patterns (Fig. 2e) both when bound ASAs are analyzed (Fig. 2e, Table 1), or circulating AMLAs and ASAs are evaluated (Table 2). Anti-myosin (Fig. 2c), anti-actin (Fig. 2d), and anti-interfibrillary (anti-mitochondrial or anti-sarcoplasmic) antibodies were also demonstrated in various degrees in DC and myocarditis (Table 2).

Immunohistochemistry

It is clear from the data in Table 1 and from Fig. 3 that bound ASAs of the IgM, IgA, and IgG class (Fig. 2e) are diagnostic serological markers in almost 90% of patients with DC. As in myocarditis, bound IgM antibodies are not infrequent in up to 50% of cases and are interpreted as indicators of a recent immune response. Complement fixation (14%–18%), in addition – as in active inflammatory heart disease – is indicative of a secondary immunopathogenesis operating in the individual patients. Extracellular matrix antibodies to primarily directed collagen constituents and fibronectin are found to a similar degree, whereas bound AEAs are present less frequently (Table 1). As can be seen in Tables 1 and 2, not only do individuals with inflammatory heart disease or DC possess these antibodies in the myocardial tissue but are also found, to a much lesser extent in healthy controls.

Circulating Antimembrane Antibodies

In DC the incidence of circulating ASAs was 31%–42% and of circulating AMLAs (Fig. 2a) 33%–42%, whereas their incidence in myocarditis of children and adults was much higher. One can conclude that AMLAs are a good disease marker in all age groups of myocarditis (Table 2). In the selected group of 28 patients with DC, 21 demonstrated homologous and heterologous AMLAs by trivalent staining. AMLAs of the IgG isotype were found in 14, of the IgM isotype in 21, and of the IgA isotope in 4 cases. Complement fixation was demonstrated in 4 patients. ASAs on homologous blood group 0 cryostat sections were positive in 24 patients (IgG 15; IgM 21; IgA 4, C3 fixation 5 patients each).

When analyzing possible subgroups of anti-membrane antibodies such as anti-laminin antibodies (Fig. 2b), one can conclude that the latter may be part of the microheterogeneity of ASAs or AMLAs but are – at least in our patients – found less frequently (35%) than positive ASAs or AMLAs (Table 2).

Anti-myosin antibodies were found in 33% [16, 19, 20, 25], increasing with the severity of heart failure (Fig. 2c). This also applies to hypertrophic cardiomyopathy [22]. Anti-actin antibodies (Fig. 2d) we are found rarely in myocarditis or its sequelae. It appears unlikely that the antibodies directed to intracellular constituents possess a primary pathogenetic pathogenetic relevance. They might best be explained either as natural antibodies or as the consequence of a limited accessibility of antigen-presenting cells and effector cells to necrotic heart cells. Remarkably, however, they do not occur to a significant extent in patients after acute myocardial infarction or with chronic angina pectoris, to which the latter situation might preferably apply.

Natural Antibodies Increase with Age

In healthy adults controls (n = 150) 31% had low-titer AMLAs and 35% demonstrated ASAs in the peripheral blood. The incidence of bound ASAs was around 20% of the IgG isotype in healthy controls (postmortem, non-cardiac death, n = 20). Analysis in healthy children demonstrated clearly that in healthy newborns neither ASAs nor AMLAs can be detected. They appeared after the first 6 months of age. Their incidence increased gradually. This property of anti-membrane antibodies fits well to the characteristics of natural autoantibodies.

Natural Antibodies Possess Extensive Cross-reactivity

Cross-reactivity is a key marker of ASAs and AMLAs. In coxsackievirus B induced myocarditis this can be clearly stated for several bands of sarcolemmal

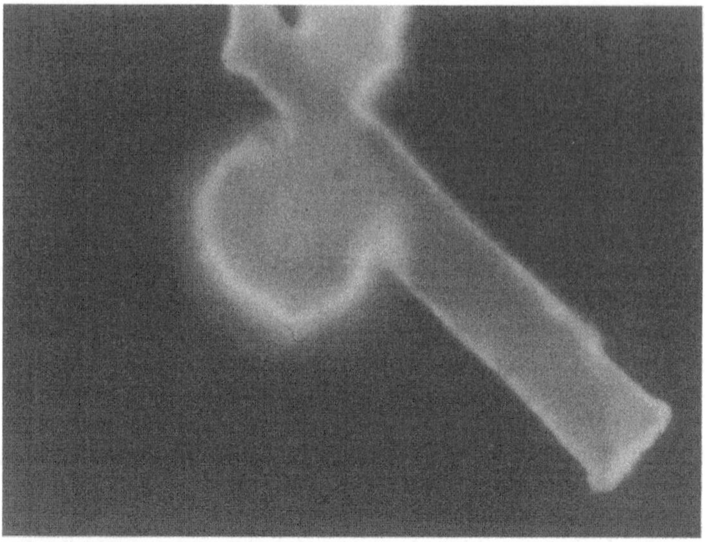

a

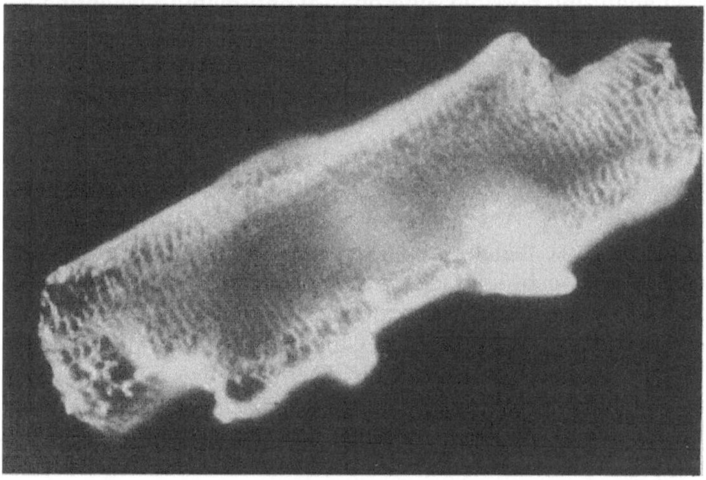

b

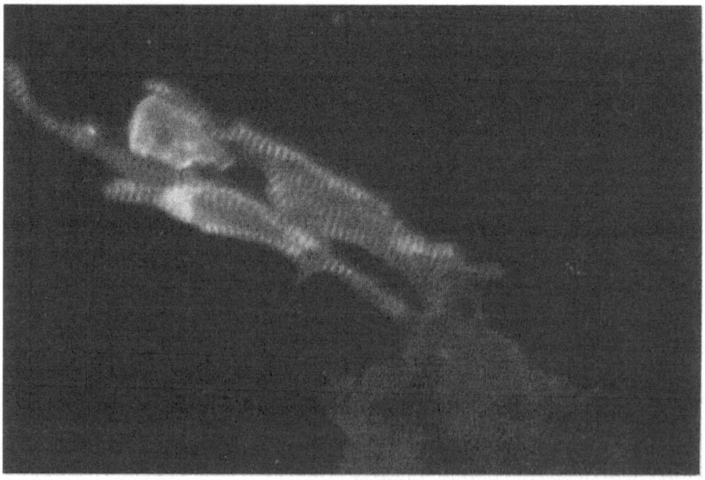

c

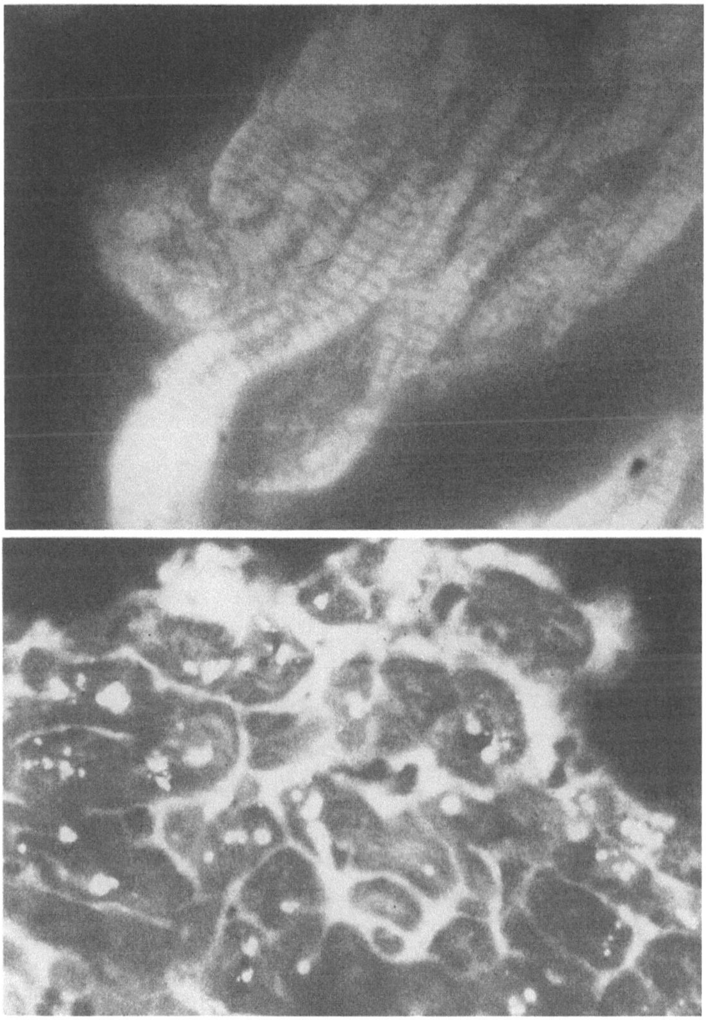

d

e

Fig. 2. a Circulating anti-myolemmal antibodies in a 27-year-old patient with dilated postmyocarditic heart muscle disease using a human adult myocyte as antigen (titer 1:160). **b** Anti-laminin antibodies were demonstrated as concomittant staining of sarcolemma and Z bands of isolated rat cardiocytes (titer 1:320) in a 31-year-old man with DC who had to undergo heart transplantation 16 months after initial diagnosis of heart failure. The anti-laminin ELISA was positive as well (from [20] with permission). **c** Antimyosin antibodies demonstrated with isolated human myocytes as characteristic A band (titer 1:40) in a 49-year-old patient with DC of NYHA class III (from [20] with permission). **d** Antiactin antibodies are found rarely in DC. In this 36-year-old woman with postmyocarditic heart muscle disease the titer reached 1:80 (from [20] with permission). **e** Bound antisarcolemmal antibodies in the endomyocardial biopsy of the same patient. The IgG, IgM, and IgA subclasses were positive

Table 2. Circulating antibodies to the sarcolemma, extracellular matrix, and intermediate filaments (percentage positive) in healthy controls, and in patients with dilated cardiomyopathy or (peri)myocarditis

Reference	*n*	AMLA (homol)	ASA (homol)	ALA	A-Fibron	Z-bands	A-Actin	A-Myosin	A-Tubulin	AIDA	A-Desmin	A-Vimentin	ANT	A-M7	AEA	A-BAR	A-Collagen			
																	I	III	IV	V
Healthy controls																				
Maisch [18, 20]	200	31	35	nd	nd	5	5	5	0	0	0	0	nd	nd	17	nd	nd	nd	nd	nd
Maisch et al. [23]	45	16	18	20	nd	nd	nd	nd	nd	nd	nd	nd	nd	nd	nd	nd	nd	nd	nd	nd
Obermayer et al. [38]	25	20	20	nd	nd	0	nd	nd	0	nd	4	0	nd	nd	nd	nd	nd	nd	nd	nd
de Scheerder et al. [5]	40	10	5	nd	nd	nd	5	3	nd	nd	nd	nd	nd	nd	12	nd	5	5	5	5
Klein et al. [12]	nd	nd	nd	nd	nd	nd	nd	nd	nd	nd	nd	nd	nd	0	nd	nd	nd	nd	nd	nd
Schultheiß and Bolte [42]	nd	nd	nd	nd	nd	nd	nd	nd	nd	nd	nd	nd	nd	nd	nd	nd	nd	nd	nd	nd
Dilated cardiomyopathy																				
Klein et al. [12]	nd	nd	nd	nd	nd	nd	nd	nd	nd	nd	nd	nd	nd	30	nd	nd	nd	nd	nd	nd
Maisch et al. [27]	79	9	10	nd	nd	nd	4	20	2	nd	nd	nd	nd	nd	13	nd	nd	nd	nd	nd
Maisch [17]	30	33	42	nd	nd	nd	10	33	2	nd	nd	nd	nd	nd	45	nd	nd	nd	nd	nd
Schultheiß and Bolte [42]	51	nd	nd	72	nd	nd	nd	nd	nd	nd	nd	nd	nd	nd	nd	nd	12	24	6	24
Obermayer et al. [38]	36	42	31	nd	nd	8	0	8	nd	nd	0	0	nd	nd	31	nd	nd	nd	nd	nd
Myocarditis and perimyocarditis																				
de Scheerder et al. [5]	12	100	12	nd	nd	nd	58	67	nd	nd	nd	nd	nd	nd	91	nd	35	40	35	35
Klein et al. [12]	nd	nd	nd	nd	nd	nd	nd	nd	nd	nd	nd	nd	nd	13	nd	nd	nd	nd	nd	nd
Maisch et al. [16, 35]																				
Viral myocarditis (adults)	44	79–90	75–90	nd	nd	15	7	10–50	0	0	nd	nd	nd	nd	80	nd	nd	nd	nd	nd
Idiopathic myocarditis	144	59	45	nd	nd	nd	0	23	9	0	nd	nd	nd	nd	40	nd	nd	nd	nd	nd
Maisch et al. in children [34]		100	100	nd	nd	0	0	0	0	0	nd	nd	nd	nd	91	nd	nd	nd	nd	nd
Maisch et al. [23]	132	nd	nd	30–35	nd	nd	nd	nd	nd	nd	nd	nd	nd	nd	nd	nd	nd	nd	nd	nd
Obermayer et al. [38]	25	64	72	nd	nd	16	0	4	0	nd	0	0	nd	nd	72	nd	nd	nd	nd	nd
Schultheiß and Bolte [42]	29	nd	nd	60	20	nd	nd	nd	nd	nd	nd	nd	nd	nd	nd	nd	30	40	35	35

AMLA, Anti-myolemmal antibody; ASA, anti-sarcolemmal antibody; ALA, anti-laminin antibody; A-, anti-; Fibron, fibronectin; AIDA, anti-intercalated disc antibody; ANT, anti-nucleotid translocator; AEA, anti-endothelial antibody; Bar, beta-receptor; nd, not done.

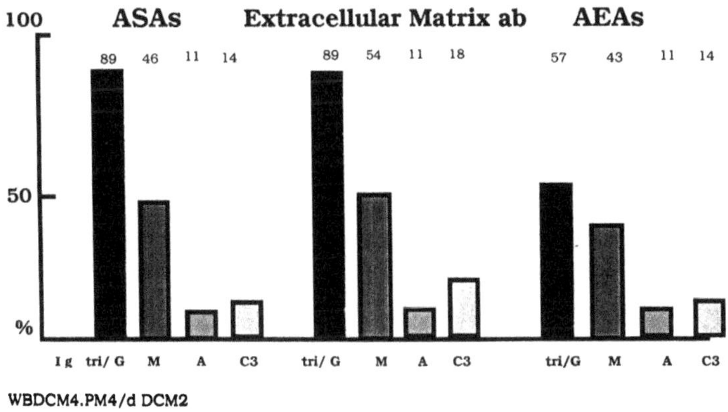

Fig. 3. Distribution of bound antibodies to the sarcolemma, endothelial cell, and the extracellular matrix in patients with dilated cardiomyopathy (n = 28). (Data taken from the Würzburg-Marburg Multicenter Trial [19]

proteins to which the patients' sera bind [23, 28]. In a quantitative analysis of ten patients with coxsackievirus B myocarditis cross-reactive antibodies fixed on the 220-, 110-, 48-, 35-, 31-, and 26-kDa bands. Virus-specific epitopes were detected as 33- and 34-kDa bands, whereas cardioselective epitopes were 90-, 78-, 72-, 67-, and 45-kDa bands [23]. Cross-reactivity to other membrane constituents was also possible. In the indirect test with human skeletal muscle tissue about one third of the sera that bound to the cardiac sarcolemma also fixed to the skeletal muscle sarcolemma.

In streptococcal disease, for example, rheumatic fever cross-reactivity between epitopes of the sarcolemma and the streptococcal membrane has become clear [4, 45]. In infective endocarditis induced by streptococci Thometzek et al. [45] also demonstrated cross-reactivity to a number of epitopes of the sarcolemma. In DC cross-reactivity to most of the enteroviral proteins could not be confirmed (unpublished) since the most relevant binding sites for ASAs and AMLAs appeared to be a 48- and 90-kDa proteins of the sarcolemma. Apart from these two cardinal proteins others are also found in up to 60% to bind circulating antibodies in DC. Preliminary analyses show that patients with little cardiac compromise (NYHA class II) more often have an IgM antibody directed to the sarcolemma which does not possess lytic activity in vitro to isolated myocytes. In contrast, DC patients with severe froms of congestive heart failure have, in addition, an IgG antibody which is often directed to the 90-kDa protein and may exert cytolytic properties (unpublished). From these data it can be concluded that ASAs and AMLAs in dilated and inflammatory heart muscle disease belong to a microheterogeneity of anti-membrane antibodies. Some ASAs and AMLAs are cross-reactive, and others are cardiospecific. Their pathogenetic relevance can be evaluated in vitro when they are incubated with vital resting [23, 27, 31, 32, 34, 35] or stimulated contracting myocytes [6].

Conclusions

These studies show that AMLAs and ASAs have specificity for conserved structures. In enteroviral myocarditis they possess extensive cross-reactivity to virus and possible to other tissues. They increase with age. These three characteristics argue for the role of AMLAs and ASAs as natural antibodies. As in myocarditis, ASAs and AMLAs may occur, however, as a result of bacterial (e.g., in infective endocarditis and rheumatic fever) and viral stimulation in coxsackievirus and influenza myocarditis [23, 35]. They are more often of the IgG or IgA than of the IgM isotype and fix complement. They exert a cytolytic and cytotoxic response in vitro that may well be operative in vivo. In DC cross-reactivity is less easily confirmed. A microheterogeneity of AMLAs and ASAs can also be demonstrated in conjunction with antibodies to mitochondrial [41, 42], myofibrillar [37], and cytoskeletal structures [20, 38] (Table 2). Recently antibodies to the beta-adrenoceptor have been described that may either inhibit [13] or stimulate [46] the receptor in vitro with isolated fetal heart cells.

The latter characteristics indicate that ASAs and AMLAs both in myocarditis and DC represent truly a microheterogeneity of pathogenetically relevant immunoglobulins occurring in the context of a polyclonal stimulation either after viral infection or other forms of postcardiac injury. It has now been possible to characterize their function in vitro with resting and stimulated isolated adult or fetal heart cells. This makes it very likely that some of the anti-cardiac antibodies have functional properties at least in vitro. It remains speculative but not unlikely that the functions assessed in vitro also represent their pathogenetic property in vivo.

References

1. Archard LC, Bowles NE, Cunningham L, Freeke CA, Olsen EGJ, Rose ML, Meany B, Why HJF, Richardson PJ (1991) Molecular probes for detection of persisting enterovirus infection of human heart and their prognostic value. Eur Heart J 12 [Suppl D]:56–59
2. Aretz HT, Bilingham ME, Edwards WD, Fallon JT, Fenoglio JJ Jr, Olsen EGJ, Schoen FJ (1987) Myocarditis-A pathologic definition and classification. Am J Cardiovasc Pathol 1:3–11
3. Boyden SV (1963) Natural antibodies and the immune response. Adv Immunol 5:1–7
4. Cunningham MW, Hall NK, Krisher KK, Spanier AM (1986) A study of anti-group a streptococcal monoclonal antibodies cross-reactive with myosin J Immunol 136:293–297
5. De Scheerder I, de Buyzere M, Algoed L, de Lange M, Delanghe J, Bogaert AM, Clement DL (1987) Characteristic anti-heart antibody patterns in postcardiac injury syndrome, endocarditis and acute myocarditis. Eur Heart J 8 [Suppl J]:237–238
6. Drude L, Wiemers F, Maisch B (1991) Impaired myocyte function in vitro incubated with sera from patients with myocarditis. Eur Heart J 12 [Suppl D]:36–38
7. Herzum M, Maisch B (1988) Humoral and cellular immune response in human myocarditis and dilated cardiomyopathy. Pathol Immunopathol Res 7:240–250
8. Huber SA, Lodge PA (1984) Coxsackie virus B3 myocarditis in BALB/c mice: evidence for autoimmunity to myocyte antigens. Am J Pathol 116:21–29
9. Huber SA, Lodge PA (1984) Coxsackie virus B3 myocarditis: identification of different pathogenic mechanisms in DBA/2 and BALB/c mice. Am J Pathol 116:21–29

10. Jerne NK (1955) The natural selection theory of antibody formation. Proc Natl Acad Sci USA 41:849–857
11. Kandolf R, Klingel K, Mertsching H, Canu A, Hohenadl C, Zell R, Reimann BY, Heim A, McManus BM, Foulis AK, Schultheiss HP, Erdmann BE, Riecker G (1991) Molecular studies on enteroviral heart disease: patterns of acute and persistent infections. Eur Heart J 12 [Suppl D]:49–55
12. Klein J, Stanek G, Bittner R, Horvat R, Holzinger C, Glogar D (1991) Lyme borreliosis as a cause of myocarditis and heart muscle disease. Eur Heart J 12 [Suppl D]:73–75
13. Limas CJ, Limas C (1991) Beta-adrenoceptor antibodies and genetics in dilated cardiomyopathy – an overview and review. Eur Heart J 12 [Suppl D]:175–177
14. Maisch B (1985) Rickettsial perimyocarditis – a follow up study. Heart Vessels 2:55–59
15. Maisch B (1985) Surface antigens of adult heart cells and their use in diagnosis. Basic Res Cardiol 80 [Suppl 1]:47–52
16. Maisch B (1986) Immunologic regulator and effector functions in perimyocarditis, postmyocarditic heart muscle disease and dilated cardiomyopathy. Basic Res Cardiol 81 [Suppl 1]:217–242
17. Maisch B (1987) Immunological mechanisms in human cardiac injury. In: Spry JE (ed) Immunology and molecular biology of cardiovascular disease. In: Shillingfort JP (ed) Current status of clinical cardiology. MTP Press, London, pp 225–252
18. Maisch B (1987) The sarcolemma as antigen in the secondary immunopathogenesis of myopericarditis. Eur Heart J 8 [Suppl J]:155–166
19. Maisch B (1988) The use of myocardial biopsy in heart failure. Eur Heart J 9 [Suppl H]:59–71
20. Maisch B (1989) Autoreactivity to the cardiac myocyte, connective tissue and the extracellular matrix in heart disease and postcardiac injury. Springer Semin Immunopathol 11:369–395
21. Maisch B (1990) Myocarditis. Curr Opin Cardiol 5:320–327
22. Maisch B (1992) Anticardiac antibodies in hypertrophic cardiomyopathy as a marker of severity. Postgrad Med J 68 [Suppl 1]:29–35
23. Maisch B, Bauer E, Cirsi M, Kochsiek K (1993) Cytolytic cross-reactive antibodies directed angainst the cardiac membrane and viral proteins in coxsackie virus B3 and B4 myocarditis – characterization and pathogenetic relevance. Circulation 1993 (in press)
24. Maisch B, Berg PA, Kochsiek K (1980) Autoantibodies and serum inhibition factors (SIF) in patients with myocarditis. Klin Wochenschr 58:219–225
25. Maisch B, Berg PA, Kochsiek K (1980) Immunological parameters in patients with congestive cardiomyopathy. Basic Res Cardiol 75:219–221
26. Maisch B, Büschel G, Izumi T, Eigel P, Regitz V, Deeg P, Pfeifer U, Schmaltz A, Herzum M, Liebau G, Kochsiek K (1985) Four years of experience in endomyocardial biopsy – an immunohistologic approach. Heart Vessels 1 [Suppl]:59–67
27. Maisch B, Deeg P, Liebau G, Kochsiek K (1983) Diagnostic relevance of humoral and cytotoxic immune reactions in primary and secondary dilated cardiomyopathy. Am J Cardiol 52:1071–1078
28. Maisch B, Drude L, Hengstenberg C, Herzum M, Hufnagel G, Kochsiek K, Schmaltz A, Schönian U, Schwab MD (1991) Are antisarcolemmal (ASAs) and antimyolemmal antibodies (AMLAs) "natural" antibodies? Basic Res Cardiol 86 [Suppl 3]:101–114
29. Maisch B, Eichstädt H, Kochsiek K (1983) Immune reactions in infective endocarditis. I. Clinical data and diagnostic relevance of antimyocardial antibodies. Am Heart J 106: 329–337
30. Maisch B, Lotze U, Schneider J, Kochsiek K (1986) Antibodies to human sinus node in sick sinus syndrome. PACE 9:1101–1109
31. Maisch B, Kochsiek K (1983) Humoral immune reactions in uremic pericarditis. Am J Nephrol 3:264–271
32. Maisch B, Maisch S, Kochsiek K (1982) Immune reactions in tuberculous and chronic constrictive pericarditis. Am J Cardiol 50:1007–1013
33. Maisch B, Outzen H, Roth D, Hiby A, Herzum M, Hengstenberg C, Hufnagel G, Schönian U, Kochsiek K (1991) Prognostic determinants in conventionally treated

myocarditis and perimyocarditis – focus on antimyolemmal antibodies. Eur Heart J 12 [Suppl D]:81–87
34. Maisch B, Schwab D, Bauer E, Sandhage K, Schmaltz AA, Wimmer M (1987) Antimyolemmal antibodies in myocarditis in children. Eur Heart J 8 [Suppl J]:167–175
35. Maisch B, Trostel-Soeder R, Stechemesser E, Berg PA, Kochsiek K (1982) Diagnostic relevance of humoral and cell-mediated immune reactions in patients with acute viral myocarditis. Clin Exp Immunol 48:533–545
36. Maisch B, Wendl I (1988) Cytomegalovirus DNA in endomyocardial biopsies of patients with (peri) myocarditis. Eur Heart J 9 [Suppl 1]:190(P1000) (abstr)
37. Neu N, Beisel K, Traystman M, Rose N, Craig SW (1987) Auto-antibodies specific for the cardiac myosin isoform and found in mice susceptible to coxsackie virus B3-induced myocarditis. J Immunol 138:2488–2495
38. Obermayer U, Scheidler J, Maisch B (1987) Antibodies against micro- and intermediate filaments in carditis and dilated cardiomyopathy – are they a diagnostic marker? Eur Heart J 8 [Suppl J]:181–186
39. Paris S, Fosset M, Samuel D, Ailhand G (1977) Chick embryo plasma membrane from cardiac muscle cells having respiratory control and tolerance to calcium. Biochem Biopsy Res Commun 72:327–33
40. Schönian U, Crombach M, Maisch B (1991) Does CMV infection play a role in myocarditis? New aspects from in-situ hybridization. Eur Heart J 12 [Suppl D]:65–68
41. Schultheiss HP (1987) The mitochondrium as antigen in inflammatory heart disease. Eur Heart J 8 [Suppl J]:203
42. Schultheiss HP, Bolte HD (1985) Immunological analysis of autoantibodies against the adenine nucleotide translocator in dilated cardiomyopathy. J Mol Cell Cardiol 17:601
43. Schwarz RS (1986) Are natural autoantibodies real? Ann Inst Pasteur Immunol 137:156–58
44. Seligman M (1986) The origin and nature of autoantibodies. Ann Inst Pasteur Immunol 137:149–151
45. Thometzek P, Maisch B (1987) Antibodies cross-reactive to streptococci in infective endocarditis. Eur Heart J 8 [Suppl J]:319–322
46. Wallukat G, Morwinski M, Kowal K, Förster A, Boewer V, Wollenberger A (1991) Autoantibodies against the β-adrenergic receptor in human myocarditis and dilated cardiomyopathy: β-adrenergic agonism without desensitization. Eur Heart 12 [Suppl D]:178–181
47. Woodruff JF (1980) Viral myocarditis. A review article. Am J Pathol 101:4225–4484

X. Therapeutic Options in IDC and Myocarditis

Animal Models for Therapeutic Trials of Viral Myocarditis

A. Matsumori

Introduction

Clinically, viral myocarditis may appear in a wide variety of forms, ranging from a total lack of clinical manifestations to sudden, unexpected death. Myocarditis may be subacute or even chronic, leading to progressive myocardial failure and death [1, 2]. We have developed animal models for viral myocarditis using encephalomyocarditis (EMC) virus in which congestive heart failure developed in a range of the acute to subacute stages [5], and dilatation and hypertrophy as seen in dilated cardiomyopathy developed in the chronic stage [6]. Histologically, myocardial necrosis became apparent on day 7, cellular infiltration was most marked on day 14, and thereafter, inflammatory reaction decreased, and fibrosis appeared; calcification persisted to day 28. Marked myocardial fibrosis and hypertrophy of myocardial cells became evident on day 90. We have investigated the natural history and pathogenesis of viral myocarditis and assessed diagnostic methods and therapeutic and preventive interventions in these models [7–20, 23, 26–28, 30, 31, 34]. This review discusses our recent therapeutic trials of viral myocarditis.

Virus Vaccine and Passive Immunization

Passive immunization with hyperimmune rabbit sera given early after virus inoculation had a protective effect in EMC virus myocarditis [10]. Virus-specific vaccine prevented development of myocarditis due to EMC virus. After subsequent challenge with the virus, all vaccinated mice survived without developing myocarditis [10]. Vaccination of mothers before pregnancy completely inhibited myocardial virus replication and had a protective effect on EMC virus infection in offspring [11]. Such vaccines may not be required for widespread use in any entire population but could be important for specifically targeted high-risk groups.

Third Division Department of Internal Medicine Kyoto University, 54 Kawaracho Shogoin, Sakyo-ku, Kyoto 606, Japan

Figulla et al. (Eds.)
Idiopathic Dilated Cardiomyopathy

Antiviral Agents

In an animal model of EMC virus myocarditis, we demonstrated that human leukocyte interferon-α A/D given 1 day before or simultaneously with inoculation with EMC virus inhibited multiplication of virus in the heart and protected mice from developing myocarditis. Prevention was dependent on dosage and on the time of initiation of treatment. When treatment was started previous to or simultaneously with virus inoculation, interferon-α A/D at a dose of 10^7 U/kg per day effectively reduced both inflammatory response and myocardial damage [12].

Ribavirin (virazole, 1-β-D-ribofuranosyl-1,2,4-triazole-3-carboxamide) is a synthetic nucleoside analogue, structurally related to inosine and guanosine, and has broad antiviral activity against RNA and DNA viruses. Clinically, its efficacy has been demonstrated in measles, influenza, and respiratory syncytial virus. Ribavirin at a dose of 200 mg/kg inhibited viral replication in the heart and improved survival and myocardial damage when treatment was started simultaneously with infection [8]. Ribavirin was also effective at a dose of 400 mg/kg beginning 1 day or 3 days after inoculation. Ribavirin also had beneficial effects on coxsackievirus B3 myocarditis. The minimal effective dose required for effect against coxsackievirus myocarditis was smaller than that for EMC virus myocarditis [13].

As described above, both interferon-α A/D and ribavirin were effective, but large quantities of these agents were necessary to obtain beneficial effects. The side effects of these drugs are usually dose related; synergistic combination of drugs may allow effective treatment at lower drug concentrations. This strategy has been highly successful in the treatment of a number of bacterial diseases and cancers. Since the mode of action of interferon may differ from that of ribavirin, we investigated the possibility that recombinant interferon-α A/D is synergistic with ribavirin. Ribavirin at 100 mg/kg or interferon at 10^6 U/kg alone did not inhibit EMC virus replication in the heart. When used together, 100 mg/kg ribavirin and 10^6 U/kg interferon per day achieved a striking effect. The use of the synergistic combination of interferon and ribavirin to treat mice infected with EMC virus produced an enhanced survival. The myocardial virus concentration, the inflammatory response, and myocardial damage were also effectively reduced in mice treated with the above combination of ribavirin and interferon.

We further investigated the effects of interferon-α and ribavirin on coxsackievirus B3 myocarditis in mice. In this experiment, 4-week-old C3H mice were inoculated with the Nancy strain of coxsackievirus B3. The combination treatment of ribavirin 100 mg/kg plus interferon 5×10^6 U/kg effectively reduced viral replication on day 4 even when the treatment was started 1 day after infection. This combination achieves greater effects at lower concentrations than either drug used alone, and there is potential for reduction in frequency of undesirable side effects of both drugs. The use of combinations of antiviral agents deserves further careful study for other serious viral infections.

Immunosuppressive Agents

In our previous study, the use of immunosuppressive therapy with corticosteroids or cyclosporine in experimental myocarditis induced by EMC virus was associated with greater mortality when administered early in illness. Beneficial effects were not seen by later administration of these agents [23, 31].

We also investigated the effect of FK506, a macrolide antibiotic obtained from *Streptomyces tsukubaensis*, collected from the soil of the Tsukuba area in northern Japan [3]. A wide variety of pharmacological and immunological studies have shown that FK506 has strong immunosuppressive activity against mixed lymphocyte reactions, and that it is a valuable drug for augmenting the acceptability of organ transplants. FK506 suppresses T cell-mediated immunity, suppressing both interleukin 2 (IL-2) and IL-2 receptor expression on T cells [3], and its immunosuppressive activity is up to 100 times stronger than that of cyclosporin A. Treatment with FK506, at a rate of 0.5 or 1.0 mg/kg, started on the same day as EMC virus inoculation, did not improve survival or reduce severity of myocardial lesions on day 14.

Activated T cells express a variety of plasma membrane receptors that are absent from the surface of resting T cells, for example, receptors for IL-2, insulin, and transferrin. De novo acquisition of membrane receptors for IL-2 marks a critical event in the course of T cell activation. M7/20, a rat anti-mouse IL-2 receptor monoclonal antibody, blocks IL-2-mediated growth and inhibits binding of IL-2 to its cellular receptor. Intraperitoneal injection with M7/20 at a daily dose of 5 μg per mouse for 10 days caused indefinite survival of murine cardiac allograft [4]. We investigated the effect of immunosuppressive therapy with the anti-IL-2 receptor antibody M7/20 in EMC virus myocarditis. (M7/20 was supplied by Drs. Terry B. Strom and Vicki Kelly.) BALB/c mice were inoculated with EMC virus and injected intraperitoneally daily with M7/20 at a rate of 5 μg per mouse for 14 days. There was no significant difference in mortality or in cardiac histopathology between treated mice and controls [19]. Thus, therapy with anit-IL-2 receptor monoclonal antibody may not have a beneficial effect on EMC virus myocarditis.

The implication of these studies with respect to inflammatory myocarditis in man is that therapy with immunosuppressive agents such as corticosteroids, cyclosporine or FK506, although still potentially useful for myocarditis secondary to primary autoimmune disorders, may be deleterious when the disease results from an acute picornavirus infection, at least in the early period of viral replication and shortly thereafter. In this setting, immunosuppressive therapy should be used only after a carefully controlled trial study, which is needed to define its potential role in the therapy of inflammatory myocarditis in man.

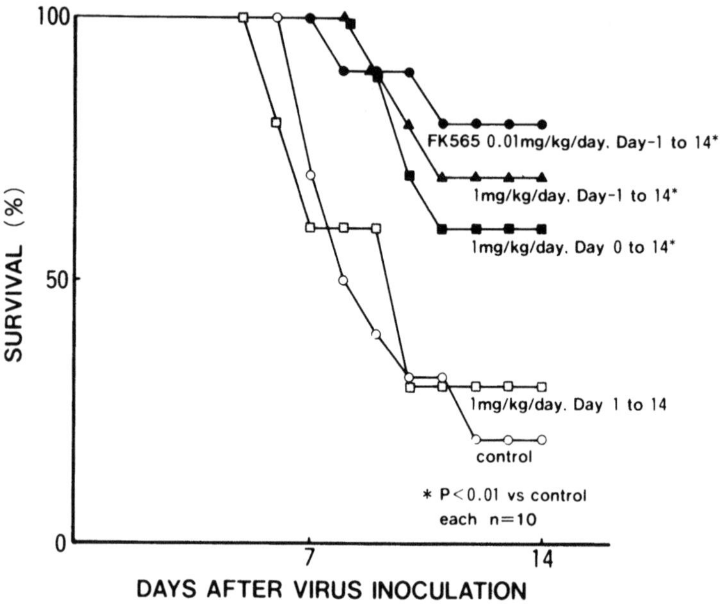

Fig. 1. When FK565 was started before or on the same day as virus inoculation, the survival was significantly higher than in the control group. (Reproduced from [26] with permission of *American Heart Journal*)

Immunostimulant Therapy Using a New Immunoactive Peptide, FK565

FK565 is a novel low molecular weight synthetic immunoactive compound (molecular weight of 503) [22]. Previous studies have reported that the host defense ability of mice was markedly enhanced when FK565 was given before induction of systemic infections such as *Escherichia coli*, *Klebsiella pneumoniae*, S. *marcescens* and *Pseudomonas aeruginosa* [22]. The effect of FK565 has been studied on some types of virus infection, such as Friend leukemia and herpes simplex viruses, and has been shown to improve survival of mice.

Although the mechanism of the effect of FK565 has not been established. FK565 has been reported to increase the number of macrophages in vivo. In vitro, FK565 enhances several macrophage functions, such as attachment to substrate, production of O_2^-, and secretion of lysosomal enzymes. In addition, FK565 increases interferon production in mice stimulated with poly (I, C) and also enhances natural killer cell activity [29].

In this study, FK565 did not demonstrate direct antiviral activity. However, peritoneal exudate cells and spleen cells harvested from mice treated with FK565 reduced EMC virus multiplication in cell culture. Therefore, the inhibitory activity against EMC virus appears to stem from its activation of host defense mechanisms such as macrophages and lymphocytes. In the in

Table 1. Effect of FK565 (1 mg/kg) on EMC virus myocarditis

	n	Virus titer of the heart ($\log_{10}$PFU/mg)	Histological scores	
			Cell infiltration	Necrosis
Started on day −1	10	2.77 ± 0.17**	1.2 ± 0.63*	1.2 ± 0.42**
Started on day 0	10	2.46 ± 0.35**	1.7 ± 0.81	1.3 ± 0.47
Control	7	3.33 ± 0.26	2.0 ± 0.81	2.0 ± 1.00

*$p < 0.05$; **$p < 0.025$ versus control.

vivo study, FK565 adminstered 1 day before or on the same day as EMC virus inoculation significantly reduced viral replication and increased survival (Fig. 1). Treatment with 1 mg/kg FK565 prior to virus inoculation was most effective in reducing inflammatory response and myocardial damage (Table 1) [26]. The effect of FK565 was also studied using coxsackievirus B3 myocarditis. FK565 at a rate of 0.01 mg/kg, started simultaneously with virus inoculation, effectively inhibited myocardial virus replication, reduced inflammatory response and myocardial damage, and improved survival. As FK565 did not show direct antiviral activity in vitro, the inhibitory activity of FK565 against coxsackievirus B3 myocarditis seems to be due to its activation of host defense mechanisms [20]. Side effects of high doses of FK565 (6 mg/kg) have been reported. Marked loss of body weight in normal DBA/2 mice may result from the toxicity of the drug itself. Therefore, in our study, we used a maximum dose of 1 mg/kg, and no apparent toxic effects were observed. The dose of 0.01 mg/kg also increased survival.

The present study suggests that immunostimulant therapy may slow the progress or viral myocarditis during the viral-mediated phase, if given early after infection. We believe that our results justify further investigations on the immunoactivating therapy of acute myocarditis.

Effect of α_1-Adrenergic Blocker

α-Adrenergic blockers are clinically effective in patients with congestive heart failure. Bunazosin, an α_1-blocker and balanced vasodilator, has been reported to be useful in the treatment of patients with congestive heart failure [21]. It has been reported to reduce mean aortic pressure, total vascular resistance index, left ventricular end-diastolic pressure, and increased cardiac index in chronic congestive heart failure after experimental myocardial infarction and in patients with congestive heart failure. In this study, bunazosin did not improve the survival rate of mice infected with the virus. This result is compatible with several double-blind clinical studies concerning the survival rate of patients with congestive heart failure treated with α_1-blocker, although there have also been some contradictory reports. However, bunazosin showed a protective effect in the subacute stage of viral myocarditis if injected from the day of viral inoculation to day 14. The

Table 2. Effect of bunazosin on EMC virus myocarditis

	Lung weight/ body weight ($\times 10^{-3}$)	Antimyosin uptake	Histological scores	
			Cell infiltration	Necrosis
Bunazosin 10 mg/kg	6.7 ± 1.5	9.1 ± 4.0	0.9 ± 1.2*	0.6 ± 0.7
Infected control	8.9 ± 1.4**	13.0 ± 3.6**	2.3 ± 1.3**	1.6 ± 0.8**
Normal	6.6 ± 0.9	7.3 ± 0.6	0.0 ± 0.0	0.0 ± 0.0

*$p < 0.05$; **$p < 0.01$ versus normal.

parameters that indicate congestive heart failure (lung weight to body weight ratio and right and left ventricular dimensions) were significantly reduced in the bunazosin group as compared with the placebo group. In this study, indium-111 labeled antimyosin antibody was used as an objective parameter of myocardial damage. In a previous study, we demonstrated that antimyosin uptake correlates well with myocardial necrosis in the same model of murine viral myocarditis [15]. Antimyosin uptake in the bunazosin group was significantly lower than that in the placebo group, indicating less myocardial damage in the bunazosin group than in the placebo group (Table 2) [34].

Effect of Angiotensin-Converting Enzyme Inhibitors

Recently it has been reported that captopril, an antihypertensive drug which inhibits angiotensin-coverting enzyme activity, is effective for severe congestive heart failure and that it has anti-inflammatory properties. We studied the effect of captopril on EMC virus myocarditis in BALB/c mice. Captopril at a rate of 10, 30, or 100 mg/kg was administered orally from day 4 to 14 after virus inoculation. Survival of mice treated with 100 mg/kg per day was significantly higher than in the placebo treatment group. The ratios of heart weight to body weight were significantly lower, and histological examination showed less extensive myocardial necrosis, cellular infiltration, and calcification in mice treated with 30 or 100 mg/kg as compared with the placebo treatment group [27].

Thus, captopril improved virus-induced myocardial injury and congestive heart failure dose-dependently in an experimental model, even if treatment was started when the rate of viral replication was at its peak. This study suggests that captopril is a promising new therapeutic agent for acute myocarditis. The most noticeable point in this study is that captopril decreased myocardial injury and improved survival of mice even when treatment was started significantly after infection and around the peak rate of virus replication. Rezkalla et al. [25] also reported beneficial effects of captopril on acute coxsackievirus B3 murine myocarditis. In their early treatment protocol, mice were treated with 100 mg/kg captopril daily on days 1–6; heart weight and histological score were significantly less in the treated group. In the late treatment protocol, they reported a reduction in heart weight of treated mice at days 20 and 30, although the degrees of inflammation, necrosis, and dystrophic calcification were similar to the infected controls. It is clear that there is a beneficial effect in early treatment. However, in treatment for chronic stage it cannot be concluded that treatment is beneficial merely because heart weight was lower than in infected controls, as there were no significant improvements in the histological changes. Our results showed that heart weight and the ratio of heart weight to body weight in the uninfected mice treated with captopril were significantly lower than those in the uninfected normal group. Thus, the effect of captopril in the later stage of coxsackievirus B3 myocarditis reported by Rezkalla et al. may merely reflect the drug effect itself and not its effects on myocarditis [18].

Recently, we have investigated the comparative effects of captopril and enalapril. We used captopril at doses of 10, 30, and 100 mg/kg and enalapril at 3, 10, and 30 mg/kg. It has been reported that enalapril is 3–11 times more potent in decreasing blood pressure than captopril. Enalapril improved congestive heart failure but did not improve survival of mice or reduce incidence or type of histopathological changes (Table 3) [28]. Thus, different beneficial effects were seen on survival and myocardial injury by captopril and enalapril. Our results suggest that unique structures such as captopril's sulfhydryl groups may play an important role in creating the beneficial effects of acute myocarditis.

Effect of β-Adrenergic Receptor Blockers

Traditionally, congestive heart failure in patients has been considered to be a contraindication for the use of β-adrenergic receptor blockers. However, recent studies suggest that these agents can have beneficial effects and can improve survival in patients with dilated cardiomyopathy [32, 33]. Although myocarditis is thought to be one of several causes of dilated cardiomyopathy [1, 2], few studies have investigated the role of β-adrenergic blockers in viral myocarditis. To our knowledge, there has been only one published report [24] on the effects of β-adrenergic blockers on myocarditis. In that study, the effect of metoprolol was examined on murine acute coxsackievirus B3 myocarditis. There was an increase in the mortality of the mice and in severe pathological changes of the hearts of mice given metoprolol at 32.5 mg/kg per day. The deleterious hemodynamic and immunological effects of metoprolol were considered to cause these effects because the myocardial virus concentration was increased.

We have recently investigated the effect of carteolol on EMC virus myocarditis. Carteolol is a nonselective β-adrenergic blocker with moderate intrinsic sympathomimetic activity and little membrane-stabilizing activity, and it has about a 20- to 30-fold stronger β-adrenergic blocking effect than propranolol. In our experiment, moderate doses of carteolol had no deleterious effect on survival, heart weight to body weight ratio, histo-

Table 3. Effect of angiotensin-converting enzyme inhibitors on EMC virus myocarditis

	n	Survival (%)	Body weight	Heart weight/ body weight ($\times 10^{-3}$)	Histological scores	
					Cell infiltration	Necrosis
Placebo	37	46	12.1 ± 1.8	5.6 ± 0.8	2.6 ± 0.6	2.5 ± 0.7
Captopril						
10 mg/kg	24	50	13.8 ± 4.1	5.2 ± 0.8	1.8 ± 0.9	1.8 ± 0.9
30 mg/kg	21	62	15.8 ± 3.1*	4.9 ± 0.6*	1.3 ± 1.0*	1.3 ± 1.0*
100 mg/kg	26	73*	15.4 ± 3.9*	4.8 ± 0.5*	1.2 ± 0.8*	1.1 ± 0.9*
Placebo	30	43	11.4 ± 1.8	7.5 ± 1.2	2.6 ± 0.5	2.6 ± 0.5
Enalapril						
3 mg/kg	24	42	12.6 ± 2.1	7.3 ± 1.3	2.6 ± 0.5	2.7 ± 0.5
10 mg/kg	23	44	12.6 ± 2.9	6.2 ± 1.2*	2.3 ± 0.6	2.4 ± 0.5
30 mg/kg	21	57	12.6 ± 3.4	6.2 ± 0.9*	2.3 ± 0.9	2.3 ± 0.9

* $p < 0.05$ versus placebo.

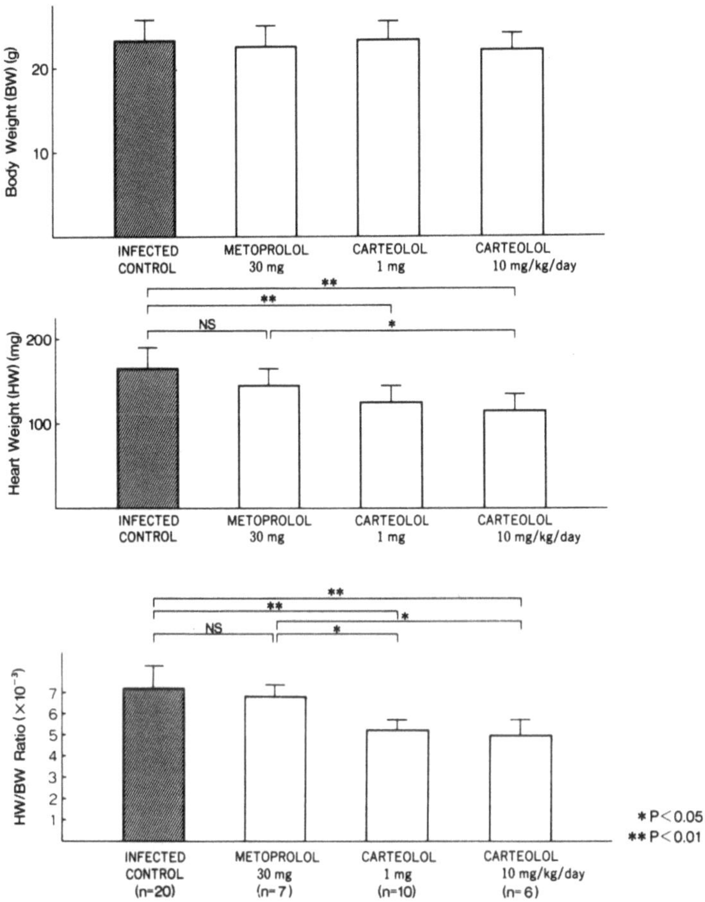

Fig. 2. Bar graphs of effects of carteolol and metoprolol on body weight, heart weight, and heart weight to body weight ratio in the chronic experiment. Heart weight and heart weight to body weight ratio of mice given carteolol at 1 and 10 mg/kg per day was significantly lower than those of the infected control group. Heart weight of mice given carteolol daily at 10 mg/kg and heart weight to body weight ratio of mice given carteolol daily at 1 and 10 mg/kg was also significantly lower than those of mice given metoprolol. On the other hand, heart weight and heart weight to body weight ratio of mice given metprolol was similar to those of the infected control group. (Reproduced from [30] with permission of the *American Heart Association*)

pathological findings, or myocardial virus replication in the acute and subacute stages [30]. Thus, carteolol did not worsen viral myocarditis in these mice. The difference between Rezkalla's work and ours may be due to differences in the strain of mice, virus, or the β-blocker. In the chronic experiment, DBA/2 mice were inoculated with EMC virus and were then given carteolol at 1 or 10 mg/kg or metoprolol at 30 mg/kg daily, starting on day 14, and mice were killed on day 104. The ratio of heart weight to body weight and histopathological scores were significantly lower in mice given carteolol than in the infected controls (Fig. 2). Furthermore, left ventricular

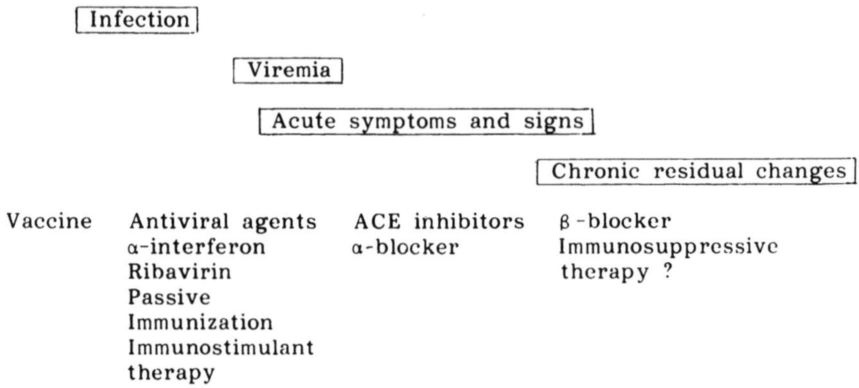

Fig. 3. Acute viral myocarditis: new treatment strategy

cavity dimension, left ventricular wall thickness, and myocardial fiber diameter of the left ventricle were significantly reduced in mice given carteolol as compared with the control group. Metoprolol did not cause any significant changes. This study suggests that carteolol prevents the development of myocardial lesions similar to those in dilated cardiomyopathy after myocarditis in the chronic stage [30].

A new strategy for the treatment of viral myocarditis is summarized in Fig. 3. Virus vaccine and passive immunization are specific preventive methods for a known viral etiology. Antiviral agents such as interferon-α and ribavirin and the new immunostimulant agent, FK565, are broad-spectrum agents and can be used for various types of viral infection, although use of these agents should be started early after infection. An α-adrenergic blocker is also helpful if started early. Angiotensin-converting enzyme inhibitors are effective even when treatment is started significantly after infection. β-Adrenergic blockers do not have beneficial effects in the acute stage, but a nonselective β-blocker with intrinsic sympathomimetic activity, carteolol, is useful for prevention of development of myocardial lesions similar to those in dilated cardiomyopathy after myocarditis.

Acknowledgements. This work was supported, in part, by a Research Grant from the Ministry of Health and Welfare, a Grant-in-Aid for Developmental Scientific Research and for General Scientific Research from the Ministry of Education, Science and Culture, Japan, and Kanazawa Research Fund. I thank Drs. Nobuyoshi Tomioka, Yoshiki Matoba, Ikutaro Okada, Makoto Tominaga, Takehiko Yamada, Hiroshi Suzuki, Yukihito Sato and Scott E. Monrad for great contributions to the above studies and Drs. Chuichi Kawai, Clyde S. Crumpacker and Walter H. Abelmann for pertinent and valued advice.

References

1. Kawai C, Matsumori A, Kitaura Y, Takatsu T (1978) Viruses and the heart: viral myocarditis and cardiomyopathy. Prog Cardiol 7:141–162
2. Kawai C, Matsumori A, Fujiwara H (1987) Myocarditis and dilated cardiomyopathy. Annu Rev Med 38:221–239
3. Kino T, Hatanaka H, Miyata S, Inamura N, Nishiyama M, Yajima T, Goto T, Okuhara M, Kohsaka M, Aoki H, Ochiai T (1987) FK-506, a novel immunosuppressant isolated from a streptomyces: immunosuppressive effect of FK-506 in vitro. J Antibiot (Tokyo) 40:1256–1265
4. Kirkman RL, Barrett LV, Gaulton GN, Kelley VE, Koltun WA, Schoen FJ, Ythier A, Strom TB (1985) The effect of anti-interleukin-2 receptor monoclonal antibody on allograft rejection. Transplantation 40:719–722
5. Matsumori A, Kawai C (1982) An experimental model for congestive heart failure following encephalomyocarditis virus myocarditis in mice. Circulation 65:1230–1235
6. Matsumori A, Kawai C (1982) An animal model of congestive (dilated) cardiomyopathy: dilatation and hypertrophy of the heart in the chronic stage in DBA/2 mice with myocarditis caused by encephalomyocarditis virus. Circulation 66:355–360
7. Matsumori A, Kawai C (1984) Animal models of congestive heart failure and congestive (dilated) cardiomyopathy due to viral myocarditis in mice. In: Bolte H-D (ed) Viral heart disease. Springer, Berlin Heidelberg New York, pp 35–56
8. Matsumori A, Wang H, Abelmann WH, Crumpacker CS (1985) Treatment of viral myocarditis with ribavirin in an animal preparation. Circulation 71:834–839
9. Matsumori A, Kawai C, Crumpacker CS, Abelmann WH (1987) The use of animal models for preventive and therapeutic trials of viral myocarditis. In: Kawai C, Abelmann WH, Matsumori A (eds) Cardiomyopathy update 1. Pathogenesis of myocarditis and cardiomyopathy. University of Tokyo Press, Tokyo, pp 37–47
10. Matsumori A, Crumpacker CS, Abelmann WH, Kawai C (1987) Virus vaccine and passive immunization for the prevention of viral myocarditis in mice. Jpn Circ J 51:1362–1364
11. Matsumori A, Crumpacker CS, Abelmann WH (1987) Pathogenesis and preventive and therapeutic trials in an animal model of dilated cardiomyopathy induced by a virus. Jpn Circ J 51:661–664
12. Matsumori A, Crumpacker CS, Abelmann WH (1987) Prevention of viral myocarditis with recombinant human leukocyte interferon-α A/D in a murine model. J Am Coll Cardiol 9:1320–1325
13. Matsumori A, Okada I, Kawai C, Crumpacker CS, Abelmann WH (1988) Animal models for therapeutic trials of viral myocarditis: effect of ribavirin and alpha interferon on coxsackievirus B3 and encephalomyocarditis virus myocarditis. In: Schultheiss HP (ed) New concepts in viral heart disease. Springer, Berlin Heidelberg New York, pp 377–384
14. Matsumori A, Kawai C (1989) Immunomodulating therapy in experimental myocarditis. Springer Semin Immunopathol 11:77–88
15. Matsumori A, Ohkusa T, Matoba Y, Okada I, Yamada T, Kawai C, Tamaki N, Watanabe Y, Yonekura Y, Endo K, Konishi J, Khaw B, Haber E (1989) Myocardial uptake of antimyosin monoclonal antibody in a murine model of viral myocarditis. Circulation 79:400–405
16. Matsumori A, Hoshino T, Kishimoto C, Kawai C (1990) Morphological features and electric disturbances in experimental viral myocarditis. In: Olsen EGJ, Sekiguchi M (eds) Proceeding of the 2nd international symposium on cardiomyopathy and myocarditis. University of Tokyo Press, Tokyo, pp 217–222
17. Matsumori A, Matoba Y, Tomioka N, Kawai C (1990) Experimental viral myocarditis: immunopathogenetic aspects. In: Baroldi G, Camerini F, Goodwin JF (eds) Advances in cardiomyopathies. Springer, Berlin Heidelberg New York, pp 181–189
18. Matsumori A, Suzuki H, Kawai C (1990) Beneficial effects of captopril in coxsackievirus myocarditis. Circulation 82:1882

19. Matsumori A, Yamada T, Kawai C (1991) Immunodulating therapy in viral myocarditis: effects of tumor necrosis factor, interleukin-2 and anti-interleukin-2 receptor antibody in an animal model. Eur Heart J 12 [Suppl D]:203–205
20. Matsumori A, Sato Y (1991) Treatment of coxsackievirus B3 myocarditis by a new immunoactive peptide, FK-565 in a murine model. Circulation 84 [Suppl II]:633
21. Matsuzaki M, Kohno M, Matsuda Y, Ogawa H, Ozaki M, Toma Y, Fujii H, Kusukawa R (1989) The effect of vasodilators on hemodynamics and renal circulation in stable chronic heart failure. Jpn Circ J 53:463–464
22. Mine Y, Yokota Y, Wakai Y, Fukada S, Nishida M, Goto S, Kuwahara S (1983) Immunoactive peptides, FK156 and FK565. I. Enhancement of host resistance to microbial infection in mice. J Antibiot (Tokyo) 36:1045–1050
23. Monrad ES, Matsumori A, Murphy JC, Fox JG, Crumpacker CS, Abelmann WH (1986) Therapy with cyclosporine in experimental murine myocarditis with encephalomyocarditis virus. Circulation 73:1058–1064
24. Rezkalla S, Kloner RA, Khatib G, Smith F, Khatib R (1988) Effect of metoprolol in acute coxsackievirus B3 murine myocarditis. J Am Coll Cardiol 12:412–414
25. Rezkalla S, Kloner RA, Khatib G, Khaib R (1990) Beneficial effects of captopril in acute coxsackievirus B3 murine myocarditis. Circulation 81:1039–1047
26. Sato Y, Maruyama S, Kawai C, Matsumori A (1992) Effect of immunostimulant therapy on acute viral myocarditis in an animal model. Am Heart J 124:428–434
27. Suzuki H, Matsumori A (1990) Myocardial injury due to viral myocarditis improved by captopril in mice. Circulation 82 [Suppl III]:674
28. Suzuki H, Matsumori A (1991) Effects of angiotensin-converting enzyme inhibitors on myocardial injury and congestive heart failure due to myocarditis in mice: comparison with captopril and enalapril. Circulation 84 [Suppl II]:634
29. Talmadge JE, Chirigos MA (1985) Comparison of immunomodulatory and immunotherapeutic properties of biologic response modifiers. Springer Semin Immunopathol 204:1–15
30. Tominaga M, Matsumori A, Okada I, Yamada T, Kawai C (1991) β-Blocker treatment of dilated cardiomyopathy: beneficial effect of carteolol in mice. Circulation 86:2021–2028
31. Tomioka N, Kishimoto C, Matsumori A, Kawai C (1986) Effects or prednisolone on acute viral myocarditis in mice. J Am Coll Cardiol 7:868–872
32. Waagstein F, Hjalmarson A, Varnauskas E, Wallentin I (1975) Effect of chronic β-adrenergic receptor blockade in congestive cardiomyopathy. Br Heart J 37:1022–1036
33. Waagstein F, Caidahi K, Wallentin I, Bergh CH, Hjalmarson A (1989) Long-term β-blockade in dilated cardiomyopathy: effects of short- and long-term metoprolol treatment followed by withdrawal and readministration of metoprolol. Circulation 80:551–563
34. Yamada T, Matsumori A (1990) Therapeutic effect of bunazosin on experimental viral myocarditis in mice evaluated by antimyosin uptake. Circulation 82 [Suppl III]:675

Immunosuppressive Therapy in Human Myocarditis

J.B. O'Connell

Introduction

The role of infection in initiating deleterious immune responses which precipitate chronic heart disease was first proposed in the early nineteenth century [28]. However, more than 150 years passed before an accurate diagnosis of myocarditis could be established during life and pharmacologic agents that alter host immune responses incorporated into clinical practice.

Shortly after the extraction of adrenocorticotrophic hormone, reports of the treatment of myocarditis with corticosteroid immunosuppression appeared [9]. The high frequency of myocarditis following influenza epidemics allowed investigators to study larger series of patients. Ainger noted that corticosteroids were a valuable adjunct in eight children who developed congestive heart failure refractory to conventional therapy following an influenzal illness [1]. Stoegmann described 25 infants and children with congestive heart failure following coxsackievirus infection who were treated with high-dose glucocorticoids [25]. Successful reversal of the cardiac decompensation was accomplished only if corticosteroids were administered early after the onset of the illness. In adult populations, sporadic case reports appeared in the literature; however, successful clinical trials could not be conducted because of lack of accurate diagnostic tools.

Mason and colleagues were first to report the use of endomyocardial biopsy (developed for cardiac allograft rejection surveillance) in patients with unexplained congestive heart failure, and they noted lymphocytic infiltration or antibody deposition in ten of the patients studied [14]. Although standards of histologic diagnosis had not yet been established, a flurry of enthusiasm for applying this technique to patients with unexplained heart failure ensued [19]. The accurate identification of myocarditis led to interest in therapeutic modalities that alter the host immune response which was postulated to be responsible for the myocardial injury. Hence, great enthusiasm for administration of immunosuppression in patients with active myocarditis resulted.

The Department of Medicine University of Mississippi Medical Center, Jackson, MS 39216-4505, USA

Figulla et al. (Eds.)
Idiopathic Dilated Cardiomyopathy

Table 1. Uncontrolled reports of immunosuppression in biopsy-confirmed myocarditis (from [18])

Investigators	Year	*n*	Improved	Therapy
Mason et al. [14]	1980	10	5	CS, AZA
Sekiguchi et al. [24]	1980	3	2	CS
Edwards et al. [6]	1982	4	2	CS
Fenoglio et al. [8]	1983	19	8	CS, AZA
Zee-Cheng et al. [29]	1984	11	5	CS, AZA, ATG
Daly et al. [4]	1984	9	7	CS, AZA
Vignola et al. [27]	1984	6	5	CS, AZA
Fenely et al. [7]	1984	2	2	CS, AZA
Dec et al. [5]	1985	9	4	CS, AZA
Mortensen et al. [16]	1985	12	8	CS, AZA, CYA
Hosenpud et al. [12]	1985	6	0	CS, AZA
Salvi et al. [23]	1989	14	8	CS, AZA
Hobbs et al. [11]	1989	34	25	
Chan et al. [3]	1990	13	6	CS, AZA
Jones et al. [13]	1991	9	4	
Total		161	91 (56%)	

CS, Corticosteroids; AZA, azathioprine; ATG, antithymocyte globulin; CYA, cyclosporine.

Uncontrolled Trials

In the initial report, Mason et al. described a favorable effect of immunosuppression in the ten patients with unexplained heart failure who had histologic evidence of myocarditis [14]. Although these investigators called for a randomized prospective trial, only uncontrolled reports appeared in the literature (Table 1) [3–8, 11–14, 16, 23, 24, 27, 29]. In some reports, a profound beneficial effect was noted, while in others the effect was far less pronounced. Because all reports before 1985 antedated the widely accepted Dallas Criteria, histologic diagnostic standards varied widely [2]. Objective signs of improvement differed among and within the studies, ranging from resolution of the histology to objective evidence of improved cardiac function. Immunosuppressive therapy was not standardized, and although all regimens incorporated corticosteroids, the addition of azathioprine, antithymocyte globulins, or cyclosporine were included in select patients. Nonetheless, the cumulative improvement rate convinced some physicians that immunosuppressive therapy was warranted in patients with active myocarditis.

When peripartal heart disease with a high frequency of active myocarditis was studied by Midei et al., a dramatic improvement was noted with immunosuppression in this uncontrolled study [15]. When patients with positive results on gallium-67 scan who did not undergo myocardial biopsy received prednisone and azathioprine, more than 30% improved in long-term follow-up [21]. However, when a larger population of patients was studied in a prospective trial of prednisone, transient improvement was detected, but by nine months no differences were found between those patients who were treated and those who were not [22]. The natural history of myocarditis has not been defined.

The recent observation that enteroviral genome may be detected in myocytes from patients with active myocarditis suggests that viral persistence may be involved in the pathogenesis of the disease [26]. Although morbidity has not been emphasized, none of the reports has noted a deterioration resulting from viral dissemination or immunosuppression-mediated exacerbation of viral-induced necrosis.

The role of immunosuppressive therapy in active myocarditis can be settled only by a prospective clinical trial with specific end-points.

Myocarditis Treatment Trial

Responding to this need for a prospective, randomized trial, investigators in the United States initiated the Myocarditis Treatment Trial [20]. Patients with ejection fractions below 45% and active myocarditis by the Dallas criteria were randomized in 23 enrollment centers to receive immunosuppression consisting of either prednisone and azathioprine, prednisone

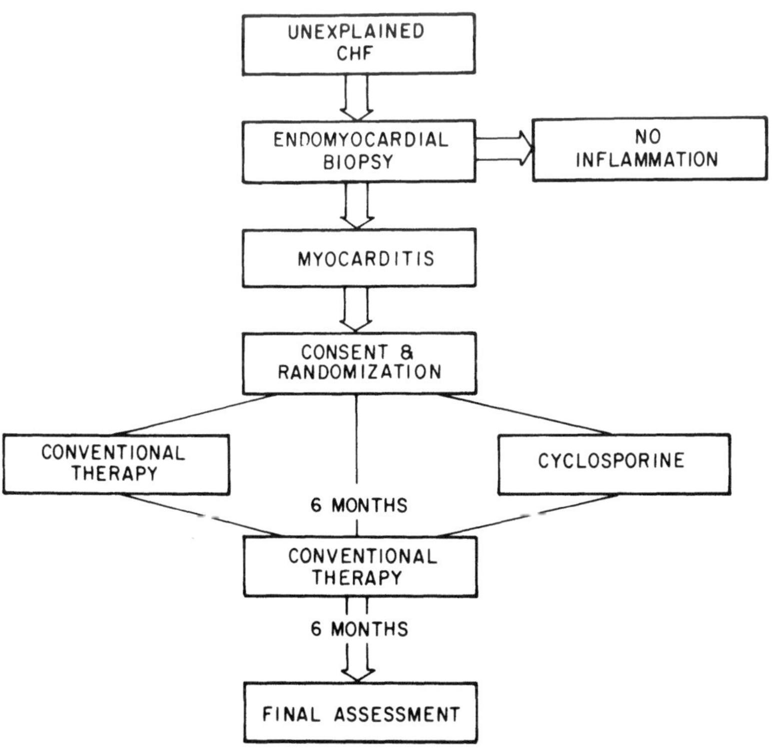

Fig. 1. Study protocol of the Myocarditis Treatment Trial (from [19])

and cyclosporine, or conventional medical management alone for 6 months following baseline studies. The primary end-points in the study included left ventricular ejection fraction by gated blood pool imaging and treadmill exercise time on a standardized exercise protocol. Secondary end-points included histologic resolution, changes in frequency of ventricular arrhythmia, echocardiographic left ventricular volume and wall thickness, hemodynamics, and survival.

Randomization commenced in October 1986. Because of the popularity of endomyocardial biopsy in patients with unexplained congestive heart failure outside formal enrollment centers, the required sample size could not be achieved during the funded period. Consequently, the prednisone and azathioprine arm was deleted, and these patients were randomized either to prednisone and cyclosporine or to conventional management alone, decreasing the required number of subjects (Fig. 1). Enrollment was discontinued in November 1991.

Of the 2305 patients who underwent biopsy for suspected myocarditis in the trial, 2227 had adequate samples. Of these, 208 (9.3%) had active myocarditis. From this group 111 patients were randomized. The baseline characteristics of this population included a mean age of 42 ± 14 years, 62% male, a left ventricular ejection fraction of 0.24 ± 0.10, and an exercise time on a modified Naughton protocol of 9.4 ± 5.3 min. Chest pain was detected in 35% of the patients, and an increase in MB fraction of the creatine kinase was detected in 12%. Influenzalike symptoms were noted in 60% of the randomized patients and an elevated erythrocyte sedimentation rate in 61%. An elevated white blood cell count was detected in 25%, and 18% had fever on examination. Of note, 20% of the patients showed no evidence of an antecedent viral syndrome. The follow-up is ongoing, and consequently results of the end-point analyses are not available.

The Myocarditis Treatment Trial will define whether the specific immunosuppressive regimen improves exercise time and ejection fraction in patients with active myocarditis. The trial cannot be considered definitive. Although the natural history of myocarditis will be defined in those randomized to conventional therapy, other immunosuppressive regimens may be more effective and should be studied prospectively. In this light, the monoclonal antibody OKT3 has been administered with success in resolution of myocarditis [10].

Because of the lack of definitive proof of the benefit of immunosuppressive therapy in active myocarditis, these patients may be selected as recipients for cardiac transplantation. The caveat of such reflex selection is that dramatic improvement in hemodynamic abnormalities may occur even when left ventricular assistance is required. However, undoubtedly some patients should be considered for cardiac transplantation. In a retrospective analysis of 12 patients undergoing cardiac transplantation for active myocarditis in the explanted heart, survival was significantly shorter than that of age-matched or age- and sex-matched control subjects [17]. These patients also experienced a 2.2 ± 1.1-fold increase in rejection compared to institutional average rejection rates. Of the first episodes of rejection 60%

occurred within 2 weeks of transplantation. Of initial rejection episodes 25% were fatal. These results suggest that the immunologic milieu of active myocarditis serves as a hostile environment for a transplanted heart and results in aggressive early rejection, further underscoring the immunologic nature of this disease. Consequently, cardiac transplantation should be considered only in patients with severe hemodynamic compromise who have been given the benefit of maximal medical and mechanical support for their condition.

Conclusions

The results of immunosuppressive trials in human subjects with active myocarditis are inconclusive. The Myocarditis Treatment Trial may resolve some of the issues. However, several new specific immunosuppressive agents are on the horizon, and clinical trials with these newer agents in patients with active myocarditis may be warranted before firm conclusions regarding the effect of immunosuppression are drawn.

References

1. Ainger LE (1964) Acute aseptic myocarditis: Corticosteroid therapy. J Pediatr 64:716–723
2. Aretz HT, Billingham ME, Edwards WD, Factor SM, Fallon JT, Fenoglio Jr JJ, Olsen EGJ, Schoen FJ (1986) Myocarditis. Am J Cardiovasc Pathol 1:3–14
3. Chan KY, Iwahara M, Benson LN, Wilson GJ, Freedom RM (1991) Immunosuppressive therapy in the management of acute myocarditis in children: a clinical trial. J Am Coll Cardiol 17:458–460
4. Daly K, Richardson PJ, Olsen EGJ, Morgan-Capner P, McSorley C, Jackson G, Jewitt DE (1984) Acute myocarditis. Role of histological and virological examination in the diagnosis and assessment of immunosuppressive treatment. Br Heart J 51:30–35
5. Dec Jr GW, Palacios IF, Fallon JT, Aretz HT, Mills J, Lee DC, Johnson RA (1985) Active myocarditis in the spectrum of acute dilated cardiomopathies. Clinical features, histologic correlates, and clinical outcome. N Engl J Med 312:885–890
6. Edwards WD, Holmes Jr DR, Reeder GS (1982) Diagnosis of active lymphocytic myocarditis by endomyocardial biopsy. Quantitative criteria for light microscopy. Mayo Clin Proc 57:419–425
7. Feneley MP, Gavaghan TP, Ralston M, Hickie JB, Baron DW (1984) Diagnosis and management of acute myocarditis aided by serial myocardial biopsy. Aust NZ J Med 14:826–830
8. Fenoglio Jr JJ, Ursell PC, Kellogg CF, Drusin RE, Weiss MB (1983) Diagnosis and classification of myocarditis by endomyocardial biopsy. N Engl J Med 308:12–18
9. Garrison RF, Swisher RC (1953) Myocarditis of unknown etiology (Fiedler's) treated with ACTH. J Pediatr 42:591–599
10. Gilbert EM, O'Connell JB, Hammond ME, Renlund DG, Watson FS, Bristow MR (1988) Treatment of myocarditis with OKT3 monoclonal antibody. Lancet 1:759
11. Hobbs RE, Pelegrin D, Ratliff NB, Bott-Silverman C, Rincon G, Sterba R, Badhwar K (1989) Lymphocytic myocarditis and dilated cardiomyopathy: treatment with immunosuppressive agents. Cleve Clin J Med 56:628–635
12. Hosenpud JD, McAnulty JH, Niles NR (1985) Lack of objective improvement in ventricular systolic function in patients with myocarditis treated with azathioprine and prednisone. J Am Coll Cardiol 6:797–801

13. Jones SR, Herskowitz A, Hutchins GM, Baughman KL (1991) Effects of immunosuppressive therapy in biopsy-proved myocarditis and borderline myocarditis on left ventricular function. Am J Cardiol 68:370–376
14. Mason JW, Billingham ME, Ricci DR (1980) Treatment of acute inflammatory myocarditis assisted by endomyocardial biopsy. Am J Cardiol 45:1037–1044
15. Midei MG, DeMent SH, Feldman AM, Hutchins GM, Baughman KL (1990) Peripartum myocarditis and cardiomyopathy. Circulation 81:922–928
16. Mortensen SA, Baandrup U, Buch J, Bendtzen K, Hvid-Jacobsen K (1985) Immunosuppressive therapy of biopsy proven myocarditis: experiences with corticosteroids and cyclosporin. Int J Immunother 1:35–45
17. O'Connell JB, Dec GW, Goldenberg IF, Starling RC, Mudge GH, Augustine SM, Costanzo-Nordin MR, Hess ML, Hosenpud JD, Icenogle TB, Menlove RL, Billingham ME (1990) Results of heart transplantation for active lymphocytic myocarditis. J Heart Transplant 9:351–356
18. O'Connell JB, Mason JW (1988) Immunosuppressive therapy in experimental and clinical myocarditis. Pathol Immunopathol Res 7:292–304
19. O'Connell JB, Mason JW (1989) Diagnosing and treating active myocarditis. West J Med 150:431–435
20. O'Connell JB, Mason JW (1990) The applicability of results of streamlined trials to clinical practice: the Myocarditis Treatment Trial. Stat Med 9:193–197
21. O'Connell JB, Robinson JA, Henkin RE, Gunnar RM (1981) Immunosuppressive therapy in patients with congestive cardiomyopathy and myocardial uptake of gallium-67. Circulation 64:780–786
22. Parrillo JE, Cunnion RE, Epstein SE, Parker MM, Suffredini AF, Brenner M, Schaer GL, Palmeri ST, Cannon III RO, Alling D, Wittes JT, Ferrans VJ, Rodriguez ER, Fauci AS (1989) A prospective, randomized, controlled trial of prednisone for dilated cardiomyopathy. N Engl J Med 321:1061–1068
23. Salvi A, Di Lenarda A, Dreas L, Silvestri F, Camerini F (1989) Immunosuppressive treatment in myocarditis. Int J Cardiol 22:329–338
24. Sekiguchi M, Hiroe M, Take M, Hirosawa K (1980) Clinical and histopathological profile of sarcoidosis of the heart and acute idiopathic myocarditis. Concepts through a study employing endomyocardial biopsy. II. Myocarditis. Jpn Circ J 44:264–273
25. Stoegmann VW (1972) Die Grippemyokarditis beim Saugling und Kleinkind. Wien Klin Wochenschr 84:502–504
26. Tracy S, Wiegand V, McManus B, Gauntt C, Pallansch M, Beck M, Chapman N (1990) Molecular approaches to enteroviral diagnosis in idiopathic cardiomyopathy and myocarditis. J Am Coll Cardiol 15:1688–1694
27. Vignola PA, Aonuma K, Swaye PS, Rozanski JJ, Blankstein RL, Benson J, Gosselin AJ, Lister JW (1984) Lymphocytic myocarditis presenting as unexplained ventricular arrhythmias: diagnosis with endomyoardial biopsy and response to immunosuppression. J Am Coll Cardiol 4:812–819
28. Wenger NK, Abelmann WH, Roberts WC (1982) Myocardial disease. In: Hurst JW (ed) The heart. McGraw-Hill, New York, pp 1278–1299
29. Zee-Cheng C-S, Tsai CC, Palmer DC, Codd JE, Pennington DG, Williams GA (1984) High incidence of myocarditis by endomyocardial biopsy in patients with idiopathic congestive cardiomyopathy. J Am Coll Cardiol 3:63–70

Interferon Therapy in Enterovirus-Associated Idiopathic Dilated Cardiomyopathy

M. Stille-Siegener[1], A. Heim[1], K. Klingel[2], R. Kandolf[2], G. Mall[3], H. Kreuzer[1], and H.R. Figulla[1]

Introduction

Enteroviruses are thought to play a pathogenetic role in the development of human idiopathic dilated cardiomyopathy (IDC). Reports of the percentage of patients with IDC or myocarditis and detection of enteroviral RNA in the myocardium range from 17% to 30% [6, 8].

Interferon Treatment

Presuming the enteroviruses to be the causative agent in the virus-positive patients with IDC, an antiviral therapy can be considered a causal treatment. We chose interferon-α for our trial because of promising results reported on the treatment of viral hepatitis [1, 4] and the simple route of subcutaneous administration by the patients themselves. As there are some case reports suspecting high doses of interferon to cause a reversible heart dilatation [2, 3, 7] and because of the relatively favorable hemodynamic course of our IDC patients (data not shown), we designed the following study protocol in agreement with our ethics committee (Fig. 1).

All patients were diagnosed by left ventricular catheterization. Four or five left ventricular biopsies were taken both for microscopy and for in situ hybridization [5]. We also measured the left ventricular end-diastolic diameter by echo, left ventricular ejection fraction at rest by gated blood pool scanning, and mean pulmonary artery pressure at maximal work load by Swan-Ganz catheterization. After 12 months echo, gated blood pool scanning, and Swan-Ganz catheterization were repeated.

According to the hemodynamic parameters there were two groups of positive patients. Those showing improvement or stabilization obtained further conventional therapy alone; the others were treated with interferon-α in addition. After another 6 months the interferon therapy was ended

[1] Department of Cardiology and Pulmonology, Division of Internal Medicine, University of Göttingen, W-3400 Göttingen, Fed. Rep. of Germany
[2] Max-Planck-Institut für Biochemie, W-8033 Martinsried, Fed. Rep. of Germany
[3] Pathologisches Institut, Städtische Kliniken Darmstadt, W-6100 Darmstadt, Fed. Rep. of Germany

Figulla et al. (Eds.)
Idiopathic Dilated Cardiomyopathy

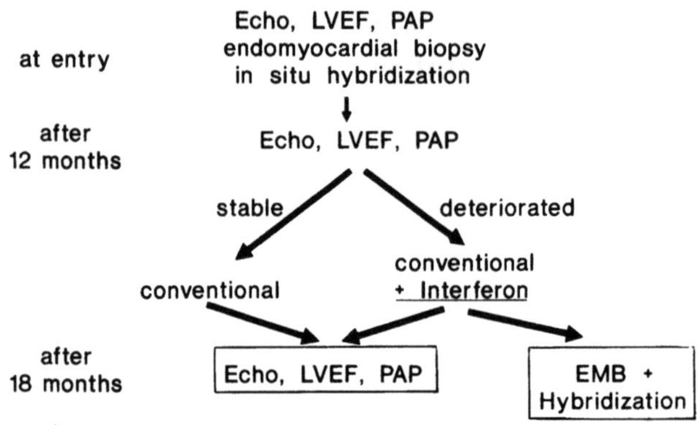

Fig. 1. Interferon treatment in virus-positive IDC: study protocol

Table 1. Characteristics of patients included in the interferon trial

Patients	Sex	Age	Date of in situ hybridization	Preceding myocarditis
G.P.	F	59	8/88	1985
F.H.	M	35	9/88	–
G.H.	M	46	11/89	–
E.M.	F	59	2/91	1986

and a reexamination was performed in both groups. In addition to the hemodynamic work-up described above, the patients treated with interferon underwent a second left ventricular catheterization to obtain material for histopathology and, in situ hybridization.

The therapeutic regimen of our interferon trial was as follows: We used interferon-α (Roferon A; kindly given by Hoffmann-La Roche). The patients themselves injected 3×10^6 U s.c. every other day. The controls during therapy consisted of echo, ECG, clinical examination, and blood tests. The duration of therapy was 6 months.

Enteroviral RNA was detectable in the myocardium of 21 patients with IDC. Up to now one virus-positive patient has died; 15 of 21 patients were reexamined. Four of these patients deteriorated (decrease in LVEF at rest $\geqslant$ 5%, increase in mean PAP at maximal work load $\geqslant$ 3 mmHg), thus being suitable for the interferon treatment (Table 1).

Interestingly there were two patients with an anamnestic history of myocarditis. The myocarditis was based on a clinical diagnosis in one case, a left ventricular catheterization was not performed at that time; in the other case the diagnosis was confirmed histologically by endomyocardial biopsy.

Table 2 presents the results of the 6-month interferon therapy. Hemodynamic improvement was found in all patients. On average the

Table 2. Hemodynamic data before and after interferon treatment

Patients	Before treatment		After treatment	
	EF (%)	PAP (mmHg)	EF (%)	PAP (mmHg)
G.P.	34	58	41	33
F.H.	25	49	31	31
G.H.	41	28	60	31
E.M.	31	43	31	31
	33 (Av.)	45 (Av.)	41 (Av.)	32 (Av.)

ejection fraction increased from 33% to 41%; mean pulmonary artery pressure at maximal work load decreased from 45 to 33 mmHg. After the interferon therapy virus was still detected in two of four patients in the subsequent biopsy by in situ hybridization. There were only mild side effects of the interferon therapy. Three patients developed fever after the first injections, two patients complained of a short influenzalike syndrome at the beginning and one showed a transient elevation of transaminases.

Summary

Less than one third of our virus-positive patients with IDC (4/15 with reexamination) showed hemodynamic deterioration about 1 year after diagnosis. No severe side effects were seen in the four patients treated with interferon. All four virus-positive patients demonstrated hemodynamic improvement after a 6-month interferon-α therapy; however, in two of the four patients enteroviral RNA was further detectable after interferon therapy.

References

1. Alexander GJM, Brahm J, Fagan EA, Smith HM, Daniels HM, Eddleston ALWF, Williams R (1987) Loss of HBsAg with interferon therapy in chronic hepatitis B virus infection. Lancet 2:66–69
2. Cohen MC, Muberman MS, Nesto RW (1988) Recombinant alpha 2 interferon-related cardiomyopathy. Am J Med 85:549–551
3. Deyton LR, Walker RE, Kovacs JA, Herpin B, Parker M, Masur H, Fauci AS, Lane HC (1989) Reversible cardiac dysfunction associated with interferon alfa therapy in AIDS patients with Kaposi's sarcoma. N Engl J Med 321:1246–1249
4. Hoofnagle JH, Peters M, Mullen KD, Jones DB, Rustgi V, DiBisceglie AM, Hallahan C, Park Y, Meschievitz C, Jones EA (1988) Randomized, controlled trial of recombinant human alpha-interferon in patients with chronic hepatitis B. Gastroenterology 95:1318–1325
5. Kandolf R, Ameis D, Kirschner P, Canu A, Hofschneider PH (1987) In situ detection of enteroviral genomes in myocardial cells by nucleic acid hybridization: an approach to the diagnosis of viral heart disease. Proc Natl Acad Sci 84:6272–6276
6. Kandolf R, Hofschneider PH (1989) Viral heart disease. Springer Semin Immunopathol 11:1–13

7. Sonnenblick M, Rosenmann D, Rosin A (1990) Reversible cardiomyopathy induced by interferon. Br Med J 300:1174–1175
8. Tracy SM, Chapman NM, McManus BM, Pallansch MA, Beck MA, Carstens J (1990) A molecular and serologic evaluation of enteroviral involvement in human myocarditis. J Mol Cell Cardiol 22:403–414

The Role of β-Blockers in Idiopathic Dilated Cardiomyopathy: Clinical Effect and Molecular Mechanism

F. Waagstein, L.-X. Fu, and Å. Hjalmarson

Heart failure is associated with activation of the sympathetic nervous system. In patients with idiopathic dilated cardiomyopathy (IDC) the adrenergic nervous system (ANS) is markedly activated, but the role of the ANS in determining the pathophysiological process and clinical outcome of patients with IDC has yet to be defined. Investigations carried out in the past 25 years have suggested that enhanced neurohormonal activity could have either a beneficial or a deleterious effect in patients with the disease. An increase in adrenergic activity could either increase the circulation by augmenting cardiac output and blood pressure or raise ventricular preload by inducing peripheral vasoconstriction, salt and water retention, and the release of other vasoconstrictor neurohormones. This could adversely affect the patient's clinical status and long-term prognosis. Whether activation of the ANS is beneficial or detrimental to IDC patients remains a heated topic of controversy.

Much progress in understanding the ANS has been achieved over the past 15 years by observing the effects of long-term administration of β-blockers in patients with IDC. Experimentally, there is a growing evidence that the increased sympathetic tone is associated with a loss of myocardial surface β-adrenoceptors with a selective loss of the myocardial contractile response to β-adrenergic stimulation. Recently, it has also been shown that there is an increase of the guanine nucleotide binding inhibitory regulatory protein (G_i) as measured by pertussis toxin induced [^{32}P]ADP ribosylation in human failing heart due to IDC. These alterations have been demonstrated to be of at least the same functional importance as β-adrenoceptor down-regulation for the regulation of myocardial contractility, providing the fundamental basis for exploring the mechanisms behind the effect of β-blocker in IDC. Despite the growing number of investigations supporting the concept of long-term treatment with β-blockers in this setting, debate continues regarding the molecular mechanisms of action of β-blockers and their clinical applications.

Wallenberg Laboratory, Sahlgren's Hospital, University of Göteborg, 413 45 Göteborg, Sweden

Figulla et al. (Eds.)
Idiopathic Dilated Cardiomyopathy

Receptor-Directed and G Protein Modulated Signal Transduction in Heart Failure with IDC

Congestive heart failure secondary to IDC is associated with a number of alterations in the ANS. The plasma norepinephrine levels of patients with heart failure are invariably elevated for a number of reasons. The degree of elevation has been shown to be predictive of mortality [19, 25]. In addition, evidence suggests that changes in plasma norepinephrine concentrations and clinical response to therapy are closely related. Decreased responsiveness of baroreceptors has been reported in patients with heart failure, resulting in potentially excessive release of neurohomones, including norepinephrine. Alternatively, hemodynamic improvement could be accompanied by a decrease in the influence of autoregulatory mechanisms – including vasoconstrictor hormone release, which could contribute to partial restoration of central baroreceptor function. Peripheral nutritive perfusion may have an effect on systemic sympathetic hyperactivity as well as central events. The role of continuously increased plasma norepinephrine levels in patients with heart failure has not been defined. However, it is generally agreed that the increase mirrors the activation of the ANS in an effort to compensate for the reduced pump performance of the failing heart. Evidence suggests that markedly increased plasma norepinephrine levels may in fact be detrimental to the heart. The deleterious effects of continuously high plasma norepinephrine levels on the myocardium could be due in part to decreased β-adrenoceptor density and ineffective coupling of receptor–G protein complex. In addition, catecholamines have direct toxic effects on the myocardium [7, 10, 25].

Since Thomas and Marks [21] speculated that myocardial β-adrenoceptors become desensitized to the elevated level of circulating catecholamine, a growing number of investigations have demonstrated that in patients with heart failure due to IDC, β-adrenoceptor density and adenylyl cyclase activity (both basal and agonist stimulated) are reduced [12], and the inotropic response to the β-adrenergic agonist is impaired. Postreceptor signal transduction, i.e., G protein–adenylyl cyclase complex, may be altered as well [3, 5]. We have previously shown the increased levels of mRNAs encoding two spliced forms of $G_{s\alpha}$ and subspecies 3 of $G_{i\alpha}$ in human failing ventricles [11–13]. Recently, Fu et al. used antisense ^{32}P- or ^{35}S-labeled $G_{s\alpha}$ RNA probes (nucleotides 883–1738), SP6 polymerase with *Bam*HI linearized pGEM2 plasmid as template and synthesizing ^{32}P- or ^{35}S-labeled $G_{i\alpha 2}$ RNA antisense probe(nucleotides 1378–1749) with T7 polymerase from linerarized plasmid (*Xba*I). They found that either $G_{s\alpha}$ mRNA or $G_{i\alpha 2}$ mRNA was unchanged in human failing left ventricles (L.X. Fu et al., to be published). To quantify subspecies of $G_{s\alpha}$ and $G_{i\alpha}$ in the human failing heart, Fu et al. raised antisera against unique, synthetic peptides corresponding to $G_{s\alpha}$, $G_{i\alpha 1-2}$, and $G_{i\alpha 3}$. $G_{s\alpha}$ antiserum was generated against a decapeptide corresponding to the carboxyl-terminal of $G_{s\alpha}$ predicted from cDNA sequences [20]. It has been shown that the antisera

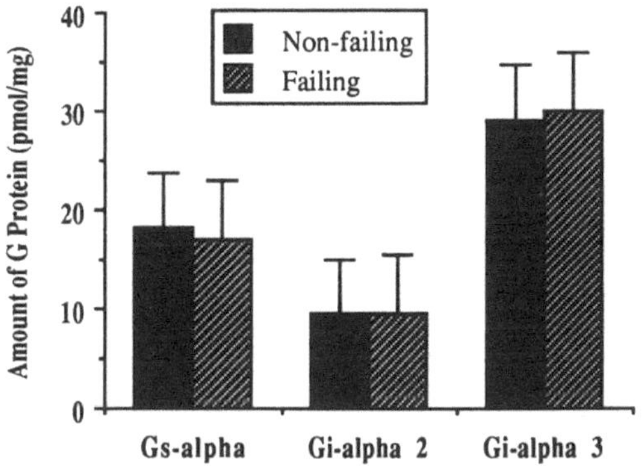

Fig. 1. Amount (peptide equivalent) of α subunits of G_s and subspecies of G_i, determined by competetive ELISA on membrane extracts. Data are the mean ±SEM of duplicate determinations from six individuals

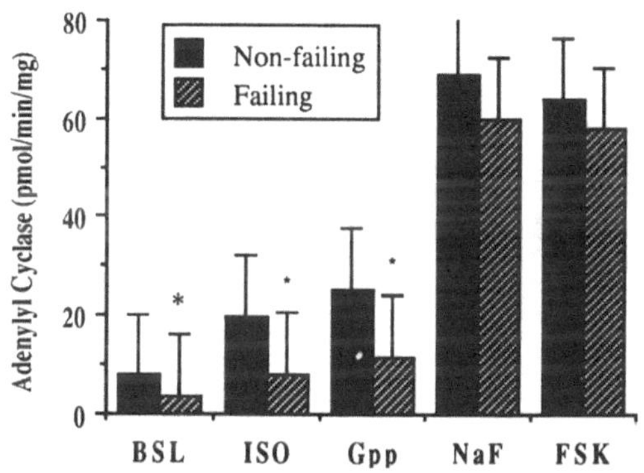

Fig. 2. Adenylyl cyclase activities were measured as amount of cAMP (pmol $mg^{-1} min^{-1}$). Stimulation was made by: *BSL*, basal activity; *ISO*, isoproterenol, 50 μM; *GPP*, Gpp(NH)p, 100 μM; NaF, sodium fluoride, 10 μM; FSK, forskolin, 100 μM.*, $p < 0.01$. Data are the mean ±SEM of duplicate determinations from six individuals

could selectively detect the two isoforms of $G_{s\alpha}$, 45 kDa and 52 kDa. $G_{i\alpha1-2}$ antiserum was raised against the carboxyl-terminal decapeptide of transducin-α. This antiserum recognized not only transducin but also $G_{i\alpha1-2}$. Because these cDNAs have the identical carboxyl-terminal sequence, differing by a single residue from transducin, this antiserum can react equivalently with $G_{i\alpha1}$ and $G_{i\alpha2}$. Since it has been demonstrated that there is lack of $G_{i\alpha1}$ in human heart (X.L. Fu et al., to be published), it may be assumed that this antisera detects mainly $G_{i\alpha2}$. $G_{i\alpha3}$ antiserum was produced against a decapeptide corresponding specifically to the carboxyl-terminal of $G_{i\alpha3}$. As shown in the present study, antisera $G_{i\alpha1-2}$ and $G_{i\alpha3}$ can reveal 40-kDa protein. Neither the amount of $G_{s\alpha}$ nor that of $G_{i\alpha3}$ in human failing hearts differed from controls (Fig. 1). The functional activity of G protein in the

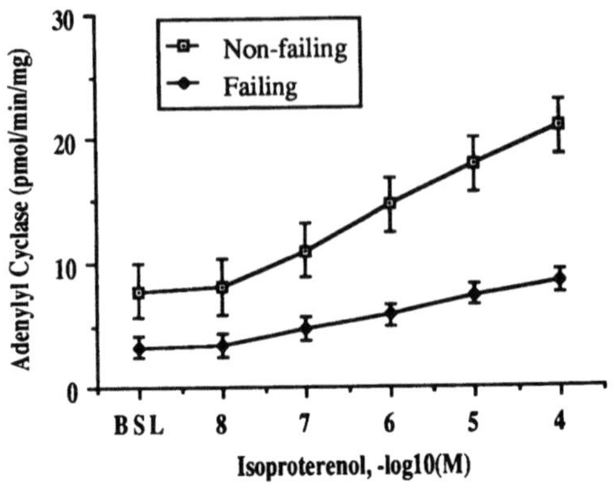

Fig. 3. Isoproterenol-stimulated adenylyl cyclase activity. *BSL*, basal activity. Data are the mean ±SEM of duplicate determination from six individuals

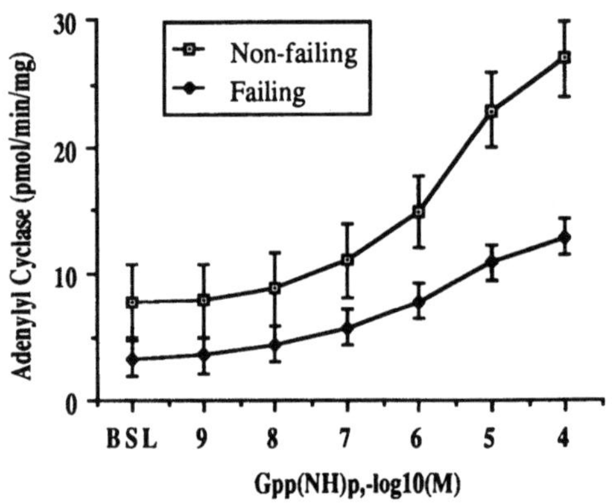

Fig. 4. Gpp(NH)p-stimulated adenylyl cyclase activity. *BSL*, Basal activity. Data are the mean ±SEM of duplicate determinations from six individuals

human failing heart has been determined pharmacologically by effector-stimulated adenylyl cyclase activity. In human failing ventricles, under basal conditions and under stimulation with isoproterenol (via the β-adrenergic pathway), adenylyl cyclase activities were significantly decreased. In addition, there was a marked decline in the adenylyl cyclase activity stimulated by Gpp(NH)p (via both G_s- and G_i-modulated pathways; Figs. 2–5). However, in the presence of fluoride ions (activation of G_s independently of G_i), there was no difference in adenylyl cyclase activity between the human failing hearts and the control hearts, suggesting normal functional activity of G_s in human heart failure. Consequently, the decreased activity of adenylyl cyclase stimulated by Gpp(NH)p as described above in the human failing heart is a balanced result from the normal function of G_s and the increased bioactivity of G_i. This substantial alteration in G_i functional activity with identical $G_{i\alpha2}$ mRNA and normal amount of G_i in human failing ventricles suggests that $G_{i\alpha}$ is modified posttranslationally.

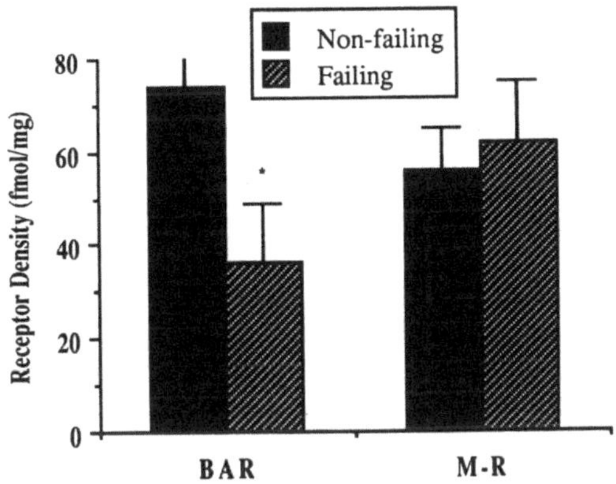

Fig. 5. Densities of β-adrenoceptors and muscarinic receptors. *BAR*, β-adrenoceptor; *M-R*, muscarinic receptor;*, $p < 0.01$. Data are the mean ±SEM of duplicate determinations from six individuals

Thus, it is clear that the β-adrenoceptor–G protein–adenylyl cyclase system plays a crucial role in the short-term preservation of cardiac function; however, the consequence of this compensatory mechanism may be the promotion of further cardiac damage [7, 10]. Eventually, adrenergic stimulation of contractile function is limited by the development of desensitization of the β-adrenoceptor–G protein–adenylyl cyclase system. The desensitization involves sequestration or down-regulation of the receptors from the myocardial cell surface as well as functionally significant covalent modifications of the receptors and/or G proteins, and is an obvious limitation on the use of β-agonists as positive inotropic agents. Consequently, the phenomenen of desensitization has provided us a hint to use β-blockers as a potential therapeutic strategy that might obviate or lessen desensitization.

Clinical Approach to β-Blockade in Patients with IDC

It has been proposed that treament with β-blockers could protect the β-adrenoceptors from chronic down-regulation and thereby effectively cause recovery or up-regulation of β-adrenoceptors. If so, the adrenergic system could again participate in inotropic support during periodic elevation of circulating catecholamines required by exercise or other activities, despite the presence of a competitive β-blocking agent [7, 10]. In our laboratory we first demonstrated that long-term therapy with a low dose of β-blocker in selected patients improves left ventricular performance and symptoms in patients with dilated cardiomyopathy [20, 22]. This observation was soon confirmed in two double-blind, placebo-controlled trials using metoprolol and bucindolol [4]. There is now no doubt that a small dose of metoprolol can have a considerable clinical effect when given to patients in NYHA functional class IV. We believe that the improvement observed in left ventricular function after metoprolol is due to a true increase in contractility and not merely to the relationship between end-systolic wall stress and rate-corrected mean velocity of circumferential fiber shortening. When all data from controlled and uncontrolled studies are pooled, these strongly indicated that metoprolol and some of the other β-blockers improve myocardial function and clinical conditions considerably. In addition, long-term β-blockade in patients with IDC may improve survival. The ongoing international multicenter trial on metoprolol in dilated cardiomyopathy is aimed at comparing long-term metoprolol versus placebo treatment in patients with IDC. The primary objective of this trial is to study mortality.

Regarding the treatment regimens, the small initial daily dose should be escalated gradually over several weeks until the final dose is reached. The maximal daily doses are comparable to those used in treating patients with hypertension or angina pectoris. In most studies, β-blockers were given in two or three divided daily doses. For example, when metoprolol treatment is initiated in a patient with heart failure due to IDC (NYHA IV), the dosage during the first week should be as low as 5 mg b.i.d. Thereafter the dosage can be increased every week for 6 weeks, to a maximal of 50 mg t.i.d. In addition, the following guidlines should be followed: (a) adjustment of concomitant treatment in hypotensive patients; (b) avoidance of co-administration of cardiodepressant class I antiarrhymic agents; (c) looking for early signs of improved diastolic function; (d) monitoring of the patient for signs of intolerance, including profound hypotension and aggravation of right-sided heart failure; (e) monitoring of patients who are withdrawn from metoprolol after long-term treatment; if severe heart failure develops, administer dobutamine temporarily as an antidote.

Several groups of investigators have studied the effects of withdrawing β-blockers after long-term treatment. In 1980 Swedberg et al. demonstrated that patients treated with metoprolol for 6–50 months experienced a significant improvement in cardiac function and exercise capacity [20]. When β-blockers were withdrawn, most patients exhibited a marked deterioration in cardiac function within 3–4 months. Within a few weeks, some patients underwent dramatic deterioration, manifested by severe tachycardia, pulmonary edema, and even death. In our 1989 study of patients who had received long-term metoprolol therapy, drug withdrawal was followed by a relapse in patients whose NYHA class had improved. Improvement, again, was seen after β-blocker treatment was resumed. These results suggest that benefit associated with β-blockade is a true pharmacological effect, not a spontaneous improvement [23, 24].

Possible Mechanism Behind Beneficial Effect of β-Blockade in Patients with Heart Failure due to IDC

Via β-Adrenoceptors Coupled to G Proteins

Human heart contains a heterogenous population of β_1-, β_2-, and even β_3-adrenergic receptors in both atria and ventricles. Both β_1 and β_2 subtypes may subserve somewhat different functions, particularly when atrium is compared with ventricles, and when normal heart is compared with cardiac tissue from individual with congestive heart failure. The techniques of contemporary molecular biology, which complement the radioligand binding experiments and represent the newest approach to the understanding of β-adrenoceptors, have been employed to great advantage in identifying the sequence, structure and structure-activity relationship of β-adrenoceptors. It has currently been suggested that the β-adrenoceptor is best thought of as a dynamic entity rather than a tight structural association that can be regulated by conformational changes in receptor structure and is mobile not only within the cell membrane but very likely in the cytoplasma as well [1, 2]. Neurally released norepinephrine, one of the most important endogenous catecholamines, has a higher affinity for the β_1 than for the β_2 receptor. This may account for the differential regulations of the two coexisting subtypes. However, down-regulation of the β-adrenoceptor occurs after prolonged exposure of either receptor subtype to an agonist and usually is preceded by a number of agonist-specific processes, including uncoupling of the receptor from G protein and sequestration or internalization of the receptor. Long-term β-blockade in patients with chronic congestive heart failure has been shown to increase myocardial β-adrenoceptor density [15]. Other studies in humans have demonstrated that long-term blockade increases the number of β-adrenoceptors in lympocytes and in right atrial tissues obtained during cardiac surgery. Thus, one of the mechanisms presumed to contribute to the beneficial effects of β-blockers in patients with IDC is up-regulation of β-adrenoceptors. Measurements of adenylyl cyclase activity suggest that β-blockers may enhance enzyme activity following incubation with a β-adrenergic agonist.

Via Partial Agonist Activity

Partial agonist activity of the β-blocker appears to enhance the degree of relative up-regulation of the β-adrenoceptors [7]. In patients who had been pretreated with pindolol, β-adrenoceptor density was 51% higher than in nontreated subjects, compared with 38% higher in patients receiving β-blockers without partial agonistic activity.

Via Other Mechanisms

A number of other mechanisms may also contribute to the favorable effects of β-blockers in patients with IDC, such as inhibition of free radicals, reduction of oxygen consumption, and modification of muscarinic receptors. Overall, much remains to be learned concerning β-blockers and their mechanism of action and subsequent effects on myocardial β-adrenoceptor and signal transduction.

References

1. Anderson JL, Lutz JR, Gilbert EM (1985) A randomized trial of low-dose β-blockade therapy for the idiopathic dilated cardiomyopathy. Am Heart J 55:471–475
2. Brodde OE, Kretsch R, Ikezono K, Zerkowski HR, Reidemeister JC (1986) Human β-adrenoceptors: relation of myocardial and lymphocyte β-adrenergic density. Science 231:1584–1585
3. Böhm M, Gierschik P, Jakobs K-H, Pieske B, Schnabel P, Ungerer M, Erdman E (1990) Increase of Gi-alpha in human hearts with dilated but not ischemic cardiomyopathy. Circulation 82:1249–1265
4. Engelmeier RS, O'Connell JB, Walsh R (1985) Improvement in symptoms and exercise tolerance by metoprolol in patients with dilated cardiomyopathy: a double-blind, randomized, placebo-controlled trial. Circulation 72:536–546
5. Feldman AM, Cates AE, Veazey WB, Hershberger RE, Bristow MR, Baughman KL, Baumgartner WA, Van Dop C (1988) Increase of the 40 000 mol wt pertussis toxin substrate (G protein) in the failing human heart. J Clin Invest 82:189–197
6. Fu LX, Waagstein F, Hjalmarson Å (1990) A new insight into adriamycin-induced cardiotoxicity. Int J Cardiol 29:15–20
7. Fu LX, Waagstein F, Hjalmarson Å (1991) An overview of β-adrenoceptor signal transduction systen-desensitization in heart failure and effect of β-blockade. Int J Cardiol 30:261–268
8. Fu LX, Sjögren KG, Liang QM, Waagstein F, Hoebeke J, Hjalmarson Å (1991) Activity of receptors coupled to guanine nucleotide binding regulatory protein in doxorubicin induced cardiomyopathy. Cardiovasc Res 25:145–150
9. Fu LX, Bergh C-H, Hoebeke J, Liang QW, Sjögren KG, Waagstein F, Hjalmarson Å (1991) Effect of metoprolol on activity of receptors coupled to G-proteins in adriamycin-induced cardiotoxicity. Basic Res Cardiol 86:117–126
10. Fu LX, Waagstein F, Hjalmarson Å (1991) Beta-adrenoceptor-G protein-adenylyl cyclase system in cardiac disease: a new insight into desensitization. Clin Physiol 11:1–7
11. Fu LX, Waagstein F, Hjalmarson Å (1990) Increased transcript of Gi-alpha 3 in human failing heart. Eur Heart J 11:1013 (abstr)
12. Fu LX, Sjögren KG, Liang QM, Waagstein F, Hjalmarson Å (1990) Distribution and function of beta-adrenoceptor-adenylyl cyclase in human failing heart. Eur Heart J 11:P1477 (abstr)
13. Fu LX, Liang QM, Waagstein F, Hjalmarson F (1990) Increases of mRNAs of two spliced forms of Gs-alpha in human failing heart. J Mol Cell Cardiol 22 [Suppl III]:PT 28
14. Gwathmey JK, Copelas L, Mackinnon, Schoen FJ, Feldman MD, Grossman W, Morgan JP (1987) Abnormal intracellular calcium handling in myocardium from patients with end-stage heart failure. Circ Res 61:70–76
15. Heibrunn SM, Shah P, Bristow MR, Valantine HA, Ginsburg R, Fowler MB (1989) Increased β-adrenoceoptor density and improved hemodynamic response to catecholamine stimulation during long-term metoprolol therapy in heart failure from dilated cardiomyopathy. Circulation 79:483–490
16. Ikonomides JS, Salerno TA, Wittnich C (1990) Calcium and the heart: an essential partnership. Can J Cardiol 6:305–316
17. Limas CJ, Olivari MT, Goldenberg IF, Levine TB, Benditt DG, Simon A (1987) Calcium uptake by cardiac sarcoplasmic recticulum in human dilated cardiomyopathy. Cardiovasc Res 21:601–605
18. Maki T, Leinoen H, Naveri H (1989) Response of the β-adrenergic system to maximal dynamic exercise in congestive heart failure secondary to dilated cardiomyopathy. Am J Cardiol 63:1348–1353
19. Packer M, Lee WH, Kessler PD (1987) Role of neurohormonal mechanisms in determing survival in patients with severe chronic heart failure. Circulation 75 [Suppl IV]:80–92
20. Swedberg K, Hjalmarson Å, Waagstein F (1980) Beneficial effects of long-term β-blockade in congestive cardiomyopathy. Br Heart J 44:117–133

21. Thomas JA, Marks BH (1978) Plasma norepinephrine in congestive heart failure. Am J Cardiol 41:233–243
22. Waagstein F, Hjalmarson Å, Varnauskas E (1975) Effect of chronic β-adrenergic receptor blockade in congestive cardiomyopaty. Br Heart J 37:1022–1036
23. Waagstein F, Hjalmarson Å, Swedberg K, Wallentin I (1983) β-Blockers in dilated cardiomyopathy: they work. Eur Heart J [Suppl A]:173–178
24. Waagstein F, Caidahl K, Wallentin I, Bergh C-H, Hjalmarson Å (1989) Long-term β-blockade in dilated cardiomyopathy: effects of short and long-term metoprolol treatment followed by withdrawal and readministration of metoprolol. Circulation 80:551–563
25. Waagstein F, Fu LX, Hjalmarson Å (1991) Defining the role of β-Blockers in dilated cardiomyopathy. Prim Cardiol 178:34–41

The Role of Calcium Channel Blockers in Idiopathic Dilated Cardiomyopathy

H.R. Figulla and H. Kreuzer

The etiology of human idiopathic dilated cardiomyopathy (IDC) is yet unclear; it has been observed, however, that some physiological abnormalities involved in the perpetuation of cardiomyopathy may be affected by calcium channel blockers. These abnormalities are discussed in this article. The theoretical considerations will be supported by clinical data of calcium channel blockers in IDC. The abnormalities observed in IDC which may be affected by calcium channel blockers are: abnormal myocardial calcium handling, inappropriate vasomotility, capillary disarray, and tachycardia with incomplete relaxation.

Abnormal Myocardial Calcium Handling

Muscles from explanted hearts of IDC patients show no alterations in contractile and regulatory proteins [2] but a diminished capacity for restoring low resting calcium levels during diastole. This may be due to increased calcium entry via voltage-dependent channels and slowed restoration due to a decreased rate of calcium resequestration by the sarcoplasmic reticulum [10]. Recent experiments demonstrated a decrease in the myocytal level of the mRNA encoding the calcium ATPase of the sarcoplasmic reticulum [19]. The administration of calcium channel blockers has been found to diminish excess calcium inflow [10] and may restore low resting calcium levels.

Furthermore, ethanol is held to be a risk factor in the development of IDC. Long-term administration of ethanol in animals is known to produce myocyte changes such as dilatation of sarcoplasmatic reticulum, swelling of mitochondria, and lipid deposition. These defects are associated with membrane damage, increased calcium influx, decreased fatty acid oxidation, impaired sodium/potassium-stimulated ATPase activity, and decreased ATP levels [1, 14, 21]. Using magnetic resonance spectroscopy, Wu and coworkers [25] showed that the phosphorylation potential decreases in an isolated heart preparation of ethanol-induced cardiac depression. The phosphorylation potential could be restored and dysfunction prevented by concomitant treat-

Department of Cardiology and Pulmonology, Division of Internal Medicine, University of Göttingen, W-3400 Göttingen, Fed. Rep. of Germany

Figulla et al. (Eds.)
Idiopathic Dilated Cardiomyopathy

ment with verapamil during long-term ethanol consumption [9, 25] probably due to prevention of excessive mitochondrial calcium loading.

The hereditary cardiomyopathic Syrian hamster suffers from severe calcium overload [16]. There are conflicting data as to whether this is due to an increased number of calcium channels [12, 15]. However, glycolysis is partially rate limited by phosphofructokinase, an enzyme which is inhibited by calcium and hydrogen ions. In isolated hearts with pyruvate as substrate shunting glycolysis, the phosphorylation potential and developed pressure increased significantly, indicating that late heart failure is associated with calcium and/or hydrogen ion induced inhibition of glycolysis [1]. Inhibition of enhanced transsarcolemmal calcium inflow by chronic treatment with calcium channel blockers may be a contributing mechanismn in preventing the development of myocardial lesions in the affected hamsters [13].

Thus, in human IDC, treatment with calcium channel blockers may affect increased transsarcolemmal calcium transients and alcohol toxicity.

Inappropriate Vasomotility

Endothelium-dependent dilatation of the coronary microvasculature is impaired in IDC. The decrease in coronary resistance elucidated by acetylcholine infusion into the coronary vaculature was only 41% in IDC as compared with controls in whom it was 232% [24]. These figures indicate that vasomotility is impaired in IDC. In the cardiomyopathic Syrian hamster the focal pattern of myocyte necrosis indicates microcirculatory disorders. Factor et al. [3] demonstrated vasospasms in these animals. In experiments with timed plasma staining we observed that the coronary circulatory transit time was delayed in myocardial areas of 200–300 μm width. The delayed coronary blood flow was reestablished by treatment with verapamil or hydralazine [6]. Both substances additionally prevented myocytolysis.

Thus, there is evidence that vasomotility in IDC patients and cardiomyopathic Syrian hamsters is impaired. Calcium channel blockers can restore this impairment in the latter.

Capillary Disarray

Human hearts with IDC exhibit hypertrophy of myocytes [17]. Intercapillary distance consequently increases, which may impede the diffusional metabolic supply. Therefore, living human hearts excised during cardiac transplantation due to intractable IDC were reperfused with a Krebs-Henseleit solution containing fluorescein isothiocyanate (FITC) bound to dextran (molecular weight 58 kDa). The FITC-dextran complex did not penetrate the tissue, and the intravascular space could therefore be precisely detected and the capillary density and distribution evaluated. The average functional capillary density was only 1245 ± 345 capillaries/mm^2, with an average muscle fiber diameter of 27.0 + 6.0 μm [7] as compared to 2483–3342 capillaries in

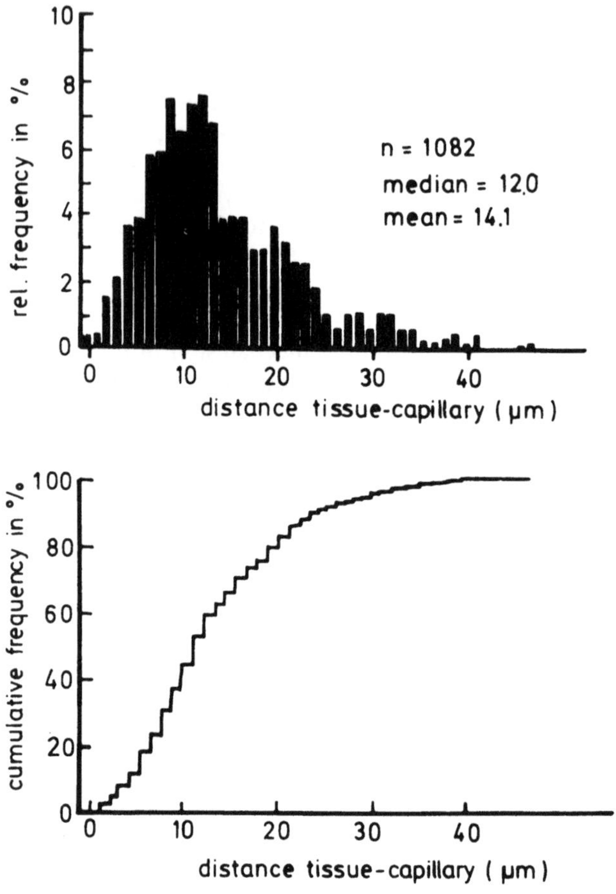

Fig. 1. The distances from 1082 test points superimposed onto tissue slices from eight human hearts with IDC to the nearest capillary were measured. These data gave a frequency distribution of the tissue-capillary distance. (From [7] by permission)

hypertrophied and normal hearts [23]. Mean tissue capillary distance was 14.1 μm (Fig. 1). Assuming a simple diffusionable model (Krogh cylinder) of oxygen transport, the tissue cylinder supplied with oxygen has an radius at the venous side of the capillary of 25 μm. By applying calcium channel blockers, the pressure rate product can be lowered by about 25% [5]. This increases the radius of oxygen diffusion in our model from 25 to 29 μm (Fig. 2) and thus improves oxygen supply in IDC hearts bordering on anoxia (Fig. 3).

Tachycardia with Incomplete Relaxation

Tachycardia is a frequent phenomenon in IDC due to baroreceptor desensitization and high levels of endogenous catecholamines. However,

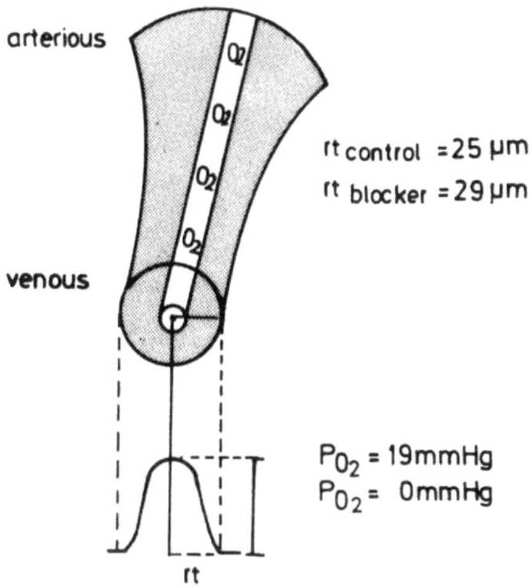

Fig. 2. A simple diffusional model of oxygen transport (Krogh cylinder), according to a simplified Krogh-Erlang diffusional equation, $P_{o2} = (AR^2)/K$, where P_{o2} is the oxygen gradient within a muscle tissue cylinder R surrounding a capillary in the center with an oxygen conductivity coefficient of $K = 2.5 \times 10^{-8}\,\mathrm{ml\,cm^{-1}\,min^{-1}\,mmHg^{-1}}$. Under the assumption of a 20% part of interstitial structured tissue with very little oxygen consumption, an oxygen consumption of the heart muscle tissue of $A = 9.7\,\mathrm{ml} \times 100\,\mathrm{ml^{-1}\,min^{-1}}$ and an oxygen pressure of 19 mmHg at the venous side of the capillary, the tissue cylinder supplied by oxygen has a radius of $rt_{\mathrm{control}} = 25\,\mu\mathrm{m}$. Lowering heart muscle oxygen consumption by 25% due to calcium channel blockers extends the tissue cylinder to $rt_{\mathrm{blockers}} = 29\,\mu\mathrm{m}$

in isolated muscles from patients with end-stage heart failure, heart rate potentiation resulted in an abbreviated action potential with a rise in intracellular resting calcium concentrations, resting tension, and reduced augmentation of peak force. Thus it was concluded that slowing the heart rate allows sufficient time for impaired calcium restitution processes to operate [11]. In patients with IDC, cardiac index, ejection fraction, and $dp/dt_{\max}$ of the pressure curve decreases during rapid pacing to 140/min as compared with normal hearts (G. Hasenfuss, personal communication). Diltiazem can significantly lower sinus heart rate in IDC [5] and thus should exhibit further beneficial effects in IDC.

Adjunctive Treatment with Calcium Channel Blockers in IDC

Although the process of cardiomyopathy is not completely understood, there are pathophysiological abnormalities involved which justify investigating the effects of calcium channel blockers in a clinical trial in IDC patients. We therefore performed a longitudinal pilot trial in 22 randomly selected IDC patients treated with diltiazem adjunctive to conventional therapy. Diltiazem was administered in increasing doses over 1 week to a final dosage of 90 mg t.i.d. orally. The study group was then compared with a historical

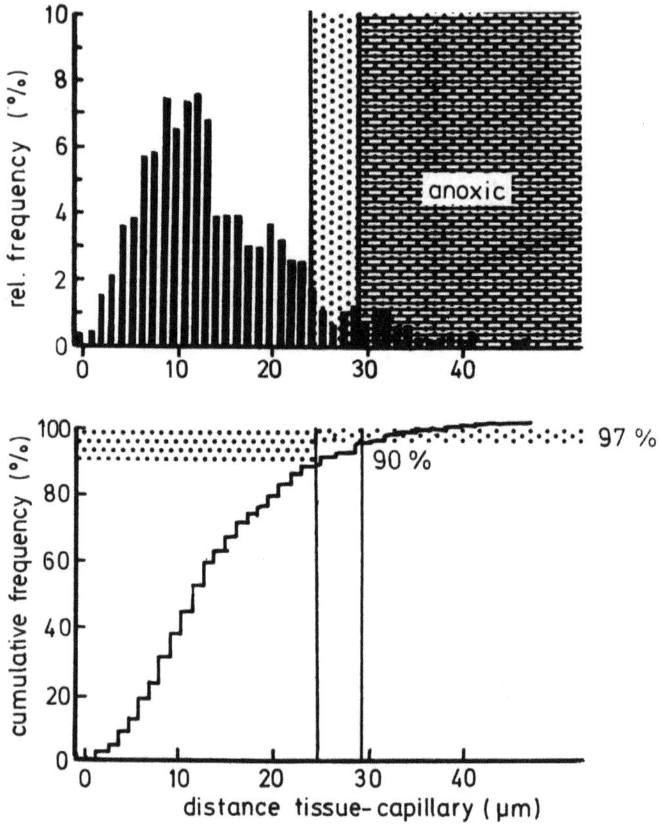

Fig. 3. According to the capillary distribution found in human cardiac tissue with IDC, at the venous side of the capillary about 10% of the tissue is anoxic. By the administration of calcium channel blockers the oxygen consumption can be lowered by 25%, which increases tissue oxygenation, reducing the anoxic area to 3%

control group from the same institution who were treated similarily but without diltiazem [8]. All patients underwent left heart catheterization, Swan-Ganz catheterization, gated blood pool scanning at rest and during exercise, and left ventricular endomyocardial biopsy. Reduced myofibrillar volume fraction was found on endomyocardial biopsy for all patients; we believe this to be the most sensitive and specific indicator for further hemodynamic deterioration or death in individual patients [4]. The patients were then seen at regular intervals for follow-up examinations. In the diltiazem groups, 15 of the 22 patients manifested hemodynamic improvement within a mean of 15.4 months; the condition of six patients remained unchanged, and one deteriorated. The mean left ventricular ejection fraction increased from 0.34 to 0.44 ($p < 0.001$) as opposed to the control group in which nine patients died prior to reexamination. The left ventricular ejection fraction of the remaining controls decreased significantly from 0.41 to 0.37 ($p < 0.05$) in a similar observation period (Fig. 4) [8]. We attributed

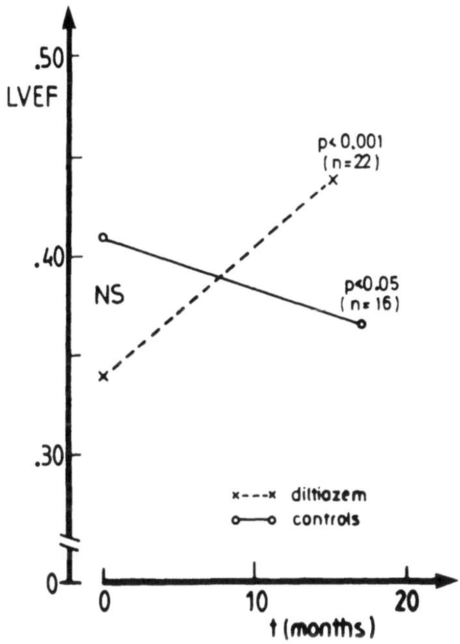

Fig. 4. Mean left ventricular ejection fraction (*LVEF*) at rest versus time, in patients who survived until subsequent reexamination. Mean ejection fraction increased in patients receiving diltiazem but decreased in the control group over a similar observation period. In the control group, nine patients with diminished left ventricular ejection fraction died before reexamination. Therefore, in this group only the course of the surviving 16 patients could be shown. (From [8], by permission)

the beneficial clinical and hemodynamic course in this trial to *specific* actions of diltiazem in IDC since in a study with diltiazem in patients with myocardial dysfunction due to infarction, those with impaired left ventricular function had a poor outcome [20].

The beneficial results of this open clinical trial initiated in 1989 a large scale, randomized double blinded clinical trial (DiDi trial = diltiazem in dilated cardiomyopathy). It has just completed its recruitment period after including 185 patients. The final results will be available in 1994.

Open Questions

Many questions in the context of calcium channel blockers and IDC remain unanswered and are presently under investigation:

- Is the L-type calcium channel density in IDC the same as compared to controls? There are conflicting findings at present [18, 22].
- Will calcium channel blocker treatment affect L-type calcium channel density?

- Do calcium channel blockers improve the balance of oxygen supply and demand in IDC?
- Do calcium channel blockers beneficially affect diastolic dysfunction in IDC during tachycardia?

Abnormal intracellular calcium handling in human end-stage heart failure has been discovered in recent years. Many facts are yet poorly understood. Therefore, biological experimentation should be extended from the cardiomyopathic Syrian hamster, in which calcium channel blockers show beneficial effects, to humans. Clinical trials such as the DiDi trial are integrative experiments with the potential for answering these questions. However, these trials do not explore the underlying pathophysiological mechanisms in IDC.

References

1. Auffermann W, Wu ST, Parmley WW, Wikman-Coffelt J (1990) Glycolysis in heart failure: a ^{31}P-NMR and surface fluorometry study. Basic Res Cardiol 85:342–357
2. D'Agnolo A, Luciani GB, Mazzucco A, Gallucci V, Salviati G (1992) Contractile properties and Ca^{2+} release activity of the sarcoplasmic reticulum in dilated cardiomyopathy. Circulation 85:518–525
3. Factor SM, Minase T, Cho S et al. (1982) Microvascular spasm in the cardiomyopathic Syrian hamster. A preventable cause of focal myocardial necrosis. Circulation 66:342–354
4. Figulla HR, Rahlf G, Nieger M, Luig H, Kreuzer H (1985) Spontaneous hemodynamic improvement or stabilization and associated biopsy findings in patients with congestive cardiomyopathy. Circulation 71:1095–1104
5. Figulla HR, Kreuzer H, Luig H (1986) Verapamil, Diltiazem oder Nifedipin bei schwerer linksventrikulärer Funktionsstörung? Dtsch Med Wochenschr 111:11–14
6. Figulla HR, Vetterlein F, Glaubitz M (1987) Inhomogenous capillary flow and its prevention by verapamil and hydralazine in the cardiomyopathic Syrian hamster. Circulation 76:208–216
7. Figulla HR, Vetterlein F, Wiegand V, Schüler S, Kreuzer H (1988) Capillary density and oxygen supply in human dilated cardiomyopathy. In: Kaltenbach M, Hopf R, Kunkel (eds) New aspects of hypertrophic cardiomyopathy. Springer, Berlin Heidelberg New York, pp 255–259
8. Figulla HR, von Rechenberg J, Wiegand V, Soballa R, Kreuzer H (1989) Beneficial effects of long-term diltiazem treatment in dilated cardiomyopathy. J Am Coll Cardiol 13:653–658
9. Garrett JS, Wikman-Coffelt J, Sievers R, Finkbeiner WE, Parmley WW (1987) Verapamil prevents the development of alcoholic dysfunction in hamster myocardium. J Am Coll Cardiol 9:1326–1331
10. Gwathmey JK, Copelas L, MacKinnon R, Schoen FJ, Feldman MD, Grossman W, Morgan JP (1987) Abnormal intracellular calcium handling in myocardium from patients with endstage heart failure. Circ Res 61:70–76
11. Gwathmey JK, Slawsky MT, Hajjar RJ, Briggs GM, Morgan JP (1990) Role of intracellular calcium handling in force-interval relationships of human ventricular myocardium. J Clin Invest 85:1599–1613
12. Howlett SE, Rafuse VF, Gordon T (1988) [^{3}H]-Nitrendipine binding sites in normal and cardiomyopathic hamsters: absence of a selective increase in putative calcium channels in cardiomyopathic hearts. Cardiovasc Res 22:840–846
13. Jasmin G, Solymoss B (1975) Prevention of hereditary cardiomyopathy in the hamster by verapamil and other agents. Proc Soc Exp Biol Med 149:193–198
14. Katz AM, Freston JW, Messineo FC, Herbette LG (1985) Membrane damage and the pathogenesis of cardiomyopathies. J Mol Cell Cardiol 17 [Suppl 2]:11–20

15. Kobayashi A, Yamashita T, Kaneko M, Nishiyama T, Hayashi H, Yamazaki N (1987) Effects of verapamil on experimental cardiomyopathy in the Bio 14.6 Syrian hamster. J Am Coll Cardiol 10:1128–1134
16. Lossnitzer K, Janke J, Hein B, Stauch M, Fleckenstein A (1975) Disturbed myocardial calcium metabolism: a possible pathogenetic factor in the hereditary cardiomyopathy of the Syrian hamster. In: Fleckenstein A, Rona G (eds) Recent advances in studies on cardiac struction and metabolism: 6. Pathophysiology and morphology of myocardial cell alteration. University Park Press, Baltimore, pp 207–217
17. Mall G, Schwarz F, Derks H (1982) Clinicopathologic correlations in congestive cardiomyopathy. Virchows Arch 397:67–82
18. Marsh JD, Allen PD, Dennis AR (1987) Decreased calcium channel expression in sarcolemmal membranes from human hearts with congestive heart failure. J Mol Cell Cardiol 19:533
19. Mercadier J-J, Lompré A-M, Duc P, Boheler KR, Fraysse J-B, Wisnewsky C, Allen PD, Komajda M, Schwartz K (1990) Altered sarcoplasmic reticulum Ca^{2+}-ATPase gene expression in the human ventricle during end-stage heart failure. J Clin Invest 85:305–309
20. Multicenter Diltiazem Postinfarction Trial Group (1988) The effect of diltiazem on mortality and reinfarction after myocardial infarction. N Engl J Med 319:385–392
21. Noren GR, Staley NA, Einzig S, Mikell FC, Asinger RW (1983) Alcohol-induced congestive cardiomyopathy: an animal model. Cardiovasc Res 17:81–87
22. Rasmussen PR, Minobe W, Bristow MR (1990) Calcium antagonist binding sites in failing and nonfailing human ventricular myocardium. Biochem Pharmacol 39:691–696
23. Roberts JT, Wearn JT (1941) Quantitative changes in the capillary muscle relationship in human hearts during normal growth and hypertrophy. Am Heart J 21:617–633
24. Treasure CB, Vita JA, Dox DA et al. (1990) Endothelium-dependent dilatation of the coronary microvasculature is impaired in dilated cardiomyopathy. Circulation 81:772–774
25. Wu S, White R, Wikman-Coffelt J, Sievers R, Wendland M, Garrett J, Higgins CB, James T, Parmely WW (1987) The preventive effect of verapamil on ethanol-induced cardiac depression: phosphorus-31 nuclear magnetic resonance and high-pressure liquid chromatographic studies of hamsters. Circulation 75:1058–1064

Gene Therapy for Cardiovascular Disease: Realistic Possibilities for the Future

J.M. Leiden

Introduction

The ability to program recombinant gene expression in cardiac and skeletal muscle would represent a significant advance in our ability to treat a variety of acquired and inherited diseases of the cardiovascular system. This report summarizes recent work that demonstrates the feasibility of expressing recombinant genes in cardiac and skeletal muscle in vivo. These studies have shown that plasmid DNA injected directly into the myocardium is taken up and stably expressed in cardiac myocytes in vivo. This technique has been used to express a number of recombinant genes in the rodent heart for periods of as long as 6 months. Preliminary experiments suggest that this approach may be useful for stimulating angiogenesis in the myocardium of patients with ischemic cardiomyopathies. In a second set of experiments, we have shown that genetically modified skeletal myoblasts that secrete human growth hormone can be introduced by intramuscular injection into the skeletal muscle of normal syngeneic mice. Such injected cells fuse into multinucleated myotubes and produce stable and physiological levels of human growth hormone in the serum of the injected mice. This approach holds promise for the treatment of a number of acquired and inheritied serum protein deficiencies.

Somatic gene therapy can be defined as the ability to introduce and express recombinant genes in nongerm line cells of a recipient organism in vivo (reviewed in [5, 17]). All approaches to somatic gene therapy can be divided into two categories: (a) ex vivo gene therapy in which cells are removed from an animal, transduced with a recombinant gene in the laboratory (in vitro), and then reimplanted or transplanted back into the recipient host, and (b) in vivo gene therapy in which a recombinant gene is introduced directly into the appropriate cells of a recipient animal in vivo. Before being considered for use in humans all gene therapy techniques must fulfill a number of basic requirements, including the initial cloning of the gene of interest, the development of an efficient method for introducing this gene into the appropriate mammalian cells, a system that programs high-level expression of this gene in the tissue of interest, and, perhaps most im-

The Departments of Medicine and Pathology, University of Chicago, Chicago, IL 60637, USA

Figulla et al. (Eds.)
Idiopathic Dilated Cardiomyopathy

portantly, a minimal risk of persistent infection or mutagenesis of the recipient host. A variety of systems have been used to introduce genes into fibroblasts [11], hepatocytes [1], bone marrow [18], keratinocytes [4], and endothelial cells [10] in vivo. These include RNA and DNA viruses, liposomes (fatty particles that surround DNA and promote fusion with cell membranes), and calcium phosphate/DNA complexes.

Until recently, cardiac and skeletal muscle were not considered good target organs for somatic gene therapy. However, several recent advances have now made gene therapy using muscle a feasible and exciting approach. This report summarizes recent work from my laboratory exploring new techniques and applications for somatic gene therapy in cardiac and skeletal muscle.

Expression of Recombinant Genes in Cardiac Myocytes in Vivo

Previous work from Wolff and coworkers [19] demonstrated that plasmid DNA is taken up and expressed by a small percentage of skeletal myocytes following direct intramuscular injection in vivo. To test whether a similar approach could be used to program recombinant gene expression in cardiac myocytes in vivo, 100 μg pRSV β-galactosidase (β-gal) plasmid DNA was injected into the apical wall of the left ventricle of 6-week-old Sprague Dawley rats [8]. The pRSV β-gal plasmid was chosen for these experiments because it programs the expression of the bacterial β-gal enzyme which stains cells blue. Thus it is possible to use standard histological techniques to directly identify the cell types expressing the recombinant gene in vivo. These studies showed that the injected plasmid DNA is taken up by a small proportion (fewer than 1%) of cardiac myocytes and expressed for periods of at least 6 months in these animals (the longest time thus far examined) [8]. Expression was limited to myocytes in the area of injection and was never observed in endothelial cells or fibroblasts. Moreover, blue staining was never observed in control rats injected with irrelevant plasmid DNAs, confirming that it was due to expression of the recombinant β-gal gene. We have subsequently expressed a number of recombinant genes in the heart using this technique, including the bacterial chloramphenicol acetyl transferase gene and the firefly luciferase gene. In addition, we have shown that this method can be used for the analysis of cardiac-specific transcriptional regulatory elements [13] in vivo, thus obviating the need for expensive and time-consuming transgenic approaches to this problem.

In preliminary studies we have recently used this approach to express the cDNA for fibroblast growth factor-5 (FGF-5) in the myocardium. Results from five animals showed a 35% increase in capillary density in the areas of FGF-5 injection 3 weeks after injection (E. Barr et al., manuscript in preparation). This increase in capillary density was observed in each of the FGF-5 injected animals, was statistically significant ($p < 0.01$), was not observed in control animals injected with the pRSV β-gal plasmid, and was not due to muscle atrophy as the myocyte density was the same in FGF-5

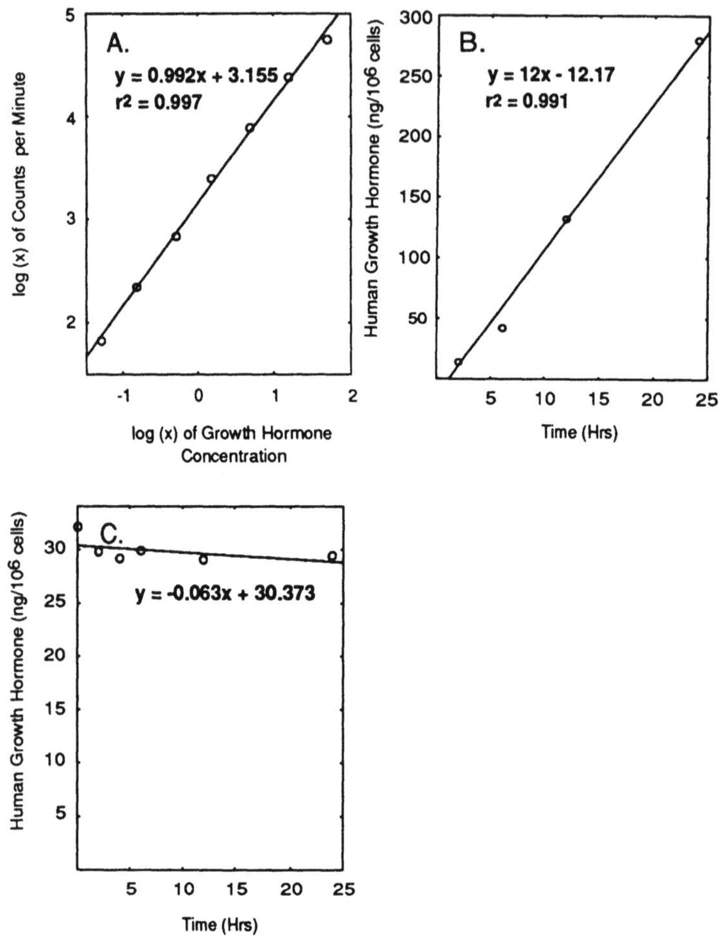

Fig. 1A–C. Human growth hormone production in vitro by the G19 clone of pRSVGH-transfected C2C12 myoblasts. C2C12 myoblasts were cotransfected with the pRSVGH and pRSVneo plasmids using calcium phosphate. After selection in 1 mg/ml G418 (GIBCO, Grand Island, NY), clones were assayed for hGH production and secretion by a commercially available radioimmunoassay (RIA; Nichols Institute Diagnostics, San Juan Capistrano, CA). One clone, G19, was expanded for further studies. **A** Validation of the RIA for hGH; dilutions of recombinant hGH containing between 0.05 and 50 ng/ml were assayed in duplicate by RIA as per the manufacturer's instructions. **B** Kinetics of hGH secretion by G19 cells in vitro; 10^6 G19 cells were seeded onto replicate 100-mm^2 dishes containing 10 ml DMEM + 20% fetal bovine serum and 1% chick embryo extract (GIBCO). Aliquots (100 μl) of culture medium were assayed for hGH by RIA at the time points shown. All assays were performed in duplicate using culture medium from two different plates, and mean values were normalized for cell number to eliminate differences due to cell division. **C** Kinetics of hGH decay during incubation of C2C12 monolayers. Cultures of 10^6 normal C2C12 cells were incubated in hGH-containing culture medium previously conditioned by exposure to G19 cells for 24 h Duplicate 100-μl aliquots of culture medium from two different plates were assayed for hGH at the time points shown. All statistical calculations were performed using Statview II software (Abacus concepts Inc., Berkeley, CA) on an Apple Macintosh IIcx computer. (By permission from [2], copyright by the AAAS, 1997)

and pRSV β-gal injected animals. Thus, this technique may hold promise for the treatment of patients with chronic ishemic cardiomyopathies. Current studies are focused on improving the efficiency of recombinant gene uptake and the development of efficient cardiac-specific expression vectors.

Genetically Modified Skeletal Myoblasts: A Novel System for the Delivery of Recombinant Proteins to the Systemic Circulation

A number of human diseases are currently treated by repeated intramuscular injection of purified or recombinant proteins. These include diabetes mellitus [16], pituitary dwarfism [15], and hemophilia A [9]. Development of a somatic gene therapy approach to stably supply such recombinant proteins to the circulation would represent a significant advance in our ability to treat these patients. Previous studies have used genetically modified fibroblasts [11], keratinocytes [4], bone marrow [18], and hepatocytes [1] for the systemic delivery of such recombinant proteins. However, it has thus far

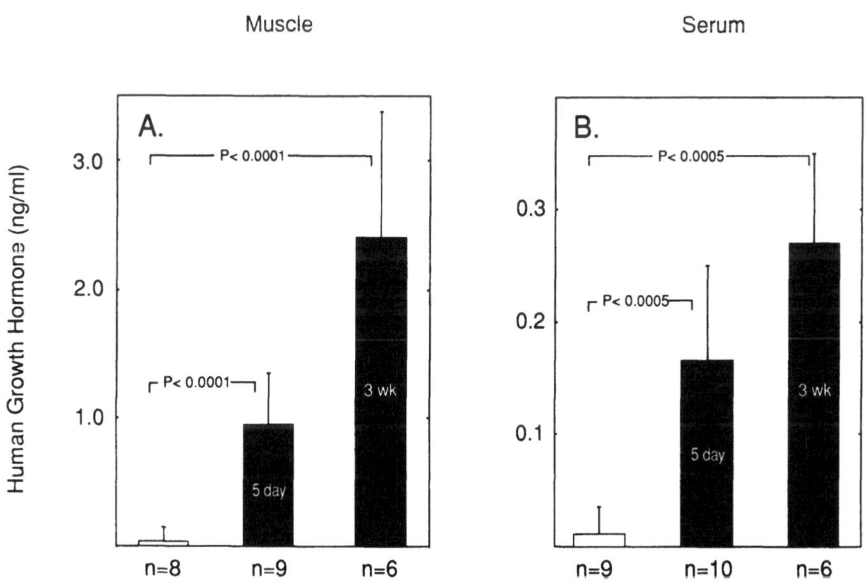

Fig. 2A,B. Human growth hormone levels in muscle and serum of C2C12 myoblast-injected mice. Four-week-old C3H mice were injected intramuscularly with 6×10^6 G19 cells (*solid bars*) or control F17 cells (*open bars*) in a total volume of 240 μl DMEM medium (GIBCO) at six separate sites using a 30-G needle. Mice were euthanized 5 days or 3 weeks after injection, and muscle samples were assayed for hGH by radioimmunoassay. **A** Human growth hormone levels in muscle homogenates from control (n = 8) and G19-injected (n = 9 at 5 days, and n = 6 at 3 weeks) mice. **B** Human growth hormone levels in serum from control (n = 9) and G19-injected (n = 10 at 5 days, and n = 6 at 3 weeks) mice. All samples were assayed in duplicate. The data are presented as mean ± S.D. (By permission from [2], copyright 1991 by the AAAS)

proven difficult to produce stable and physiological levels of recombinant serum proteins in normal animals.

Skeletal myoblasts display a number of unique features which suggest that they might be useful as a vehicle for the delivery of recombinant proteins to the systemic circulation. Skeletal myoblasts or satellite cells can be easily isolated from routine muscle biopsies, grown to large numbers in vitro, and transfected with recombinant genes using standard approaches [3, 12]. Previous studies have demonstrated that such myoblasts can be introduced back into skeletal muscle in vivo by simple intramuscular injection. Following implantation, a significant fraction of these injected cells fuse into multinucleated myotubes without apparent disruption of the normal muscle structure or function [6, 7, 14].

To assess the feasibility of using genetically modified myoblasts for the somatic gene therapy of acquired and inherited serum protein deficiencies murine C2C12 skeletal myoblasts were transfected with an expression vector for human growth hormone (hGH). A single clone (G19) was selected that expressed high levels of hGH in vitro (12 ng/106 cells per hour; Fig. 1). Normal syngeneic mice were injected intramuscularly with 6×10^6 cells of this clone, and both muscle and serum were assayed for hGH using a commercially available and sensitive radioimmunoassay that was specific for the human form of the hormone (Fig. 2) [2]. As expected, no hGH was detected in the serum or muscle of control mice injected with C2C12 cells that did not express hGH. In contrast, mice injected with the G19 hGH-expressing cells contained significant levels of hGH in muscle and serum at both 5 days and 3 weeks following implantation (Fig. 2). Moreover, the levels of hGH in the serum of the G19-injected animals were one to four times as high as those observed in serum from normal human volunteers [2].

To determine the fate of the injected, genetically modified myoblasts, C2C12 cells expressing the bacterial β-gal gene were injected intramuscularly into normal syngeneic C3H mice, and muscle biopsies were processed for β-gal activity 5–7 days following injection. These experiments showed that the injected mononuclear myoblasts fused into multinucleated myotubes following intramuscular injection with little or no disruption of the muscle architecture [2]. Taken together these studies demonstrate the feasibility of using genetically modified myoblasts to stably deliver physiological levels of recombinant proteins to the systemic circulation. The results are especially encouraging given the relatively short half-life of hGH in mice (less than 20 min). Ongoing studies are focused on using this approach to treat canine hemophilia A using genetically modified primary canine myoblasts.

References

1. Armentano D, Thompson AR, Darlington G, Woo SL (1990) Expression of human factor IX in rabbit hepatocytes by retrovirus-mediated gene transfer: potential for gene therapy of hemophilia. Proc Natl Acad Sci USA 87:6141–6145
2. Barr E, Leiden JM (1991) Systemic delivery of recombinant proteins by genetically modified myoblasts. Science 254:1507–1509
3. Blau HM, Webster C (1981) Isolation and characterization of human muscle cells. Proc Natl Acad Sci USA 78:5623–5627
4. Flowers ME, Stockschlaeder MA, Schuening FG, Niederwieser D, Hackman R, Miller AD, Storb R (1990) Long-term transplantation of canine keratinocytes made resistant to G418 through retrovirus-mediated gene transfer. Proc Natl Acad Sci USA 87:2349–2353
5. Friedmann T (1989) Progress toward human gene therapy. Science 244:1275–1281
6. Hughes SM, Blau HM (1990) Migration of myoblasts across basal lamina during skeletal muscle development. Nature 345:350–353
7. Law PK, Goodwin TG, Wang MG (1988) Normal myoblast injections provide genetic treatment for murine dystrophy. Muscle Nerve 11:525–533
8. Lin H, Parmacek MS, Morle G, Bolling S, Leiden JM (1990) Expression of recombinant genes in myocardium in vivo after direct injection of DNA. Circulation 82:2217–2221
9. Mammen EF (1983) Seminars in thrombosis and hemostasis. Semin Thromb Hemost 9:1–72
10. Nabel EG, Plautz G, Nable GJ (1990) Site-specific gene expression in vivo by direct gene transfer into the arterial wall. Science 249:1285–1288
11. Palmer TD, Thompson AR, Miller AD (1989) Production of human factor IX in animals by genetically modified skin fibroblasts: potential therapy for hemophilia B. Blood 73:438–445
12. Parmacek MS, Bengur AR, Vora AJ, Leiden JM (1990) The structure of regulation of expression of the murine fast skeletal troponin C gene. Identification of a developmentally regulated, muscle-specific transcriptional enhancer. J Biol Chem 265:15970–15976
13. Parmacek MS, Vora AJ, Shen T, Barr E, Jung F, Leiden JM (1992) Identification and characterization of a cardiac-specific transcriptional regulatory element in the slow/cardiac troponin C gene. Mol Cell Biol (in Press)
14. Partridge TA, Morgan JE, Coulton GR, Hoffman EP, Kunkel LM (1989) Conversion of mdx myofibres from dystrophin-negative to -positive by injection of normal myoblasts. Nature 337:176–179
15. Rudman D, Kutner MH, Blackston RD, Cushman RA, Bain RP, Patterson JH (1981) Children with normal-variant short stature: treatment with human growth hormone for six months. N Engl J Med 305:123–131
16. Schiffrin A, Belmonte MM (1982) Comparison between continuous subcutaneous insulin infusion and multiple injections of insulin. A one-year prospective study. Diabetes 31:255–264
17. Swain JL (1989) Gene therapy: a new approach to the treatment of cardiovascular disease. Circulation 80:1495–1496
18. Wilson JM, Danos O, Grossman M, Raulet DH, Mulligan RC (1990) Expression of human adenosine deaminase in mice reconstituted with retrovirus-transduced hematopoieic stem cells. Proc Natl Acad Sci USA 87:439–443
19. Wolff JA, Malone RW, Williams P, Chong W, Acsadi G, Jani A, Felgner PL (1990) Direct gene transfer into mouse muscle in vivo. Science 247:1465–1468

Subject Index

Springer-Verlag and the Environment

We at Springer-Verlag firmly believe that an international science publisher has a special obligation to the environment, and our corporate policies consistently reflect this conviction.

We also expect our business partners – paper mills, printers, packaging manufacturers, etc. – to commit themselves to using environmentally friendly materials and production processes.

The paper in this book is made from low- or no-chlorine pulp and is acid free, in conformance with international standards for paper permanency.

MIX
Papier aus verantwortungsvollen Quellen
Paper from responsible sources
FSC® C105338

If you have any concerns about our products,
you can contact us on
ProductSafety@springernature.com

In case Publisher is established outside the EU,
the EU authorized representative is:
Springer Nature Customer Service Center GmbH
Europaplatz 3, 69115 Heidelberg, Germany

Printed by Libri Plureos GmbH
in Hamburg, Germany